Biofilms
Infection,
and
Antimicrobial
Therapy

Biofilms, Infection, and Antimicrobial Therapy

edited by
John L. Pace
Mark E. Rupp
Roger G. Finch

CRC Press
Taylor & Francis Group
Boca Raton London New York

CRC Press is an imprint of the
Taylor & Francis Group, an **informa** business
A TAYLOR & FRANCIS BOOK

CRC Press
Taylor & Francis Group
6000 Broken Sound Parkway NW, Suite 300
Boca Raton, FL 33487-2742

First issued in paperback 2019

© 2006 by Taylor & Francis Group, LLC
CRC Press is an imprint of Taylor & Francis Group, an Informa business

No claim to original U.S. Government works

ISBN-13: 978-0-8247-2643-0 (hbk)
ISBN-13: 978-0-367-39219-2 (pbk)
Library of Congress Card Number 2005044047

Library of Congress Cataloging-in-Publication Data

Biofilms, infection, and antimicrobial therapy / edited by John L. Pace, Mark Rupp, and Roger G. Finch.
     p. ; cm.
  Includes bibliographical references and index.
  ISBN 0-8247-2643-X (alk. paper)
  1. Biofilms. 2. Pathogenic microorganisms. 3. Anti-infective agents. I. Pace, John L. II. Rupp, Mark. III. Finch, R. G. (Roger G.)
  [DNLM: 1. Biofilms. 2. Anti-Bacterial Agents. 3. Infection. 4. Prosthesis-Related Infections. QW 50 B6147 2005]

QR100.8.B55B565 2005
616.9′041--dc22
                                              2005044047

**Visit the Taylor & Francis Web site at
http://www.taylorandfrancis.com**

**and the CRC Press Web site at
http://www.crcpress.com**

Dedication

To
Yvonne, Sarah, and Katherine,
who suffered by my absence during
the many hours that I spent at
the laboratory bench — JLP

Preface

One of the greatest medical achievements of the 20th century undoubtedly was the rapid and effective introduction of antiinfective chemotherapy subsequent to Fleming's discovery of penicillin. Clearly untold numbers of lives have been saved by the application of natural product, semisynthetic, and synthetic antimicrobial compounds to the treatment of infectious disease. Academic, industrial, and government scientists, physicians and associated healthcare workers, and administrators all contributed to these efforts that have served humankind so well.

This success in some way belies the effectiveness of the agents that were discovered and developed into powerful therapies because we have often overlooked the complexity of the pathogenic microorganisms and their interactions with human and animal hosts. The clearest example of this was our very slow grasp of the significance of microbial biofilms. Von Leeuwenhoek alluded to biofilms in his writings (circa 1650), and later reports by Zobell (1942) and others clearly described multicellular prokaryotic communities from natural environments. Another 30 years passed before Costerton and colleagues began to emphasize the very significant role of adherent microbial communities in human infectious disease. In the 20 years subsequent to that declaration, many advances have been made, and yet much of the antimicrobial research continues to focus on methods utilizing planktonic bacteria. This phenomenon is due in part to the often good correlations of minimal inhibitory concentration (MIC) breakpoints for antibiotics with clinical outcomes, and takes into account exposure at the site of infection as well as activity against the target. However, the successful use of the MIC value may also have been due to the nature of infections and patient's status. As the infectious disease landscape continues to evolve such that biofilm-related infections become more predominant, the correlation between *in vitro* susceptibility results and clinical outcomes may likely decline. The erosion of the basis for effective antimicrobial therapy has arisen from the nature of the patient population due to underlying debilitation and impaired immune status, but also from the increasing emphasis on the use of indwelling medical devices (IMD). This trend at least is well known and has been recognized by governmental agencies including the Food and Drug Administration and the Centers for Disease Control and Prevention, and behooves the scientific and medical community to reflect on future approaches to addressing this problem.

In this book we have attempted to develop a general understanding of the problems for a broad audience. Thomas and colleagues have contributed an excellent overview of economic burden associated with treating biofilm-related infections. Schinabeck and Ghannoum have described the central role of IMD in the infections, and Gorman and Jones have provided us with detailed descriptions of how implant surface characterists contribute to the adherence of pathogens and initiation of infections. In addition to the major principles, discussion of cutting edge aspects have been provided by Lewis and colleagues in the area of antibacterial tolerance and persisters, which is central to the problem of treating biofilm-related infections. Animal models have been described by Handke and Rupp essential for identification of new effective therapeutic regimens. Discussion of the effectiveness of various antimicrobial agents for therapy of associated infections and extensive clinical protocols have been included as well. Finch and Gander have described how an understanding of antimicrobial pharmacodynamics can lead to improved efficacy, and the chapter by Dodge et al. on beta-lactam therapy is insightful. The detailed protocols provided by Lewis and Raad, Stryjewski and Corey, and Antonios et al. should aid physicians in the treatment of these refractory and often difficult-to-treat infections. It is our hope that this book will provoke thought and lead to more research resulting in improved therapies for infections where biofilms play a substantial role.

Table of Contents

SECTION III   Emerging Issues, Assays, and Models

SECTION IV Overview of Antiinfective Agents and Clinical Therapy

Contributors

Vera Antonios, M.D.
Mayo Clinic, Rochester, Minnesota

Steven L. Barriere, Pharm.D.
Theravance, Inc., South San Francisco, California

Elie Berbari, M.D.
Mayo Clinic, Rochester, Minnesota

Karen Bush, Ph.D.
Johnson & Johnson Pharmaceutical Research & Development,
LLC, Raritan, New Jersey

Howard Ceri, Ph.D.
The Biofilm Research Group, Department of Biological Sciences,
University of Calgary, Calgary, Alberta, Canada

G. Ralph Corey, M.D.
Department of Medicine, Division of Infectious Diseases, Duke University
Medical Center and Duke Clinical Research Institute, Durham,
North Carolina

Ingrid L. Dodge, Ph.D.
Johnson & Johnson Pharmaceutical Research & Development,
LLC, La Jolla, California

Corinne Dorel, Ph.D.
Unite de Microbiologie et Genetique, Composante INSA de Lyon,
Villeurbanne, France

Paul D. Fey, Ph.D.
Departments of Internal Medicine and Pathology and Microbiology,
University of Nebraska Medical Center, Omaha, Nebraska

Roger Finch, M.D.
Division of Microbiology and Infectious Diseases, Nottingham City Hospital,
University of Nottingham, Nottingham, United Kingdom

Steven M. Frey, M.S.
ICOS Corporation, Bothell, Washington

Sarah Gander, M.D.
Division of Microbiology and Infectious Diseases,
Nottingham City Hospital, University of Nottingham,
Nottingham, United Kingdom

Nafsika H. Georgopapadakou, Ph.D.
Methylgene, Inc., Montreal, Canada

Mahmoud A. Ghannoum, Ph.D.
Center for Medical Mycology, Department of Dermatology,
University Hospitals of Cleveland and Case Western Reserve University,
Cleveland, Ohio

Sean P. Gorman, Ph.D.
School of Pharmacy, The Queen's University of Belfast,
Medical Biology Centre, Belfast, United Kingdom

Luke D. Handke, Ph.D.
Department of Pathology and Microbiology, University of
Nebraska Medical Center, Omaha, Nebraska

Daniel J. Hassett, Ph.D.
Departments of Molecular Genetics, Biochemistry and Microbiology,
University of Cincinnati College of Medicine, Cincinnati, Ohio

Stephen Hawser, Ph.D.
Arpida AG, Muenchenstein, Switzerland

Julie M. Higashi, M.D., Ph.D.
Division of Infectious Diseases, Department of Medicine,
University of California and Veterans Affairs Medical Center,
San Francisco, California

Matthias A. Horstkotte, M.D.
Institut für Infektionsmedizin, Zentrum für Klinisch-Theoretische Medizin,
Universitätsklinikum Hamburg-Eppendorf, Hamburg, Germany

Khalid Islam, Ph.D.
Arpida AG, Muenchenstein, Switzerland

Xiuping Jiang, Ph.D.
Department of Food Science and Human Nutrition, Clemson University,
Clemson, South Carolina

David S. Jones, Ph.D.
School of Pharmacy, The Queen's University of Belfast,
Medical Biology Centre, Belfast, United Kingdom

Gregory Jubelin, Ph.D.
Unite de Microbiologie et Genetique, Composante INSA de Lyon,
Villeurbanne, France

Niilo Kaldalu, Ph.D.
Institute of Technology, Tartu University, Tartu, Estonia

Iris Keren, Ph.D.
Northeastern University, Boston, Massachusetts

Johannes K.-M. Knobloch, M.D.
Institut für Infektionsmedizin, Zentrum für Klinisch-Theoretische Medizin,
Universitätsklinikum Hamburg-Eppendorf, Hamburg, Germany

Philippe Lejeune, Ph.D.
Unite de Microbiologie et Genetique, Composante INSA de Lyon,
Villeurbanne, France

Kim Lewis, Ph.D.
Northeastern University, Boston, Massachusetts

Russell E. Lewis, Pharm.D.
University of Houston College of Pharmacy, Houston, Texas

Isaiah Litton, Pharm.D.
School of Pharmacy, West Virginia University, Morgantown, West Virginia

Dietrich Mack, M.D.
Chair of Medical Microbiology and Infectious Diseases, The Clinical School,
University of Wales Swansea, Swansea, United Kingdom

Susan Meier-Davis, DVM-Ph.D.
Corebio Group, Inc., Belmont, California

Douglas W. Morck, Ph.D.
The Biofilm Research Group, Department of Biological Sciences,
University of Calgary, Calgary, Alberta, Canada

Merle E. Olson, Ph.D.
The Biofilm Research Group, Department of Biological Sciences,
and Microbiology and Infectious Diseases, University of Calgary,
Calgary, Alberta, Canada

Douglas Osmon, M.D.
Mayo Clinic, Rochester, Minnesota

John L. Pace, Ph.D.
Protez Pharmaceuticals, Inc., Malvern, Pennsylvania

Luciano Passador, Ph.D.
Department of Microbiology and Immunology
University of Rochester Medical Center, Rochester, New York

Issam I. Raad, M.D.
Department of Infectious Diseases, Infection Control, and Employee Health,
University of Texas M.D. Anderson Cancer Center, Houston, Texas

Harald Rinde, M.D. LLD.
BioBridged Strategies, San Diego, California

Holger Rohde, M.D.
Institut für Infektionsmedizin, Zentrum für Klinisch-Theoretische Medizin,
Universitätsklinikum Hamburg-Eppendorf, Hamburg, Germany

Mark E. Rupp, M.D.
Department of Internal Medicine, Division of Infectious Diseases,
University of Nebraska Medical Center, Omaha, Nebraska

Matthew K. Schinabeck, M.D.
Division of Infectious Diseases and Center for Medical Mycology, University
Hospitals of Cleveland and Case Western Reserve University, Cleveland, Ohio

Devang Shah, Ph.D.
Northeastern University, Boston, Massachusetts

Karen Joy Shaw, Ph.D.
Johnson & Johnson Pharmaceutical Research & Development,
LLC, La Jolla, California

Andrea Smiley, Ph.D.
Departments of Molecular Genetics, Biochemistry and Microbiology,
University of Cincinnati College of Medicine, Cincinnati, Ohio

Amy L. Spoering, Ph.D.
Northeastern University, Boston, Massachusetts

Douglas G. Storey, Ph.D.
The Biofilm Research Group, Department of Biological Sciences,
University of Calgary, Calgary, Alberta, Canada

Martin E. Stryjewski, M.D., MHS
Department of Medicine, Division of Infectious Diseases,
Duke University Medical Center and Duke Clinical Research Institute,
Durham, North Carolina

Paul M. Sullam, Ph.D.
Division of Infectious Diseases, Department of Medicine,
University of California and Veterans Affairs Medical Center,
San Francisco, California

John G. Thomas, Ph.D.
Department of Pathology, School of Medicine, West Virginia
University, Morgantown, West Virginia

Kenneth D. Tucker, Ph.D.
National Cancer Institute at Frederick, Frederick, Maryland

Jan Verhoef, M.D.-Ph.D.
Eijkmann-Winkler Institute of Medical Microbiology, University of Utrecht,
Utrecht, The Netherlands

Rosaire Verna, M.D.
Georgetown University School of Medicine, Washington, D.C.

Section I

Biofilms: Background,
Significance, and Roles of
Catheters and Indwelling
Devices

1 Microbial Biofilms

Xiuping Jiang and John L. Pace

CONTENTS

1.1 INTRODUCTION

Bacteria exist in two basic states, planktonic or sessile cells. It is believed that planktonic cells are important for rapid proliferation and spread into new territories, whereas sessile or slow-growing populations are focused on perseverance. Studies have revealed that the adherent bacteria, growing in consortia known as biofilms, are present in virtually all natural and pathogenic ecosystems (1,2). These biofilms are defined as structured communities of microbial species embedded in a biopolymer matrix on either biotic or abiotic substrata.

Zobell (3) published the first scientific study of biofilms. However, it was not until the 1970s that researchers realized that biofilms are universally present (1). Both abiotic and biotic surfaces such as mineral, metal, animal or plant surfaces, lung and intestine, and all types of medical implants are subject to bacterial colonization and biofilm formation. On the one hand, beneficial properties of biofilms have been put to use in industrial processes, but on the other hand they pose substantial challenges including chronic biofilm-related infections (4–6). Most importantly, the biofilm is characterized by its resistance to biocides, antibiotic chemotherapy, and clearance by humoral or cellular host defense mechanisms (5,7,8). Therefore, treatments with traditional concentrations of biocides or antibiotics are ineffective at eradicating the biofilm populations. In order to control unfavorable biofilm formation on a variety of surfaces important to the medical practices and industry, and to develop preventive strategies and methods for biofilm control, we need to fully understand the mechanisms involved in initial attachment, development of the biofilm phenotype, maturation and detachment, and the related regulatory processes at the molecular level.

1.2 BIOFILM FORMATION

Biofilm formation is a complex dynamic process. Earlier studies in environmental and industrial microbiology have already examined biofilm formation in various ecosystems, and concluded that bacteria form biofilms in essentially the same manner regardless of which environment they inhabit (1). Surfaces are normally conditioned with water, lipids, albumin, extracellular polymer matrix, or other nutrients from the surrounding environment (8). Bacteria adhere to the surfaces, initially in a reversible association and then through irreversible attachment, and eventually develop into an adherent biofilm of highly structured and cooperative consortia (2,9,10). Mature biofilms typically consist of differentiated mushroom- and pillar-like structures of cells embedded in copious amounts of extracellular polymer matrix or glycocalyx, which are separated by water-filled channels and voids to allow convective flows that transport nutrients and oxygen from the interface to the interior parts of the biofilm, and remove metabolic wastes.

The process of microbial adherence to surfaces is largely dictated by a number of variables, including the species of bacteria, cell surface composition, nature of surfaces, nutrient availability, hydrodynamics, cell-to-cell communication, and global regulatory networks (2,9,11–13). There are at least three phases involved in the biofilm formation process (2). The first one is characterized by the redistribution of attached cells by surface motility. A second phase is from the binary division of attached cells, and the third is aggregation of single cell or cell flocs from the bulk fluid to the developing biofilm. This accumulation step requires coordinated efforts from the biofilm community to produce a well-organized mature biofilm. The following are major stages involved in the process of biofilm formation.

1.2.1 Reversible Attachment

Once at the surface, different physical, chemical, and biological processes take place during this initial interaction between bacterial cell and the surface. On the abiotic surface, primary attachment between bacteria and the surface is generally mediated by non-specific interactions such as electrostatic, hydrophobic, or van der Waals forces, whereas adhesion to biotic surface such as tissue is through specific molecular (lectin or adhesin) docking mechanisms (8). Planktonic cells are thought to initiate the contact with a surface either randomly or in a directed fashion via chemotaxis and motility. Earlier studies have shown that the rate of bacterial adhesion to a wide variety of surfaces is affected by some physical characteristics such as hydrophobicity of surfaces, which is determined by bacterial surface-associated proteins (14,15). Other studies suggest that motility is very important for the planktonic cells to make initial contacts with an abiotic surface, and for bacteria to spread across the surface (16,17). Flagella-mediated motility can bring the cell within close proximity of the surface to overcome repulsive forces between bacterium and the surface where bacterium will be attached. O'Toole and Kolter (16) used the surface attachment defective (*sad*) mutants of *Pseudomonas aeruginosa* PA14 to elucidate the role of flagella and type IV pili in early stages of biofilm formation on the abiotic surface. Type IV pili are responsible for a form of surface-associated movement known as twitching motility, and have been shown to play an important role in bacterial adhesion to eukaryotic cell surfaces and pathogenesis (13). Mutants defective in the production of flagellum attached poorly to polyvinylchloride (PVC) plastic, whereas mutants defective in biogenesis of the polar-localized type IV pili formed a monolayer of cells on the PVC plastic but didn't develop microcolonies over the course of experiment. As suggested by the authors, twitching motility may enable the cells to migrate along the surface to form multicellular aggregates characteristic of the wild-type strain. These results suggest that flagella-mediated motility is important for the formation of a bacterial monolayer on the abiotic surface, whereas type IV pili appear to play a role in subsequent microcolony formation.

Contrary to the conventional wisdom, bacteria form biofilms preferentially in very high-shear environments as compared with low shear environments (5). One of the explanations is that the high-shear flow aids in organization and strengthens the biofilm, making it more resistant to mechanical breakage.

1.2.2 IRREVERSIBLE ATTACHMENT

After binding to the surface through exopolymeric matrix, bacterial cells start the process of irreversible adhesion, proliferation, and accumulation as multilayered cell clusters. These extracellular matrices, composed of a mixture of materials such as polysaccharides, proteins, nucleic acids, and other substances, are considered to be essential in cementing bacterial cells together in the biofilm structure, in helping to trap and retain nutrients for biofilm growth, and in protecting cells from dehydration and the effects of antimicrobial agents. The role of slime production in *P. aeruginosa* biofilm has been extensively studied (18,19). This microorganism produces alginic acid as an exopolysaccharide glycocalyx under the control of *algACD* gene cluster (19). In the biofilm, *algC is* expressed at levels approximately 19 times greater than that in planktonic cells. High molecular weight exopolysaccharide polymers, possibly microbial DNA, and surface proteins are used to efficiently retain the bacterial cells within the biofilm matrix. Many genes with related function are regulated at the transcriptional level, permitting microorganisms to switch from planktonic to sessile forms under different environmental conditions.

1.2.3 MATURATION OF BIOFILM FORMATION

Once having irreversibly attached to a surface, bacterial cells undergo phenotypic changes, and the process of biofilm maturation begins. Bacteria start to form micro-colonies either by aggregation of already attached cells, clonal growth (cell division) or cell recruitment of planktonic cells or cell flocs from the bulk liquid. The attached cells generate a large amount of extracellular components which interact with organic and inorganic molecules in the immediate environment to create the glycocalyx. Confocal laser scanning microscopy revealed a novel and complex three-dimensional structure of the biofilm community (20). Costerton et al. (9) suggested that the microcolony is the basic unit of biofilm growth in the same way as the tissue makes up the more complex organisms. Analogously the water channels inside the biofilm represent a primitive circulatory system, mimicking that of higher organisms.

Microbial biofilms display discrete temporal and spatial structure. The basic "design" of the mushroom-like microcolonies with intervening water channels is optimal from a nutrient access point of view, because the nutrients are transported to the bacteria via the water channels at low flow rate (10,21). There are many microenvironments within a biofilm—each varying because of differences in local conditions such as nutrient availability, pH, oxidizing potential (redox), and so on. Cells near the surface of the biofilm microcolony are exposed to high concentrations of O_2, while near the center oxygen is rapidly depleted to near anaerobic levels (22). The steep oxygen gradients are paralleled by gradients for either nutrients or metabolites from the biofilm, which creates a heterogenic environment even for the single-species of biofilm (23,24). Apparently, biofilms display both structural and metabolic heterogeneity. In adapting to these niches, bacteria within a biofilm display many types of phenotypes, with broad metabolic and replicative heterogeneity, providing the community as a whole with enormous capability to resist stresses, whether from host defense systems or antimicrobial agents (10).

1.2.4 ROLE OF QUORUM SENSING IN BIOFILM FORMATION

It is now well established that bacterial cells communicate through the secretion and uptake of small diffusible molecules. Recent evidence indicates that biofilm formation might be regulated at the level of population density-dependent gene expression controlled by cell-to-cell signaling, or quorum sensing (QS) (25). A large number of bacterial species are known to possess this communication mechanism, through which bacteria can sense changes in their environment and coordinate gene expression in favor of the survival for the entire community (13,25). These cell–cell communication systems regulate various functions as diverse as motility, virulence, sporulation, antibiotic production, DNA exchange, and development of multicellular structures such as biofilm and fruiting body formation (26–28).

Dozens of putative bacterial signals have been discovered. Among them are acylhomoserine lactones (AHLs) known as autoinducer-1 (AI-1) signals in gram-negative bacteria (29), amino acids and short cyclic peptide signals in gram-positive bacteria (30), and a furanosyl borate diester known as autoinducer-2 (AI-2) signals of both groups (31,32). AHLs have been detected in naturally occurring biofilms (33). During bacterial growth, these signal molecules accumulate in the surrounding environment, and are transported among bacterial cells through diffusion across the bacterial membrane or active efflux and influx mechanisms (28). Until a critical threshold concentration is reached, these AHLs molecules start to bind to their cognate receptors, which in turn become activated, and stimulate or repress the transcription of target genes (32).

An involvement of QS in the regulation of biofilm formation was originally reported for *P. aeruginosa*, a gram-negative opportunistic human pathogen, which grows primarily as biofilms in the lungs of cystic fibrosis patients (25). At least two extracellular signal systems have been identified in *P. aeruginosa*, and they are involved in either cell-to-cell communication or cell density-dependent gene expression of various virulence factors and biofilm differentiation. Since QS mechanism requires high cell density, these signals may not be produced in the early stages of biofilm formation; however they can play an important role on mature biofilm differentiation. A study of a *P. aeruginosa lasI-rhiI* signaling mutant revealed that the mutant strain formed a thin, flat, and undifferentiated biofilm as compared with wild type bacterium, which formed characteristic microcolonies separated by water channels and void spaces (25). A signal molecule, N-3-oxo-dodecanoyl homoserine lactone (OdDHL), was involved in the differentiation of the biofilm into the normal characteristic structure. A recent study suggests that environmental factors such as nutrients and hydrodynamic conditions may play a similar role to QS in the regulation of biofilm structural development (34). As a matter of fact, in faster flowing system, signal molecules may be washed out of the biofilm by the flowing bulk liquid, which may diminish the impact of QS on the biofilm formation. QS concept was initially discovered from studies on planktonic bacteria in suspension. As for the biofilm sessile cells, the cell communication between signal-producing and signal-receiving cells is dependent on the proximity and the diffusion path between them. Therefore, QS signals may regulate gene expression in biofilms much more individually and subtly. Further study in this field is warranted.

1.2.5 DETACHMENT OF BIOFILMS

Bacterial cells detached from the biofilm reenter the planktonic state, and may reattach to virgin areas and initiate a new round of biofilm formation. Therefore, biofilm formation may be cyclical in nature. Hydrodynamic flow of liquid over and through the biofilm would likely break parts of the biofilm and carry them away from the surface (10). Nutrient starvation, overexpression of alginate lyase, loss of EPS, and perhaps QS may be involved in the control of biofilm detachment (18,35). Further study is needed in elucidating the environmental factors, and functions or regulatory pathways involved in the detachment of bacterial cells or cell clumps from biofilms.

1.3 THE UBIQUITOUS CHALLENGE

Indwelling medical devices (IMDs) include urinary and vascular catheters, shunts, prosthetic joints, heart valves, pacemakers, stents, endotracheal tubes, breast, and miscellaneous fixed and removable (contact lens, denture) implants (5,36).

With medical advancement over the past few decades, increased use of these devices has been favored by both physicians and patients. As an example, about 200 million catheters of all types are being used to treat various human diseases in the United States (36,37). IMDs are normally surrounded by tissues and body fluids such as blood, saliva, urine, and synovial fluid, which can coat the implant surface with platelets, plasma, body fluids, and proteins such as fibrins, fibronectin, collagen, and laminin, forming a glycol-proteinaceous conditioning film (38). Numerous studies have demonstrated that medical devices composed of a wide range of biomaterials are prone to microbial colonization and biofilm formation by a variety of bacterial and fungal species (5,38–40). Normal commensals of humans are a potential source for colonizing and forming biofilms on the IMDs. The skin surrounding the catheter insertion site has been implicated as the most common source of central venous catheters (CVC) colonization (41). Skin commensals such as *Staphylococcus epidermidis*, can migrate from the catheter insertion site along the catheter outer surface into the subcutaneous portion, forming a biofilm and causing the catheter-related infection (42). However, some medical procedures conducted by healthcare providers prior to or during IMD implantation can introduce other types of pathogens as well. Predominant microorganisms responsible for biofilm formation on devices include coagulase-negative staphylococci, *Staphylococcus aureus*, *Enterococcus faecalis*, and *Streptococcus* spp., and gram-negative *Escherichia coli*, *Klebsiella pneumoniae*, *Acinetobacter* spp., *Proteus mirabilis*, and *P. aeruginosa*, and *Candida* spp. (36,38). Coagulase-negative staphylococci, *S. aureus* and *Candida* spp. are commonly associated with intracardiac prosthetic devices, leading to pros-thetic valve endocarditis. As for artificial joints, both *S. epidermidis* and *S. aureus* account for more than 90% of prosthetic joint infections (43).

Early biofilm studies focused primarily on natural ecosystems such as marine, river, and other aqueous environments (44). Both natural and industrial aquatic systems contain sufficient nutrients for bacterial growth and metabolism, and available surface area for the bacteria to attach and form biofilms. In pristine alpine streams, the sessile population is about 3 to 4 logs greater than that of planktonic cells (1). In general, submerged surfaces tend to accumulate more nutrients than in the water column facilitating biofilm formation. However, in the extremely oligotrophic aquatic environment, bacteria generally do not adhere to surfaces in nutrient-deprived ecosystems such as deep-ocean and deep-ground water, where bacteria change to ultramicrobacteria as their primary starvation-survival response (45).

Biofilm formation on industrial aquatic systems plugs filters and injection faces, generates harmful metabolites (e.g., H_2S), reduces heat exchange efficiency, and causes the corrosion of metals on ship hulls or other equipment (4,46).

In food systems, the formation of microbial biofilms on either food or food processing surfaces is detrimental and undesirable. The most common places for biofilm accumulation are floors, bends in water pipes, processing surfaces, conveyor belts, slicer blades, rubber seals, and utensils for holding and carrying the food products (47–49). These abiotic surfaces are made of diversified materials such as cement, polypropylene, polyethylene, rubber, aluminum, glass, and stainless steel. In addition, both chicken and beef skin can also serve as a biotic surface for bacteria to attach and form biofilms (50). Foodborne pathogens such as *E. coli* O157:H7,

Listeria monocytogenes, Yersinia enterocolitica, Salmonella spp., and *Campylobacter jejuni* can form either single species or multi-species biofilms on food surfaces and food processing or contact equipment (49,51–53). Biofilm formation of the important foodborne pathogen *L. monocytogenes* and spoilage microorganism *Pseudomonas* spp. have been studied extensively in the food processing environment (54–57). Both microorganisms can attach to surfaces of various materials through hydrophobic, electrostatic, or exopolymer interactions. The pathogenic biofilms on food processing surfaces act as a major source for cross contamination of food products, which lead to a serious problem in food safety and economic losses. Biofilm formation is also an important issue in food sanitary programs especially considering the high resistance of biofilms to biocides, drying, heat, nutrient deprivation, antibiotics, sanitizers, and other stresses.

1.4 MECHANISMS OF BIOFILM RESISTANCE TO ANTIBIOTICS

Inherent resistance of biofilm bacteria to antibiotics has been found to be a general phenomenon. Biofilms may exhibit antibiotic resistance three or more orders in magnitude greater than those displayed by planktonic bacteria of the same strain depending on the species–drug combination (58). After exposure to the killing effects of antibiotics, a small surviving population of persistent bacteria can repopulate the surface immediately, and become more resistant to further antibiotic treatment (see Chapter 12). Paradoxically, once dispersed from the biofilm, those bacterial cells typically revert to an antibiotic susceptible form (1). A number of additional factors have been considered for the resistance of biofilm cells, including the presence of a diffusion barrier to the chemicals posed by the glycocalyx, interaction of exopolymer with the antibiotics, slow growth mode of sessile cells, hypermutation, multicellular nature of the biofilm, and possible genetic expression of certain resistance genes.

1.4.1 SLOW PENETRATION OF ANTIBIOTICS BY BIOFILM EXOPOLYMERS

Biofilm extrapolymeric matrix such as exopolysaccharides (EPS) has the potential to reduce the penetration of antibiotics and biocides either by physically slowing diffusion or chemically reacting with these compounds. EPS acts as an ion exchanger, and sequesters hydrophilic and positively charged antibiotics such as aminoglycosides. In *P. aeruginosa* biofilms, both tobramycin and gentamicin (aminoglycosides) penetrate more slowly due to interaction with extracellular polymers such as alginate (59). In contrast, fluoroquinolones such as ofloxacin and ciprofloxacin were shown to readily diffuse into biofilms (24). Tetracycline was able to reach all constituent cells of uropathogenic *E. coli* biofilms after 7.5 to 10 minutes of exposure, and there were no pockets within the biofilms where the antibiotics failed to reach (60). Walters et al. (24) compared antibiotic penetration, oxygen limitation, and metabolic activity effects on tolerance of *P. aeruginosa* biofilms to ciprofloxacin and tobramycin. Their results suggest that oxygen limitation and low metabolic activity in the biofilm interior, rather than the poor antibiotic penetration, are responsible for the antibiotic tolerance.

Apparently, poor antibiotic penetration is not the most important protective mechanism in biofilms.

1.4.2 SLOW GROWTH RATE OF BIOFILM CELLS

Due to nutrient limitation, biofilm cells have slow growth rates comparable with that of stationary phase cells. Moreover, bacterial cells in biofilms constitute a heterogeneous population with varied growth rates in different compartments of the biofilm, and varied sensitivity to antibiotics (61). In a *K. pneumoniae* biofilm model, the average growth rate for biofilm cells was only 0.032 h^{-1} as compared to that of planktonic bacteria of 0.59 h^{-1} (62). Transmission electron microscopy revealed that biofilm bacteria were affected by ampicillin near the periphery of the biofilm but were not affected in the interior, where bacterial cells probably grew slowly or entered in a stationary-phase state due to the limitation for nutrients and oxygen there. Slow growing cells generally have much reduced metabolic activity, thereby resulting in reduction of antimicrobial susceptibility.

1.4.3 INCREASED RATE OF GENETIC TRANSFER IN THE BIOFILM

Bacteria may acquire antibiotic resistance through either horizontal gene transfer (such as genes encoded on plasmid, transposon, or integron) or through mutation in different chromosomal loci. Since most bacteria found in nature live in biofilms on surfaces or at interfaces, it is likely that gene transfer by conjugation plays an important role for spreading antibiotic resistance among different bacterial species. In fact, biofilms are ideally suited to the exchange of genetic material of various origins due to the close contact and relative spatial stability of bacteria within biofilms (2). Tetracycline resistance encoded by Tn*916*-like elements was transferred readily from four tetracycline-resistant *Streptococcus* spp. to other streptococci within a model oral biofilm (63). Another study demonstrated that conjugation rates in biofilms were 1,000-fold higher than those determined by conventional plating techniques, indicating that gene transfer occurs far more frequently in biofilms than previously thought (64). Furthermore, genetic transfer rates in biofilms are orders of magnitude higher than those between planktonic cells in suspension (65). Ghigo (65) studied the role of conjugative pili in biofilm formation produced in *E. coli* carrying the F-factor, and discovered that natural conjugative plasmids express factors that induce planktonic bacteria to form or join biofilm communities. Thus antibiotic resistance and virulence factor expression may have been selected for in bacteria, bearing conjugative plasmids and that readily form biofilms, by the use of antibiotics and biocides in medical practice and industry.

1.4.4 EXPRESSION OF RESISTANCE GENES IN BIOFILMS

During stationary phase, Gram-negative bacterial cells develop stress-resistance response by expressing a number of genes under the regulation of a stationary-phase sigma factor known as RpoS (66). It was reported that RpoS regulated quorum sensing in *P. aeruginosa*, suggesting the role of RpoS in the later stages of biofilm formation (67). Due to the physiological similarity of biofilm cells and stationary phase cells, it is presumable that the expression of RpoS-regulated or biofilm-specific genes occurs in the biofilm (62). Genetic studies indicated that, in general, more genes are expressed in biofilms than in single-phase batch cultures of planktonic cells, and these up-regulated genes include many open reading frames of unknown function (19).

Beta-lactam antibiotics are routinely used for treating chronic *P. aeruginosa* infections in the lungs of cystic fibrosis patients. By using a green fluorescent protein (GFP) reporter consisting of a fusion of the *ampC* promoter to *gfp*, both the dynamic and spatial distribution of beta-lactamase induction in *P. aeruginosa* cells in biofilm were investigated (68). The study demonstrated that sub-MICs of imipenem significantly induced the expression of *ampC* in the peripheries of the microcolonies but not in the centers of the microcolonies.

Sub-lethal concentrations of antibiotic select resistance in microorganisms (69). As discussed above, the varied antibiotic concentrations in the different compartments of the biofilm may exert different selective pressure on the biofilm bacteria. In considering the heterogeneous populations inside the biofilm due to different concentrations of oxygen, nutrients, pH and other environmental factors, biofilms cells are expected to possess different levels of antibiotic resistance.

1.4.5 Hypermutation in Biofilms

The high antibiotic resistance of biofilms may be explained in part by the hypermutation phenomenon as observed in stressed bacterial cells. Both environmental and physiological stress conditions, such as starvation and antibiotic treatment, can transiently increase the mutation rates in sub-populations of bacteria allowing the bacteria to evolve faster (66,70). Upon exposure to stresses, bacterial cells undergo transient, genome-wide hypermutation (also called adaptive mutation). Approximately 1% of pathogenic *E. coli* and *Salmonella* isolates from both food-related outbreaks of disease and the natural environment, and 20% of *Pseudomonas* isolates from the lungs of cystic fibrosis patients are strong mutators with very high mutation rates (70–72). These hypermutable strains are mainly defective in methyl-directed mismatch repair (MMR) genes, a DNA repair and error-avoidance system (70). Antibiotics, as stress producers, not only select for resistance to themselves but may also increase the mutation rate, thus indirectly selecting for resistance to unrelated antibiotics. For example, fluoroquinolones induce the SOS system, a global response to DNA damage, whereas streptomycin treatment results in mistranslation and induction of a *recA*- and *umuDC*-independent mutator (70,73,74). Oliver et al. (72) compared the multiple antibiotic resistance levels between hypermutator *P. aeruginosa* and nonmutator strains, and found that the percentage of *P. aeruginosa* mutator strains carrying multi-drug resistance is significantly higher than that of nonmutators. However, to confirm the link of antibiotic resistance in biofilms with hypermutation, further study is needed.

1.4.6 Multicellular Nature of the Biofilm Community

Interestingly, bacterial survivors from antibiotic-treated biofilms remain susceptible to antibiotics, indicating that resistance in the biofilm state is not entirely due to mutation or acquisition of a resistance gene but rather to persister variants (75). These persisters are responsible for biofilm regrowth when the treatment is discontinued. When exposed to environmental stresses, the persisters arise at a considerably higher rate (10 to 10,000-fold) than mutants (75).

Biofilm cells have been recognized as multicellular organisms, using the sophisticated signal transduction networks to regulate gene expression and cell differentiation (12,76). This multicellular behavior permits biofilm cells to efficiently utilize resources for cell growth and provide collective defense against clearance by humoral or cellular host defense mechanisms and killing by biocides or antibiotic chemotherapy. As an example, penicillin-susceptible *S. aureus* formed a penicillin-resistant biofilm on a pacemaker, and caused recurring septicemia that could not be completely eradicated by antibiotic treatment (77). One possible explanation might be that bacteria in the surface layer of biofilms degrade or modify antibiotics, long after they have lost their viability, shielding their more deeply embedded neighbor cells from harmful antimicrobial agents.

1.5 CONTROLLING BIOFILMS

Likely effective strategies for biofilm treatment include prevention of bacterial cell adhesion to the substratum, reduction of polysaccharide production, and disruption of cell-to-cell communication involved in biofilm formation through physical, chemical, and biological approaches.

1.5.1 QUORUM SENSING ANALOG USED FOR BIOFILM TREATMENT

Currently, much attention is focused on developing new ways to prevent biofilm formation on both industrial and medical surfaces. Since QS plays such a substantial role in biofilm formation, one strategy considered for preventing biofilm formation is to coat or embed surfaces with compounds capable of interfering with related signaling mechanisms. Furanone compounds, produced by Australian red macroalga *Delisea pulchra*, and structurally similar to AHLs molecules, have been demonstrated to inhibit AHL-regulated phenotypes, such as bacterial swarming in *Serratia liquifaciens* and production of bioluminescence in *Vibrio* spp. (78). They hypothesized that furanone compounds may competitively bind to the AHL receptor sites of putative regulatory proteins, and interfere with AHL regulatory systems. A green fluorescent protein-based-AHL biosensor, in combination with laser confocal scanning microscopy, revealed that a synthetic derivative of natural furanone compounds penetrates the biofilm matrix and inhibits cell signaling in most sessile cells (79). As a consequence, the furonone-analog interfered with the biofilm structure and enhanced bacterial detachment from the surface. However, the QS-analog did not affect the initial attachment of *P. aeruginosa* to the surface (79). Recently, using a GeneChip® microarray technology, Kristoffersen et al. (80) was able to identify furanone target genes and to map the QS regulon. In a mouse pulmonary infection model, they also demonstrated that the synthetic furanone-analog compound inhibited the expression of several QS-controlled virulence factors of infecting

P. aeruginosa PAO1. Another study also reported that RNA III inhibiting peptide was able to disrupt QS mechanisms in biofilm-producing *S. aureus* (81). Apparently the discovery of the QS communication systems in microorganisms opens up many possibilities for designing novel non-antibiotic drugs which aim at preventing biofilm formation rather than killing the microorganisms. In combination with traditional antibiotics, these novel QS analogs may extend the effectiveness of currently-used antibiotics for controlling biofilm-related bacterial infections.

1.5.2 COMBINATION TREATMENTS OF BIOFILMS

Since no single method or chemical completely eliminates biofilm microorganisms, a combination of various treatments have been tested for controlling biofilm formation. These can include combinations of antibacterial agents such as the common practice of administering rifampin with another antibacterial, or the combination of both chemical and physical intervention (82–85). Raad et al. (83) reported that the combination of minocycline and rifampin was highly effective in preventing the colonization of catheters with slime-producing *S. epidermidis* and *S. aureus*. Both *in vitro* and *in vivo* (animal model) studies have indicated that low-frequency and low-power-density ultrasound combined with aminoglycoside antibiotics significantly reduce *E. coli* biofilms (86). Combination of low electrical currents with antibiotic treatment also appears more effective in controlling biofilms by increasing the diffusion of charged molecules and antibiotics through the biofilm matrix (82,87). However, it is currently unknown whether all of these approaches can be utilized in the patient.

Most recently, Ehrich et al. (43) described a scheme for an intelligent implant to fight biofilm-related infection. The intelligent implant would respond to bacterial communication signals through integral, gated reservoirs that release compounds to prevent biofilm formation along with high concentrations of antibiotics to wipe out the nearby planktonic bacterial cells. With further understanding of mechanisms involved in biofilm formation, more and more novel methods are expected to emerge for enhancing the killing of harmful bacteria and eradicating biofilms from surfaces.

1.5.3 CHEMICAL AND PHYSICAL TREATMENTS FOR BIOFILMS

1.5.3.1 What Can Be Learned from Industrial Environments?

Peracetic acid and chlorine are considered as the most efficient industrial disinfectants to remove biofilms from surfaces. However, the effect is only obtained if disinfectants are used at high concentrations for long reaction times and after pre-treatment with detergent (48,49). As expected, bacteria in suspension are more sensitive to disinfectants than bacteria in biofilms (88,89). Among the disinfectants tested, peracetic acid is more effective than the aldehydes, hydrogen peroxide, or chlorine against biofilm bacteria (90). Bacteria can become resistant to these biocides as well, and to avoid a buildup of resistant pathogens rotation of disinfectants should be considered.

In order to control biofilm formation it is recommended that various preventive and control strategies be implemented such as hygienic layout, design of equipment, choice and coating of materials, correct use and selection of detergents and disinfectants in combination with physical methods and use of bacteriocins and enzymes (11,48). In addition, appropriate sanitation practices must be adopted, i.e., critical sites should be monitored periodically and undergo rotation sanitation. These approaches apply to the medical arena as well.

1.6 CONCLUSIONS

Bacterial biofilms are ubiquitous both in natural environment and pathogenic ecosystems. The formation of biofilms can be beneficial or detrimental, especially in regard to IMDs. A number of variables, including the species of bacteria, cell surface composition, nature of surfaces, nutrient, hydrodynamics, cell-to-cell communication, and global regulatory networks have been studied for their impact on the development of biofilms. Most recently, with the further understanding of cell-to-cell communication in microorganisms, quorum sensing has emerged as one of the most important mechanisms for controlling the development of highly structured and cooperative biofilm consortia on both biotic and abiotic substrata.

In general, the biofilms are highly resistant to various environmental stresses, biocides, antibiotic chemotherapy, and clearance by humoral or cellular host defense mechanisms. Resistance of biofilms to antibiotic treatment often results in the failure of the chemotherapy and further recalcitrant infection associated with IMDs. As a matter of fact, biofilms are associated with more than 65% of all medical infections (Centers for Disease Control and Prevention, Atlanta, GA). Further study on the mechanisms of antibiotic resistance by biofilms will certainly expedite the development of preventive strategies and methods for biofilm control, including physical, chemical, cell-to-cell communication, or molecular approaches.

REFERENCES

1. Costerton, J.W., Cheng, K.J., Geesey, G.G., Ladd, T.I., Nickel, J.C., Dasgupta, M., and Marrie, T.J., Bacterial biofilms in nature and disease. *Ann. Rev. Microbiol.* 41, 435–464, 1987.
2. Davey, M.E., and O'Toole, G.A., Microbial biofilms: from ecology to molecular genetics. *Microbiol. Mol. Biol. Rev.* 64, 847–867, 2000.
3. Zobell, C.E., The effects of solid surfaces upon bacterial activity. *J. Bacteriol.* 46, 39–56, 1943.
4. Characklis, W.G., and Cooksey, K.E., Biofilms and microbial fouling. *Adv. Appl. Microbiol.* 29, 93–137, 1983.
5. Donlan, R.M., and Costerton, J.W., Biofilms: survival mechanisms of clinically relevant microorganisms. *Clin. Microbiol. Rev.* 15, 167–193, 2002.
6. Eisenmann, H., Letsiou, I., Feuchtinger, A., Beisker, W., Mannweiler, E., Hutzler, P., and Arnz, P., Interception of small particles by flocculent structures, sessile ciliates, and the basic layer of a wastewater biofilm. *Appl. Environ. Microbiol.* 67, 4286–4292, 2001.
7. Costerton, J.W., Stewart, P.S., and Greenberg, E.P., Bacterial biofilms: a common cause of persistent infections. *Science* 284, 1318–1322, 1999.
8. Dunne, W.M., Bacterial adhesion: seen any good biofilm lately? *Clin. Microbiol. Rev.* 15, 155–166, 2002.

9. Costerton, J.W., Lewandowski, Z., Caldwell, D.E., Korber, D.R., and Lappin-Scott, H.M., Microbial biofilms. *Annu. Rev. Microbiol.* 49, 711–745, 1995.

10. Stoodley, P., Sauer, K., Davies, D.G., and Costerton, J.W., Biofilms as complex differentiated communities. *Annu. Rev. Microbiol.* 56, 187–209, 2002.

11. Kumar, C.G., and Anand, S.K., Significance of microbial biofilms in the food industry: a review. *Int. Food Microbiol.* 42, 9–27, 1998.

12. Shapiro, J.A., Thinking about bacterial populations as multicellular micro-organisms. *Annu. Rev. Microbiol.* 52, 81–104, 1998.

13. Shirtliff, M.E., Mader, J.T., and Camper, A.K., Molecular interactions in biofilms. *Chem. Biol.* 9, 859–871, 2002.

14. Carpentier, B., and Cerf, O., Biofilms and their consequences, with particular references to hygiene in the food industry. *J. Appl. Bacteriol.* 75, 499–511, 1993.

15. Fletcher, M., and Loeb, G.I., Influence of substratum characteristics on the attachment of a marine pseudomonad to solid surfaces. *Appl. Environ. Microbiol.* 37, 67–72, 1979.

16. O'Toole, G., and Kolter, R., Flagellar and twitching motility are necessary for *Pseudomonas aeruginosa* biofilm development. *Mol. Microbiol.* 30, 295–304, 1998.

17. Pratt, L.A., and Kolter, R., Genetic analysis of *Escherichia coli* biofilm formation: roles of flagella, motility, chemotaxis and type I pili. *Mol. Microbiol.* 30, 285–293, 1998.

18. Boyd, A., and Chakrabarty, A.M., Role of alginate lyase in cell detachment of *Pseudomonas aeruginosa*. *Appl. Environ. Microbiol.* 60, 2355–2359, 1994.

19. Davies, D.G., and Geesey, G.G., Regulation of the alginate biosynthesis gene *algC in Pseudomonas aeruginosa* during biofilm development in continuous culture. *Appl. Environ. Microbiol.* 61, 860–867, 1995.

20. Lawrence, J.R., Korber, D.R., Hoyle, B.D., Costertion, J.W., and Caldwell, D.E., Optical sectioning of microbial biofilms. *J. Bacteriol.* 173, 6558–6567, 1991.

21. Rasmussen, K., and Lewandowski, Z., Microelectrode measurements of local mass transport rates in heterogeneous biofilms. *Biotechnol. Bioeng.* 59, 302–309, 1998.

22. Lewandowski, Z., Dissolved oxygen gradients near microbically colonized surfaces. In: Geesey, G.G., Lewandowski, Z., and Flemming, H.C., eds., *Biofouling and Biocorrosion in Industrial Water Systems*. Florida: Lewis, 1994:175–188.

23. de Beer, D., Stoodley, P., Roe, F., and Lewandowski, Z., Effects of biofilm structure on oxygen distribution and mass transport. *Biotechnol. Bioeng.* 43, 1131–1138, 1994.

24. Walters, M.C. III., Roe, F., Bugnicourt, A., Franklin, M.J., and Stewart, P.S., Contributions of antibiotic penetration, oxygen limitation, and low metabolic activity to tolerance of *Pseudomonas aeruginosa* biofilms to ciprofloxacin and tobramycin. *Antimicrob. Agents Chemother.* 47, 317–323, 2003.

25. Davies, D.D., Parsek, M.R., Pearson, J.P., Iglewski, B.H., Costerton, J.W., and Greenberg, E.P., The involvement of cell-to-cell signals in the development of a bacterial biofilm. *Science* 280, 295–298, 1998.

26. De Kievit, T.R., and Iglewski, B.H., Quorum sensing and microbial biofilms. In: Wilson, M., and Devine, D., eds., *Medical Implications of Biofilms*. Cambridge: University Press, 2003:18–35.

27. Hentzer, M., Givskov, M., and Eberl, L., Quorum sensing in biofilms: gossip in slime city. In: Ghannoum, M., and O'Toole, G.A., eds., *Microbial biofilms*. Washington, D.C.: ASM Press, 2004:118–140.

28. Smith, J.L., Fratamico, P.M., and Novak, J.S., Quorum sensing: a primer for food microbiologists. *J. Food Prot.* 67, 1053–1070, 2004.

29. Fuqua, C., and Greenberg, E.P., Self perception in bacteria: quorum sensing with acylated homoserine lactones. *Curr. Opin. Microbiol.* 1, 183–189, 1999.

30. Novick, R.P., Regulation of pathogenicity in *Staphylococcus aureus* by a peptide-based density-sensing system. In: Dunny, G.M., and Winans, S.C., eds., *Cell–Cell Signaling in Bacteria*. Washington, D.C.: ASM Press, 1999:175–192.

31. Chen, X., Schauder, S., Potter, N., van Dorsslaer, A., Pelczer, I., Bassler, B.L., and Hughson, F.M., Structural identification of a bacterial quorum-sensing signal containing boron. *Nature* 415, 545–549, 2002.

32. Miller, M.B., and Bassler, B.L., Quorum sensing in bacteria. *Annu. Rev. Microbiol.* 55, 165–199, 2001.

33. McLean, R.J.C., Whiteley, M., Stickerler, D.J., and Fuqua, W.C., Evidence of autoinducer activity in naturally occurring biofilms. *FEMS Microbiol. Lett.* 154, 259–263, 1997.

34. Purevdorj, B., Costerton, J.W., and Stoodley, P., Influence of hydrodynamics and cell signaling on the structure and behavior of *Pseudomonas aeruginosa* biofilms. *Appl. Environ. Microbiol.* 68, 4457–4464, 2002.

35. O'Toole, G., Kaplan, H.B., and Kolter, R., Biofilm formation as microbial development. *Annu. Rev. Microbiol.* 54, 49–79, 2000.

36. Thomas, J.G., Ramage, G., and Lopez-Ribot, J.L., Biofilms and implant infections. In: Ghannoum, M., and O'Toole, G.A., eds., *Microbial Biofilms*. Washington, D.C.: ASM Press, 2004:269–293.

37. Maki, D.G., and Mermel, L.A., Infections due to infusion therapy. In: Bennett, J.V., and Brachman, P.S., eds., *Hospital Infections*. Philadelphia: Lippincott-Raven, 1998:689–724.

38. Hanna, H.A., and Raad, I., Intravascular-catheter-related infections. In: Wilson, M., and Devine, D., eds., *Medical Implications of Biofilms*. Cambridge: University Press, 2003:86–109.

39. Gorman, S.P., and Jones, D.S., Biofilm complications of urinary tract devices. In: Wilson, M., and Devine, D., eds., *Medical Implications of Biofilms*. Cambridge: University Press, 2003:136–172.

40. Raad, I.I., and Hanna, H., Nosocomial infections related to use of intravascular devices inserted for long-term vascular access. In: Mayhall, C.G., ed., *Hospital Epidemiology and Infection Control*. Philadelphia: Lippincott Williams & Wilkins, 1999:165–172.

41. Raad, I.I., Costerton, W., Sabharwal, U., Sacilowski, M., Anaissie, E., and Bodey, G.P., Ultrastructural analysis of indwelling vascular catheters: a quantitative relationship between luminal colonization and duration of placement. *J. Infect. Dis.* 168, 400–407, 1993.

42. Maki, D.G., Pathogenesis, prevention, and management of infections due to intravascular devices used for infusion therapy. In: Bisno, S.L., and Waldvogal, F.A., eds., *Infections Associated with Indewelling Medical Devices*. Washington, D.C.: ASM Press, 1989:161–177.

43. Ehrlich, G.D., Hu, F.Z., Lin, Q., Costerton, J.W., and Post, J.C., Intelligent implants to battle biofilms: self-diagnosing, self-treating, self-monitoring artificial joints could combat postimplant infections attributable to biofilms. *ASM News* 70, 127–133, 2004.

44. Costerton, J.W., A short history of the development of the biofilm concept. In: Ghannoum, M., and O'Toole, G.A., eds., *Microbial Biofilms*. Washington, D.C.: ASM Press, 2004:4–19.

45. Kjelleberg, S., *Starvation in Bacteria*. New York: Plenum, 1993.

46. Rittmann, B.E., Biofilms in the water industry. In: Ghannoum, M., and O'Toole, G.A., eds., *Microbial Biofilms*. Washington, D.C.: ASM Press, 2004.

47. Farrell, B.L., Ronner, A.B., and Wong, A.C., Attachment of *Escherichia coli* O157: H7 in ground beef to meat grinders and survival after sanitation with chlorine and peroxyacetic acid. *J. Food Prot.* 61, 817–822, 1998.

48. Jessen, B., and Lammert, L., Biofilm and disinfection in meat processing plants. *International Biodeterioration and Biodegradation* 51, 265–269, 2003.
49. Chmielewski, R.A.N., and Frank, J.F., Biofilm formation and control in food processing facilities. *Comp. Rev. Food Sci. Food Safety* 2, 22–32, 2003.
50. Chung, K.T., Dickson, J.S., and Crouse, J.D., Attachment and proliferation of bacteria on meat. *J. Food Prot.* 52, 173–177, 1989.
51. Chae, M.S., and Schraft, H., Comparative evaluation of adhesion and biofilm formation of different *Listeria monocytogenes* strains. *Int. J. Food Microbiol.* 62, 103–111, 2000.
52. Dykes, G.A., Sampathkumar, B., and Korber, D.R., Planktonic or biofilm growth affects survival, hydrophobicity and protein expression patterns of a pathogenic *Campylobacter jejuni* strain. *Int. J. Food Microbiol.* 89, 1–10, 2003.
53. Stopforth, J.D., Samelis, J., Sofos, J.N., Kendall, P.A., and Smith, G.C., Influence of organic acid concentration on survival of *Listeria monocytogenes* and *Escherichia coli* O157:H7 in beef carcass wash water and on model equipment surfaces. *Food Microbiol.* 20, 651–660, 2003.
54. Aarnisalo, K., Salo, S., Miettinen, H., Suihko, M.L., Wirtanen, G., Autio, T., Lunden, J., Korkeala, H., and Sjoberg, A.M., Bactericidal efficiencies of commercial disinfectants against *Listeria monocytogenes* on surfaces. *J. Food Safety* 20, 237–250, 2000.
55. Barnes, L.M., Lo, M.F., Adams, M.R., and Chamberlain, A.H., Effect of milk proteins on adhesion of bacteria to stainless steel surfaces. *Appl. Environ. Microbiol.* 65, 4543–4548, 1999.
56. Frank, J., and Koffi, R., Surface-adherence growth of *Listeria monocytogenes* is associated with increased resistance to surfactant sanitizers and heat. *J. Food Prot.* 53, 550–554, 1990.
57. Hassan, A.N., Birt, D.M., and Frank, J.F., Behavior of *Listeria monocytogenes* in a *Pseudomonas putida* biofilm on a condensate-forming surface. *J. Food Prot.* 67, 322–327, 2004.
58. Ceri, H., Olson, M.E., Stremick, C., Read, R., Morck, D., and Buret, A., The Calgary biofilm device: new technology for rapid determination of antibiotic susceptibilities of bacterial biofilms. *J. Clin. Microbiol.* 37, 1771–1776, 1999.
59. Nichols, W.W., Dorrington, S.M., Slack, M.P.E., and Walmsley, H.L., Inhibition of tobramycin diffusion by binding to alginate. *Antimicrob. Agents Chemother.* 32, 518–523, 1988.
60. Stone, G., Wood, P., Dixon, L., Keyhan, M., and Matin, A., Tetracycline rapidly reaches all the constituent cells of uropathogenic *Escherichia coli* biofilms. *Antimicrob. Agents Chemother.* 46, 2458–2461, 2002.
61. Mah, T.F., and O'Toole, G.A., Mechanisms of biofilm resistance to antimicrobial agents. *Trends Microbiol.* 9, 34–39, 2001.
62. Anderl, J.N., Zahller, J., Roe, F., and Stewart, P.S., Role of nutrient limitation and stationary-phase existence in *Klebsiella pneumoniae* biofilm resistance to ampicillin and ciprofloxacin. *Antimicrob. Agents Chemother.* 47, 1251–1256, 2003.
63. Roberts, A.P., Cheah, G., Ready, D., Pratten, J., Wilson, M., and Mullany, P., Transfer of Tn*916*-like elements in microcosm dental plaques. *Antimicrob. Agents Chemother.* 45, 2943–2946, 2001.
64. Hausner, M., and Wuertz, S., High rates of conjugation in bacterial biofilms as determined by quantitative in situ analysis. *Appl. Environ. Microbiol.* 65, 3710–3713, 1999.
65. Ghigo, J., Natural conjugative plasmids induce bacterial biofilm development. *Nature* 412, 442–445, 2001.
66. Velkov, V.V., Stress-induced evolution and the biosafety of genetically modified microorganisms released into the environment. *J. Biosci.* 26, 667–683, 2001.

67. Whiteley, M., Parsek, M.R., and Greenberg, E.P., Regulation of quorum sensing by RpoS in *Pseudomonas aeruginosa. J. Bactetriol.* 182, 4356–4360, 2000.
68. Bagge, N., Hentzer, M., Andersen, J.B., Ciofu, O., Givskov, M., and Høiby, N., Dynamics and spatial distribution of β-lactamase expression in *Pseudomonas aeruginosa* Biofilms. *Antimicrob. Agents Chemother.* 48, 1168–1174, 2004.
69. Khachatourians, G.G., Agricultural use of antibiotics and the evolution and transfer of antibiotic-resistant bacteria. *CMAJ* 159, 1129–1136, 1998.
70. Blazquez, J., Hypermutation as a factor contributing to the acquisition of antimicrobial resistance. *Clin. Infect. Dis.* 37, 1201–1209, 2003.
71. LeClerc, J.E., Li, L.B., Payne, W.L., and Cebula, T.B., High mutation frequencies among *Escherichia coli* and *Salmonella* pathogens. *Science* 274, 1208–1211, 1996.
72. Oliver, A., Canton, R., Campo, P., Baquero, F., and Blazquez, J., High frequency of hypermutable *Pseudomonas aeruginosa* in cystic fibrosis lung infection. *Science* 288, 1251–1253, 2000.
73. Phillips, I., Culebras, E., Moreno, F., and Baquero, F., Induction of the SOS response by new 4-quinolones. *J. Antimicrob. Chemother.* 20, 631–638, 1987.
74. Ysern, P., Clerch, B., Castano, M., Gilbert, I., Barbe, J., and Liagostera, M., Induction of SOS genes in *Escherichia coli* and mutagenesis in *Salmonella* Typhimurium by fluoquinolones. *Mutagenesis* 5, 63–66, 1990.
75. Lewis, K., Programmed death in bacteria. *Microbiol. Mol. Bio. Rev.* 64, 503–514, 2000.
76. Stewart, P.S., Multicellular resistance: biofilms. *Trends Microbiol.* 9, 204, 2001.
77. Marrie, T.J., and Costerton, J.W., A scanning and transmission electron microscopic study of an infected endocardial pacemaker lead. *Circulation* 66, 1339–1343, 1982.
78. Givskov, M., de Nys, R., Manefield, M., Gram, L., Maximilien, R., Eberl, L., Molin, S., Steinberg, P.D., and Kjelleberg, S., Eukaryotic interference with homoserine lactone-mediated prokaryotic signaling. *J. Bacteriol.* 178, 6618–6622, 1996.
79. Hentzer, M., Riedel, K., Rasmussen, T.B., Heydorn, A., Andersen, J.B., Parsek, M.R., Rice, S.A., Eberl, L., Molin, S., Høiby, N., Kjelleberg, S., and Givskov, M., Inhibition of quorum sensing in *Pseudomonas aeruginosa* biofilm bacteria by a halogenated furanone compound. *Microbiology* 148, 87–102, 2002.
80. Kristoffersen, P., Hentzer, M., Manefield, M., Wu, H., Costerton, J.W., Andersen, J.B., Riedel, K., Molin, S., Eberl, L., Rasmussen, T.B., Steinberg, P., Bagge, N., Kjelleberg, S., Kumar, N., Høiby, N., Schembri, M.A., Song, Z., and Givskov, M., Attenuation of *Pseudomonas aeruginosa* virulence by quorum sensing inhibitors. *EMBO J.* 22, 3803–3815, 2003.
81. Giacometti, A., Cirioni, O., Gov, Y., Ghiselli, R., Del Prete, M.S., Mocchegiani, F., Saba, V., Orlando, F., Scalise, G., Balaban, N., and Dell'Acqua, G., RNA III inhibiting peptide inhibits *in vivo* biofilm formation by drug-resistant *Staphylococcus aureus. Antimicrob. Agents Chemother.* 47, 1979–1983, 2003.
82. Costerton, J.W., Ellis, B., Lab, K., Johnson, F., and Khoury, A.E., Mechanism of electrical enhancement of efficacy of antibiotics in killing biofilm bacteria. *Antimicrob. Agents Chemother.* 38, 2803–2809, 1994.
83. Raad, I.I., Darouiche, R., Hachem, R., Sacilowski, M., and Bodey, G.P., Antibiotics and prevention of microbial colonization of catheters. *Antimicrob. Agents Chemother.* 39, 2397–2400, 1995.
84. Rediske, A.M., Roeder, B.L., Brown, M.K., Nelson, J.L., Robison, R.L., Draper, D.O., Schaalje, G.B., Robison, R.A., and Pitt, W.G., Ultrasonic enhancement of antibiotic action on *Escherichia coli* biofilms: an *in vivo* model. *Antimicrob. Agents Chemother.* 43, 1211–1214, 1999.

85. Soboh, F., Khoury, A.E., Zamboni, A.C., Davidson, D., and Mittelman, M.W., Effects of ciprofloxacin and protamine sulfate combinations against catheter-associated Pseudomonas aeruginosa biofilms. *Antimicrob. Agents Chemother.* 39, 1281–1286, 1995.

86. Rediske, A.M., Hymas, W.C., Wilkinson, R., and Pitt, W.G. Ultrasonic enhancement of antibiotic action on several species of bacteria. *J. Gen. Appl. Microbiol.* 44, 283–288, 1998.

87. Davis, C., Wagle, N., Anderson, M.D., and Warren, M.M., Iontophoresis generates an antimicrobial effect that remains after iontophoresis ceases. *Antimicrob. Agents Chemother.* 36, 2552–2555, 1992.

88. Harkonen, P., Salo, S., Mattila-Sanholm, T., Wirtanen, G., Allison, D.G., and Gilbert, P., Development of a simple *in vitro* test system for the disinfection of bacterial biofilm. *Wat. Sci. Technol.* 39, 219–225, 1999.

89. Stopforth, J.D., Samelis, J., Sofos, J.N., Kendall, P.A., and Smith, G.C., Influence of extended acid stressing in fresh beef decontamination runoff fluids on sanitizer resistance of acid-adapted *Escherichia coli* O157:H7 in biofilms. *J. Food Prot.* 20, 2258–2266, 2003.

90. Exner, M., Tuschewitzki, G.J., and Scharnegel, J., Influence of biofilms by chemical disinfectants and mechanical cleaning. *Zbl. Bakteriol. Hyg. B* 183, 549–563, 1987.

2 Economic Impact of Biofilms on Treatment Costs

John G. Thomas, Isaiah Litton, and Harald Rinde

CONTENTS

2.1   INTRODUCTION

The world of diagnostic microbiology has lived a rather sheltered life. It had been assumed that free floating planktonic microorganisms were the primary causes of infectious diseases (1–12). Little consideration was given to the manner by which these pathogens lived in the host environment. However, with aging of the patient population and increases in chronic diseases, indwelling medical devices (IMD) and inert surfaces have become a significant part of medical practice (4–7,13–19). Subsequently it was realized that the microorganisms were capable of forming attached sessile communities called biofilms on these surfaces, and the key principle of "To attach is to survive" is now readily recognized (1–7,9,20–22).

While the definitions of biofilms and the associated diseases have grown significantly since 2000, key uniform characteristics have been retained. In essence, biofilms are composed of microbial cells: (a) irreversibly attached to a substratum, interface, or to each other; (b) imbedded in a matrix of extracellular polymeric substance of pathogen origin; and (c) exhibiting altered phenotypes from their planktonic counterparts. Biofilms, whether found at fluid/solid interfaces in contact with the blood or urine, or gas/solid interfaces as found in the respiratory tract, exhibit 3-D spatial organization and physiology regulated by responses to environmental conditions (1–6,20). In fact, biofilms can be considered a primitive type of developmental biology of cells differentiated and striving within the matrix to optimize utilization of available nutritional resources and maintain steady state.

It is postulated that there are at least three spatial regions within the biofilm, each contributing to the phenotype of increased relative antimicrobial susceptibility and survival. Level 1 is a glycocalyx layer that may act as a protective barrier. Organisms above the shield of slime might be expendable, although sloughed cells may act as a means of dissemination analogous to the metastasis of cancer cells. Level 2 is a layer where resistance transfer is likely facilitated due to proximity of the organisms. Cells found in this layer may also be responsible for enzymatic activities within the biofilm. Finally, Level 3, the innermost layer, is consistent with organisms that are metabolically "inactive," having reduced their size, and physiology is primarily focused on housekeeping maintenance functions. This may be the layer where the recently described "persister" cells are found, providing the nucleus for additional survival capabilities (see Chapter 12).

In 2000, the Centers for Disease Control and Prevention cited biofilm-associated diseases as two of the seven major healthcare safety challenges facing the medical community (Table 2.1) (21). The single most important aspect of these challenges is the remarkably increased antibiotic resistance of the biofilm ensconced microorganisms. The biofilm phenotype can reduce antimicrobial susceptibility and increase tolerance up to 1,000-fold substantially diminishing antimicrobial efficacy, and leading to clinical failures. Consequences of biofilms and the associated loss of susceptibility are described in Tables 2.2 and Figure 2.1 (22–24). The data derived from the minimal biofilm eradication concentration (MBEC) clearly contrasts with that determined by the classical minimal inhibitory concentration (MIC) assay. This has provided an entirely new focus for infectious diseases therapy.

TABLE 2.1
CDCs Seven Healthcare Safety Challenges

In 5 years, the CDCs Division of Healthcare Quality Promotion plans on accomplishing these challenges:

Challenge 1   Reduce catheter-associated adverse events by 50% among patients in healthcare settings (Biofilms).

Challenge 2   Reduce hospitalizations and mortality from respiratory tract infections among long-term care patients by 50% (Biofilms).

Note: www.cdc.gov/ncidod/hip.challenges.htm

TABLE 2.2
Clinical Consequences of Biofilm Architecture

- Sessile bacteria within the biofilms have an increased resistance up to 4000-fold higher than planktonic phenotype, "Colonization Resistance"
- Biofilms increase the opportunity for gene transfer, ESBLs
- Bacteria express new and sometimes more virulent Phenotypes or Life Forms within the biofilm
- Bacteria are resistant to both immunologic and non-specific defense mechanisms
- Sessile bacteria reach a much higher density (10^{11} CFUs/ml) than planktonic (10^8 CFUs/ml)
- A positive blood culture is a "failed organism," i.e. not attached

The number of diseases associated with biofilms has also continued to evolve as further understanding of their pathophysiologies is unraveled. Significant resources are still focused on the original six upon which much information has been gathered, and can be clearly defined by the infectious disease (ID) burden. These included: otitis media, periodontal disease, cystic fibrosis, endocarditis, indwelling medical devices, and prostatitis. Additional suspected diseases more recently associated with biofilms but requiring clarification are shown in Table 2.3. Considerable focus on IMDs is consistent with their substantial current and projected use. The wide varieties of IMDs are listed in Table 2.4. Complementing information for device-related infections in the United States is presented in Table 2.5. In each case, it is important to recognize the consequences relative to cost associated with the resistance and the ID burden for the health care system relating to the tripartate scheme that will be described in subsequent sections.

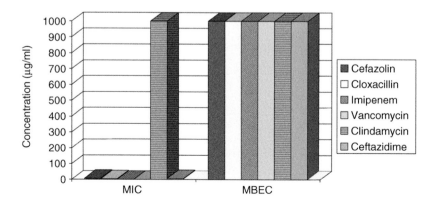

FIGURE 2.1 (See color insert following page 270) *Staphylococcus epidermidis* sensitivity.

TABLE 2.3
Partial List of Human Infections Involving Biofilms

Infection or disease	Common bacterial species involved
Dental caries	Acidogenic Gram-positive cocci (*Streptococcus* spp.)
Periodontitis	Gram-negative anaerobic oral bacteria
Otitis media	Nontypeable *Haemophilus influenzae*
Chronic tonsillitis	Various species
Cystic fibrosis pneumonia	*Pseudomonas aeruginosa, Burkholderia cepacia*
Endocarditis	Viridans group streptococci, staphylococci
Necrotizing fascitis	Group A streptococci
Musculoskeletal infections	Gram-positive cocci
Osteomyelitis	Various species
Biliary tract infections	Enteric bacteria
Infectious kidney stones	Gram-negative rods
Bacterial prostates	*Escherichia coli* and other Gram-negative bacteria

TABLE 2.4
Diversity of Indwelling Medical Devices (IMDs)

A wide array of IMDs
- Prosthetic heart valves
- Prosthetic joints
- Orthopedic implants
- Shunts
- Pacemaker and defibrillator
- Endotracheal intubation
- Hemodialysis/Peritoneal dialysis devices
- Dental implants
- Intravascular catheters
- Intrauterine devices (IUDs) (inert and chemically modified plastic)

TABLE 2.5
Device-Related Infections in the U.S.

Device	Usage/Year	Infection Risk
Cardiac assist devices	700	50–100%
Bladder catheters	Tens of millions	10–30%
Fracture fixators	2 million	5–10%
Dental implants	1 million	5–10%
Central venous catheters	5 million	3–8%
Vascular grafts	450,000	2–10%
Penile implants	15,000	2–10%
Cardiac pacemakers	400,000	1–5%
Prosthetic heart valves	85,000	1–3%
Joint prostheses	600,000	1–3%

2.2 ECONOMIC ANALYSIS

As the evolution in clinical microbiology has begun to unfold, so too has the evolution of cost assessment. The reality is that outcomes are becoming a major feature for the economic evaluation of infectious disease impact and care. Donabedian (25,26) gave us the first definition of outcomes; it had two components: clinical (quality of life (QOL)); and financial (direct and indirect). It was described as an assessment of the patient's management following completion of the hospital stay and brought into focus another key factor, length of stay (LOS). A major feature of the Donabedian theory was the optimization of value of services defined as:

$$Value = \frac{Quality}{Cost\ (\$)}.$$

Key to the value enhancement theme is the concept that "An investment in quality should lead to eventual decrease in cost which ultimately improves the value of services provided." Subsequently, the ECHO model was developed, which was a further attempt to define outcomes. Various outcomes are listed below:

ECHO
Economic: Costs (Laboratory, pharmacy, total) LOS
Clinical: ± Cultures, body temperature, other values
Humanistic: Patient, physician, nursing, satisfaction, health related
OUTCOMES quality of life (QOL)

Today, we have evolved the health economic formula into a Tripartate Formulation, which involves cost, mortality, and QOL. There is a reason. COST can be quantified using the national, regional and individual hospital values mandated by the federal government, and analyzed for comparison. It can be linked to clinical diseases via the Diagnostic Related Groups (DRG). Quality of life is composed of Quality Adjusted Life Years (QAL) which is a composite of mortality/life years lost and reduced quality of life. The total of the tripartate can be mapped as the "Burden of a specific infectious disease":

Infectious diseases burden = Cost/Mortality/QAL.

This can be further stratified to rank the total cost for specific infectious diseases (Table 2.6). Ultimately, the consequence of strategies (diagnostics, therapeutics, clinical medicine) of the health care system can be defined.

2.3 COST PER BIOFILM ASSOCIATED DISEASE (DRG BASED, ICD-9 CODES, ETC.)

As described earlier, the consequences of the ID burden can best be quantified using outcomes assessment. This assessment has grown considerably more complex, but still can be broken down into the tripartarte described earlier: cost, mortality and QOL. Obviously, each of these has significant subsets. Table 2.7 lists components of the outcomes assessment in the patient's cost analysis. Direct non-medical costs and

TABLE 2.6
Three Main Components of Burden of Infectious Disease

- Mortality/life years lost
- Reduced quality of life (QOL) QALY, Quality Adjusted Life Year
- Cost of disease ($)

DRG 415 Surgical site infection hip replacement	QALY, Mostly QoL/Cost of disease
DRG 416 Sepsis	QAL, Mostly life years/Cost of disease
DRG 421 Hepatitis	QALY, Mixed/Cost of disease

indirect costs could be described as components of the QOL. Direct medical costs are listed and are provided as a guide. The QOL, which is hard to quantify, also has individual subsets. These could include personal perception of health, mobility, ability to function in their living, ability to work, level of emotional well-being, and energy and social interactions. All of these are part of a Short Form-36 (SF-36) survey and continuum in comparative analysis of the ID Burden.

2.3.1 MEDICAL COST

Tier I. In order to standardize and allow for comparison between various ID burdens, we used as our primary resource DRGs. DRGs are signature costs of cumulative institutions, tabulated by the federal agencies and are available as Medpar (Medicare charge/reimbursement: http://www.cms.hhs.gov/statistics/medpar/default.asp), and NIS (available through HCUPMET: www.ahrq.gov/data/hcup/hcupmet.htm).

Tier II. Secondarily, we supplemented this information with published literature focusing on specific infectious diseases and biofilms and cost per patient. We focused

TABLE 2.7
Three Areas of Cost for Infectious Diseases

Direct Non-Medical Costs	Direct Medical Costs	Indirect Costs
Home help	Hospitalization	Reduced work productivity
Transportation	Primary drug cost	Absenteeism
	Physician visits	Unemployment
	Diagnostic procedures	
	Disease complications	
	ADEs	
	Physiotherapy	
	Occupational therapy	
	Complementary/alternative remedies	

TABLE 2.8
DRGs for Biofilm Associated Diseases

DRG #	Description
475	Respiratory system diagnosis with ventilator support
415	OR procedure for infectious and parasitic diseases
79	Respiratory infections and inflammations age >17
416	Septicemia age >17
89	Simple pneumonia and pleurisy age? 17 w cc
68	Otitis media and URI age >17 w cc
126	Acute and subacute endocarditis
418	Postoperative and post-traumatic infections

---

on eight DRGs (Table 2.8), given the significance of 33 biofilm-associated DRGs (Table 2.9) and related costs of the to the approximately 212 total cited.

Tier III. Our third level of information was the most difficult to obtain, but more pertinent given the antibiotic resistance consequence of biofilms. We wanted to establish the costs of antibiotic resistance and the failure of antibiotic therapy relative to the three features that make up outcomes and define ID burden: QOL, mortality, and costs. *Tier IV* had the least amount of data available, but reflected another attempt to quantify cost/incident: Tabulated by organism, particularly yeast biofilms (Table 2.10).

Figures 2.2 (Mean Charge per DRG) and Figure 2.3 (Loss per Incidence for Medicare Charges) were primary features in stratifing the eight infectious diseases via

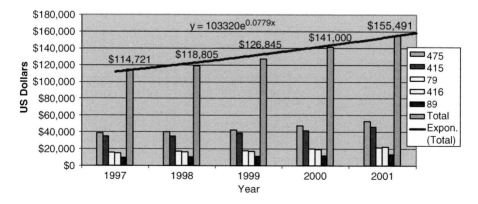

FIGURE 2.2 (See color insert)   Mean charge per DRG.

TABLE 2.9
DRGs and MCDs for 33 Infectious Disease Categories

DRG	MDC	Description
20	1	Nervous system infection except viral meningitis
211	1	Viral meningitis
68	3	Otitis Media and URI age? 17 w cc
69	3	Otitis Media and Uri age ? 17 w/o cc
70	3	Otitis Media and URI age 0–17
79	4	Respiratory infections and inflammation age >17 w cc
80	4	Respiratory infections and inflammation age >17 w/o cc
81	4	Respiratory infections and inflammation age 0–17
89	4	Simple pneumonia and pleurisy age >17 w cc
90	4	Simple pneumonia and pleurisy age >17 w/o cc
91	4	Simple pneumonia and pleurisy age 0–17
126	5	Acute and subacute endocarditis
238	8	Osteomyelitis
242	8	Septic arthritis
277	9	Cellulitis age >17 w cc
278	9	Cellulitis age >17 w/o cc
279	9	Cellulitis age 0–17
320	11	Kidney and urinary tract infections age >17 w cc
321	11	Kidney and urinary tract infections age >17 w/o cc
322	11	Kidney and urinary tract infections age 0–17
415	18	OR procedure for ID
416	18	Septicemia age >17
417	18	Septicemia age 0–17
418	18	Postoperative and post-traumatic infections
419	18	FUO age >17 w cc
420	18	FUO age >17 w/o cc
421	18	Viral illness age >17
422	18	Viral illness and FUO age 0–17
423	18	Other infections and parasitic diseases
475	4	Respiratory diagnosis with ventilator support
488	25	HIV with OR procedure
489	25	HIV with major related condition
490	25	HIV w or w/o other condition

outcomes assessments. We were particularly interested in the change over years and reviewed the NIS/Medpar data from 1997–2001. In calculations and preparation of the figure, we looked at eight key cost features. These included: MCD (Main Charge per DRG per National Inpatient Sample), TCMD (Total Charge for Medicare per DRG), TRMD (Total Medicare Reimbursement per DRG), TMDD (Total Medicare Discharge per DRG), MCMD (Main Charge to Medicare per DRG), MRMD (Main Reimbursement by Medicare per DRG) and LTI (Loss per Incidence Medicare).

TABLE 2.10
Organisms Commonly Forming Biofilms Arranged by IMD

Device	Principal Organism[a]	Other Organisms[a]
Urinary bladder catheter	*E. coli*	*Canadida* spp.
		CoNS
		E. faecalis
		P. mirabilis
		K. pneumoniae
		Other gram-negative species
Endotracheal tube	Enteric Gram-negative species	*Candida* spp.
		P. aeruginosa
		Enterococcus
		Staphylococcus and *Streptococcus* spp.
		Diphtheroids
Central venous catheter or Swan–Ganz catheter	CoNS	*S. aureus*
		Enterococci
		K. pneumoniae
		Candida spp.
		P. aeruginosa
		E. faeacalis
		Microorganisms selected for by local antibiotic prescribing practices
Orthopedic prostheses	*Staphylococcus* spp.	*S. pneumoniae*
		Other streptococcal species
Mechanical heart valves	CoNS in association with contamination at the time of surgery	*S. aureus*
		Streptococcus spp.
		Gram-negative bacilli
		Enterococci
		Diphtheroids
		Candida spp.
Periotoneal dialysis catheters	*S. aureus*	*P. aeruginosa*
		Other gram-negative species
		Candida spp.
Vascular grafts	CoNS	*S. aureus*
		P. aeruginosa
		Flora indigenous to the graft site (e.g., Gram-negative bacilli in AAA grafts developing a fistula to the gut)

[a] CoNS, coagulase-negative staphylococci; AAA, abdominal aorta aneurysm. Courtesy of Jason A. Bennett, West Virginia University.

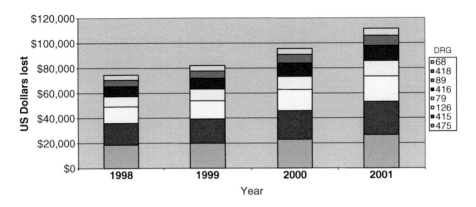

FIGURE 2.3 (See color insert) Loss per incidence for Medicare charges.

The data is stunning. Increased mean charge per DRG from 1997 to 2001 increased from $114,721 per case to $155,491. Furthermore, the contributors of those charges did not change but rather remain the most expensive, DRG 475, DRG 415 and DRG 79. By far, the two highest charges per DRG were DRG 475 and DRG 415.

The loss per incidence for medicare charges was equally as dramatic, again, emphasizing both DRG 475 and DRG 415. Of interest is the relative proportion of diseases. In the late 1990s, diseases associated with pneumonia were very high, although they did not appear in later years of our database, again a reflection of the change in population and kind of diseases emerging most often in the ICU. Still, a significant amount of published information concerned the cost associated with infectious disease related to IMDs. This reflected the growing increase in the number of ICU patients, the LOS, and the fact that today a preponderance of patients spend a significant amount of time in the ICU (ICU – LOS/TOTAL – LOS).

2.3.1.1 Ventilator Associated Pneumonia (VAP) and Endotrachs (DRG 475)

Ventilator associated pneumonia (VAP) is a huge expenditure for institutions and today is associated with the number one nosocomial infection (27). Table 2.11 lists significant facts about VAP and the costs associated therein. VAP in the ICU occupies about 7 to 8% of total healthcare. Prolonged mechanical ventilation in ICU consumes 50% of the ICU resources. This is significant in that an average of 30% of patients in the ICU are on a ventilator, while it ranges from 18 to 60%. Ventilator associated pneumonia increases hospital LOS from four to nine days. VAP is responsible for about $1.5 billion additional healthcare dollars and the ICU demands 20 to 34% of all hospital costs.

Complications of VAP clearly increase the cost. Simultaneous fungemia has been described as the most expensive complication of VAP (28). Table 2.12 shows the

TABLE 2.11

U.S. VAP Trivia

- 250K cases/yr of nosocomial pneumonias
- 23K deaths/yr
- Second most common of all nosocomial infections
- ET intubation and MV represent the single greatest risk for developing pneumonia
- Risk of VAP is 1–3%/day on MV
- Overall mortality of patients with VAP is 20–50%
- VAP increases hospital stay by 4 to 9 days
- HAP increases hospital cost by $5K/episode
- VAP is responsible for $1.5 billion in additional healthcare expenditures

outcome measurements for two target populations: blood culture positive vs. suspected fungemia. In either case the consequences are significant. The average length of stay is 44 days for proven fungemics with a DRG 475 vs. 44 with "suspected." The average costs for the positive fungemic is $68,000, and for the suspected is $81,000 given the impact of accurate diagnosis from the laboratory. Total reimbursement of $52,689 for the positive fungemic vs. $49,205 for the suspected again reflects the loss per incidence as described earlier for DRG 475.

2.3.1.2 Sepsis/Catheter Related Sepsis (DRG 416)

Foley catheters and the potential for catheter related sepsis (CRS) are incredible resource consuming events in the United States (3,7,9,10–12,14,15,17,22,28–30). There are three to four million hospital admissions annually, and over two million have nosocomial infections. Over 40%, or 900,000, are nosocomial urinary tract infections or NUTIs. Of these, 4,500 patients die each year due to nosocomial urinary tract infections. The cost to treat is between $680 and $3,800 per patient. Total cost is somewhere between $0.65 and $3.5 billion. Catheter related sepsis, a biofilm presentation, increases costs by approximately $28,000 per case.

2.3.1.3 Dental/Periodontitis

Oral health disparities in the U.S. were recognized by the Surgeon General's Report in 2000 which focused on particular populations including those in rural areas such as Appalachia. West Virginia's population of 1.9 million exhibited dental disease with one-third of adult population under 35 missing six or more teeth, while 31% were endentulous. While there were a variety of contributing factors such as smokeless tobacco, drinking, and smoking, the predominant feature was poor oral hygiene focusing on periodontal pathogens. Here, the QOL was the primary feature of ID Burden, which is difficult to quantify.

TABLE 2.12
Summary of Outcome Measurements of Two Target Populations: Blood Culture Positive (A) vs. Suspected Fungemia (B)

	Target Population					
	A			B		
	Average	Minimum	Maximum	Average	Minimum	Maximum
Demographics						
Number	13			13		
Age	38	0	73	46	0	82
Gender (M/F)	6/7			6/7		
Hospital service						
Non-ICU	9			10		
ICU	4			3		
Outcomes						
LOS (days)	44	5	146	42	21	101
Death during hosp	31			15		
Hospitalization						
Charges ($)	117,104	9,696	372,921	96,480	31,087	200,848
Costs ($)	68,652	6,435	226,445	81,004	23,998	152,757
Total reimbursement ($)	52,689	4,837	151,094	49,205	3,175	92,626
Pharmacy						
Charges ($)	28,348	3,262	74,311	34,974	12,626	55,245
Costs ($)	15,252	1,755	39,993	18,599	6,714	29,379
Microbiology						
Charges ($)	2,720	492	9,048	2,774	1,141	5,238
Costs ($)	1,029	1,886	3,465	1,055	428	2,003

2.3.1.4   Other DRGs

Other significant DRGs included otitis media (DRG 68), endocarditis (DRG 126), and cystic fibrosis (DRG 298).

2.4   ANTIBIOTIC RESISTANCE

The emergence of antibiotic resistance, particularly in ICUs, has been well documented over the last ten years. Emergence of the multi-drug resistant (MDR) Gram negative rods in the latter part of 2000, the re-emergence of the Gram positive cocci, particularly vancomycin-resistant enterococci (VRE) in 2001, and presently, the addition of "colonization resistance" via biofilms, has increased the awareness that past strategies had significant limitations (10,20). Today, the selective pressure of antibiotics has led to such strategies as de-escalation, cycling, and computer-assisted algorithms for the management of antibiotic prescribing. There is recognition of the incredible cost of resistance and its consequence on LOS, outcomes, and ID burden.

Only recently, however, has it begun to be a cost feature, arising from analyses of laboratory data and the importance of tabulating the contribution of resistance to overall outcomes.

It is estimated that the between US$100 million and US $300 million are spent each year on associated resistant bacteria. In general, it is calculated that resistant organisms cost twice as much as those that are susceptible to treatment. For example, blood culture costs alone associated with therapy of resistant *Pseudomonas aeruginosa* infection in the ICU adds an estimated $7,340 per case.

Table 2.13 lists the ID burden of three selected infectious diseases in the United States. This is related to the DRGs chosen earlier which highlight sepsis (DRG 416), pneumonia (DRG 79), and VAP (DRG 475). Based on calculations for hospitalizations, mortality rate, and cost per patient, an incidence of sepsis costs the patient between $22,000 and $70,000, pneumonia costs $12,000 to $22,000, and VAP $41,000. Infections associated with indwelling medical devices may be managed by removal of the device, and replacement of a central venous line can be as high as $14,000 per incidence.

In an average, 500-bed hospital, the data indicate that 40 patients per year with hospital acquired sepsis, and between 20 and 30 patients per year with ventilator-associated pneumonia, will receive inadequate initial antimicrobial therapy (31). Table 2.14 describes the burden for three selected infectious diseases for an average 500-bed hospital with the cost per year increased to that institution. Sepsis cost would increase between $7.5 million to $24 million for sepsis, $8 million to $15 million for pneumonia, and for VAP approximately $2.8 million. A cost of $4.3 million was calculated for therapy of cervical site infections often associated with grafts and indwelling support vehicle. Incredibly, the annual addition or burden of hospital acquired infections from sepsis, pneumonia, and cervical site infections in an average 500-bed hospital would be $14 million to $21 million, and approximately 120 patients would die from these infections.

TABLE 2.13
Infective Disease Burden of Three Selected Infectious Diseases in the U.S. Focusing on Antibiotic Resistance

	Hospitalizations	Mortality Rate	Mortality	$ Per Patient
Severe sepsis	660,000	23%	150,000	$22,000–$70,000
Community acquired	395,000	13%	50,000	
Hospital acquired	265,000	38%	100,000	
Adequate initial Rx (70%)	185,000	28%	51,000	
Adequate initial Rx (30%)	80,000	62%	49,000	
Pneumonia	1,300,000	9	115,000	$12,000–$22,000
Community acquired	1,000,000	2.4%	24,000	
Hospital acquired	300,000	30%	90,000	
Ventilator associated pneu.	135,000	45%	61,000	$41,000
Adequate initial Rx (56–75%)	75–100,000	10–20%		
Inadequate initial Rx (25–44%)	35–60,000	40–60%		

TABLE 2.14
Burden of Three Selected Infectious Diseases in an "Average" 500 Focusing on Antibiotic Resistance

	Hospitalizations	Mortality Rate	Mortality	$ Per Patient
Severe sepsis	350	23%	80	$7.5 million–$24 million
Community acquired	200	13%	26	
Hospital acquired	140	38%	55	$3 million–$10 million
Adequate initial Rx (70%)	100	28%	28	
Adequate initial Rx (30%)	40	62%	25	
Pneumonia	670	9	60	$8 million–$15 million
Community acquired	500	2.4%	12	
Hospital acquired	170	30%	50	$7 million
Ventilator associated pneu.	70	45%	30	$2.8 million
Adequate initial Rx (56–75%)	40–50	10–20%		
Inadequate initial Rx (25–44%)	20–30	40–60%		
Surgical site infections (2.6%)	360	4.3%	15	$4.3 million

2.5 STRATEGIES TO ADDRESS THE INCREASE IN BIOFILM DISEASES, PARTICULARLY ANTIBIOTIC ASSOCIATED RESISTANCE

A number of strategies may be initiated to improve treatment outcomes and reduce costs associated with infectious diseases. Focused teams for catheter care have clearly shown benefit (23). The infection control team is an integral and important part in maintaining clean sites and recognizing the involvement of multi-species biofilms that originate from microbiota tissue location. Increased awareness of the consequences of biofilms and the loss of susceptibility is paramount. Fortunately, there is a growing awareness of the "universality" of biofilms and their impact on the dental as well as the medical community.

For years laboratories have employed standardized methodologies described by National Committee for Clinical Laboratory Standards (NCCLS) guidelines which imply evaluation of single species planktonic populations. Obviously, biofilms present a whole new spectrum of problems for the diagnostic laboratory, none of which have been standardized. However, Ceri et al. (12) published an article describing evaluation of the MBEC which utilizes monospecies biofilms prepared in a 96-well format (see also Chapter 13). It is a method that resembles, in many features,

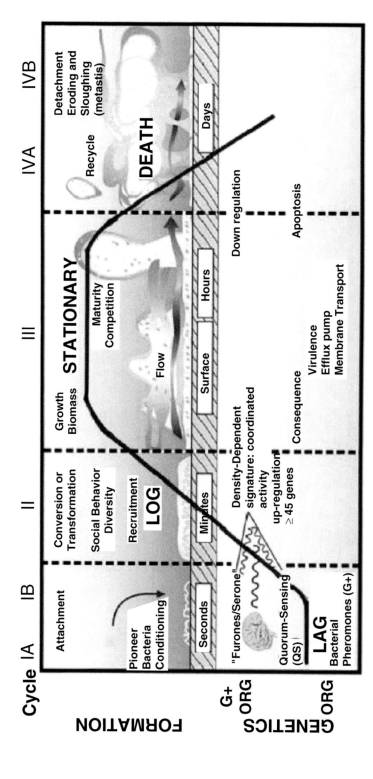

FIGURE 2.4 Biofilm cycle: formation/genetics.

the standardized broth microdilution minimal inhibitory concentration (MIC) assay as defined by NCCLS for the planktonic population. At the very least, it is important for laboratories to increase the awareness that susceptibilities presently done are based on the single species and a free-floating or planktonic population. In any management of biofilm associated infections, it is apparent that early management decisions are critical, particularly in VAP, where early antibiotic selection is key. Later changes in inadequate therapy have almost no impact upon the consequences. Hence, laboratory participation is critical. Rapid susceptibility reporting is paramount in providing good decisions on early antibiotic therapy, even if it is determined against a planktonic population. It may be very important for the laboratory to understand the stage the biofilm disease (Figure 2.4). Additionally, culturing of indwelling medical devices needs to be evaluated. Lumenal cultures may be preferable over external cultures from catheters, and vortexing and sonication may be a requirement to release the matrix-ensconced microorganisms. This is particularly important in the case of infections by yeast.

Ultimately, the best treatment/management may be the combination of prevention, antiinfectives and host modulatory therapy. A number of recent publications have described immune modulators or immunologic response modifiers that have impact upon the biofilm. This included low dose doxycyclines and periodontal diseases, and the macrolides, particularly azithromycin in cystic fibrosis, both of which may exhibit some adjunctive anti-inflammatory activity. Further, a number of publications have described combined usage of an anti-inflammatory and an antiinfective in otitis media. Although the optimization of the treatment of biofilm associated diseases is at an early stage, long-standing methods are clearly not the answer.

2.6 CONCLUSIONS

New technologies may enhance clinical outcomes measurements and decrease the overall cost of patient management. These approaches and services need to demonstrate the reduction of overall costs, reduction in mortality, and improvement of QOL. Well-constructed trials would be expected to validate and quantify the savings/ID burden. Decisions need to be informed and evidence-based. This brings us back to the concept of value:

$$\text{Value} = \frac{\text{Improvement in clinical outcomes, reduced mortality, improved quality of life}}{\text{Cost of new technologies and services} - \text{Cost savings}}.$$

Finally, improved outcomes will be associated with better communication via the internet. Information about hospital quality of performance will become available to patients. Hospital competition will become dependent upon quality of performance, and economic performance will depend upon quality performance. Lastly, all of these will be associated with ID (bio)burdens, particularly those associated with biofilms. Biofilms will continue to be the growing reservoir of infections that are untreatable. The resistance associated with these and the cost therein will become unbearable for those institutions which do not comprehend and fully implement successful strategies to address biofilm-related infections.

REFERENCES

1. Costerton, J.W., Lewandowski, Z., Caldwell, D.E., Korber, D.R., and Lappin-Scott, H.M., Microbial biofilms. *Annu. Rev. Microbiol.* 49, 711–745, 1995.
2. Costerton, J.W., Stewart, P.S., and Greenberg, E.P., Bacterial biofilms: a common cause of persistent infections. *Science* 284, 1318–1322, 1999.
3. Donlan, R.M., Biofilm formation: a clinically relevant microbiological process. *Clin. Infect. Dis.* 33, 1387–1392, 2001.
4. Donlan, R.M., Biofilms and device-associated infections. *Emerg. Infect. Dis.* 7, 277–281, 2001.
5. Donlan, R.M., and Costerton, J.W., Biofilms: survival mechanisms of clinically relevant microorganisms. *Clin. Microbiol. Rev.* 15, 167–193, 2002.
6. Hall-Stoodley, L., Costerton, J.W., and Stoodley, P., Bacterial biofilms: from the natural environment to infectious diseases. *Mature Reviews Microbiology* 2, 95–108, 2004.
7. Fux, C.A., Stoodley, P., Hall-Stoodley, L., and Costerton, W.J., Bacterial biofilms: a diagnostic and therapeutic challenge. *Expert Rev. Anti-Infect. Ther.* 1(4), 667–683, 2003.
8. Ceri, H., Olson, M.E., Stremick, C., Read, R.R., Morck, D., and Buret, A., The Calgary Biofilm Device: new technology for rapid determination of antibiotic susceptibilities of bacterial biofilms. *J. Clin. Microbiol.* 37, 1771–1776, 1999.
9. Habash, M., and Reid, G., Microbial biofilms: their development and significance for medical device-related infections. *J. Clin. Pharmacol.* 39, 887–898, 1999.
10. Hanna, H.A., Radd, I.I., Hackett, B., Wallace, S.K., Price, K.J., Coyle, D.E., and Parmley, C.L., Antibiotic-impregnated catheters associated with significant decrease in nosocomial and multidrug-resistant bacteremias in criticall ill patients. *Chest* 124, 10030–11038, 2003.
11. Mermel, L.A., Farr, B.M., Sherertz, R.J., Raad, I.I., O'Grady, N., Harris, J.S., and Craven, D.E., Guidelines for the management of intravascular catheter-related infections. *Clin. Infect. Dis.* 32, 1249–1272, 2001.
12. O'Grady, N.P., Alexander, M., Dellinger, E.P., Gerberding, J.L., Heard, S.O., Maki, D.G., Masur, H., McCormick, R.D., Mermel, L.A., Pearson, M.L., Raad, I.I., Randolph, A., and Weinstein, R.A., Guidelines for the prevention of intravascular catheter-related infections. *Infect. Control Hosp. Epidemiol.* 23, 759–769, 2002.
13. Raad, I.I., Vascular catheters impregnated with antimicrobial agents: present knowledge and future direction. *Infect. Control Hosp. Epidemiol.* 18(4), 1–5, 1999.
14. Ryder, M., Peripheral access options. *Surgical Oncology Clinical of North America,* 4(3), 395–427, 1995.
15. Ryder, M., Peripherally inserted central venous catheters. *Nursing Clinical of North America,* 28, 937–992, 1993.
16. Weinstein, R.A., MD Adding insult to injury: device-related infections. *Infectious Diseases Conference Summaries* – 2000, Medscape, Inc., 2000.
17. Ryder, M., The role of biofilm in vascular catheter-related infections. *New Dev. Vasc. Dis.* 2(2), 15–25, 2001.
18. Chandra, J., Kuhn, D.M., Mukherjee, P.K., Hoyer, L.L., McCormick, T., and Ghannoum, M.A., Biofilm formation by the fungal pathogen *Candida albicans,* development, architecture, and drug resistance. *J. Bacteriol.* 183, 5385, 2001.
19. Kuhn, D.M., Chandra, J., Mukherjee, P.K., and Ghannoum, M.A., Comparison of biofilms formed by *Candida albicans* and *Candida parapsilosis* on bioprosthetic surfaces. *Infect. Immun.* 70, 878–888, 2002.

20. Murga, R., et al., Biofilm formation by Gram-negative bacteria on central venous catheter connectors: effect of conditioning films in a laboratory model. *J. Clin. Microbiol.* 39, 2294–2297, 2001.
21. CDC. Seven healthcare safety challenges. 2000.
22. Tighe, M.J., Kite, P., Thomas, D., et al., Rapid diagnosis of catheter-related sepsis using the acridine orange cytospin test and an endoluminal brush. *J. Nutr.* 20, 215–218, 1996.
23. Kite, P., Tighe, M.J., Thomas, D., et al., Gram/acridine orange leucocyte cytospin (G/AOLC) rapid method rapid diagnosis of catheter related sepsis using the acridine orange leucocyte cytpspin test and an endoluminal brush. *J. Parenteral. Nutr.* 3, 215–218, 1996.
24. Ceri, H., Olson, M.E., Stremick, C., Read, R.R., Morck, D., and Buret, A., The Calgary Biofilm Device, new technology for rapid determination of antibiotic susceptibilities of bacterial biofilms. *J. Clin. Microbiol.* 37, 1771–1776, 1999.
25. Donabedian, A., The role of outcomes in quality assessment and assurance. *QRB Qual. Rev. Bull.* 18, 356–360, 1992.
26. Donabedian, A., The effectiveness of quality assurance. *Int. J. Qual. Health Care* 8, 401–407, 1996.
27. Adair, C.G., Gorman, S.P., Feron, B.M., Byers, L.M., Jones, D.S., Goldsmith, C.E., Moore, J.E., Kerr, J.R., Curran, M.D., Hogg, G., Webb, C.H., McCarthy, G.J., and Milligan, K.R., Implications of endotracheal tube biofilm for ventilator-associated pneumonia. *Intensive Care Med.* 25, 1072–1076, 1999.
28. Wilson, L.S., Reyes, C.M., Stolpman, M., Speckman, J., Allen, K., and Beney, J., The direct cost and incidence of systemic fungal infections. *Value Health* 5, 26–34, 2002.
29. O'Grady, N.P., et al., and working group, Guidelines for the Prevention of Intravascular Catheter-Related Infections. *J. Infusion Nursing* 25(65), S37–S64, 2002.
30. Raad, I., Intravascular-catheter-related infections. *Lancet* 351, 893–899, 1998.
31. Siegman-Izra, Y., Anglim, A.M., Shapiro, D.E., et al., Diagnosis of catheter-related bloodstream infection: A meta-analysis. *J. Clin. Microbiol.* 35, 928–936, 1997.

3 Biofilm-Related Indwelling Medical Device Infections

Matthew K. Schinabeck and
Mahmoud A. Ghannoum

CONTENTS

3.1 INTRODUCTION

Anton van Leeuwenhoek unknowingly was the first person to discover bacterial biofilms causing human disease when he described the presence of "animalcules" while microscopically examining scrapings of his own dental plaque in the seventeenth century (1). Yet it wasn't until the 1970s that it became widely accepted that bacteria in all natural ecosystems lived in the biofilm state. Biofilms, however, were not acknowledged as an important cause of indwelling medical device (IMD) infections until the early 1990s when electron microscopic examination of explanted IMDs, believed to be the foci of infection, revealed large numbers of bacteria encased in a thick extracellular matrix (2). These findings initiated a rapid increase in the number of investigators studying biofilm-related IMD infection.

The classic characteristics of biofilm-related infections were the same as those that had been observed for years in IMD-related infections without a clear etiology. These similarities included the following: (a) biofilms form on an inert surface or dead tissue, (b) they grow slowly with a delayed onset of symptoms, (c) biofilm infection is not resolved by host defense mechanisms, (d) planktonic cells released from biofilms (programmed detachment) act as a nidus of infection, and (e) antibiotic therapy does not kill mature biofilms (1). Now that biofilms have been accepted as a major cause of IMD-related infections, a better understanding of the diagnosis, pathogenesis, and resistance mechanisms of biofilm related infections will be critical to the reduction of morbidity and mortality associated with IMD infections. In this chapter, we will discuss the magnitude and pathogenesis of IMD-related infections and also review the proposed mechanisms by which biofilms adherent to IMDs cause systemic disease in humans.

3.2 MAGNITUDE OF IMD-RELATED INFECTIONS

Over the past several decades there have been rapid advances in medical technology associated with a dramatic increase in the use of IMDs. Biofilm related infection is a serious complication associated with use of these devices. The ability of biofilms to evade the host immune responses and their enhanced antimicrobial resistance phenotype make biofilm-related IMD infections very difficult to manage (1,3–5). Often, the only reliable treatment for biofilm-associated IMD infections is the removal of the infected device. However, this can be associated with increased morbidity and mortality, prolonged hospitalization, and increased healthcare costs (6). It has been estimated that the cost for treating a device-related infection can range from 5–7 times the cost of the original implantation (7). Those unfortunate patients who cannot have their infected device removed face a life of suppressive antimicrobial therapy to prevent recurrent blood stream infection.

A list of indwelling devices that have been associated with biofilm related infection is provided in Table 3.1. Despite the large number of devices implanted, it has been difficult to accurately determine incidence and prevalence figures for IMD-related infections (8). In 2001, for example, greater than 40 million IMDs were placed in patients in the United States with infection rates ranging from 1% to 50% (9).

TABLE 3.1
Indwelling Medical Devices Associated with Biofilm Related Infection

Intravenous catheters	Contact lenses
Endotracheal tubes	Intrauterine devices
Mechanical heart valves	Pacemakers/automated intracardiac devices
Peritoneal dialysis catheters	Prosthetic joints
Urinary catheters	Tympanostomy tubes
Voice prosthetsis	Breast implants
Penile implants	Vascular grafts
Biliary stents	Orthopedic devices (fixators, nails, screws)

TABLE 3.2
Estimated Number of IMDs Inserted in the U.S. Per Year with Estimated Infection Rates and Attributable Mortality

Device	Estimated No. of Devices Placed/Year	Infection Rate (%)	Attributable Mortality
Urinary catheters	>30,000,000	10–30	Low
Central venous catheters	5,000,000	3–8	Moderate
Fracture fixation devices	2,000,000	5–10	Low
Dental implants	1,000,000	5–10	Low
Joint prostheses	600,000	1–3	Low
Vascular grafts	450,000	1–5	Moderate
Cardiac pacemakers	300,000	1–7	Moderate
Breast implants (pairs)	130,000	1–2	Low
Mechanical heart valves	85,000	1–3	High
Penile implants	15,000	1–3	Low
Heart assist devices	700	25–50	High

Scale for attributable mortality: low, <5%; moderate, 5–25%; high, >25%.
Source: Donlan, R.M., Biofilm formation: a clinically relevant microbiological process. *Clin. Infect. Dis.* 33(8), 1837–1392, 2001. With permission.

Table 3.2 lists the estimated number of IMDs placed per year in the U.S. with the associated infection rate and attributable mortality. However, the true rates of biofilm-related IMD infection are probably higher than those reported in Table 3.2 for the following reasons: (a) the re-infection rate of re-implanted devices is several fold higher than in those with their first IMD implantation, (b) there are increased rates of infection in patients with only partially explanted IMDs (i.e. pacemakers), (c) poor diagnostic methods for biofilm-associated IMD infections lead to under diagnosis, (d) failure of current microbiology techniques to reliably culture biofilm encased organisms may lead to false negative cultures, and (e) antibiotics are often started prior to obtaining the appropriate culture specimens (9).

The data illustrated in Table 3.2 depicts clear differences in severity of infection associated with the various IMDs. For example, six times more urinary catheters are placed per year compared to central venous catheters (CVCs) yet the attributable mortality is <5% for urinary catheters compared to 5–25% in CVC-associated bloodstream infections (9). On the other hand, intravascular IMD infections (CVC, vascular grafts, pacemakers/AICDs, mechanical heart valves, and left ventricular assist devices) are the most serious with attributable mortality rates >25% for infected prosthetic heart valves and heart assist devices (9). Other IMD infections may not be associated with high mortality rates, but patients may still suffer from significant morbidity. For example, the surgical removal of infected breast implants, penile implants, and prosthetic joints can have profound cosmetic and psychological effects while not putting the patients life in danger (9). In summary, the rapid increase in medical technology has led to a dramatic rise in the use of IMDs with a resultant increase in IMD-related infection. The true incidence of these infections is probably underestimated and the morbidity and attributable mortality associated with these infections varies greatly by the type and location of the device.

3.3 PATHOGENESIS OF IMD-RELATED INFECTIONS

Adherence of microorganisms to a device surface is the initial step in the pathogenesis of IMD-related infection and is dependent upon the following: (a) device related factors, (b) host factors, (c) microorganism–biomaterial interactions, and (d) microorganism related factors. In this section, we will discuss in detail the role of the first three factors in the pathogenesis of biofilm associated IMD infections. The individual microorganism related factors are beyond the scope of this chapter and can be found elsewhere (9,10).

3.3.1 DEVICE RELATED FACTORS

The *in vitro* study of microbial adherence to biomaterials has revealed several device related factors that effect biofilm formation. Certain materials used in the design of IMDs are more conducive to microbial adherence/biofilm formation than others. In vitro studies performed by many laboratories have determined that microbial adherence to biomaterials occurs in the following order: latex > silicone > PVC > Teflon > polyurethane > stainless steel > titanium (9,11). Surface characteristics determining the adherence properties of specific materials include: (a) surface texture, (b) surface charge, and (c) hydrophobicity (9).

Materials with irregular or rough surfaces tend to have enhanced microbial adherence compared to smooth surfaces (9). Locci et al. (12) documented that surface irregularities in central venous catheters (CVC) varied with different polymer materials and that bacteria preferentially adhered to surface defects within minutes after infusing the catheters with a contaminated buffer solution. Another study examined the surface topography of five commercially available polyurethane CVCs by scanning electron microscopy (13). The catheters with the most surface irregularities had significantly more adherent bacteria compared to catheters with smoother surfaces (13).

Bacterial cells, which tend to have hydrophobic cell surfaces, are attracted to the hydrophobic surfaces of many of the biomaterials currently used in IMDs (14). This hydrophobic interaction between the microorganism and the biomaterial leads to increased adherence and subsequent biofilm formation. An increase in the surface hydrophilicity of the polymers used in IMDs leads to weakened hydrophobic interactions between the infecting organism and the polymer and decreased adherence (15). Tebbs et al. (13) confirmed this by comparing the adherence of five *Staphylococcus epidermidis* strains to polyurethane CVC and a commercially produced hydrophilic-coated polyurethane CVC. Adhesion of the bacteria was significantly reduced in the hydrophilic-coated CVC.

Biomaterial surface charge greatly influences adherence of microorganisms. Most microorganisms exhibit a negative surface charge in an aqueous environment. Therefore, a negatively charged biomaterial surface should lead to decreased adherence of microorganisms due to a repulsion effect between both negatively charged surfaces (15,16). This has been documented by several groups. Hogt et al. (17) showed that coagulase-negative staphylococci adherence to MMA-copolymers was higher in positively charged polymers compared to negatively charged polymers. Kohnen et al. (16) also showed a reduction in the adherence of *S. epidermidis* adhesion to polyurethane, which had been surface-grafted with acrylic acid leading to a negatively charged surface.

Overall, several device related factors appear to play an important role in the *in vitro* adherence and subsequent biofilm formation on IMDs. Biomaterials with rough, hydrophobic, and positively charged surfaces tend to have increased microorganism adherence compared to biomaterials that are smooth, hydrophilic, and negatively charged. Surface modification of biomaterials, in an attempt to prevent microorganism adherence, has been the primary approach to decrease the number of IMD-related infections. Two main strategies have been employed in the prevention of IMD-related infection: (a) the development of biomaterials with anti-adhesive properties using physico-chemical methods, and (b) incorporation or coating biomaterials with antimicrobial agents (11). The physico-chemical modification of biomaterials has significantly reduced the adherence of microorganisms *in vitro*, as described above, however these modified materials have not performed well *in vivo*. This suggests that the adhesion of bacteria to a biomaterial *in vivo* is influenced by other factors besides the nature of the synthetic material (16).

3.3.2 The Role of Conditioning Films

The *in vivo* role of device related factors are further complicated by the formation of conditioning films on the surface of implanted IMDs. Shortly after an IMD is inserted into the body, the surface of the biomaterial is rapidly covered by a layer of proteins, fibrin, platelets, and other constituents referred to as the conditioning film, which alters the surface properties of the biomaterial (15). There are many components that make up the conditioning film. The majority of the molecules are proteinaceous including fibronectin, fibrinogen, fibrin, albumin, collagen, laminin, vitronectin, elastin, and von Willebrand factor (8,11). Of these proteins, fibronectin, fibrinogen, and fibrin are the most important for adherence of microorganisms and have been shown to enhance the adherence of Gram positive cocci, Gram negative rods, and *Candida albicans* (11,18,19).

The adherence of microorganisms to IMDs *in vivo*, therefore, may be dependent upon specific interactions between organism specific structures (i.e., proteins and pili, etc.) and receptors on the conditioning film coated biomaterial rather than the device specific factors and/or the bacteria–biomaterial interactions which seem to predominate *in vitro* (15). For example, Hawser and Douglas reported that *C. albicans* biofilm formation *in vitro* is significantly decreased when grown on smooth, hydrophilic polyurethane discs compared to silicone elastomer (20). However, our group has recently published data using a unique animal model of catheter-associated *C. albicans* biofilm infection showing that there is no significant difference in *in vivo* biofilm formation when polyurethane or silicone catheters were used (Figure 3.1) (21). These findings confirm that there is a difference in adherence to various

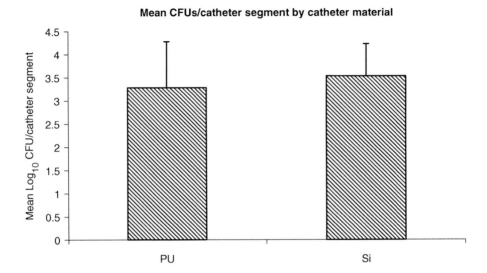

FIGURE 3.1 *In vivo* biofilm formation on polyurethane and silicone catheters. *C. albicans* $\log_{10}$ CFUs/catheter segment 7 days post-infection. There is no significant difference in biofilm formation on polyurethane versus silicone catheters *in vivo*.

biomaterials in the *in vitro* vs. *in vivo* setting. This has been attributed to the ability of *in vivo* conditioning films to mask the underlying properties of the biomaterials used in the device.

3.3.3 BACTERIA–BIOMATERIAL INTERACTIONS

In vitro adhesion of microorganisms to biomaterial surfaces has been extensively studied and it has been shown to require both non-specific reversible interactions and highly specific irreversible interactions (15). Initial attachment, or reversible adhesion, of microorganisms to a biomaterial surface is dependent upon the physical characteristics of the microorganism, the biomaterial, and the surrounding environment (22). Microorganisms randomly arrive near the surface of the implanted IMD by several mechanisms: (a) direct contamination, (b) contiguous spread, or (c) hematogenous spread (22). Once near the surface of the IMD, initial adherence of the microorganism depends upon microorganism–biomaterial interactions including van der Waals forces and hydrophobic interactions (11). The cell surface of most microorganisms is negatively charged as are the majority of biomaterials used in IMDs. The common charges of the microorganism and the IMD surface will repel each other, however the effect of van der Waals forces overcome this repulsion beginning about 10 nm from the IMD surface effectively keeping the microorganism near the biomaterial surface (22).

Many microorganisms and biomaterial surfaces are hydrophobic and, therefore, hydrophobic forces play an important role in the early attachment of microorganisms to IMDs. Pashley et al. (23) has shown that hydrophobic forces are 10 to 100 times stronger than van der Waals forces at 10 nm from the biomaterial surface and that these forces may be exerted up to 15 nm from the IMD surface. The hydrophobic forces easily overcome electrostatic repulsion and position the microorganisms 1–2 nm from the biomaterial surface (22). The close proximity of the microorganisms to the IMD surface then allows irreversible adhesion to occur.

Irreversible adhesion of the microorganism to biomaterials occurs with the binding of specific microorganism adhesions to receptors expressed by the conditioning film (22). The specific mechanisms involved in this irreversible adhesion have been examined the closest in *Staphylococcus aureus* and *S. epidermidis*, which are the most common microorganisms causing IMD-related infection. *S. aureus* relies on specific cell surface proteins called "microbial surface components recognizing adhesive matrix molecules" (MSCRAMM) which bind to specific host ligands that are found in the conditioning films. The most important MSCRAMMs are the fibronectin-binding proteins (FnBPs), the fibrinogen-binding proteins (clumping factors, Clf), and the collagen adhesin, which binds to collagen (9). However, the role of MSCRAMMs in the pathogenesis of *S. aureus* IMD-related infections remains controversial due to studies with conflicting results (24,25).

Cell surface proteins also play an important role in *S. epidermidis* adhesion to IMDs. Proteinaceous autolysin (26) and polysaccharide adhesin (PSA) (27) are two surface proteins that play an early role in the irreversible adhesion of *S. epidermidis* to IMD surfaces (28). Once adherent to the biomaterial surface, cell accumulation and early biofilm formation are dependent upon the polysaccharide intercellular adhesin (PIA), which promotes intercellular adhesion (10).

3.4 IMD-RELATED BIOFILMS AND HUMAN DISEASE

From an epidemiological standpoint, biofilms play an important role in IMD-related infections. Quantitative culture and various imaging techniques performed on explanted IMDs have confirmed the presence of biofilms on these infected devices. The rapid clinical improvement in patients following the removal of infected IMDs is highly suggestive that biofilms are causing the systemic illness. However, the exact process by which biofilm-associated microorganisms attached to IMDs lead to disease in the human host remains poorly understood (29). Proposed mechanisms include the following: (a) detachment of cells or cell aggregates from biofilm coated IMDs leading to systemic bloodstream infection, (b) production of endotoxins from biofilm encased microorganisms, and (c) evasion of the host immune response (29). In this section, we will review the current literature supporting these possible mechanisms.

3.4.1 Cell Detachment

In natural environments, detachment of cells from a mature biofilm is considered a survival and propagation strategy. However, in IMD-related infections, detachment of microorganisms from biofilms may represent a fundamental mechanism leading to the dissemination of infection in the human host (22,30). Although detachment has not been specifically studied in relation to IMD-related infections, some information can be gained from the small number of *in vitro* studies that have previously been performed. Detachment has been determined to be a function of growth phase, biofilm size, nutrient conditions, and local environmental factors such as hemodynamic or mechanical shear forces (22). Increases in shear stress, as may be expected with the changes in direction or rate of flow in an intravenous catheter, have been shown to increase cell detachment from *in vitro* biofilms (29). The detachment of bacterial cells from biofilms has been categorized into the following two processes based on the magnitude and frequency of the detachment: (a) erosion—the continual detachment of single cells and small portions of the biofilm, and (b) sloughing—the rapid, massive loss of biofilm (30).

A recent study by Stoodley et al. (30) examined the role of detachment in mixed species biofilms grown in flow cells using digital time-lapse microscopy. Direct evidence of detachment and the size of the detached aggregates were determined by microscopic examination of effluent from the flow cells. Although single cells and aggregates of 10 cells or less accounted for 90% of the biomass particles visualized, large cell clusters (>1000 μm^2) represented over 60% of the total detached biofilm mass. The largest cluster identified in the effluent contained 1.6×10^3 cells, which is greater than the mean infective dose for many microorganisms that frequently cause biofilm related IMD infections. Although these results have yet to be confirmed *in vivo*, it is not difficult to imagine the detachment of large cell clusters leading to subsequent bloodstream and/or secondary infections by seeding distant sites. Since these detached cell clusters will most likely possess the antimicrobial resistance and ability to evade the host immune responses characteristic of the original biofilm, the resultant secondary infections caused by these detached cells may be particularly difficult to treat (30). Preliminary work in a rabbit model of *C. albicans* biofilm-related catheter infection by our group has confirmed that biofilm-related IMD infections can lead to seeding of distant sites (unpublished data). In these experiments, biofilm infections are established in surgically placed intravenous catheters using a *C. albicans* isolate tagged with green fluorescent protein (21). The animals are connected to a constant infusion of normal saline with 5% dextrose for 7 days prior to being sacrificed. Quantitative culture of kidneys removed from the animals consistently grew *C. albicans*, which was confirmed as the original isolate infecting the catheters by identifying the green fluorescent protein in the organ samples using confocal scanning laser microscopy (unpublished data). This new data confirms earlier *in vitro* studies suggesting that detachment of individual cells or cell aggregates from biofilms adherent to implanted IMDs plays an important role in the systemic complications associated with these infections. Use of *in vivo* animal models to demonstrate this phenomenon in other microbial biofilms should be encouraged.

3.4.2 ENDOTOXIN PRODUCTION

Gram negative organisms embedded in biofilms on the surface of IMDs will produce endotoxins that can stimulate a host immune response *in vivo* in the infected patient leading to a pyogenic reaction (29,31). Although no studies have defined the levels or kinetics of endotoxins secreted by biofilms and no studies have been performed to date, Vincent et al. (32) has shown that endotoxin levels correlated with bacterial counts in biofilms coating dialysis tubing. Further studies are required to clarify the role of biofilm produced endotoxin in human disease.

3.4.3 RESISTANCE TO THE HOST IMMUNE RESPONSE

The ability of IMD-related biofilms to persist until the infected device is removed suggests that the host immune response is ineffective against biofilms. This has become an intense area of interest for many researchers, but to date there has been very little published on this topic. There appears to be two main mechanisms by which IMD-related biofilms are able to evade the host defense: (a) the generation of a local zone of immunosuppression surrounding the IMD increases the risk for infection to develop (14), and (b) inherent properties of the biofilm appear to protect it from the host immune response. To date, only the role of polymorphonuclear leukocytes and opsinizing antibodies against biofilms have been studied in depth. The role of cellular immunity against biofilms is virtually unknown.

The biomaterials used in IMDs result in the activation of the host immune response leading to local tissue damage and the development of an immunoincompetent, fibroinflammatory zone that increases the susceptibility of the IMD to infection (14). Local stimulation of macrophages and the cellular immune response by the IMD surface results in superoxide radical and cytokine-mediated tissue damage increasing the risk for device-related infection (14). Zimmerli et al. (33) showed that polymorphonuclear leukocytes (PMN) adjacent to an IMD had low ingestion and bactericidal activities associated with low levels of granular enzymes and a decreased ability to mount a respiratory burst. The reduced activity of local PMNs against IMD-related biofilm infections is believed to be due to the prolonged stimulation of local PMNs by the nonphagocytosable IMD surface (frustrated phagocytosis) (33) and may also be related to PMN interactions with the extracellular matrix (34).

In addition to frustrated phagocytosis, the inability of PMNs to penetrate the biofilm extracellular matrix has been hypothesized to be a mechanism by which biofilms are able to elude the host immune response. However, recent *in vitro* evidence produced by Leid et al. (35) has shown that PMNs are able to penetrate the ECM of *S. aureus* biofilms grown under shear conditions. However, the PMNs that penetrated the ECM were unable to engulf bacteria within the biofilm suggesting that other unknown mechanisms inhibiting PMN function must exist in the biofilm environment (35).

Several investigators have evaluated the role of opsonic antibodies against biofilms. Ward et al. (36) showed that vaccinated rabbits, with a 1,000 fold-higher titer of antibodies, were unable to mount an immune response (phagocytosis) to peritoneal implants infected with biofilms. It was hypothesized that the biofilm ECM prevented the opsonic antibodies from reaching the surface of bacteria embedded in the biofilm (36). This hypothesis has subsequently been confirmed by Meluleni and colleagues (37) who showed that cystic fibrosis patients with high titers of mucoid *P. aeruginosa* opsonic antibodies were unable to mediate opsonic killing of biofilm, but not planktonic cells. However, when the ECM was degraded with alginate lyase or specific anti-alginate antibodies were added, comparable killing of *P. aeruginosa* in the biofilm and planktonic state occurred (37). These findings confirm the importance of the protective effects of the biofilm ECM against the host humoral immune response.

In summary, biofilms lead to disease in humans by detachment of cells, persistent secretion of endotoxins, and evasion of the host immune response. Although some literature exists to support these mechanisms, much more work is required to fully comprehend the relationship between biofilm formation and human disease. A true understanding of these mechanisms may one day lead to the development of novel preventive or therapeutic strategies against IMD-related biofilm infections.

3.5 CONCLUSIONS

Recent advances in medical technology have led to a drastic rise in the number of implanted IMDs, which has been accompanied by an increase in the number of IMD-related infections. Prior to the 1990s these infections were poorly understood. However, the recent discovery that biofilms adherent to the biomaterial surfaces of IMDs are responsible for these infections has led to a rapid increase in research and knowledge. We now have a basic understanding of how microorganisms adhere to IMDs leading to biofilm formation which include specific device related factors, the role of conditioning films, and specific microorganism–biomaterial interactions. We have also begun to understand how IMD-related biofilm infections cause human disease through cell detachment, secretion of endotoxins, or evasion of the host immune defenses. Overall, research in the area of IMD-related biofilm infections has progressed rapidly, but much more work is needed in order to develop effective preventive and therapeutic treatment strategies for these extremely difficult to treat infections.

REFERENCES

1. Costerton, J.W., Stewart, P.S., and Greenberg, E.P., Bacterial biofilms: a common cause of persistent infections. *Science* 284(5418), 1318–1322, 1999.
2. Costerton, J.W., Lewandowski, Z., Caldwell, D.E., Korber, D.R., and Lappin-Scott, H.M., Microbial biofilms. *Annu. Rev. Microbiol.* 49, 711–745, 1995.
3. Mukherjee, P.K., Chandra, J., Kuhn, D.M., and Ghannoum, M.A., Mechanism of fluconazole resistance in *Candida albicans* biofilms: phase-specific role of efflux pumps and membrane sterols. *Infect. Immun.* 71(8), 4333–4340, 2003.
4. Kuhn, D.M., George, T., Chandra, J., Mukherjee, P.K., and Ghannoum, M.A., Antifungal susceptibility of *Candida* biofilms: unique efficacy of amphotericin B lipid formulations and echinocandins. *Antimicrob. Agents Chemother.* 46(6), 1773–1780, 2002.
5. Kuhn, D.M., and Ghannoum, M.A., *Candida* biofilms: antifungal resistance and emerging therapeutic options. *Current Opinion in Infectious Diseases* 5(2), 186–197, 2004.
6. Donlan, R.M., Biofilm formation: a clinically relevant microbiological process. *Clin. Infect. Dis.* 33(8), 1387–1392, 2001.
7. Bandyk, D.F., and Esses, G.E., Prosthetic graft infection. *Surg. Clin. North Am.* 74(3), 571–590, 1994.
8. Reid, G., Biofilms in infectious disease and on medical devices. *Int. J. Antimicrob. Agents* 11(3–4), 223–226, 1999.

9. Darouiche, R.O., Device-associated infections: a macroproblem that starts with microadherence. *Clin. Infect. Dis.* 33(9), 1567–1572, 2001.

10. Dunne, W.M. Jr., Bacterial adhesion: seen any good biofilms lately? *Clin. Microbiol. Rev.* 15(2), 155–166, 2002.

11. Pascual, A., Pathogenesis of catheter-related infections: lessons for new designs. *Clin. Microbiol. Infect.* 8(5), 256–264, 2002.

12. Locci, R., Peters, G., and Pulverer, G., Microbial colonization of prosthetic devices. I. Microtopographical characteristics of intravenous catheters as detected by scanning electron microscopy. *Zentralbl. Bakteriol. Mikrobiol. Hyg.* [B] 173(5), 285–292, 1981.

13. Tebbs, S.E., Sawyer, A., and Elliott, T.S., Influence of surface morphology on *in vitro* bacterial adherence to central venous catheters. *Br. J. Anaesth.* 72(5), 587–591, 1994.

14. Schierholz, J.M., and Beuth, J., Implant infections: a haven for opportunistic bacteria. *J. Hosp. Infect.* 49(2), 87–93, 2001.

15. Jansen, B., Peters, G., and Pulverer, G., Mechanisms and clinical relevance of bacterial adhesion to polymers. *J. Biomater. Appl.* 2(4), 520–543, 1988.

16. Kohnen, W., and Jansen, B., Polymer materials for the prevention of catheter-related infections. *Zentralbl. Bakteriol.* 283(2), 175–186, 1995.

17. Hogt, A.H., Dankert, J., and Feijen, J., Adhesion of *coagulase-negative staphylococci* to methacrylate polymers and copolymers. *J. Biomed. Mater. Res.* 20(4), 533–545, 1986.

18. Raad, I.I., Luna, M., Khalil, S.A., Costerton, J.W., Lam, C., and Bodey, G.P., The relationship between the thrombotic and infectious complications of central venous catheters. *JAMA* 271(13), 1014–1016, 1994.

19. Murga, R., Miller, J.M., and Donlan, R.M., Biofilm formation by gram-negative bacteria on central venous catheter connectors: effect of conditioning films in a laboratory model. *J. Clin. Microbiol.* 39(6), 2294–2297, 2001.

20. Hawser, S.P., and Douglas, L.J., Biofilm formation by *Candida* species on the surface of catheter materials *in vitro*. *Infect. Immun.* 62(3), 915–921, 1994.

21. Schinabeck, M.K., Long, L.A., Hossain, M.A., Chandra, J.M., Mukherjee, P.K., Mohammad, S., and Ghannoum, M.A., Development of a rabbit model of *Candida albicans* biofilm infection: evaluation of liposomal amphotericin B antifungal lock therapy. *Antimicrob. Agents Chemother.* 48(5), 1727–1732, 2004.

22. Gristina, A.G., Biomaterial-centered infection: microbial adhesion versus tissue integration. *Science* 237(4822), 1588–1595, 1987.

23. Pashley, R.M., McGuiggan, P.M., Ninham, B.W., and Evans, D.F., Attractive forces between uncharged hydrophobic surfaces: direct measurements in aqueous solution. *Science* 229(4718), 1088–1089, 1985.

24. Darouiche, R.O., Landon, G.C., Patti, J.M., Nguyen, L.L., Fernau, R.C., and McDevitt, D., Role of *Staphylococcus aureus* surface adhesins in orthopaedic device infections: are results model-dependent? *J. Med. Microbiol.* 46(1), 75–79, 1997.

25. Greene, C., McDevitt, D., Francois, P., Vaudaux, P.E., Lew, D.P., and Foster, T.J., Adhesion properties of mutants of *Staphylococcus aureus* defective in fibronectin-binding proteins and studies on the expression of fnb genes. *Mol. Microbiol.* 17(6), 1143–1152, 1995.

26. Tojo, M., Yamashita, N., Goldmann, D.A., and Pier, G.B., Isolation and characterization of a capsular polysaccharide adhesin from *Staphylococcus epidermidis*. *J. Infect. Dis.* 157(4), 713–722, 1988.

27. Heilmann, C., Gerke, C., Perdreau-Remington, F., and Gotz, F., Characterization of Tn917 insertion mutants of *Staphylococcus epidermidis* affected in biofilm formation. *Infect. Immun.* 64(1), 277–282, 1996.

28. Rupp, M.E., Ulphani, J.S., Fey, P.D., Bartscht, K., and Mack, D., Characterization of the importance of polysaccharide intercellular adhesin/hemagglutinin of *Staphylococcus epidermidis* in the pathogenesis of biomaterial-based infection in a mouse foreign body infection model. *Infect. Immun.* 67(5), 2627–2632, 1999.

29. Donlan, R.M., and Costerton, J.W., Biofilms: survival mechanisms of clinically relevant microorganisms. *Clin. Microbiol. Rev.* 15(2), 167–193, 2002.

30. Stoodley, P., Wilson, S., Hall-Stoodley, L., Boyle, J.D., Lappin-Scott, H.M., and Costerton, J.W., Growth and detachment of cell clusters from mature mixed-species biofilms. *Appl. Environ. Microbiol.* 67(12), 5608–5613, 2001.

31. Rioufol, C., Devys, C., Meunier, G., Perraud, M., and Goullet, D., Quantitative determination of endotoxins released by bacterial biofilms. *J. Hosp. Infect.* 43(3), 203–209, 1999.

32. Vincent, F.C., Tibi, A.R., and Darbord, J.C., A bacterial biofilm in a hemodialysis system. Assessment of disinfection and crossing of endotoxin. *ASAIO Trans.* 35(3), 310–313, 1989.

33. Zimmerli, W., Lew, P.D., and Waldvogel, F.A., Pathogenesis of foreign body infection: evidence for a local granulocyte defect. *J. Clin. Invest.* 73, 1191–1200, 1984.

34. Johnson, G.M., Lee, D.A., and Regelmann, W.E., Interference with granulocyte function by *staphylococcus epidermidis* slime. *Infect. Immun.* 54, 13–20, 1986.

35. Leid, J.G., Shirtliff, M.E., Costerton, J.W., and Stoodley, A.P., Human leukocytes adhere to, penetrate, and respond to *Staphylococcus aureus* biofilms. *Infect. Immun.* 70(11), 6339–6345, 2002.

36. Ward, K., Olson, M.E., Lam, K., and Costerton, J.W., Mechanism of persistent infection associated with peritoneal implants. *J. Med. Microbiol.* 36, 406–413, 1992.

37. Meluleni, G.J., Grout, M., Evans, D.J., and Pier, G.B., Mucoid *Pseudomonas aeruginosa* growing in a biofilm *in vitro* are killed by opsonic antibodies to the mucoid exopolysaccharide capsule but not by antibodies produced during chronic lung infection in cystic fibrosis patients. *J. Immunol.* 155(4), 2029–2038, 1995.

4 Medical Device Composition and Biological Secretion Influences on Biofilm Formation

Sean P. Gorman and David S. Jones

CONTENTS

4.1   INTRODUCTION

Medical device-related infection has become a significant problem for the increasing numbers of patients who otherwise benefit from temporarily inserted or long-term implanted devices (1). The reason for this may be illustrated by a relatively simple analogy whereby the insertion of a urethral catheter is visualized as comparable to building a bridge between the outside world and the sterile bladder, along which bacteria can travel (2). The biofilm mode of growth on medical device biomaterials is of particular importance in device-related infection as this confers a number of advantages on bacteria, not least of which is extreme resistance to therapy with consequent patient detriment (3). These aspects are further detailed elsewhere in this book. Microbial biofilms can form on any medical device ranging from those that are relatively easily inserted and removed such as catheters, endotracheal and naso-gastric tubes, and contact lenses to those that are long-term implants such as hip joints, cardiac valves, and intraocular lenses. The greatly increased resistance of microbial biofilms to antimicrobial agents means these device-related infections can often only be treated after removal of the medical device, when this is a viable option, thereby increasing the trauma to the patient and the cost of the treatment (4).

The area of medical device technology has grown rapidly in recent years reflecting the advances made in the development of biocompatible materials (biomaterials) suitable for medical device construction. The medical device industry has, in turn, been guided and driven by the demands of the medical profession seeking constant improvements and innovation in the devices available for an increasingly elderly and, often, affluent population. These developments include biomaterial surface coatings, surface treatments, drug-incorporated bulk polymers and coatings, drug-polymer conjugates, and the availability of novel biodegradable (sometimes also referred to as bioresorbable or bioerodable) polymers that can mimic the natural turnover of mucus and cells in the body. The latter materials are increasingly referred to as biomimetic biomaterials as they have the ability to prevent microbial adherence and biofilm formation by shedding their surface (5,6). Consequently, there is a very broad range of materials employed in device composition with an attendant broad range of surface influences on the development of device microbial biofilm. Furthermore, medical devices will encounter many types of biological secretion or body fluid (urine, saliva, synovial fluid, blood, cerebrospinal fluid, gastrointestinal secretions) of varying composition depending on the clinical area. The deposition of these fluids on the device surface will also be a determining factor in the ability of a micro-bial pathogen to adhere to and subsequently colonize the surface to form a biofilm. The influence of these key elements, namely medical device composition and body fluid contact, on the development of biofilm will be the subject of this chapter.

4.2   MEDICAL DEVICE BIOCOMPATIBILITY

The great variety of biomaterials currently employed in medical device construction implies that at least some acceptable degree of success has been attained in respect to the acceptability of these by the body. Biocompatibility has been defined as "the utopian state where a biomaterial presents an interface with a physiologic environment without the material adversely affecting the environment or the environment adversely affecting the material" (7). This definition relates mainly to the cellular effects of the biomaterial. All biomaterials have some form of reactive effect on tissue. It was generally thought that biomaterials designed for implantation were chemically inert and, therefore, had no major effect on host physiological processes. However, many materials induce a variety of inflammatory and healing responses (8).

The effect of a medical device on the urothelium, for example, depends on the chemical composition and smoothness of the biomaterial, its surface characteristics and coatings, and the coefficient of friction (9). Materials with a smooth surface and low coefficient of friction appear to be more biocompatible by reducing the mechanical trauma and the shear forces at the biomaterial–tissue interface (10,11). Inappropriate biomaterial surface degradation and particle shedding can also lead to tissue toxicity and overall loss of biocompatibility (12,13). Materials that produce a significant inflammatory reaction in the host are likely to reduce the host's ability to resist infection. In addition, materials that least disrupt the mucosal defence mechanisms are better tolerated and result in lower rates of long-term sequelae (14). In the blood, the nature of the adsorbed protein layer determines a number of adverse events; thrombus formation can arise as a result of platelet adhesion, platelet activation, and initiation of coagulation (15). The most widely used method to prevent thrombus formation at a device surface is the administration of heparin. Loss of biocompatibility may arise through the leaching of substances from device biomaterials in situ. Latex has been shown to produce a greater inflammatory cell reaction than that induced by silicone catheters (16). In all these situations the immediate, normal environment of the medical device is potentially altered. The nature of the biomaterial–body fluid interaction may, therefore, also be different from that observed in the laboratory situation. The ability of a pathogen to enter the vicinity of these interactions, adhere and successfully form a biofilm on the device is obviously a complex process.

4.3 MEDICAL DEVICE BIOMATERIALS

A biomaterial can be defined as any substance, natural or synthetic, used in the treatment of a patient that at some stage, interfaces with tissue (17). Hundreds of polymers are now used in part or whole construction of medical devices therefore a brief description only is provided of some of the biomaterials most frequently encountered.

4.3.1 SILICONE

A commercially viable synthetic process to produce silicones was first developed in response to the war needs of the 1940s. In 1950, silicones were first used in medical applications. Silicone has become the standard against which other materials must be compared for biocompatibility. It is soft, nonirritating and clinically stable, making it ideal for long-term use in the urinary tract. It has good surface properties, which allow easy insertion, and lower rates of encrustation and bacterial adherence (10). Recently, novel, naturally lubricious silicones with the ability to deliver drugs have been developed (18). These novel silicones overcome the pain associated with device

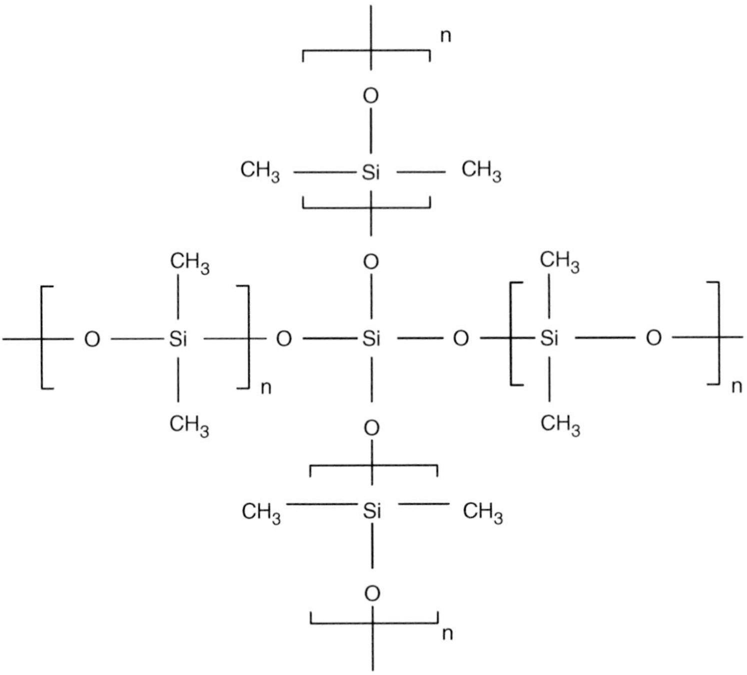

FIGURE 4.1 Structure of silicone.

insertion and tissue trauma experienced with device removal, due to the inherent disadvantages of a normally high coefficient of friction. Figure 4.1 shows the structure of silicone (polydimethylsiloxane).

4.3.2 POLYURETHANE

Polyurethanes are polymers containing the urethane linkage –OC(O)NH–. The term polyurethane refers to a broad variety of elastomers that are usually formed by the addition of a polyglycol to an isocyanate. By changing the chemical constituents, they can be readily tailored for many applications. Polyurethanes have good mechanical properties, are relatively inexpensive, and are commonly used in clinical practice.

4.3.3 LATEX

Latex was first discovered by Columbus and is composed of water and 30 to 35% natural rubber (poly cis-1,4-isoprene) (Figure 4.2). Latex catheters are inexpensive and have good elasticity, but tend to be more prone to bacterial adherence and have a greater potential for allergic sensitization than other materials (10). Many people now suffer from latex allergy which can cause symptoms ranging from allergic rhinitis, urticaria and wheezing to anaphylactic shock, brain damage, or death. Latex catheters are suitable for short-term use only and frequently are coated with additional polymers.

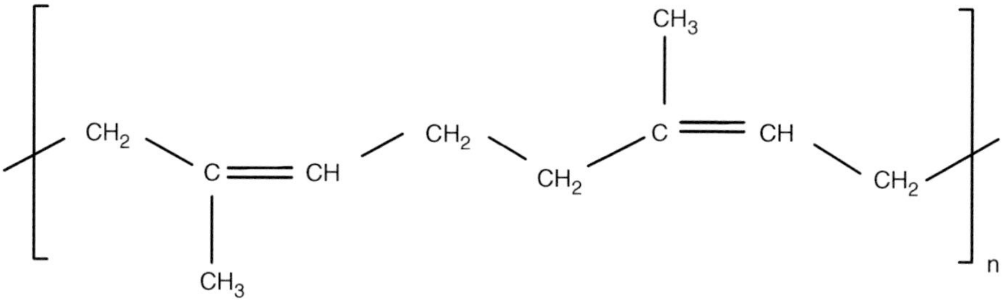

FIGURE 4.2 Structure of poly *cis*-1,4-isoprene.

4.3.4 HYDROGELS

Hydrogels are hydrophilic polymers that swell on contact with water, retaining a significant amount of water within their polyanionic structure (12). As the water content increases, medical devices become softer, more flexible, and more slippery, resulting in a greater ease of insertion. Hydrogels used for device coatings may include polyvinyl alcohol, hydroxy ethyl methacrylate, n-vinyl pyrollidone and a variety of other hydrophilic polymers. Hydrogel devices usually have a confluent coating of hydrogel on the surface and exhibit lower coefficients of kinetic friction and lower bacterial adhesion than most hydrophobic polymers (19).

4.3.5 POLY(TETRA FLUORO ETHYLENE)

Poly(tetra fluoro ethylene), also known as Teflon® or PTFE, is used largely as a surface coating. It is usually applied by dip-coating the substrate polymer into a solution containing PTFE particles and a binder polymer (often a polyurethane). After dip-coating, the binder polymer is cured. The resulting surface has a low coefficient of friction, but is not a continuous coating of PTFE (19). Latex urinary catheters coated with Teflon are smoother than plain latex catheters, reducing the incidence of urethritis and encrustation.

4.3.6 POLY(VINYL CHLORIDE)

Poly(vinyl chloride) (PVC) can be used for a wide range of medical device applications. It is strong, transparent, smooth and, most important, very inexpensive. In order to make it flexible, plasticizers are incorporated. Plasticized PVC catheters have a wide lumen but tend to be more rigid than latex or silicone catheters and are often found to be uncomfortable. Therefore, these catheters are more suitable for short-term, intermittent catheterization. Endotracheal tubes are also manufactured from PVC.

4.3.7 COPOLYMERS

Many copolymers have been developed for device construction or coatings. For example, C-Flex™ (Concept Polymer Technologies) is a proprietary silicone-modified block polymer, poly(styrene)-polyoefin, which provides an intermediate material between silicone and polyurethane (19). Silitek™ (Medical Engineering

Corporation) is another silicone-based copolymer and Percuflex™ (Boston
Scientific Corporation), an olefinic based copolymer (10).

4.4 CONDITIONING FILM FORMATION
ON MEDICAL DEVICE BIOMATERIALS

Two major problems arise in the use of implantable medical devices, microbial
adhesion to the constituent biomaterials and failure of the biomaterial to successfully
integrate with host tissue (20). Contact of an implanted medical device with biological
secretions or body fluids allows the deposition of a conditioning film. All surfaces
exposed to a biological environment will acquire a proteinaceous or glycoproteina-
ceous conditioning film. In the urinary tract, for example, this film is produced
host urinary components and has been shown to be composed of proteins, electrolytes
and other unidentified organic molecules. The protein composition of the conditioning
film on ureteral stents shows different protein profiles between encrusted and nonencrusted stents. Although both types were found to contain albumin, Tamm–Horsfall
Protein and α1-microglobulin absorption was limited to nonencrusted devices.
The deposition of a conditioning film markedly enhanced crystal precipitation and
aggregation events leading to blocking encrustations on the device surface (21).

   Protein adsorption onto biomaterial surfaces is a complex phenomenon that
directs subsequent biological responses to the surface (22). A conditioning film is
dynamic in the sense that its content likely changes with time, due to adsorption,
desorption and replacement of components and conformational alterations being
made at the liquid–surface interface (23). The formation of a conditioning film is a
very important event because it alters the surface of the biomaterial and may provide
receptor sites for bacterial adhesion (10). Conditioning films from saliva and urine
on poly(vinylchloride) and polyurethane reduced the measured contact angles
of both materials, thereby rendering them significantly more hydrophilic (24).
In respect to biofilm formation, the initial adhesion of bacterial cells to a conditioned
surface is considered a random event (25).

   In theory, the ideal device biomaterial should possess surface properties that
will interact favorably with the host environment. A conditioning film of body fluid
origin may contain a wide range of proteins such as fibronectin, laminin, fibrin,
collagen and immunoglobulins, and some of these molecules can act as receptors
promoting colonization by microorganisms (26). The fate of the medical device
surface has been described justifiably as "a race for the surface" between macromolecules, microorganisms and tissue cells (27). Potential colonizers first encounter
the conditioning film originating from the surrounding body fluid. Colonization of
the surface by tissue cells leads to development of a strong monolayer. The device
becomes fully integrated and microorganisms are confronted by living host cells
rather than by an artificial acellular substratum. An example of this is illustrated
in Figure 4.3.

   The integrated cell surface overcomes microbial colonization via its viability,
intact cell surface, and regular host defences. However, microbial colonization
potential is higher than that of the host tissue cells. Furthermore, host cells are

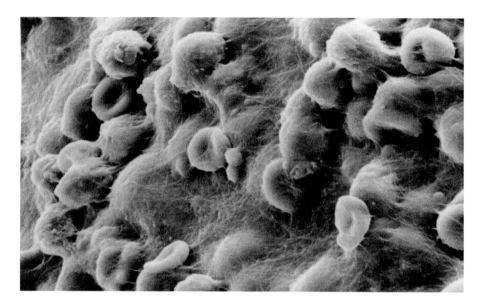

FIGURE 4.3 An extensive conditioning film on a medical device showing presence of fibrin and erythrocyte deposits.

unable to displace adherent microbial invaders and the device biomaterial often exhibits a poor host tissue integration performance. The microorganism, therefore, has a distinct advantage over the host tissue and device-related infection often ensues.

There have been a number of studies seeking to characterize conditioning films with the application of such techniques as electron spectroscopy for chemical analysis, time-of-flight secondary ion mass spectrometry, and radiolabeling (22).

4.5 MICROBIAL ADHERENCE TO MEDICAL DEVICES

As the process of microbial adhesion is the initial event in the pathogenesis of infection, failure to adhere will result in removal of the microorganism from the surface of an implanted medical device and avoidance of device-related infection (28). The adherence of a microorganism to a medical device surface involves several stages (29). Following device implantation and the deposition of a conditioning film, microorganisms must be transported to the surface and progress from an initial reversible attachment, to an irreversible adhesion and finally, proceed to the development of a microbial biofilm. This process can be influenced by a number of parameters such as bacterial cell wall characteristics, the nature of the fluid interface, and the biomaterial composition and characteristics. The development of the biofilm on the device surface may proceed as a succession of adhesion and multiplication events (30). The first organisms attaching are the primary colonizers and this is mediated through specific or non-specific physicochemical interactions with the components of an adsorbed, organic conditioning film. If the conditions are suitable, the primary colonizers can then multiply. A process of coaggregation may then occur wherein bacterial cells from the planktonic population specifically adhere to cells in the biofilm in a process known as coadhesion. These secondary colonizers may provide for a multispecies biofilm community (30).

4.5.1 STAGE 1. MICROBIAL TRANSPORT TO THE SURFACE

Microorganisms arrive at a surface more or less at random as a result of diffusion by Brownian motion, convection arising from currents in the surrounding medium, or by the chemotactic ability of the microorganism. Attachment, on a distance and time scale, may be divided into an initial, long-range, non-specific, reversible adherence stage and a subsequent, close-range, essentially irreversible, specific adherence stage (27).

4.5.2 STAGE 2. INITIAL ADHESION

When a microorganism arrives at a distance of approximately 50 nm from a medical device surface with a conditioning film, adhesion is initiated via long- and short-range forces. Microorganisms can be regarded as nonideal, living colloidal particles. This is the basis of the Derjaguin–Landau–Verwey–Overbeek (31,32) hypothesis on initial adherence wherein charged particles in an aqueous environment are surrounded by a diffuse layer of opposite charge. An electrical double layer is created and electrostatic interactions arise when two of these double layers overlap, e.g. when a microorganism approaches the surface of a biomaterial. This interaction is either repulsive, where both entities possess the same charge, or attractive if the charges are opposite. The energy of such interactions is determined by the zeta potential of each surface. The distance at which the electrostatic interactions become relevant is dependent on the thickness of the double layers, which in turn is influenced by the surface ionic charge and ionic concentration of the surrounding medium. The latter is, therefore in the context of an implanted medical device, determined by the surrounding body fluid.

When the microorganism initially approaches the device surface, long-range van der Waals forces come into play with an increasing contribution from long-range electrostatic forces as distances become shorter. When the separation distance exceeds 1 nm, the total interaction is given by the sum of these van der Waals and electrostatic forces. When a negatively charged organism and a negatively charged device surface co-exist, the total interaction energy, or total Gibbs energy (G_{tot}), is a function of separation and the ionic strength of the surrounding medium. The forces acting on an organism seeking to adhere to a device surface in a body fluid of low, medium, and high ionic strength may be visualized in Figure 4.4.

If the suspending fluid is of low ionic strength, the G_{tot} profile presents a barrier to adhesion in the form of a positive maximum. At a distance of less than 2 nm from the device surface, G_{tot} changes to a steep (primary) minimum where irreversible adhesion occurs. When the suspending fluid is of medium ionic strength, as is the case with saliva (affecting devices in the respiratory and upper gastrointestinal tracts), the positive maximum is reduced due to a decrease in the range of repulsive electrostatic forces. The approaching microorganism encounters a secondary minimum, resulting

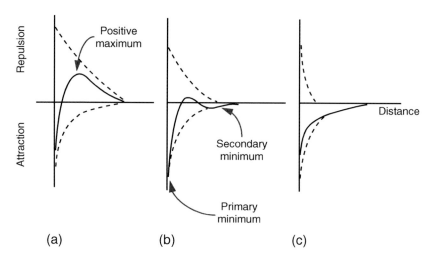

FIGURE 4.4 DLVO hypothesis as applied to the adherence of a microorganism to an implanted medical device. Attractive and repulsive forces are represented within a suspending media, such as a body fluids of (a) low, (b) medium or (c) high ionic strength.

from the decreased maximum, which offers an opportunity to adhere reversibly. A proportion of the microorganisms will possess sufficient thermal energy to overcome the reduced barrier, reach the primary minimum and hence adhere irreversibly. If the surrounding body fluid or medium is of high ionic strength, G_{tot} is entirely negative, allowing all of the microorganisms to potentially reach the primary minimum and adhere irreversibly.

The surface charge of microorganisms and medical device materials is generally negative and so a long-range interaction with primary and secondary minima usually arises. It is possible to determine interaction energy when a microorganism makes direct contact with a surface assuming that the interfaces between the microorganism/liquid (ml) and solid/liquid (sl) are replaced by a solid/microorganism (sm) interface. The interfacial free energy of adhesion (ΔG_{adh}) is given by the equation:

$$\Delta G_{adh} = \gamma_{sm} - \gamma_{sl} - \gamma_{ml}$$

where γ is the interfacial free energy between the particular components in the system. If ΔG_{adh} is negative, adhesion is thermodynamically favorable and will occur. Microorganisms may make the transition from secondary to primary minimum by overcoming the positive maximum barrier or alternatively, by bridging the gap using microbial surface structures such as fibrils and fimbriae. In order to complete the adhesion process, removal of the water between interacting bodies is necessary. This is made possible by hydrophobic groups on the microbial cell surface and may help explain the role of cell surface hydrophobicity in microbial adhesion.

The importance of the nature of the surrounding body fluid in determining microbial adherence to a device can be observed most readily in the urinary tract where urine can range in composition. For example, the balance of electrolytes, pH, urea and creatinine concentration can all affect the extent to which microbes can adhere to a urinary device (33). A urine with a low urea concentration and a pH 5–8 range increased adherence (34). When magnesium and calcium were present, as in a urinary encrusting environment, the adherence of *Staphylococcus epidermidis* was also significantly increased (34). A radical decrease in adherence was observed when ethylenediaminetetraacetic acid was present. The presence of 5% CO_2 in the environment also increases adherence of a number of bacteria to different types of polymeric device (35,36).

4.5.3 Stage 3. Microbial Attachment to Device Surface

Following initial reversible adhesion, irreversible attachment of the microorganism to the device surface will occur through specific covalent, ionic or hydrogen bonding. These interactions arise either through direct contact or by means of extracellular filamentous appendages. Irreversible attachment is mediated by adhesins of microbial origin and complementary receptors on the biomaterial surface. Figure 4.5 shows the appearance of a device surface in the early stages of microbial colonization.

4.5.4 Stage 4. Colonization

Under suitable conditions, microorganisms adherent to the device surface multiply, adhere to one another, and elaborate biofilm, leading to the formation of microcolonies and a mature biomaterial-based infection.

FIGURE 4.5 Staphylococcal cells adhering to an endotracheal tube.

4.6 FACTORS INFLUENCING ADHERENCE

The key factors influencing microbial adherence to a medical device, as related above in the equation predicting the energy involved, are the microorganism, the biomaterial surface, and the suspending medium.

4.6.1 THE MICROORGANISM: SURFACE CHARACTERISTICS

The cell surface of a bacterium possesses many structures and properties that contribute to bacterial adhesion including fimbriae (pili), the cell wall (techoic acid in Gram-positive bacteria), and outer cell membrane (liposaccharides in Gram-negative pathogens). These characteristics influence the surface charge and hydrophobicity of the bacterial cell, thereby directly affecting adherence.

Environmental conditions affecting the microorganism such as the surrounding medium (body fluid, gaseous atmosphere, and temperature) will influence the nature of the cell surface and therefore adhesion. Growth of *Enterococcus faecalis* and *Escherichia coli* in human urine results in altered cell surface hydrophobicity and surface charge, which is associated with an increased adherence to polyurethane devices (36). In contrast, growth of *S. aureus* and *P. aeruginosa* in human urine leads to reduced adherence of these bacteria to glass and polystyrene. In addition, when these isolates are cultured in an atmosphere of 5% CO_2/95% air, as opposed to air alone, their adherence to polystyrene is significantly altered (37). *Staphylococcus epidermidis* exhibits an altered cell surface chemistry following growth in an atmosphere enriched with 5% CO_2, and this is translated into a modified adherence to polystyrene and silicone (38). Adherence by coagulase-negative staphylococci to these device materials was altered by growth in a 5% CO_2-enriched atmosphere (39). However, the influence of CO_2 on the adherence of *S. epidermidis* to polystyrene appears to be dependent on the growth medium (40). The adherence of different strains of *Candida albicans* to vaginal epithelial cells fluctuates depending on atmospheric CO_2 levels and the pH of the medium (41). Temperature also has an effect on microbial cell surfaces, with *C. albicans* grown at room temperature being more hydrophobic than those grown at 37°C, while yeasts grown at 25°C, rather than 37°C, exhibit an increased adherence to human buccal epithelial cells (42,43).

The physico-chemical character of microbial cell surfaces, i.e., hydrophobicity and charge, will influence adherence to biomaterial surfaces since the process is strongly governed by hydrophobic and electrostatic interactions. Hydrophobic strains of *P. aeruginosa* exhibited greater adherence to PVC, polyurethane and siliconized latex than hydrophilic strains (44). Hydrophobic bacteria adhered in greater numbers than hydrophilic cells to sulfated polystyrene (45). These relatively hydrophobic cells were observed to have high negative electrokinetic potential, a measure of surface charge. This apparent contradiction could be explained by charged groups occupying a minor fraction of the total cell surface area. Hydrophobic strains of *E. coli* have been shown to migrate along solid surfaces at a faster rate than hydrophilic strains, and shorter migration times were noted for organisms that were highly negatively charged (46). Microbial adherence to silicone rubber, with and without a salivary conditioning film, revealed that organisms with the most negative zeta potential adhered most slowly to the negatively charged silicone surface (47).

4.6.2 THE BIOMATERIAL SURFACE

Factors relating to the biomaterial that have been found to be important in the adherence of microorganisms include surface hydrophobicity, charge, and roughness (microrugosity). The role of hydrophobicity as a non-specific binding parameter in bacterial adhesion is broadly accepted as the hydrophobicity of a biomaterial surface dictates the composition of the conditioning film (48). The influence of biomaterial surface hydrophobicity was demonstrated in a study involving biomaterial surface free energy, a parameter inversely related to hydrophobicity (49). The adherence of *S. epidermidis* was found to be greater when the biomaterial surface free energy was close to that of the bacteria. The reverse was true when a hydrophilic strain of *E. coli* was employed. This pattern of behavior in adherence was confirmed when the adherence of a hydrophilic isolate of *E. coli* to co-polymers of poly (methylmethacrylate) and poly(hydroxyethylmethacrylate) increased as the hydrophobicity of the biomaterials decreased (50). In a similar fashion, the adherence of *S. epidermidis* and *S. saprophyticus* to a homologous series of methacrylate polymers and co-polymers was found to vary depending on the biomaterial surface charge. Increased adherence to positively charged surfaces was observed (51). It was suggested that microbial adherence rates onto positively charged surfaces were diffusion limited whereas adherence rates onto negatively charged surfaces were more surface-reaction controlled due to a potential energy barrier. Therefore, one way to reduce bacterial adhesion to a device might be to modify biomaterial surfaces to reduce their hydrophobicity. Coating polymeric materials with mucin results in decreased hydrophobicity of the material and decreases the adherence of *S. aureus* and *S. epidermidis* (52). Further description of approaches that may be taken to reduce biofilm on urinary devices is provided by Tunney et al. (53).

Wilkins et al. (54) proposed a relationship between biomaterial surface microrugosity and bacterial adherence while attempting to explain why *S. aureus* and *E. coli* failed to maximally adhere to the most hydrophobic material investigated. In a subsequent study, microrugosity of polymeric threads was identified as a factor influencing bacterial adherence. It was suggested that rougher, grooved areas, observed via scanning electron microscopy (SEM), offered a larger surface area for contact and so increased adherence (55). Employing SEM and confocal laser scanning microscopy (CLSM), the surface of used and unused continuous peritoneal ambulatory dialysis catheters was examined (56). CLSM revealed that the surface roughness of used catheters was greater than that of new devices. Following incubation of *S. epidermidis* with both types of catheter, an increased adherence to sections of used catheter indicated that microbial adherence was promoted by a surface exhibiting a greater microrugosity. Surface roughness was put forward as the dominant factor, ahead of surface free energy, in determining the adherence of oral bacteria leading to supra-gingival plaque formation (29). CLSM and atomic force microscopy are increasingly useful techniques for evaluating the surface characteristics of medical devices in relation to biofilm formation (57).

A series of studies examining the influence of conditioning fluid deposition on microbial adherence and biofilm formation on the surface of a range of medical devices produced very similar outcomes (24,58–60). A wide range of device materials

TABLE 4.1
The Effects of Saliva Treatment on the Advancing and Receding Contact Angles and Surface Roughness (Microrugosity) of PVC (Mean ± S.D.)

Treatment of PVC	Advancing Contact Angle[a]	Receding Contact Angle[a]	Microrugosity Z_{rms} (nm)[b]
Untreated	92.39 ± 0.27	65.06 ± 0.41	36.02 ± 0.35
PBS-treated	92.12 ± 0.13	67.33 ± 0.93	35.92 ± 0.31
Saliva-treated	86.45 ± 0.55	56.32 ± 0.36	19.55 ± 0.40

[a] Determined using a dynamic contact angle analyser.
[b] Determined using atomic force microscopy.
Source: Jones, D.S., McGovern, J.G., Woolfson, A.D., and Gorman, S.P., The role of physiological conditions in the oropharynx on the adherence of respiratory isolates to endotracheal tube polyvinylchloride. *Biomaterials* 18, 503–510, 1997. With permission.

was examined in the studies including silicone, polypropylene and PVC. Typical results for PVC are summarized in Table 4.1.

The adherence of bacterial and candidal species treated with peritoneal dialysate or saliva in an environment of 5% CO_2 was decreased to device materials treated in a similar manner. The contact angles and the rugosity of the treated surfaces were decreased.

4.6.3 THE SUSPENDING MEDIUM

The absorption of components from the suspending fluid can profoundly affect the adhesive properties of microorganisms. The ionic strength, osmolality, and pH of urine all influence the initial attachment of bacteria, as can the urinary concentration of urea, creatinine and proteins (10). Exposure of urinary pathogens to ascorbic acid or cranberry juice produces uropathogen surfaces that are more positively charged due to pH changes (61,62). Urine also contains urinary inhibitors that may protect the uroepithelium. Tamm–Horsfall Protein, the most abundant protein found in human urine, acts as a host defense mechanism by binding to type 1 fimbriated *E. coli* and removing organisms from the bladder (63).

As described above, in the process of adherence of microorganisms to an implanted medical device, one or both entities will be exposed to a biological secretion or body fluid of host origin. The subsequent conditioning of the microbial cell and/or biomaterial surface will inevitably modify the nature of both surfaces, thereby determining the outcome of the adherence process. There are numerous reports recounting the influence of conditioning fluids on biomaterial surface character and subsequent microbial adherence. Poisson et al. (64) observed that prior to colonization of endotracheal (ET) tubes, microorganisms preferentially adhere to a biological film of human origin rather than to the constituent biomaterial itself. Treatment of catheter materials with body fluids, such as plasma and serum albumin, reduced bacterial adherence, an outcome attributed to the adsorption of body fluid components onto the biomaterial surface (65). Bonner et al. (36) found that treatment of polyurethane with human urine led to a decrease in biomaterial advancing contact angle, indicating a more hydrophilic surface. Adherence of *E. coli* and *E. faecalis,* grown in Mueller–Hinton broth, was shown to increase after the biomaterial was exposed to human urine. In contrast, conditioning of silicone rubber surfaces with saliva generally reduces the rate of adherence by bacteria and yeasts and leads to an overall decrease in adherence (47). In agreement with the latter finding, Gorman et al. (56) reported a reduction in microbial adherence to peritoneal catheters following treatment of the constituent material with spent peritoneal dialysate.

4.7 EXEMPLAR: ENDOTRACHEAL TUBE BIOFILM FORMATION IN RELATION TO A SALIVARY CONDITIONING FILM

Microorganisms adhering in the oropharynx, to either epithelial tissue or the endo-tracheal tube (ET), will encounter a salivary liquid medium. Likewise, the surface of the ET tube will be exposed to the body fluid. The following illustrates, in a clinical context, the relationship between a medical device and adhering pathogens when both are in contact with a body fluid.

Hospital acquired (nosocomial) infection causes significant patient morbidity and mortality with associated high costs of hospitalization (5). It is accepted that of all nosocomial infections in the ICU, pneumonia is the most commonly reported in mechanically ventilated patients (66). There is a two- to ten-fold increase in mortality among ICU patients with ventilator-associated pneumonia (VAP) compared to ICU patients whom do not experience VAP (67–69). Reported crude mortality of patients with VAP ranges from 24 to 71%, depending on pathogen, type of ICU, diagnostic method utilized, and extent of underlying disease (70,71). Invasive medical devices such as the ET tube and nasogastric (NG) tube are an important factor in the patho-genesis of VAP. The likelihood of aspiration is increased when an NG tube is insert-ed, predisposing the patient to gastric reflux. ET tubes assist the entry of bacteria into the tracheobronchial tree and aspiration into the lower airway of contaminated secretions by weakening the natural barrier between the oropharynx and trachea. The pooling and leakage of subglottic secretions around the tracheal cuff and ET tube induced mucosal injury further predispose patients to the development of VAP (72). Biofilm formation on the endotracheal tube (Figure 4.6) has also been implicated in the pathogenesis of VAP (73–75).

4.7.1 THE GASEOUS ATMOSPHERE IN THE OROPHARYNX

The microbial population in the oral cavity is predominantly comprised of anaerobic organisms, so it is reasonable to suggest that the mucous membranes represent an anaerobic environment. The air taken into the lungs during inspiration consists of approximately 21% oxygen; this level falls to between 12 and 14% in the oral cavity and declines further to 1 or 2 % in the periodontal pocket. Measurement of oxidation-reduction potential, where positive values indicate aerobic conditions, confirms that the oral cavity comprises a gaseous environment in which there is a reduced oxygen tension (76). In addition, expired gas passing from the lungs through

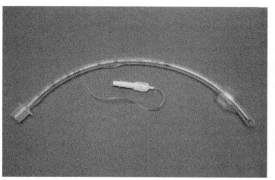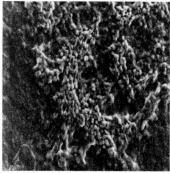

FIGURE 4.6 A PVC endotracheal tube (left) with a scanning electron micrograph of a typical lumenal staphylococcal biofilm (right).

the oropharynx creates an atmosphere in which there is a carbon dioxide (CO_2) concentration of 5% compared to a level of 0.05% in air (77).

4.7.2 THE SALIVARY PELLICLE AND ADHERENCE

Microorganisms resident in the oral cavity and oropharynx are continually bathed in saliva and the microbial cell surface can become coated with specific salivary proteins. These proteins may provide additional receptor sites and modify subsequent microbial adherence in the oropharynx. In addition, saliva can aggregate bacteria, improving microbial removal from the oral cavity by swallowing. The adherence of streptococci, for example, is increased by saliva constituents such as low molecular weight salivary mucin, proline-rich glycoproteins, α amylase, and proline-rich peptides. Some species bind salivary amylase and as a result are found exclusively in hosts exhibiting salivary amylase activity. Mucin-like glycoproteins present in saliva interact with many streptococcal species such that a mechanism for clearing the oral cavity of these organisms is established. On the other hand, salivary agglutinin bound to buccal epithelial surfaces can promote streptococcal adherence (78). The buffering system in saliva is controlled primarily by bicarbonate ions but phosphates, peptides and proteins also help in pH maintenance. Mean salivary pH lies in the range 6.75–7.25, however this will vary with flow rate that in turn is subject to circadian rhythms. We should not always consider the body fluid conditioning film to be disadvantageous as, for example, saliva also possesses antimicrobial elements important in controlling microbial colonization of the oral cavity. Among such components are antibodies, lysozyme, lactoferrin, the sialperoxidase system, and antimicrobial peptides such as histidine-rich polypeptides. On the other hand, saliva can act as a source of nutrition for the microorganisms that encounter it in the oral environment (76). Despite the antimicrobial nature of whole saliva constituents, and the body's own immune system, the mouth and oropharynx exist as an ecological niche allowing microbial populations to survive. Tactics adopted by such microorganisms to enable survival include continuous antigenic variation and antigenic masking by, e.g., slime capsules or adsorption of host macromolecules on to the microbial cell surface. Microorganisms may possess antigens that are sufficiently similar to those of the host so as to go unnoticed. It is also possible to inactivate host defences by microbial production of enzymes such as proteases.

REFERENCES

1. Rapp, R.P., Adair, C.G., and Gorman, S.P., Nosocomial infections. In: DiPiro, J., Talbert, R., Yee, G., Matzke, G., Wells, B., Posey, L., eds., *Pharmacotherapy: A Pathophysiological Approach,* 3rd Edition. Appleton & Lange, New York, 1996: 2387–2400.
2. Nickel, J.C., Catheter-associated urinary tract infection: New perspectives on old problems. *Canad. J. Infec. Control* 6(2), 38–42, 1991.
3. Tunney, M.M., Gorman, S.P., and Patrick, S., Infection associated with medical devices. *Rev. Med. Microbiol.* 74, 195–205, 1996.
4. Mah T.F.C., and O'Toole G.A., Mechanisms of biofilm resistance to antimicrobial agents. *Trends in Microbiology* 9(1), 34–39, 2001.
5. Gorman, S.P., and Jones, D.S., Antimicrobial biomaterials for medical devices. World Markets Research Centre, Medical Device Manufacturing and Technology, 2002: 97–101.
6. Gorman, S.P., and Jones, D.S., Complications of urinary devices. In: Wilson, M., ed., *Medical Implications of Biofilms.* Cambridge University Press, 2003:136–170.
7. Mardis, H.K., and Kroeger, R.M., Ureteral stents. *Urologic Clinics of North America* 15(3), 471–479, 1988.
8. Belanger, M.C., and Marois, Y., Hemocompatibility, biocompatibility, inflammatory and *in vivo* studies of primary reference materials low-density polyethylene and polydimethylsiloxane: A review. *Journal of Biomedical Materials Research (Applied Biomaterials)* 58, 467–477, 2001.
9. Sofer, M., and Denstedt, J.D., Encrustation of biomaterials in the urinary tract. *Current Opinion in Urology* 10, 563–569, 2000.
10. Denstedt, J.D., Wollin, T.A., and Reid, G., Biomaterials used in urology: current issues of biocompatibility, infection and encrustation. *J. Endourol.* 12(6), 109–112, 1998.
11. Jones, D.S., Garvin, C.P., and Gorman, S.P., Design of a simulated urethra model for the quantitative assessment of urinary catheter lubricity. *J. Mat. Sci. Materials in Medicine* 12, 15–21, 2001.
12. Denstedt, J.D., Reid, G., and Sofer, M., Advances in ureteral stent technology. *World Journal of Urology* 18, 237–242, 2000.
13. Keane, P.F., Bonner, M., Johnston, S.R., Zafar, A., and Gorman, S.P., Characterisation of biofilm and encrustation on ureteral stents *in vivo. Br. J. Urol.* 73, 687–692, 1994.
14. Burrows, L.L., and Khoury, A.E., Issues surrounding the prevention and management of device—related infections. *World Journal of Urology* 17, 402–409, 1999.
15. Keuren, J.F.W., Wielders, S.J.H., Willems, G.M., Morra, M., Cahalan, L., Cahalan, P., and Lindhout, T., Thrombogenicity of polysaccharide-coated surfaces. *Biomaterials* 24, 1917–1924, 2003.
16. Talja, M., Korpela, A., and Jarvi, K., Comparison of urethral reaction to full silicone, hydrogen-coated and siliconised latex catheters. *Br. J. Urol.* 66, 652–657, 1990.
17. Wollin, T.A., Tieszer, C., Riddell, J.V., Denstedt, J.D., and Reid, G., Bacterial biofilm formation, encrustation, and antibiotic adsorption to ureteral stents indwelling in humans. *J. Endourol.* 12(2), 101–111, 1998.

18. Woolfson, A.D., Malcolm, R.K., Gorman, S.P., Jones, D.S., Brown, A.F., and McCullagh, S.D., Self-lubricating silicone elastomer biomaterials. *J. Mater. Chem.* 13, 2465–2470, 2003.
19. Wironen, J., Marotta, J., Cohen, M., and Batich, C., Materials used in urological devices. *Journal Of Long-Term Effects of Medical Implants* 7(1), 1–28, 1997.
20. Gristina, A.G., Giridhar, G., Gabriel, B.L., Naylor, P.T., and Myrvik, Q.N., Cell biology and molecular mechanisms in artificial device infections. *Int. J. Artificial Organs* 16, 755–763, 1993.
21. Santin, A., Wassall, M.A., Peluso, G., and Denyer, S.P., Adsorption of α-1microglobulin from biological fluids onto polymer surfaces. *Biomaterials* 18, 823–827, 1997.
22. Wagner, M.S., Horbett, T.A., and Castner, D.G., Characterising multicomponent adsorbed protein films using electron spectroscopy for chemical analysis, time-of-flight secondary ion mass spectrometry, and radiolabelling: capabilities and limitations. *Biomaterials* 24, 1897–1908, 2003.
23. Tieszer, C., Reid, G., and Denstedt, J., Conditioning film deposition on ureteral stents after implantation. *J. Urology* 160, 876–881, 1998.
24. McGovern, J.G., Garvin, C.P., Jones, D.S., Woolfson, A.D., and Gorman, S.P., Modification of biomaterial surface characteristics by body fluids *in vitro*. *Int. J. Pharmaceutics* 149, 251–254, 1997.
25. Jenkinson, H.F., and Lappin-Scott, H.M., Biofilms adhere to stay. *Trends in Microbiology* 9(1), 9–10, 2001.
26. Dickinson, G.M., and Bisno, A.L., Infections associated with indwelling devices: concepts of pathogenesis; infections associated with intravascular devices. *Antimicrobial Agents and Chemotherapy* 33, 597–601, 1989.
27. Gristina, A.G., Naylor, P., and Myrvik, Q., Infections from biomaterials and implants: a race for the surface. *Medical Progress Through Technology* 14, 205–224, 1988.
28. Ofek I., and Beachey E.H., General concepts and principals of bacterial adherence in animals and man. In *Bacterial Adherence*, Ed. E.H. Beachey. Chapman and Hall, London, 1980:1–29.
29. Quirynen, M., and Bollen, C., The influence of surface roughness and surface free energy on supragingival and subgingival plaque formation in man – a review of the literature. *J. Clin. Periodont.* 22, 1–14, 1995.
30. Rickard, A.H., Gilbert, P., High, N.J., Kolenbrander, P.E., and Handley, P.S., Bacterial coaggregation: an integral process in the development of multi-species biofilms. *Trends in Microbiology* 11, 94–100, 2003.
31. Derjaguin, B.V., and Landau, V., Theory of the stability of strongly charged lyophobic sols and the adhesion of strongly charged particles in solutions of electrolytes. *Acta. Physicochemica. USSR* 14, 633–662, 1941.
32. Verwey, E.J.W., and Overbeek, J.T.G., *Theory of the Stability of Lyophobic Colloids.* Amsterdam, Elsevier, 1948.
33. Reid, G., Lam, D., Policova, Z., and Neumann, A.W., Adhesion of two uropathogens to silicone and lubricious catheters: influence of pH, urea and creatinine. *J. Mat. Sci. Mad. Med.* 4, 17–22, 1993.
34. Dunne, W.M., and Burd, E.M., The effects of magnesium, calcium, EDTA, and pH on the *in vitro* adhesion of Staphylococcus epidermidis to plastic. *Microbiol. Immunol.* 36(10), 1019–1027, 1992.
35. Gorman, S.P., McGovern, J.G., Woolfson, A.D., Adair, C.G., and Jones, D.S., The concomitant development of poly (vinyl chloride)-related biofilm and antimicrobial reistance in relation to ventilator-associated pneumonia. *Biomaterials* 22, 2727–2741, 2001.

36. Bonner, M.C., Tunney, M.M., Jones, D.S., and Gorman, S.P., Factors affecting *in vitro* adherence of ureteral stent biofilm isolates to polyurethane. *Int. J. Pharmaceutics* 151, 201–207, 1997.

37. Wassal, M.A., McGarvey, A, and Denyer, S.P., Influence of growth conditions on adherence potential and surface hydrophobicity of urinary tract isolates. *J. Pharm. Pharmacol.* 46, 1044, 1994.

38. Denyer, S.P., Davies, M.C., Evans, J.A., Finch, R.G., Smith, D., Wilcox, M.H., and Williams P., Influence of carbon dioxide on the surface characteristics and adherence potential of coagulase-negative staphylococci. *J. Clin. Microbiol.* 28, 1813–1817, 1990.

39. Wilcox, M.H., Finch, R.G., Smith, D.G.E., Williams, P., and Denyer, S.P., Effects of carbon dioxide and sublethal levels of antibiotics on adherence of coagulase-negative staphylococci to polystyrene and silicone rubber. *J. Antimicrob. Chemother.* 27, 577–587, 1991.

40. Hussain, M., Wilcox, M.H., White, P.J., Faulkner, M.K., and Spencer, R.C., Importance of medium and atmosphere type to both slime production and adherence by coagulase-negative staphylococci. *J. Hospital Infection* 20, 173–184, 1992.

41. Persi, M.A., Burnham, J.C., and Duhring, J.H., Effects of carbon dioxide and pH on adhesion of *Candida albicans* to vaginal epithelial cells. *Infection and Immunity* 50, 82–90, 1985.

42. Hazen, K.C., Plotkin, B.J., and Klimas, D.M., Influence of growth conditions on cell surface hydrophobicity of *Candida albicans* and *Candida glabrata*. *Infection and Immunity* 54, 269–271, 1986.

43. Kennedy, M.J., and Sandin, R.L., Influence of growth conditions on *Candida albicans* adhesion, hydrophobicity and cell wall ultrastructure. *J. Medical and Veterinary Mycology* 26, 79–92, 1988.

44. Martinez-Martinez, L., Pascual, A., and Perea, E.J., Kinetics of adherence of mucoid and non-mucoid *Pseudomonas-aeruginosa* to plastic catheters. *J. Med. Microbiol.* 34, 7–12, 1991.

45. van Loosdrecht, M.C.M., Lyklema, J., Norde, W., Schraa, G., and Zehnder, A.J.B., Electophoretic mobility and hydrophobicity as a measure to predict the initial steps of bacterial adhesion. *Appl. Environ. Microbiol.* 53, 1898–1901, 1987a.

46. Harkes, G., Dankert, J., and Feijen, J., Bacterial migration along solid surfaces. *Appl. Environ. Microbiol.* 58, 1500–1505, 1992.

47. Busscher, H.J., Geertsema-Doornbusch, G.I., and van der Mei, H.C., Adhesion to silicone rubber of yeasts and bacteria isolated from voice prostheses: influence of salivary conditioning films. *J. Biomed. Mater. Res.* 34, 201–209, 1997.

48. Brunsima, G.M., van der Mei, H.C., and Busscher, H.J., Bacterial adhesion to surface hydrophilic and hydrophobic contact lenses. *Biomaterials* 22, 3217–3224, 2001.

49. Ferreiros, C.M., Carballo, J., Criado, M.T., Sainz, V., and Delrio, M.C., Surface free energy and interaction of *Staphylococcus epidermidis* with biomaterials. *FEMS Microbiol. Lett.* 60, 89–94, 1989.

50. Tunney, M.M., Jones, D.S., and Gorman, S.P., Methacrylate polymers and copolymers as urinary tract biomaterials: resistance to encrustation and microbial adhesion. *Int. J. Pharmaceut.* 151, 121–126, 1997.

51. Hogt, A.H., Dankert, J., and Feijen, J., Adhesion of *Staphylococcus epidermidis* and *Staphylococcus saprophyticus* to a hydrophobic biomaterial. *J. General Microbiol* 131, 2485–2491, 1985.

52. Shi, L., Ardehali, R., Caldwell, K.D., and Valint, P., Mucin coating on polymeric material surfaces to suppress bacterial adhesion. *Colloids and Surfaces B: Biointerfaces* 17, 229–239, 2000.

53. Tunney, M.M., Jones, D.S., and Gorman, S.P., Biofilm and biofilm-related encrustation of urinary tract devices. *Methods in Enzymology.* Ed. Doyle R. Academic Press, 1999: 558–565.

54. Wilkins, K.M., Hanlon, G.W., Martin, G.P., and Marriott, C., The role of adhesion in the migration of bacteria along intra-uterine contraceptive device polymer monofilaments. *Int. J. Pharmaceut.* 58, 165–174, 1990.

55. Wilkins, K.M., Martin, G.P., Hanlon, G.W., and Marriot, C., The influence of critical surface tension and microrugosity on the adhesion of bacteria to polymer monofilaments. *Int. J. Pharmaceut.* 57, 1–7, 1989.

56. Gorman, S.P., Mawhinney, W.M., Adair, C.G., and Issoukis, M., Confocal laser scanning microscopy of peritoneal catheter surfaces. *J. Med. Microbiol.* 38, 411–417, 1993.

57. Adair, C.G., Gorman, S.P., Byers, L.M., Gardiner, T., and Jones, D.S., Confocal laser scanning microscope examination of microbial biofilms. In: An, Y.H., and Friedman, R.J., eds., *Handbook of Bacterial Adhesion.* Humana Press, New Jersey, USA, 2000:249–259.

58. Gorman, S.P., Jones, D.S., Adair, C.G., McGovern, J.G., and Mawhinney, M.W., Conditioning fluid influences on the surface properties of, and adherence of Staphylococcus epidermidis to silicone and polyurethane peritoneal catheters. *J. Mater. Sci.: Materials in Medicine* 8, 631–635, 1997.

59. Jones, D.S., McGovern, J.G., Adair, C.G., Woolfson, A.D., and Gorman, S.P., Conditioning film and environmental effects on the adherence of *Candida* spp. to silicone and poly(vinylchloride) biomaterials. *J. Mat. Sci.: Materials in Medicine* 12, 399–405, 2001.

60. Jones, D.S., McGovern, J.G., Woolfson, A.D., and Gorman, S.P., The role of physiological conditions in the oropharynx on the adherence of respiratory isolates to endotracheal tube polyvinylchloride. *Biomaterials* 18, 503–510, 1997.

61. Habash, M., and Reid, G., Microbial biofilms: Their development and significance for medical device-related infections. *J. Clin. Pharmacol.* 39, 887–898, 1999.

62. Habash, M.B., van der Mei, H.C., Busscher, H.J., and Reid, G., Absorption of urinary components influences the zeta potential of uropathogen surfaces. *Colloids and Surfaces B: Biointerfaces* 19, 13–17, 2000.

63. Hawthorne, L., and Reid, G., The effect of protein and urine on uropathogen adhesion to polymer substrata. *J. Biomed. Mater. Res.* 24, 1325–1332, 1990.

64. Poisson, D.M., Arbeille, B., and Laugier, J., Electron microscope studies of endotracheal tubes used in neonates: do microbes adhere to the polymer? *Res. Microbiol.* 142, 1019–1027, 1991.

65. Carballo, J., Ferreiros, C.M., and Criado, M.T., Importance of experimental design in the evaluation of the influence of proteins in bacterial adherence to polymers. *Med. Microbiol. Immunol.* 180, 149–155, 1991.

66. George, D.L., Epidemiology of nosocomial pneumonia in intensive care unit patients. *Clin. Chest Med.* 16, 29–44, 1995.

67. Cross, A.S., and Roup, B., Role of respiratory assistance devices in endemic nosocomial pneumonia. *Am. J. Med.* 70, 681–685, 1981.

68. Craven, D.E., Kunches, L.M., Kilinsky, V., Lichtenberg, D.A., Make, B.J., and McCabe, W.R., Risk factors for pneumonia and fatality in patients receiving continuous mechanical ventilation. *Am. Rev. Respir. Dis.* 133, 792–796, 1986.

69. Torres, A., Aznar, R., Gatell, J., Jimenez, P., Gonzalez, J., Ferrer, A., Celis, R., and Rodriguez-Roisin, R., Incidence, risk, and prognosis factors of nosocomial pneumonia in mechanically ventilated patients. *Am. Rev. Respir. Dis.* 142, 523–528, 1990.

70. Jimenez, P., Torres, A., Rodriguez-Roisin, R., Delabelacasa, J.P., Aznar, R., Gatell, J.M., and Agustividal A., Incidence and aetiology of pneumonia acquired during mechanical ventilation. *Crit. Care. Med.* 17, 882–885, 1989.

71. Ferrer, M., Torres, A., Gonzalez, J., Puig, J., El-Ebiary, M., Roca, M., Gatell, J.M., and Rodriguezroisin R., Utility of selective digestive decontamination in mechanically ventilated patients. *Ann. Intern. Med.* 120, 389–395, 1994.

72. Craven, D.E., and Steger, K.A., Epidemiology of nosocomial pneumonia: new perspectives on an old disease. *Chest* 108, 1S–16S, 1995.

73. Adair, C.G., Gorman, S.P., O'Neill, F.B., McClurg B., Goldsmith E.C., and Webb C.H., Selective decontamination of the digestive tract (SDD) does not prevent the formation of microbial biofilms on endotracheal tubes. *J. Antimicrob. Chemother.* 31, 689–697, 1993.

74. Adair, C.G., Gorman, S.P., Feron, B.M., Byers, L.M., Jones, D.S., Goldsmith, C.E., Moore, J.E., Kerr, J.R., Curran, M.D., Hogg, G., Webb, C.H., McCarthy, G.J., and Milliganm, K.R., Implications of endotracheal tube biofilm for ventilator-associated pneumonia. *Intensive Care Med.* 25, 1072–1076, 1999.

75. Adair, C.G., Gorman, S.P., Byers, L.M., Jones, D.S., Feron, B., Crowe, M., Webb, H.C., McCarthy, G.J., and Milligan, K.R., Eradication of endotracheal tube biofilm by nebulised gentamicin. *Intensive Care Med.* 25, 1072–1076, 2002.

76. Loesche, W.J., Ecology of the oral flora. In: Nisengard, R.J., and Newman, M.G., eds., *Oral Microbiology and Immunology*, 2nd Edition. Saunders, Philadelphia, 1994: 307–319.

77. Bowman, W.C., and Rand, M J., The respiratory system and drugs affecting respiratory function. In: *Textbook of Pharmacology*. Blackwell Scientific Publications, Oxford, 1980:1–24.

78. Whittaker, C.J., Klier, C.M., and Kolenbrander, P.E., Mechanisms of adhesion by oral bacteria. *Ann. Rev. Microbiol.* 50, 513–552, 1996.

Section II

Biofilm-Forming Pathogens

5 Role of Biofilms in Infections Caused by *Escherichia coli*

Grégory Jubelin, Corinne Dorel, and Philippe Lejeune

CONTENTS

5.1   INTRODUCTION

Escherichia coli, probably the best known living organism, is one of the most abundant species of the normal aerobic intestinal flora of humans (about 100 bacteria per gram of faeces) and warm-blooded animals. Unfortunately, *E. coli* is also a pathogenic organism responsible for numerous infections, causing a range of illnesses from neonatal diarrhea to cystitis and bacteremia. Uropathogenic strains of *E. coli* account for 70 to 95% of urinary tract infections, one of the most common bacterial diseases. These infections are especially frequent in cases of catheterization. Due to biofilm development on the indwelling catheters, the incidence of infection increases 5 to 10% per day (1). As the general duration is between 2 and 4 days, 15 to 30% of catheterized patients will acquire urinary tract infections. Within 3 weeks of use, 100% of patients with catheters will become infected. A detrimental property of such abiotic surface-associated growth is the expression of biofilm specific characters, such as increased resistance to antibiotics and immunological defenses (2). In order to develop drugs and surface coatings able to delay the contamination of catheters (and therefore to maintain the bacterial sensitivity to antimicrobial agents), it is necessary to understand the physiological bases of the colonization process and search for molecules capable of interfering with this process (3).

In addition, gaining knowledge in biofilm formation by *E. coli* could be exploited against infections not related to catheterization. A large part of the pathogenic strategies developed by *E. coli* strains during interaction with eucaryotic tissues involve bacterial structures and sensing mechanisms which have been described as determinants of biofilm formation. Bacterial functions, such as positive and negative chemotaxis, motility and gliding properties, synthesis of fimbriae and adhesins, secretion of polysaccharides and other polymers, and quorum sensing mechanisms, are often of crucial importance in initiating efficient attachment to both living tissues and abiotic surfaces.

Certain bacterial structures involved in biofilms formation can also play a fundamental role in cell invasion and intracellular proliferation. Anderson and coworkers (4) have recently described that uropathogenic *E. coli* strains are able to invade into bladder superficial cells and maturate into intracellular biofilms, creating pod-like bulges on the bladder surface. This particular differentiation could explain how bladder infections can persist in the face of robust host defenses.

5.2 COLONIZATION OF ABIOTIC MATERIALS BY *E. coli*

Bacteria present, or introduced, into the human body can reach the surface of an indwelling medical device by three different modes (5): passive transport due to the flow of air or fluids, diffuse transport resulting from Brownian motion, and active movement requiring flagella. As transposon mutations leading to the suppression of the adherence properties of *E. coli* W3110 (6) have been found in genes responsible for chemotaxis and flagellar motility, chemotactic processes could be of great importance to catheter colonization, for instance, in response to concentration gradients of contaminating ferric ions released from the implant.

Gliding and swarming movements on surfaces also seem to be involved in the initial stage of attachment: time-lapse microscopic observations of *Pseudomonas aeruginosa* adhesion showed that the bacteria move along the surface, almost as if they are scanning for an appropriate location for attachment (7). In accordance with these observations, a non-adherent phenotype was reported after transposon inactivation of swarming motility due to overflagellation in *Salmonella enterica* Serovar Typhimurium (8).

In the next step, individual bacteria have to interact with the surface in a sufficiently strong manner to prevent disruption by convective forces or Brownian motion. Specialized cell surface structures, such as fimbriae and adhesins, seem to be required to establish the first link between the bacterium and the surface (or molecules adsorbed on its surface). Using transposon mutagenesis followed by screening for non-adherent clones, Pratt and Kolter (6) identified type I pili as structures able to mediate the adhesion of *E. coli*. The role of another type of fimbriae in surface colonization was revealed after isolation of a point mutation in the regulatory gene *ompR* of a laboratory *E. coli* K-12 strain unable to form biofilms (9).

This mutation resulted in the overproduction of curli, a particular type of thin and flexible fimbrium, and allowed the mutant strain to stick to any type of material. Further studies with curlin antibodies demonstrated the constitutive synthesis of these fimbriae and their presence at the surface of bacteria isolated from patients with catheter-related infections (10). In addition, transduction of knock-out mutations in the curli-encoding genes of the clinical strains demonstrated their role in adhesion to biomaterials (9). A third type of pili able to mediate adhesion was identified by Ghigo (11): all the conjugative fimbriae encoded by transferable plasmids (including F) of several incompatibility groups could act as adhesion structure and promote biofilm formation. Furthermore, transfer frequencies of conjugative plasmids seemed to be increased in biofilms (11,12). These observations are of high medical importance because they raise the question of the role of biofilms in the spreading of antibiotic resistance genes, not only on contaminated prosthetic devices, but also on hospital equipment.

When individual bacteria come into contact with a material, a cascade of precise physiological changes is triggered. Prigent-Combaret and coworkers showed that the expression of about 40% of the *E. coli* genome was significantly modified during the first hours of the colonization process (13,14). They observed that the synthesis of the flagella, the determinants of motility, was repressed in the attached bacteria by downregulation of the gene encoding the flagellar structural protein, whereas production of colanic acid, a matrix exopolysaccharide, is induced. Recent studies addressed the role of colanic acid during adhesion of laboratory and clinical strains of *E. coli* (10,15). Results from these studies indicate that overexpression of the exopolysaccharide did not enhance bacterial adhesion to abiotic surfaces but rather decrease the establishment of specific binding. However, significant differences in biofilm architecture and in biofilm thickness were observed between colanic acid producing and nonproducing strains. In addition, production of cellulose by commensal strains of *E. coli* was recently demonstrated (16). As the exopolysaccharides are structural components of the extracellular matrix and lend stability to the cell–cell interconnections, all these data strongly suggest that the location and timing of exopolysaccharide production are major determinants of efficient biofilm development.

Mature biofilms are characterized by their ordered architecture consisting of mushroom-shaped colonies interspersed among less dense channels in which gas and liquid flows have been detected (17). Such a cellular organization is very complex and the construction processes obviously require a large number of genes, functions, and regulatory processes (7). In addition, quorum sensing, an intercellular signalling mechanism, has been described in *P. aeruginosa* biofilms and recognized as a determinant of biofilm structural organization, and acquirement of resistance to detergent (18). As quorum sensing molecules, typically acylated homeserine lactones, have been detected in natural biofilms developed on urethral catheters removed from patients (19), there is little doubt that cell-to-cell signals are involved in situ in biofilm construction and in the expression of specific characters, including one of the most detrimental properties of surface-associated contamination: increased resistance to antibiotics and immunological defenses.

5.3 ROLE OF CURLI IN PATHOGEN–HOST INTERACTIONS

Besides mediating adhesion to abiotic surfaces, curli confer the ability to autoaggregate (9), to bind to laminin and fibronectin (20), plasminogen (21), human contact phase proteins (22), and major histocompatibity complex class I molecules (23), and to induce internalization of *E. coli* by eukaryotic cells (24).

In human disease 55% of *E. coli* isolates from urinary tract infections were positive for curli production, and 50% of *E. coli* isolates from blood cultures of patients with sepsis were capable of producing curli *in vitro* (25,26). Normark's group observed that all serum samples from convalescent patients with *E. coli* sepsis, but not serum from healthy controls, contained antibodies against the major curli subunit (26). These data demonstrated that curli are expressed *in vivo* in human sepsis. It is conceivable that only strains producing high levels of curli could be identified as curli producing strains by *in vitro* staining procedures. In addition, the same group demonstrated that curli were responsible to a significant degree for the induction of TNF-α, IL-6, and IL-8 when human macrophages were infected with a curli producing strain of *E. coli* (26). Strikingly, they also observed that a mutant strain, which secreted soluble curli subunits, acted as a potent cytokine inducer, suggesting that it was not curli-mediated attachment to macrophages that was responsible for the induction of cytokines, but rather it was the curli subunit itself, irrespective of whether it was present as a polymer on the bacterial surface or as a secreted soluble monomer. Although the role of curli could be indirect, for instance by carrying a sufficient amount of bound LPS, these data suggest a possible role for curli in the induction of proinflammatory cytokines during *E. coli* sepsis.

5.4 SURFACE SENSING AND BIOFILM DEPENDENT GENE EXPRESSION

At present no efficient drug or coating are available to prevent biofilm formation at the surface of indwelling medical devices. However, it seems conceivable to find treatments able to hinder the contamination process. Any delay, even a few hours, in surface colonization could successfully extend the capacities of both the preventive antibiotherapy and the immunological defenses to limit the infection. Two different strategies can be adopted to reach this objective: interacting with the surface sensing mechanisms to repel the pioneering cells and disorganizing the structure of mature biofilms. In both cases, it is of pivotal importance to understand the physiological transitions of colonizing bacteria in terms of intracellular and cell-to-cell signalling processes, in order to identify potential targets for anti-biofilm molecules.

At the very first stages of colonization, individual bacteria reaching the surface have to sense their contact with the material and transmit the information to their genome in order to trigger the appropriate physiological response. To date, the studies of these signalling processes have mainly concerned *P. aeruginosa* and *E. coli* (for recent reviews, see Refs 3 and 27). In *E. coli*, Prigent-Combaret and coworkers constructed a library of reporter gene fusions by random insertions of a transposon (carrying a promoterless lacZ gene) in the chromosome of a mutant *E. coli* K12 strain able to form biofilms (13). By using a simple screen (14), they observed that after 24 hours, about 40% of the *E. coli* genes were differentially expressed in the attached and free-floating cells. To answer the question of what physicochemical parameter was sufficiently different to enable the bacteria to discriminate the liquid phase and the interface, the same authors compared the intracellular K^+ concentrations of planktonic and biofilm cell populations (13). As it is well established that in *E. coli* this internal parameter varies in accordance with the osmolarity of the external medium, a difference in intracellular K^+ concentration would indicate that the two-cell populations were present in microenvironments with different osmolarity. Ten hours after inoculation, it was observed that the attached bacteria displayed a significantly higher internal K^+ concentration. This result strongly suggested that physicochemical conditions acting on local osmolarity (or, more precisely, on water activity) could allow the bacteria to differentiate between the liquid medium and the solid material.

Other indications of the involvement of the local water activity in surface sensing by *E. coli* came from the recognition of two sensor-regulator systems in signal transmission. Laboratory strains, as well as clinical isolates, of *E. coli* were unable to adhere to abiotic surfaces after inactivation of the EnvZ/OmpR (9) and CpxA/CpxR (28) two-component systems. These sensors are known to enable bacteria to respond to external variations of osmotic pressure and pH by phosphorylation of a cytoplasmic response regulator after activation of an inner membrane sensor. Given that EnvZ/OmpR-mediated osmotic regulation of flagella, curli, and colanic acid has been demonstrated (9,13,29), its implication in surface sensing is highly probable. On the other hand, it has been shown that the CpxA/CpxR system is involved in curli regulation (28,29) and activated upon adhesion of *E. coli* to solid surfaces (30).

As two-component sensing systems have also been related to biofilm formation by other bacterial species, a model of surface sensing was recently proposed (3). The principal claim of this model is a difference in convective forces around the bacteria, when they are attached or not. Ions and molecules excreted by swimming and free-floating bacteria are dissipated by convective forces resulting from diffusion, Brownian motion, passive transport due to the flow of the liquid and active movement involving flagella. Attachment to a solid surface by specialized bridging structures, such as curli and other fimbriae, creates micro-compartments between the bacterial and material surfaces where the convective forces are reduced. In these compartments, organic molecules already present (as in any natural fluid) and excreted ions and molecules, such as amino acids and monomers, could be confined by weak chemical interactions with the abiotic surface, the bacterial appendages, and the surface of the cell. This situation could result in a reduction of water activity sufficient to activate signalling systems, such as EnvZ/OmpR and CpxA/CpxR, and induce precise changes in gene expression.

5.5 CONCLUDING REMARKS: WHICH WAY TOWARDS ANTI-BIOFILM TREATMENTS?

Escherichia coli cells, as well as pseudomonads, staphylococci, and other pathogens, are equipped with specialized structures able to interact with abiotic material and to sense their own contact with the solid surface. In nature, most bacteria are living in biofilms or consortia. The construction of such cellular organizations obviously requires a great number of functions which have to be very precisely connected, in terms of time and place. At present, only the tip of the iceberg has been described. Within natural microbial communities, there likely exist many molecules responsible for consortial functions such as cellular movement, attraction or repulsion of other organisms, growth control, and induction of particular physiological states. Potentially, some of these molecules or analogs possess anti-biofilm, or perhaps antimicrobial activity, that could be incorporated into implanted medical devices. In our opinion, the knowledge of biofilm formation has now reached a level which could allow the discovery of such molecules, by using laboratory model biofilms in appropriate screening procedures.

ACKNOWLEDGMENTS

Research in the authors laboratory was funded by grants from the French Defense Ministry (96/048 DRET) and the Centre National de la Recherche Scientifique (Réseau Infections Nosocomiales).

REFERENCES

1. Warren, J.W., Clinical presentations and epidemiology of urinary tract infections. In: Mobley, H.L.T, and Warren, J.W., eds., *Urinary Tract Infections. Molecular Pathogenesis and Clinical Management.* Washington: American Society for Microbiology, 1996:3–27.
2. Stewart, P.S., and Costerton, J.W., Antibiotic resistance of bacteria in biofilms. *Lancet* 358, 135–138, 2001.
3. Lejeune, P., Contamination of abiotic surfaces: what a colonizing bacterium sees and how to blur it. *Trends Microbiol.* 11, 179–184, 2003.
4. Anderson, G.G., Palermo, J.J., Schilling, J.D., Roth, R., Heuser, J., and Hultgren, S.J., Intracellular bacterial biofilm-like pods in urinary tract infections. *Science* 301, 105–107, 2003.
5. Van Loosdrecht, M.C.M., Lyklema, J., Norde, W., and Zehnder, A.J.B., Influence of interfaces on microbial activity. *Microbiol. Rev.* 54, 75–87, 1990.
6. Pratt, L.A., and Kolter, R., Genetic analysis of *Escherichia coli* biofilm formation: roles of flagella, motility, chemotaxis and type I pili. *Mol. Microbiol.* 30, 285–293, 1998.
7. O'Toole, G., Kaplan, H.B., and Kolter, R., Biofilm formation as microbial development. *Ann. Rev. Microbiol.* 54, 49–79, 2000.
8. Mireles, J.R., Togushi, A., and Harshey, R.M., *Salmonella enterica* Serovar Typhimurium swarming mutants with altered biofilm-forming abilities: surfactin inhibits biofilm formation. *J. Bacteriol.* 183, 5848–5854, 2001.
9. Vidal, O., Longin, R., Prigent-Combaret, C., Dorel, C., Hooreman, M., and Lejeune, P., Isolation of an *Escherichia coli* mutant strain able to form biofilms on inert surfaces: involvement of a new *ompR* allele that increases curli expression. *J. Bacteriol.* 180, 2442–2449, 1998.
10. Prigent-Combaret, C., Prensier, G., Le Thi, T.T., Vidal, O., Lejeune, P., and Dorel, C., Developmental pathway for biofilm formation in curli-producing *Escherichia coli* strains: role of flagella, curli, and colanic acid. *Environ. Microbiol.* 2, 450–464, 2000.

11. Ghigo, J.M., Natural conjugative plasmids induce bacterial biofilm development. *Nature* 412, 442–445, 2001.
12. Hausner, M., and Wuertz, S., High rates of conjugation in bacterial biofilms as determined by quantitative in situ analysis. *Appl. Environ. Microbiol.* 65, 3710–3713, 1999.
13. Prigent-Combaret, C., Vidal, O., Dorel, C., and Lejeune, P., Abiotic surface sensing and biofilm-dependent gene expression in *Escherichia coli*. *J. Bacteriol.* 181, 5993–6002, 1999.
14. Prigent-Combaret, C., and Lejeune, P., Monitoring gene expression in biofilms. *Meth. Enzymol.* 310, 56–79, 1999.
15. Hanna, A., Berg, M., Stout, V., and Razatos, A., Role of capsular colanic acid in adhesion of uropathogenic *Escherichia coli*. *Appl. Environ. Microbiol.* 69, 4474–4481, 2003.
16. Zogaj, X., Nimtz, M., Rohde, M., Bokranz, W., and Römling, U., The multicellular morphotypes of *Salmonella typhimurium* and *Escherichia coli* produce cellulose as the second component of the extracellular matrix. *Mol. Microbiol.* 39, 1452–1463, 2001.
17. Costerton, J.W., Lewandowski, Z., De Beer, D., Caldwell, D., Korber, D., and James, C., Biofilms, the customized microniche. *J. Bacteriol.* 176, 2137–2142, 1994.
18. Davies, D.G., Parsek, M.R., Pearson, J.P., Iglewski, B.H., Costerton, J.W., and Greenberg, E.P., The involvement of cell-to-cell signals in the development of bacterial biofilms. *Science* 280, 295–298, 1998.
19. Stickler, D.J., Morris, N.A., McLean, R.J.C., and Fuqua, C., Biofilms on indwelling urethral catheters produce quorum sensing molecules in situ. *Appl. Environ. Microbiol.* 64, 3486–3490, 1998.
20. Olsén, A., Jonsson, A., and Normark, S., Fibronectin binding mediated by a novel class of surface organelles on *Escherichia coli*. *Nature* 338, 652–655, 1989.
21. Sjobring, U., Pohl, G., and Olsén, A., Plasminogen, absorbed by *Escherichia coli* expressing curli or by *Salmonella enteridis* expressing thin aggregative fimbriae, can be activated by simultaneously captured tissue-type plasminogen activator (t-PA). *Mol. Microbiol.* 14, 443–452, 1994.
22. Ben Nasr, A., Olsén, A., Sjobring, U., Muller-Esteri, W., and Björck, L., Assembly of human contact phase proteins and release of bradykinin at the surface of curli-expressing *Escherichia coli*. *Mol. Microbiol.* 20, 927–935, 1996.
23. Olsén, A., Wick, M.J., Mörgelin, M., and Björck, L., Curli fibrous surface proteins of *Escherichia coli*, interact with major histocompatibility complex class I molecules. *Infect. Immun.* 66, 944–994, 1998.
24. Gophna, U., Barlev, M., Seijffers, R., Oelschlager, T.A., Hacker, J., and Ron, E.Z., Curli fibers mediate internalization of *Escherichia coli* by eukaryotic cells. *Infect. Immun.* 69, 2659–2665, 2001.
25. Römling, U., Bokranz, W., Gerstel, U., Lünsdorf, H., Nimtz, M., Rabsch, W., Tschäpe, H., and Zogaj, X., Dissection of the genetic pathway leading to multicellular behaviour in *Salmonella enterica* serotype Typhimurium and other Enterobacteriaceae. In: Wilson, M., and Devine, D., eds., *Medical Implications of Biofilms*. Cambridge: Cambridge University Press, 2003:231–261.
26. Bian, Z., Brauner, A., Li, Y., and Normark, S., Expression of and cytokine activation by *Escherichia coli* curli fibers in human sepsis. *J. Inf. Dis.* 181, 602–612, 2000.
27. Lejeune, P., Biofilm-dependent regulation of gene expression. In: Wilson, M., and Devine, D., eds. *Medical Implications of Biofilms*. Cambridge: Cambridge University Press, 2003:3–17.

28. Dorel, C., Vidal, O., Prigent-Combaret, C., Vallet, I., and Lejeune, P., Involvement of the Cpx signal transduction pathway of *E. coli* in biofilm formation. *FEMS Microbiol. Lett.* 178, 169–175, 1999.

29. Prigent-Combaret, C., Brombacher, E., Vidal, O., Ambert, A., Lejeune, P., Landini, P., and Dorel, C., Complex regulatory network controls initial adhesion and biofilm formation in *Escherichia coli* via regulation of the *csgD* gene. *J. Bacteriol.* 183, 7213–7223, 2001.

30. Otto, K., and Silhavy, T.J., Surface sensing and adhesion of *Escherichia coli* controlled by the Cpx-signalling pathway. *Proc. Natl. Acad. Sci. USA* 99, 2287–2292, 2002.

6 Staphylococcus aureus Biofilms

Julie M. Higashi and Paul M. Sullam

CONTENTS

6.1 INTRODUCTION

The Gram-positive bacterium *Staphylococcus aureus* is among the five most common isolates in the clinical microbiology laboratory (1,2). One of the reasons that it is such a ubiquitous pathogen is that it colonizes the anterior nasopharynx in 10 to 40% of humans and can be easily transferred to the skin (3). This colonization predisposes to infection. If trauma, medical procedures, percutaneous devices, or injections disrupt the natural skin or mucous membrane barrier, colonizing *S. aureus* can invade, producing a broad spectrum of clinical disease. A recent study showed that in 82% of *S. aureus* bacteremias, the blood and anterior nares isolates were identical, suggesting that these bacteremias were endogenous in origin (4).

In the community, the diseases most commonly associated with *S. aureus* are skin, soft tissue, and bone infections. Among hospitalized patients, *S. aureus* infections are more invasive and include nosocomial pneumonia, bacteremias, and endovascular infections like endocarditis and septic thrombophlebitis, which can embolize to distant organs (5). *S. aureus* is also a frequent cause of medical device and implant-related infections, with increased medical device utilization from the 1980s to 1990s being directly linked to an increased incidence in nosocomial *S. aureus* bacteremias (6). Thus, the burden of disease caused by this organism continues to grow.

S. aureus infections are difficult to manage, often needing weeks of intravenous antimicrobial therapy. Severe infections, particularly those involving implants, require surgical removal of the implant for successful cure. If not treated aggressively, the infections can recur and develop into chronic problems that require prolonged and even lifelong antimicrobial suppression. The sequelae can be very costly. A study that modeled estimates of the incidence, deaths, and direct medical costs of *S. aureus* associated hospitalizations from hospital discharges in the New York metropolitan area in 1995 found that *S. aureus* infections caused approximately twice the length of stay, deaths, and medical costs than other typical hospitalizations (7). *S. aureus* infections comprised about 1% of all hospital discharges with total estimated direct medical costs of $435.5 million. For each *S. aureus* related hospitalization, the estimated average length of stay was 20 days, direct cost $32,100, and death rate 10%. In contrast, the average length of stay for all discharges in 1995 was 9 days, direct cost $13,263, and death rate 4.1%. Clearly, *S. aureus* related infections have a significant negative impact on patient outcomes and overall costs.

The recognition that *S. aureus* forms biofilms on both native tissue and medical devices and implants has been a critical step towards understanding its pathogenesis and the challenges clinicians face with these serious infections. *Staphylococcus epidermidis*, a skin commensal that rarely causes disease without the presence of foreign material, has always been the prototypical Gram-positive biofilm organism. Early studies of device centered infections showed *S. epidermidis* attached to explanted medical devices encased in slime (8). This provided early evidence that the biofilm concept applied to human disease. Subsequently, studies have also demonstrated *S. aureus* biofilms on intravascular catheters, explanted pacemaker leads, within bone, and as vegetations on heart valves (9–13). Moreover, investigators have shown that *S. aureus* can produce the same slime as *S. epidermidis* (14). The fact that *S. aureus,* unlike *S. epidermidis,* is virulent enough to routinely produce biofilms on native tissue alone without the presence of foreign material makes elucidating its particular biofilm physiology important for devising strategies to combat its biofilms more effectively.

The overall goal of this chapter is to review the current knowledge pertaining to *S. aureus* biofilms. It begins by discussing basic *S. aureus* biofilm pathogenesis and physiology. It then reviews *S. aureus* biofilm pathogenesis, focusing on the virulence factors and molecular processes that biofilm researchers have defined as critical to *S. aureus* biofilm development and function. The clinical relevance of these processes is assessed by reviewing molecular epidemiology studies from clinical specimens. Finally, it presents approaches to the prevention and management of these difficult infections.

6.2 *S. aureus* BIOFILM PATHOGENESIS AND PHYSIOLOGY

Biofilm pathogenesis in all microorganisms is at least a three-step process. The initial stage of biofilm pathogenesis is the adhesion of microorganisms to a surface. This surface can be either biological tissue or an inorganic material. The second stage involves cell proliferation on the surface and the production of an exopolysaccharide matrix that facilitates intercellular aggregation. The third and final stage is the organization of encased bacteria into three-dimensional structures that look like mushroom shaped towers interspersed by channels through which nutrients and waste products can flow (15). Little is known about how this structural organization evolves, but confocal microscopy has enabled better visualization of this stage (Figure 6.1) (16–18). In this fully developed biofilm, it is postulated that the bacteria function like an integrated community (15).

Since the process of biofilm formation starts with bacterial adhesion, one might expect that the physiology of adherent and planktonic (non-adherent) forms of *S. aureus* differs considerably. Investigators have studied *S. aureus* in the early stages

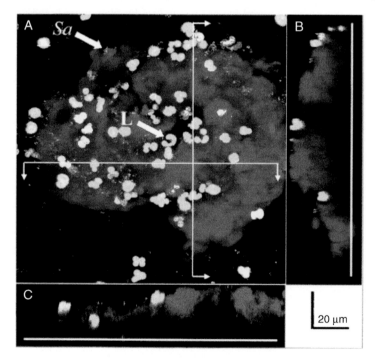

FIGURE 6.1 (See color insert following page 270) Scanning laser confocal microscopy image showing leukocytes (L) attached to and within a *S. aureus* biofilm. Note the three-dimensional structure of the biofilm with its grey mushroom tower containing bacteria interspersed with channels represented by the black areas within the biofilm. Leukocytes are present within the biofilm. (B) x and (C) y cross-sections showing that the leukocytes have not fully penetrated the biofilm but are lodged within the natural topography of the biofilm. Bar, 30 μm. Reprinted from Leid, J.G., Shirtliff, M.E., Costerton, J.W., and Stoodley, A.P., Human leukocytes adhere to, penetrate, and respond to *Staphylococcus aureus* biofilms. *Infect. Immun.* 70, 6339–6345, 2002. With permission.

of attachment to silicone surfaces and shown that differences in physiology between adherent and planktonic cells can be detected even before cellular aggregation and production of the polysaccharide matrix. Adherent bacteria grow at about half the rate of bacteria in planktonic culture, but fully retain their viability. They also have decreased membrane potentials relative to planktonic bacteria, indicating a decreased metabolism. However, *S. aureus* in the exponential phase of growth generally exhibits increased respiratory activity. Of particular note is a significant increase in antimicrobial resistance evident within 2 hours of initial adhesion and increasing over 7 days. Some of the resistance, particularly to the cell wall active β lactam and glycopeptide antimicrobials, can be attributed to the decreased growth rate of the adherent bacteria. Taken together, these differences in physiology between the adherent and planktonic *S. aureus* support the existence of two different phenotypes: biofilm and planktonic (19,20).

Two main aspects of the biofilm phenotype are particularly relevant to clinical disease. The first feature is the dramatic increase in antimicrobial resistance. For *S. aureus* biofilms, the *in vitro* minimum bactericidal concentrations (MBCs) of most antimicrobial agents averages 2 to 1,000-fold higher than their identical planktonic forms (21,22). The second feature of the biofilm phenotype is the inability of the host immune cells to kill the biofilm bacteria. For example, investigators recently showed that leukocytes were able to bind and penetrate a *S. aureus* biofilm, but were unable to phagocytose any of the bacteria within the biofilm structure (Figure 6.1) (18). These features explain why antimicrobial therapy alone for biofilm related infections frequently fails. Successful surgical debridement or replacement of an implant provides the mechanism for the mechanical removal of the bacteria.

It is still unclear how bacteria regulate the change from one phenotype to another, or what environmental cues trigger the switch. To better understand the molecular basis of the *S. aureus* planktonic and biofilm phenotypes, investigators have compared the expression of bacterial genes in these two states. In a PCR based, micro-representational difference analysis, investigators found five genes that were upregulated during biofilm growth. Three of these genes each encode an enzyme of the glycolysis or fermentation pathway, which might reflect decreased oxygen availability. The other two genes encode an enzyme that may help *S. aureus* adapt to nutrient limitation, specifically threonine, and a general stress protein that may have a homolog in *Pseudomonas fluorescens* biofilms (23). These genes all seem to help the biofilm bacteria adapt to their surface adherent state, but do not offer insights into the way the bacteria detect these environmental changes. However, the recent development of *S. aureus* DNA microarrays that allow the simultaneous evaluation of the differential gene expression for the entire genome should help identify genes that respond to environmental cues (24,25).

6.3 MOLECULES CONTRIBUTING TO
S. aureus ADHESION

Bacterial adhesion to a surface, which is the initial step of biofilm pathogenesis, is determined by a combination of interactions between the bacterial surface, substrate surface, and surrounding environment (26). The physicochemical properties of the bacterial and substrate surfaces determine the non-specific interactions like ionic, hydrophobic, and van der Waals interactions that result in overall attraction or repulsion. Molecules on both surfaces can also participate in specific ligand receptor interactions that promote adhesion. Finally, other factors like hydrodynamic flow, nutrient conditions, oxygen tension, and pH contribute to the overall process. Since these individual factors continuously change, the net sum of them, reflecting the probability of bacterial adhesion, changes as well.

It follows then, that dramatic changes in bacterial surface properties will affect adhesion to surfaces. Teichoic acids are anionic polymers that are uniformly distributed over the entire *S. aureus* peptidoglycan wall, contributing a large amount of negative charge to the bacterial surface (27). They contain numerous sites that can be glycosylated or amino acid esterified with D-alanine. While some of the genes involved in teichoic acid synthesis are essential for growth, the physiological role of these molecules is still not well understood (27,28). Investigators have shown that elimination of the D-alanine esters of teichoic acids in a *S. aureus* strain resulted in a significant decrease in initial attachment and ability to form a biofilm on polystyrene and glass, surfaces that are hydrophobic and hydrophilic, respectively (28). This mutant was created by the disruption of the *dlt*ABCD operon, which is responsible for D-alanine incorporation into teichoic acid. It is speculated that lack of D-alanine esterification reduces the attractive hydrophobic interactions between the *dltA* mutant and polystyrene. In addition, the corresponding relative increase in negative surface charge may increase the repulsive interaction between *dltA* mutant and glass. Surface property characterization of the wild type and *dltA* mutant strains could confirm the physicochemical basis of the observed changes in adhesion, but the results of such studies have not been reported. However, the mutation did not affect exopolysaccharide production, a feature necessary for the second aggregation phase of biofilm development. This finding provides additional support for teichoic acid structure contributing to initial *S. aureus* adhesion on surfaces.

Other cell wall constituents that affect primary attachment to polystyrene are autolysins. These proteins are peptidoglycan hydrolases, enzymes that cleave either the peptide or glycan moiety in the cell wall, and therefore participate in cell lysis and division (27). Autolysins were first identified as important contributors to biofilm development in *S. epidermidis*, where disruption of the autolysin gene *atlE* by transposon insertion significantly reduced initial bacterial adhesion to polystyrene. AtlE-negative strains have reduced cell surface hydrophobicity and form large aggregates, since cell separation after division is greatly affected. It is thought that the mechanism of decreased bacterial adhesion in these strains results from a reduction in hydrophobic interactions between bacteria and polystyrene, but it is also possible that the large clusters are more likely to detach during rinsing, secondary to greater drag forces (29,30). The homologue gene *atl* in *S. aureus* also encodes an autolysin, and the mutagenesis of *atl* produces bacteria in large clusters such as those seen with mutagenesis of *atlE* in *S. epidermidis* (31,32). Unfortunately, the surface properties and biofilm forming capabilities of *atl* mutants have not been characterized, so the contribution of Atl to *S. aureus* adhesion is unknown.

Modification of other *S. aureus* autolysin activity has been studied by disruption of a two component sensor-response regulatory system, ArlS–ArlR (33). Mutagenesis of *arlS* resulted in increased primary attachment to polystyrene. In contrast to mutagenesis of *atl*, where autolysis was significantly diminished, mutation of *arlS* significantly increased autolytic activity. In fact, the proposed mechanism of increased autolysis is that these strains secrete many fewer extracellular proteases. These proteases seem to mitigate the activity of the autolysins in the wild type strains. Thus, autolysins appear to generally promote staphylococcal attachment to polystyrene, but it is unclear whether these proteins affect adhesion by affecting physicochemical properties of the cell wall, or by serving as direct adhesins to polystyrene.

The adhesion of bacteria to a bare polymer surface like polystyrene is relevant to a few clinical situations. For example, bacteria may adhere to bare polymer in the immediate period before placement of device or implant into the patient. In addition, they can also attach to percutaneous devices at the exit site of the device, as with intravascular catheter hubs. However, when a medical device or implant is placed in a patient, the proteins and cells in the milieu adsorb onto its surface within seconds (34). In these cases, the bacteria will be interacting with the host modified surface rather than the bare material itself. Although physicochemical interactions are still important contributions to adhesion, the presence of proteins and cells offers the possibility of specific ligand interactions that promote adhesion.

Many *S. aureus* surface proteins participate in specific interactions, recognizing extracellular matrix (ECM) and plasma proteins. The ECM proteins are plentiful in wounds, bone, damaged endothelium, as well as orthopedic implant and soft tissue implant sites. Plasma proteins are found in not only blood, but are also immobilized in blood clots and on surfaces of blood contacting devices and implants. These proteins are ubiquitous and abundant in living organisms, offering *S. aureus* multiple targets for gaining access to establish infections.

A large number of well characterized *S. aureus* surface proteins belong to the microbial surface components recognizing adhesive matrix molecules (MSCRAMM) family (27,35). The MSCRAMM family of proteins, most of which are anchored to the peptidoglycan cell wall, bind to a range of ECM components. The MSCRAMMs typically have a long N terminal region that contains the ligand binding domain. The carboxy terminal segment has a wall spanning region that is proline or glycine rich, or composed of Ser-Asp dipeptide repeats. The wall spanning region is followed by an LPXTG motif, a hydrophobic membrane spanning domain, and finally, a positively charged tail. The LPXTG motif, highly conserved amongst Gram-positive organisms, is recognized by a membrane associated enzyme, sortase, that covalently attaches the protein to the cell wall. At least one *S. aureus* MSCRAMM is not anchored to the cell wall. The elastin binding protein (Ebp) receptor is instead an integral cytoplasmic membrane protein with two hydrophobic membrane spanning domains (36). This difference in receptor structure probably reflects the diversity of bacterial surface proteins that are putative mediators of adhesion. In fact, many other *S. aureus* surface components exist that bind ECM and plasma proteins that have yet to be fully identified or characterized.

Table 6.1 is a list of known *S. aureus* surface adhesins and their ligands (35–40). These relationships have been defined by demonstrating at least two of the three following results: (a) generation of isogenic mutant strains lacking these proteins shows decreased adhesion to their respective ligands on surfaces (49,50), (b) complementation of a number of these genes into bacterial strains that naturally lack the proteins

TABLE 6.1
S. aureus Surface Adhesins and Their Ligands

Adhesin	Adhesin Name	Ligand(s)	References
Anchored to Cell Wall			
FnBPA and FnBPB	Fibronectin binding protein A and B	Fibronectin	(35,37,48)
ClfA and ClfB	Clumping factor A and B	Fibrinogen	(35,38,39)
Cna	Cna	Collagen	(35,40)
Protein A	Protein A	von Willebrand factor, Fc domain of IgG	(41,42)
Bbp	Bone sialoprotein binding protein	Bone sialoprotein	(43)
Integral Membrane Proteins			
EbpS	Elastin binding protein	Elastin	(44)
Secreted Proteins			
Eap/Map	Extracellular adherence protein/MHC II analogous protein	Fibrinogen, fibronectin, prothrombin	(45)
Emp	Extracellular matrix protein binding protein	Fibrinogen, collagen, fibronectin, vitronectin	(46)
Uncharacterized Receptors			
Laminin binding protein		Laminin	(47)

results in increased ligand binding (49), and (c) antibodies against the adhesin or peptide fragments of either the ligand or adhesin competitively inhibit the interaction and decrease overall bacterial adhesion (44,51,52). Since these interactions can be selectively inhibited, the peptide fragments and antibodies are attractive candidate drugs for preventing biofilm infections.

An example of a well-characterized specific ligand interaction is that of *S. aureus* Phillips strain 110 kDa surface protein Cna and collagen (53). Investigators studied *S. aureus* adhesion to Type II collagen under dynamic flow conditions and quantified the effects of Cna density, collagen density and shear rates on adhesion. Higher Cna densities on the *S. aureus*, which are present during the exponential growth phase, increased *S. aureus* adhesion, while increasing collagen density had no affect on adhesion. Adhesion rates were dependent upon shear rate, generally following first-order kinetics. Interestingly, in a range of lower shear rates, 50 to 300 s^{-1}, adhesion rates increased, possibly because the attractive specific ligand interaction overcame the drag force produced by shear, and higher shear rates in this range delivered a greater number of bacteria to their ligands. At shear rates >500 s^{-1}, adhesion rates steadily decreased, as the specific ligand interaction was overwhelmed by the drag force. This study quantifies the individual contribution of several factors to the adhesion rate and illustrates how, under different sets of conditions, the primary determinants of adhesion can change.

Finally, *S. aureus* secretes proteins that are neither anchored to the cell wall nor tethered to the cytoplasmic membrane (Table 6.1) (45,46). For example, the extracellular adherence protein (EAP), also known as major histocompatibility complex class II analogous protein (MAP), binds a wide range of ECM and plasma proteins. The EAP also binds to the *S. aureus* surface and enhances its adhesion to fibroblasts and epithelial cells. Thus, EAP can act as a bridge between the ligand on the substrate surface, and *S. aureus*, through secretion of EAP, can "prime" the substrate surface for its adhesion. Moreover, EAP oligomerizes and promotes *S. aureus* aggregation, so in addition to participating in the adhesion stage of biofilm formation, can also facilitate the second stage of biofilm formation, interbacterial aggregation (45).

6.4 MOLECULAR FACTORS CONTRIBUTING TO INTERCELLULAR AGGREGATION AND EXOPOLYSACCHARIDE PRODUCTION

6.4.1 POLYSACCHARIDE INTERCELLULAR ADHESIN/POLYMERIC N-ACETYL GLUCOSAMINE

Polysaccharide intercellular adhesin (PIA), or polymeric N-acetyl glucosamine (PNAG) is the formal name for slime, the exopolysaccharide matrix produced by both *S. epidermidis* and *S. aureus* that mediates cellular accumulation into multilayer clusters in the second stage of biofilm development (29). It is predominantly comprised of linear β-1,6-linked N-acetylglucosamine and has a mean molecular mass of 460 kDa (54). About 15 to 20% of the glucosamine residues in each homoglycan molecule are nonacetylated and confer an overall positive charge. A small fraction of homoglycan molecules contains the same N-acetylglucosamine backbone, but also have phosphate or ester linked succinate groups, resulting in negatively charged molecules. The chemical composition of PIA/PNAG has been challenging to define because of difficulty distinguishing it from contaminants in nutrient media and staphylococcal wall components (54). Another polysaccharide, poly-N-succinyl β-1,6-glucosamine (PNSG), was isolated from the slime of *S. epidermidis*. PNSG has the same homoglycan backbone as PIA, but has all its glucosamine residues succinylated, thus giving it an overall negative charge (55). Interestingly, PNSG cross reacts with antibodies raised against PIA. It is now believed that PIA/PNAG and PNSG are the same molecule and the observed differences in structure are artifact. More recently, the term polymeric N-acetyl glucosamine (PNAG) has replaced PIA (56).

The genes responsible for PNAG expression were first identified in *S. epidermidis* by generating a mutant that could attach to surfaces, but was unable to form multicellular clusters and did not produce PNAG (57). The mutation producing the biofilm negative phenotype was within the intercellular adhesion (*ica*) operon. The *ica* operon consists of the *icaR* (regulatory) gene and *icaADBC* (polysaccharide biosynthesis) genes. Greater discussion of the individual function of the icaADBC proteins is available in the chapter discussing biofilms of *S. epidermidis* (Chapter 10).

The *icaR* gene is located upstream and is transcribed in the opposite orientation of the *icaADBC* genes (29). Investigators have shown that *S. aureus* and several other staphylococcal species also contain the *ica* locus, and sequence comparison of the *S. aureus* and *S. epidermidis icaADBC* and *icaR* genes had a 59 to 78% amino acid identity (14). Transcription of the *ica* locus in *S. aureus* is not constitutive, and many strains that contain the *ica* locus do not form biofilms. However, in *S. aureus* biofilm forming strains, deletion of the *ica* locus eliminates the ability to form a biofilm and make PNAG.

Understanding the regulation of *ica* locus transcription awaits further study. *In vitro*, several known stimuli induce PNAG production in staphylococci: supplemental glucose or glucosamine in nutrient media, as well as stressful conditions such as exposure to ethanol, high osmolarity, iron restriction, and oxygen deprivation (29,58). These environmental factors mediate their influence on the *ica* locus through accessory gene regulators. The IcaR protein is a repressor of *ica* locus transcription, as an internal deletion in *icaR* increased production of PNAG in a *S. epidermidis* strain (29). In *S. aureus*, investigators have shown that IcaR is a DNA binding protein that binds immediately 5′ to the *icaA* start codon, and that it may exert its repressor activity by obscuring the binding sites of the *ica* promoter (56). In addition to IcaR, a five nucleotide motif TATTT in the *ica* promoter region is important for transcriptional control of the *ica* locus. The TATTT motif is regulated by a staphylococcal DNA binding protein other than IcaR. Loss of the TATTT motif in a *S. aureus* mutant strain results in overexpression of *ica* transcripts and hypersecretion of PNAG. These insights represent significant advances in our understanding of the regulation of the *ica* locus.

6.4.2 ALPHA-TOXIN

Another molecule that also affects intercellular adhesion and aggregation of *S. aureus* is α-toxin (59). Alpha-toxin, encoded by the *hla* gene, is a secreted, multimeric toxin. It causes host cell lysis by assembling into a heptamer that functions as a pore in eukaryotic cell membranes (60). *In vitro*, mutation of *hla* results in strains that do not form a biofilm. These strains can attach to surfaces, but do not form the multilayer clusters required for the second stage of biofilm formation (59). Complementation of the *hla* gene into strains with the *hla* mutation results in hypersecretion of α-toxin and restores the ability to form biofilm. Interestingly, *in vivo* studies in a guinea pig tissue cage model showed that mRNA transcripts of *hla* in fluid extracted from infected tissue cages reaches a maximum within 48 hours after inoculation of *S. aureus*. These results provide supporting evidence that α-toxin may indeed participate in the early stages of biofilm related infections (61).

It is unclear whether α-toxin requires PNAG to promote intercellular adhesion in *S. aureus*. For example, both *S. aureus* wild type and strains with the *hla* mutation produced similar amounts of PNAG, and introduction of a multicopy dose of *hla* into an *ica* deletion mutant strain did not restore its ability to form biofilm. A study that evaluated the incidence of virulence determinants in clinical isolates showed that >99% of 334 isolates contained *hla*, while only 83% of the isolates contained the *ica* locus, indicating that *hla* is ubiquitous in *S. aureus* (62). Thus, without evidence of biofilm production in strains which are α-toxin+/PNAG− and α-toxin−/PNAG+, it is quite possible that α-toxin and PNAG are both needed for intercellular aggregation and therefore, biofilm formation.

6.4.3 Biofilm Associated Protein

The biofilm associated protein (Bap), a surface protein first identified in a bovine mastitis *S. aureus* isolate V329, affects both the first and second stages of biofilm formation *in vitro* (63). A mutant strain with a transposon insertion into the *bap* gene did not form biofilm on polystyrene, as both its initial attachment to polystyrene and intercellular adhesion were significantly reduced. Mutation of *bap* results in bacteria with decreased surface hydrophobicity relative to their wild type strains, suggesting that Bap promotes attachment to polystyrene via non-specific hydrophobic interactions (64). Complementation of the *bap* gene back into the transposon mutant and other *S. aureus* biofilm negative strains naturally lacking *bap* conferred the ability to form biofilms, further demonstrating its role as a biofilm determinant (63).

The relationship of Bap and PNAG is unclear. A PNAG producing, *bap* complemented *S. aureus* strain produced PNAG in much greater amounts than the wild type strain suggesting a biofilm potentiating relationship between Bap and PNAG (63). However, V329 and other natural Bap-positive strains are not strong PNAG producers, and the transposon mutant Bap-negative strain does not produce PNAG at all. From these studies it is difficult to discern whether Bap and PNAG production is truly independent, but the molecules themselves are distinct entities.

The Bap protein consists of 2,276 amino acids and has a molecular mass of about 240 kDa (63). Its putative carboxy terminal sequence contains the LPXTG motif, hydrophobic membrane spanning domain, and positively charged residues similar to the MSCRAMM adhesins. Its most remarkable features are an extensive repeat region, and domains that enable its own dimerization, thus suggesting a mechanism for promoting intercellular adhesion.

Despite its homology to the MSCRAMM family, Bap has no known ligands. In fact, studies have demonstrated that the presence of Bap decreases *S. aureus* binding to not only fibrinogen and fibronectin on surfaces, but also to connective tissue sections of mammary glands (65). In a murine catheter infection model, the V329 wild type strain was able to establish persistent infection more efficiently than isogenic Bap-negative m556 strain (63). However, in an intramammary gland infection model, the ability for Bap-positive strains to establish chronic infection over Bap-negative strains has been inconsistent (63,65). Moreover, Bap was present in only 5% of bovine mastitis isolates and absent in 75 *S. aureus* human clinical isolates (63). Thus, the relevance of Bap as it relates to *in vivo* human biofilm infections is unclear.

However, Bap has significant structural homology with other Gram-positive surface proteins that might provide clues to its function. The enterococcal surface protein (Esp) of *Enterococcus faecalis* also promotes *in vitro* biofilm formation (64). Interestingly, both Bap and Esp are encoded by composite transposons inserted into pathogenicity islands (66,67). Pathogenicity islands are mobile carriers of virulence associated gene clusters enabling horizontal gene transfer. Bap also has homology with a *S. aureus* surface protein that is lysed by plasmin, hence its name, plasmin sensitive protein (Pls) (68). Pls is produced by MRSA isolates, as *pls* is closely associated with *mecA*. It also prevents adhesion of *S. aureus* to ECM and host plasma proteins. Perhaps surface proteins Bap and Pls provide a mechanism for established biofilm infections to spread to other sites by discouraging attachment to host tissue, or actually promoting detachment of bacteria from existing biofilm. Despite these uncertainties regarding its function, there is evidence for the involvement of Bap at several stages of biofilm development.

6.5 *S. aureus* GLOBAL GENE REGULATORS

After reviewing the collection of *S. aureus* biofilm determinants, it becomes apparent that coordinating the expression and production of these factors is required to orchestrate the process of biofilm formation. However, it is important to recognize that existence in a biofilm is just one of the ways which *S. aureus* causes disease. Therefore, the genes that are crucial for biofilm formation are a subset of the genes involved in pathogenesis. Pathogenesis genes are, for the most part, nonessential for growth and viability under all conditions. Thus, they are accessory genes, and are utilized only if nutrients are available. Accessory genes include both pathogenesis genes as well as genes that allow adaptation to hostile environments, and are collectively called the virulon (69). Global regulator genes sense environmental signals and implement the necessary adjustments in gene expression that optimize the pathogenic potential of the virulon.

In vitro the production of virulence factors follows a temporal sequence that correlates with the *S. aureus* growth cycle. Typically, surface adhesins are expressed in stationary and early exponential phase, and then are followed by extracellular proteins like hemolysins and proteases in late and post-exponential phase (69). *S. aureus* microarray data have been able to confirm these trends by quantifying the transcription levels of accessory genes and the influence of global regulators on their expression (24,70).

In biofilms, the transition from expression of surface adhesins to extracellular proteins can be postulated to occur between the first and second steps of biofilm formation (initial adhesion through the multilayer accumulation). Sections of 4-day-old mature *S. aureus* biofilms containing bacteria with the global gene regulator, staphylococcal accessory gene regulator A (SarA) fused to a *lacZ* reporter construct, revealed greater numbers of bacteria within the deeper two thirds of the biofilm. More importantly, the percent bacteria expressing SarA was greatest in the deepest third of the biofilm, e.g., 86% vs. 65% and 20% in the more superficial portions, confirming that global regulators are active in biofilms (71). Genomic studies in biofilms have not yet been completed. However, *in vivo* studies of biofilm infections that quantified single gene transcripts showed that expression of the surface adhesin clumping factor A (ClfA) in biofilm bacteria increased steadily over 6 days, while transcripts of the extracellular protein α-toxin peaked at 2 days after inoculation (61,72). This disparity between *in vitro* and *in vivo* data reflects the complexities that typify the interactions between environment and the regulation of accessory gene expression.

Because transcription profiling studies have demonstrated that accessory gene regulators control numerous processes that appear unrelated to virulence factor expression, they are also called global regulators. The *S. aureus* global regulators are listed in Table 6.2, and their effect on individual biofilm determinants is summarized in Table 6.3 (24,61,69–70,73–87). *S. aureus* has three types of global regulators. First, there are two component systems (TCSs) consisting of a sensor protein that binds factors from the environment and subsequently activates a response regulator protein. This protein can then initiate a cascade of intracellular events that result in changes in gene expression. The *agr* locus is the best characterized TCS global regulator of *S. aureus*. Second, an interregulatory network of transcription factors mediate gene expression by binding to DNA in the promoter regions of accessory genes or other transcription factor genes. SarA was the first identified in this large family of homologous proteins (88). The third type, alternative sigma factor B (σ^B), is a transcription factor that also responds to environmental stresses like reduced ADP/ATP ratio, EtOH, and salicylate, via a phosphatase/kinase cascade (84). The interaction between these global regulators is extensive and very complex. Detailed reviews are available that provide excellent discussions (69,89).

6.5.1 Agr Locus

The *agr* locus is driven by two divergent promoters, P2 and P3. P2 drives the transcription of *agrACDB,* the products of which comprise the TCS and its autoinducing ligand, AIP. AIP is a 7–9 amino acid peptide with a unique thiolactone ring. *AgrD* encodes the propeptide of AIP, which AgrB processes to the mature form and secretes extracellularly. AgrC, the receptor for the *agr* system, binds AIP. Through activation via a phosphorelay that is not yet fully understood, AgrC initiates the activity of AgrA, the response regulator. AgrA~P then activates the transcription of P2 and P3, completing the autoinduction circuit. The transcript of P3 is RNAIII, which is the intracellular effector molecule of the *agr* system. RNAIII transcription occurs in mid to late exponential phase, stimulating the expression of extracellular protein genes, while down regulating surface protein genes (Table 6.3) (24). The autoinduction circuit is also a mechanism by which *S. aureus* tracks its own population density, a phenomenon called quorum sensing. The AIP–AgrC receptor interaction is highly specific, such that AIPs between different strains of *S. aureus* may act as inhibitors rather than activators of *agr* autoinduction (90).

Transcription profiling studies have finally confirmed after many years of speculation that the *agr* system coordinates the switch from *S. aureus* surface protein to extracellular protein expression in mid to late exponential phase *in vitro* (24). However, despite its influence on virulence factor expression, mutation of the *agr* locus does not seem to significantly affect *S. aureus* biofilm forming capabilities (91). Exceptions are *S. aureus* strains that lack *rsbU*, which activates σ^B in response to stress (92). Mutagenesis of *agr* in these strains resulted in enhancement of biofilm formation relative to their parent strains, an effect that appeared independent of both Atl and PNAG. Incubation of exogenous inhibitor peptides similar to AIP with *agr* intact parent strains also enhanced biofilm formation. These results can be attributed to the loss of *agr* and/or σ^B, making their individual contributions to these data

TABLE 6.2
Known *S. aureus* Global Gene Regulators

Regulatory Locus	Description	Role in Biofilm Formation	Interaction With		References
			agr Locus	SarA	
agrACDB/RNA III	TCS[a], autoinduced	Regulates surface proteins and extracellular proteins, population density sensing		SarA ↑ *agr*	(24,69,73)
sarA	Transcription factor	Repressor of surface proteins, proteases	Sar A ↑ *agr*		(74)
sarS	Transcription factor	Activates transcription of *spa* and possibly other surface protein genes		SarA ↓ SarS	(75)
sarT	Transcription factor	Represses transcription of *hla* and possibly other exoprotein genes	*agr*/RNAIII ↓ SarT	SarA ↓ SarT	(76)
saePQRS[b]	TCS, autoinduced	Regulates many extracellular protein genes, responds to environmental signals	RNAIII may regulate	Possible, but may be via *agr*	(69,77,87)
arlRS	TCS	Regulates autolysis and surface proteins	Counters *agr* autoinduction	Possible, but may be via *agr*	(78)
σ^B	Rpo Sigma factor	Activated by environmental stress, regulates accessory genes, generally antagonistic to *agr*. Upregulates *ica* transcription.	May decrease RNAIII transcription	Binds to SarA, SarS promoters	(79,80)
rot[c]	Transcription factor	Major transcription factor for *hla* and other proteins, influence on genes is opposite to *agr*	RNAIII probably regulates		(70,81)
mgr[d]	Transcription factor	Regulates transcription of *hla*, *spa*			(70,82)

[a] Two component system. [b] *S. aureus* exoprotein expression. [c] Repressor of transcription. [d] Multiple gene regulator.

TABLE 6.3

S. aureus Biofilm Determinants and Their Interaction with Global Gene Regulators

Gene	Timing	agr	sae	rot	sarA	σ^B	mgr	References
				Action of Regulatory Genes				
				Adhesion				
spa	exp[a]	−	?	+	−[c]		−	(24,69,70)
cna	pxp[b]	0			−			(83)
fnbA	exp				+			(24)
fnbB	exp				+	+		(24,84)
clfA	exp	0						(85)
clfB	exp	+		+	0			(24,70)
atl	exp				−			(24)
dtlD	exp	−		+				(24,70)
eap	exp				+			(24)
				Intercellular Adhesion				
hla	pxp	+	+	−	+[c]		−	(24,61,81)
ica	exp				+	+		(70,86)

[a] Exponential phase. [b] Post-exponential phase. [c] agr related. +, upregulated, −, downregulated, 0, no effect, ?, controversial.

difficult to interpret, but the evidence with *rsbU* intact strains does not suggest a dominant role for *agr* in biofilm formation *in vitro*. However, a *rsbU* intact *agr* mutated strain is significantly less virulent in a model of murine septic arthritis, so *agr* may still be very relevant to biofilm pathogenesis *in vivo* (93).

6.5.2 SARA

SarA and its homologues, e.g., SarR, SarS, SarT, are part of a growing family of winged helix-turn-helix transcription factors. These molecules act as an interdependent network, also modulating the activity of the other types of global gene regulators, *agr*/RNAIII and σ^B, to control virulence factor expression in *S. aureus*. SarA, is the most extensively studied of this family, and is a 14.7 kDa protein that binds as a dimer to AT-rich sequences called Sar boxes scattered throughout the staphylococcal genome. This suggests that SarA requires little sequence specificity for binding (69). The *sarA* locus is transcribed from three promoters *sar*P1, *sar*P2, and *sar*P3. The transcripts from these promoters overlap, each of which has a common 3′ end encoding SarA. SarA directly affects gene expression of several biofilm determinants by binding to Sar boxes in their promoter regions, resulting in target gene activation or repression (Table 6.3) (94,95). Moreover, it influences these proteins indirectly as well by binding to the intragenic region between the *agr* P2 and P3 promoters, augmenting RNAII and RNAIII transcription (95).

Transcription profiling studies have shown that SarA does not conform to the same temporal program of protein expression as the *agr* system (24). Thus, it is just as likely to increase or decrease the expression of surface or extracellular proteins in mid to late exponential phase. This may be a result of its interactions with other transcription factors. For example, its upregulation of *hla* is mediated through binding to the *sarT* promoter, as SarT is a repressor of *hla* transcription (76). Interestingly, mutagenesis of *sarA* results in a significant increase in protease production, indicating that SarA is a potent repressor of protease synthesis (96).

In contrast with the *agr* system, SarA has a profound effect on *S. aureus* biofilm formation. Mutagenesis of *sarA* in several genetically different strains resulted in the loss of biofilm forming capabilities, while complementation of *sarA* restores the biofilm phenotype (86,91). Deletion of the *agr* operon in the corresponding wild type strains did not affect biofilm formation at all, indicating that, for biofilm formation, *sarA* acts independently of *agr* (86). The effects of SarA occur in both the first and second stages of biofilm development. Loss of *sarA* results in decreased ability to adhere to fibronectin coated surfaces, as SarA up regulates both FnBPA and FnBPB adhesins on *S. aureus*. In addition, *sarA* mutagenesis, results in less PNAG production at all phases of the growth cycle relative to the wild type strains, with *ica* transcription significantly decreased in the strains lacking *sarA* (86). Proteases did not appear to contribute significantly to biofilm degradation, as incubation of *S. aureus* strains lacking *sarA* with protease inhibitors, and mutagenesis of protease genes *ssp* and *aur* in conjunction with *sarA*, did not restore biofilm forming capabilities in these strains. Clearly, the effects of *sarA* on *S. aureus* biofilm formation are pleiotropic, but these effects are influenced by σ^B as well.

6.5.3 σ^B

σ^B is activated by a complex post-translational pathway that, upon conditions of environmental stress, releases it to bind to promoter sequence GTTT(N$_{14-17}$)GGGTAT. In *S. aureus*, σ^B acts directly on at least 23 separate promoters, including one of the SarA and one of the SarS promoters. σ^B also affects genes without σ^B promoters via its action on σ^B dependent transcription factors (69). Its effects on virulence factor expression appear to be present early in the growth phase through late exponential growth phase (80).

σ^B appears to be important, but not essential for biofilm formation. For example, strains with deletion of *sig*B retain their ability to form a biofilm and produce levels of PNAG similar to their parent strains with a slight decrease in *ica* transcription (86). Concomitant mutagenesis of *sig*B and *sarA* produces strains that form better biofilms than those produced from mutagenesis of *sarA* alone. Of note, *ica* transcripts in these strains are significantly reduced compared to wild type strains. Biofilm forming capabilities in strains lacking *sarA* or both *sarA* and *sig*B are restored only with complementation of *sarA*, and not *sig*B, or *icaADBC* (86). The discrepancy between the *ica* transcript (decreased) and biofilm forming abilities (increased) observed after mutagenesis of *sig*B and *sarA* suggests the involvement of other regulator(s) in PNAG production and biofilm formation. For example, σ^B and SarA do seem to upregulate *ica* transcription. However, SarA may upregulate PNAG by repressing a factor that degrades PNAG or inhibits PNAG synthesis. σ^B might oppose the effects of SarA by upregulating expression of the factor.

6.6 CLINICAL SIGNIFICANCE OF BIOFILM DETERMINANTS

6.6.1 *IN VIVO* STUDIES

The biofilm determinants discussed in this chapter were originally identified using *in vitro* screens that demonstrated deficiencies in adhesion to test surfaces and intercellular aggregation. Table 6.4 shows the evidence confirming that these determinants contribute to virulence, as measured by animal models of biofilm infections (35,49,50,55,63,65,97–108). For the most part, mutagenesis of the genes encoding these factors will reduce, but do not completely eliminate, the capacity of *S. aureus* to establish infection. These data reflect the redundancy of the biofilm determinants within their functional groups. For example, the impact of FnBPs on virulence *in vivo* has been controversial, as isogenic strains lacking FnBPs did not consistently demonstrate a reduction in pathogenicity (97,98). Presumably, these discrepancies exist because other surface adhesins expressed by FnBP negative *S. aureus* strains were able to mediate surface attachment. However, expression of FnBPA on nonpathogenic *Lactococcus lactis* reduced its infecting inoculum to levels comparable to wild type *S. aureus*, providing compelling evidence that FnBPA is indeed a virulence factor (99).

6.6.2 MOLECULAR EPIDEMIOLOGY OF *S. aureus* BIOFILM DETERMINANTS

Table 6.4 also lists the prevalence of biofilm determinants in clinical isolates (55,62,63,109–117). These data demonstrate that α-toxin, Clfs, Eap/Map, Protein A, and FnBPs are fairly ubiquitous amongst *S. aureus* strains, while Bap, Bbp, and collagen BP are less much prevalent. Of note is the high prevalence of the *ica* locus, but variable expression of PNAG among clinical isolates (14,55). This may reflect methodological inconsistencies in the detection of PNAG production *in vitro* as well as the tighter regulation of PNAG production in *S. aureus* relative to *S. epidermidis* (14,116). Interestingly, while presence of the *ica* locus in *S. epidermidis* clinical isolates correlates with greater pathogenicity, this is not the case with *S. aureus* (62,114,115).

The diversity and variable prevalence of virulence factor genes in clinical isolates suggests that combinations of genes may be advantageous for establishing diseases in a site and host specific manner. A study examining combinations of surface adhesin and toxin genes in *S. aureus* clinical isolates showed that seven of thirty-three virulence determinants were more common in invasive disease isolates than nasal carriage isolates (62). Three of these seven determinants are also associated with biofilm formation: FnBPA, collagen binding protein, and PNAG. Moreover, it was determined that the proportion of isolates causing infections increased linearly with the proportion of the seven virulence determinants carried by individual isolates. That is, >95% of the strains that carried all seven virulence factors caused invasive disease, while only 10 to 20% of strains that carried one or two of the virulence

TABLE 6.4

S. aureus Biofilm Determinants and Global Regulators, Evidence for Virulence and Therapeutic Immunogenic Potential

Biofilm Determinant/ Global Regulator	Virulence Factor Evidence		Vaccine Protection	References
	In vivo	Presence of Gene or Phenotype in Clinical Isolates (%)		
FnBPA and FnBPB	Rat endocarditis model +	77–98	Rat endocarditis model +	62,97–99, 109–110,128
ClfA and ClfB	Rat endocarditis model +	98–100	Mouse intraperitoneal model–	50,62,99,129
Collagen BP	Mouse septic arthritis model + Rat endocarditis model +	29–56	Mouse sepsis model +	35,49,62,100, 111,130
Protein A	Mouse septic arthritis +	90–94	Bovine mastitis model +	62,101–102,131
Bbp		38–43		62
Eap/Map		93–98		62,113
EbpS		62–68		62
Alpha toxin	Mouse brain abscess model + Mouse intraperitoneal model + Mouse septic arthritis +	79–99	Mouse intraperitoneal model +	62,102–103, 110,132
PNAG/ica	Guinea pig tissue cage model ±	17–83	Mouse renal abscess model +	55,62,104–105, 114–117
Bap	Rabbit endocarditis model +[c] Mouse catheter model + Goat mastitis model ±	65–96[a] 5 bovine mastitis isolates 0 human isolates		63,65
Agr/RNAIII	Mouse septic arthritis +	1–42[b]		62,106
SarA	Rabbit endocarditis model +			107
σ[B]	Rabbit endocarditis model +			108

[a] Presence of ica in clinical isolates (1). [b] Depends on agr subgroup. [c] In S. epidermidis.

factors caused invasive disease. The study controlled for the effect of clonality on the genes, and concluded that while underlying clonality affected the distribution of certain gene combinations, it did not affect the overall virulence associated with the identified determinants. Because the nature of the invasive diseases were not reported in this study, it is impossible to assess how these data relate to biofilm infections.

6.6.3 SMALL COLONY VARIANTS IN *S. aureus*

Another group of *S. aureus* clinical isolates from infections that have similar features to biofilm infections are small colony variants (SCVs). *S. aureus* SCV infections are characterized by relapses occurring after extended disease free intervals or prolonged persistence despite appropriate parenteral antimicrobial therapy (118). The diseases with which they are associated overlap significantly with biofilm infections and include endocarditis, osteomyelitis, pneumonia, soft tissue infections, and severe bacteremia (119). In the clinical laboratory, SCVs are often either overgrown or lose their phenotype upon subculture, making their clinical relevance difficult to assess (118,119).

The SCV phenotype is that of small nonpigmented, nonhemolytic colonies that grow slowly on routine media. They are unable to utilize complex carbohydrates and rely heavily on glucose as their main nutrient source. In addition, SCVs are menadione and hemin auxotrophs, both of which are required for effective transfer of electrons and efficient generation of ATP in the citric acid cycle. These metabolic constraints are responsible for the observed phenotypic differences between SCVs and wild type *S. aureus*, which include antimicrobial resistance and decreased virulence factor production (119). For example, decreases in cell wall synthesis arising from restricted quantities of ATP results in β-lactam antimicrobial resistance. The reduction in the transmembrane electrochemical gradient, the result of impaired electron transport, limits uptake and internalization of positively charged aminoglycosides. Predictably, supplementation of nutrient media with menadione or hemin reverts the phenotype to wild type *S. aureus*, but this reversion can also occur spontaneously (118,119).

The relationship between the SCV and *S. aureus* biofilm phenotypes is unclear, but their shared characteristics suggest that they may have similar underlying physiology. Recall that biofilm *S. aureus* also grow slowly and are resistant to antimicrobials. Of note, SCVs attached to fibronectin coated surfaces also show substantial decreases in their antimicrobial sensitivity relative to that observed during planktonic growth. This suggests that SCVs may form biofilms, intimating that the SCV and biofilm phenotypes are not mutually exclusive (120). An additional similarity is that growth of *S. aureus* under anaerobic conditions encourages the emergence of both the SCV and biofilm phenotypes (58,119). In SCVs the important electron transport chain element menaquinone, which requires menadione for its synthesis, is not produced under anaerobic conditions. Anaerobiosis in the biofilm phenotype results in augmented upregulation of *ica* locus transcription and concomitant increased PNAG production, a process which is probably influenced by the global regulator σ^B (58,86). Perhaps there are similar mechanisms by which SCVs and *S. aureus* biofilm bacteria sense anaerobic stress and limit their utilization of resources.

A major difference between SCVs and *S. aureus* in biofilms is the ability of SCVs to exist within host cells, thereby averting the host immune response and antimicrobials. The mechanism by which SCVs persist intracellularly is thought to be related to the downregulation of virulence factors like α-toxin and protein A (121). Lack of α-toxin prevents SCVs from lysing their host cells, unlike their wild type counterparts. In established biofilms, where active virulence factor expression may be less important, it can be speculated that SCVs might serve as persistors, that is, rare cells resistant to killing which ensure population survival (122). Better understanding of the molecular processes that produce these phenotypes will provide answers to these enticing questions.

6.7   APPROACHES TO PREVENTION AND THERAPY

The conventional approach to management of *S. aureus* biofilm infections remains antimicrobial therapy and proper surgical resection and debridement. However, the emergence of new antimicrobial resistance typified by vancomycin intermediate and resistant *S. aureus* strains is a driving force for developing alternative approaches, among them the surface modification of intravascular catheters with antimicrobial agents. While there have been indications that these catheters reduce *S. aureus* infections, their efficacy remains controversial (123). These aspects of managing biofilm infections are addressed extensively elsewhere in this book.

Disruption of global gene regulator autoinduction circuits offers a new target to affect the expression of biofilm determinants and possibly improve the efficacy of standard therapy. This tactic has shown potential in *Pseudomonas aeruginosa* biofilm infections, where studies have shown that autoinduction circuit inhibitors were able to attenuate virulence *in vivo* and improve the susceptibility of the biofilm bacteria to tobramycin (124). In *S. aureus*, an RNAIII inhibiting heptapeptide (RIP) has been described that leads to an overall decrease in RNAIII (125). RIP appears to prevent biofilm formation on dialysis catheters *in vitro* and work synergistically with standard antimicrobial agents in a subcutaneous biofilm infection rat model (126,127). However, the amounts of RIP used in these studies were ~10,000 to 100,000 times the concentration of native concentrations of AIP. This, as well as the difficulty confirming the regulatory mechanism involving RIP inhibition of RNAIII make its significance very controversial (69).

Table 6.4 shows the evidence of the immunogenic potential for several biofilm determinants (50,55,128–132). As a rule, immunization against a single determinant results in partial protection from severe *S. aureus* disease. Interestingly, a conjugate vaccine containing capsular polysaccharides type 5 and 8 conferred partial immunity against *S. aureus* bacteremias in hemodialysis patients (133). In this extremely high risk group of patients, the incidence of bacteremias decreased by about 57%. Unfortunately, this effect was short lived, as the benefits of vaccination waned after 40 weeks. Nevertheless, the results are encouraging, and the vaccine may have practical use in high risk patients under certain circumstances, like high risk surgical procedures. As the steps of biofilm development can be carried out by a several molecules that perform the same function, the development of a multivalent vaccine against enough biofilm determinants to eliminate this redundancy seems to be a rational approach.

6.8 CONCLUSIONS

Staphylococcal disease is protean in its manifestations, reflecting the diversity of virulence determinants in its arsenal. Biofilms represent a large subset of disease manifestations caused by *S. aureus* and have an enormous detrimental impact on patient outcomes and costs. We have gained significant understanding of the molecular processes that govern *S. aureus* biofilm formation, particularly those involving the first two stages of biofilm development, surface adhesion and intercellular aggregation. However, the processes that regulate the development of the three dimensional structure of the mature biofilm and characterization of the interactions in these multicellular biofilm communities are not well understood at all. Moreover, other processes may regulate detachment of organisms from the biofilm to transport *S. aureus* to new sites of infection, another exciting area open to further investigation.

However, while virulence factors are vital for *S. aureus* pathogenicity, host immune status is equally important. A 1994 review of *S. aureus* nosocomial infections determined that over 90% of the patients had either immunocompromising medical conditions such as diabetes, lung disease, liver disease, or malignancy, and/or other predisposing factors like an intravascular medical device or trauma (5,134). Our knowledge regarding the genetic basis of host immunity as it relates to host susceptibility in *S. aureus* diseases is very limited. Improved understanding of the mechanisms of biofilm resistance to the host immune response is also critical. Real progress in these areas is crucial for designing the next generation of therapeutic and preventative modalities, especially since *S. aureus* has continued to adapt to existing therapies, thereby remaining a challenging pathogen.

ACKNOWLEDGMENTS

This work was supported in part by the Department of Veterans Affairs and the NIH grants T32 and R01AI41513.

REFERENCES

1. Jarvis, W.R., and Martone, W.J., Predominant pathogens in hospital infections. *J. Antimicrob. Chemother.* 29 Suppl A, 19–24, 1992.
2. Richards, M.J., Edwards, J.R., Culver, D.H., and Gaynes, R.P., Nosocomial infections in medical intensive care units in the United States. National Nosocomial Infections Surveillance System. *Crit. Care. Med.* 27, 887–892, 1999.
3. Williams, R., Healthy carriage of *Staphylococcus aureus*: its prevalence and importance. *Bacteriol. Rev.* 27, 56–71, 1963.
4. von Eiff, C., Heilmann, C., Herrmann, M., and Peters, G., Basic aspects of the pathogenesis of staphylococcal polymer-associated infections. *Infection* 27 Suppl 1, S7–S10, 1999.
5. Musher, D.M., Lamm, N., Darouiche, R.O., Young, E.J., Hamill, R.J., and Landon, G.C., The current spectrum of *Staphylococcus aureus* infection in a tertiary care hospital. *Medicine (Baltimore)* 73, 186–208, 1994.

6. Steinberg, J.P., Clark, C.C., and Hackman, B.O., Nosocomial and community-acquired *Staphylococcus aureus* bacteremias from 1980 to 1993: impact of intravascular devices and methicillin resistance. *Clin. Infect. Dis.* 23, 255–259, 1996.

7. Rubin, R.J., Harrington, C.A., Poon, A., Dietrich, K., Greene, J.A., and Moiduddin, A., The economic impact of *Staphylococcus aureus* infection in New York City hospitals. *Emerg. Infect. Dis.* 5, 9–17, 1999.

8. Tenney, J.H., Moody, M.R., Newman, K.A., et al., Adherent microorganisms on lumenal surfaces of long–term intravenous catheters. Importance of *Staphylococcus epidermidis* in patients with cancer. *Arch. Intern. Med.* 146, 1949–1954, 1986.

9. Marrie, T.J., Cooper, J.H., and Costerton, J.W., Ultrastructure of cardiac bacterial vegetations on native valves with emphasis on alterations in bacterial morphology following antibiotic treatment. *Can. J. Cardiol.* 3, 275–280, 1987.

10. Marrie, T.J., and Costerton, J.W., Morphology of bacterial attachment to cardiac pacemaker leads and power packs. *J. Clin. Microbiol.* 19, 911–914, 1984.

11. Marrie, T.J., and Costerton, J.W., Scanning and transmission electron microscopy of in situ bacterial colonization of intravenous and intraarterial catheters. *J. Clin. Microbiol.* 19, 687–693, 1984.

12. Marrie, T.J., Nelligan, J., and Costerton, J.W., A scanning and transmission electron microscopic study of an infected endocardial pacemaker lead. *Circulation* 66, 1339–1341, 1982.

13. Power, M.E., Olson, M.E., Domingue, P.A., and Costerton, J.W., A rat model of *Staphylococcus aureus* chronic osteomyelitis that provides a suitable system for studying the human infection. *J. Med. Microbiol.* 33, 189–198, 1990.

14. Cramton, S.E., Gerke, C., Schnell, N.F., Nichols, W.W., and Gotz, F., The intercellular adhesion (*ica*) locus is present in *Staphylococcus aureus* and is required for biofilm formation. *Infect. Immun.* 67, 5427–5433, 1999.

15. Costerton, J.W., Stewart, P.S., and Greenberg, E.P., Bacterial biofilms: a common cause of persistent infections. *Science* 284, 1318–1322, 1999.

16. Sanford, B.A., de Feijter, A.W., Wade, M.H., and Thomas, V.L., A dual fluorescence technique for visualization of *Staphylococcus epidermidis* biofilm using scanning confocal laser microscopy. *J. Ind. Microbiol.* 16, 48–56, 1996.

17. Wood, S.R., Kirkham, J., Marsh, P.D., Shore, R.C., Nattress, B., and Robinson, C., Architecture of intact natural human plaque biofilms studied by confocal laser scanning microscopy. *J. Dent. Res.* 79, 21–27, 2000.

18. Leid, J.G., Shirtliff, M.E., Costerton, J.W., and Stoodley, A.P., Human leukocytes adhere to, penetrate, and respond to *Staphylococcus aureus* biofilms. *Infect. Immun.* 70, 6339–6345, 2002.

19. Williams, I., Paul, F., Lloyd, D., et al., Flow cytometry and other techniques show that *Staphylococcus aureus* undergoes significant physiological changes in the early stages of surface-attached culture. *Microbiology* 145(Pt 6), 1325–1333, 1999.

20. Williams, I., Venables, W.A., Lloyd, D., Paul, F., and Critchley, I., The effects of adherence to silicone surfaces on antibiotic susceptibility in *Staphylococcus aureus*. *Microbiology* 143(Pt 7), 2407–2413, 1997.

21. Amorena, B., Gracia, E., Monzon, M., et al., Antibiotic susceptibility assay for *Staphylococcus aureus* in biofilms developed *in vitro*. *J. Antimicrob. Chemother.* 44, 43–55, 1999.

22. Ceri, H., Olson, M.E., Stremick, C., Read, R.R., Morck, D., and Buret, A., The Calgary Biofilm Device: new technology for rapid determination of antibiotic susceptibilities of bacterial biofilms. *J. Clin. Microbiol.* 37, 1771–1776, 1999.

23. Becker, P., Hufnagle, W., Peters, G., and Herrmann, M., Detection of differential gene expression in biofilm-forming versus planktonic populations of *Staphylococcus aureus* using micro-representational-difference analysis. *Appl. Environ. Microbiol.* 67, 2958–2965, 2001.

24. Dunman, P.M., Murphy, E., Haney, S., et al., Transcription profiling-based identification of *Staphylococcus aureus* genes regulated by the *agr* and/or *sarA* loci. *J. Bacteriol.* 183, 7341–7353, 2001.

25. Mongodin, E., Finan, J., Climo, M.W., Rosato, A., Gill, S., and Archer, G.L., Microarray transcription analysis of clinical *Staphylococcus aureus* isolates resistant to vancomycin. *J. Bacteriol.* 185, 4638–4643, 2003.

26. Wang, I.W., Anderson, J.M., Jacobs, M.R., and Marchant, R.E., Adhesion of *Staphylococcus epidermidis* to biomedical polymers: contributions of surface thermodynamics and hemodynamic shear conditions. *J. Biomed. Mater. Res.* 29, 485–493, 1995.

27. Navarre, W.W., and Schneewind, O., Surface proteins of Gram-positive bacteria and mechanisms of their targeting to the cell wall envelope. *Microbiol. Mol. Biol. Rev.* 63, 174–229, 1999.

28. Gross, M., Cramton, S.E., Gotz, F., and Peschel, A., Key role of teichoic acid net charge in *Staphylococcus aureus* colonization of artificial surfaces. *Infect. Immun.* 69, 3423–3426, 2001.

29. Gotz, F., Staphylococcus and biofilms. *Mol. Microbiol.* 43, 1367–1378, 2002.

30. Heilmann, C., Hussain, M., Peters, G., and Gotz, F., Evidence for autolysin-mediated primary attachment of *Staphylococcus epidermidis* to a polystyrene surface. *Mol. Microbiol.* 24, 1013–1024, 1997.

31. Sugai, M., Komatsuzawa, H., Akiyama, T., et al., Identification of endo-beta-N-acetylglucosaminidase and N-acetylmuramyl-L-alanine amidase as cluster-dispersing enzymes in *Staphylococcus aureus*. *J. Bacteriol.* 177, 1491–1496, 1995.

32. Yamada, S., Sugai, M., Komatsuzawa, H., et al., An autolysin ring associated with cell separation of *Staphylococcus aureus*. *J. Bacteriol.* 178, 1565–1571, 1996.

33. Fournier, B., and Hooper, D.C., A new two-component regulatory system involved in adhesion, autolysis, and extracellular proteolytic activity of *Staphylococcus aureus*. *J. Bacteriol.* 182, 3955–3964, 2000.

34. Cottonaro, C.N., Roohk, H.V., Shimizu, G., and Sperling, D.R., Quantitation and characterization of competitive protein binding to polymers. *Trans. Am. Soc. Artif. Intern. Organs* 27, 391–395, 1981.

35. Foster, T.J., and Hook, M., Surface protein adhesins of *Staphylococcus aureus*. *Trends Microbiol.* 6, 484–488, 1998.

36. Downer, R., Roche, F., Park, P.W., Mecham, R.P., and Foster, T.J., The elastin–binding protein of *Staphylococcus aureus* (EbpS) is expressed at the cell surface as an integral membrane protein and not as a cell wall-associated protein. *J. Biol. Chem.* 277, 243–250, 2002.

37. Jonsson, K., Signas, C., Muller, H.P., and Lindberg, M., Two different genes encode fibronectin binding proteins in *Staphylococcus aureus*. The complete nucleotide sequence and characterization of the second gene. *Eur. J. Biochem.* 202, 1041–1048, 1991.

38. McDevitt, D., Francois, P., Vaudaux, P., and Foster, T.J., Molecular characterization of the clumping factor (fibrinogen receptor) of *Staphylococcus aureus*. *Mol. Microbiol.* 11, 237–248, 1994.

39. Ni Eidhin, D., Perkins, S., Francois, P., Vaudaux, P., Hook, M., and Foster, T.J., Clumping factor B (ClfB), a new surface-located fibrinogen-binding adhesin of *Staphylococcus aureus*. *Mol. Microbiol.* 30, 245–257, 1998.

40. Patti, J.M., Jonsson, H., Guss, B., et al., Molecular characterization and expression of a gene encoding a *Staphylococcus aureus* collagen adhesin. *J. Biol. Chem.* 267, 4766–4772, 1992.
41. Forsgren, A., and Sjoquist, J., "Protein A" from *S. aureus*. I. Pseudo-immune reaction with human gamma-globulin. *J. Immunol.* 97, 822–827, 1966.
42. Hartleib, J., Kohler, N., Dickinson, R.B., et al., Protein A is the von Willebrand factor binding protein on *Staphylococcus aureus*. *Blood* 96, 2149–2156, 2000.
43. Tung, H., Guss, B., Hellman, U., Persson, L., Rubin, K., and Ryden, C., A bone sialoprotein-binding protein from *Staphylococcus aureus*: a member of the staphylococcal *Sdr* family. *Biochem. J.* 345(Pt 3), 611–619, 2000.
44. Park, P.W., Rosenbloom, J., Abrams, W.R., and Mecham, R.P., Molecular cloning and expression of the gene for elastin–binding protein (*ebpS*) in *Staphylococcus aureus*. *J. Biol. Chem.* 271, 15803–15809, 1996.
45. Palma, M., Haggar, A., and Flock, J.I., Adherence of *Staphylococcus aureus* is enhanced by an endogenous secreted protein with broad binding activity. *J. Bacteriol.* 181, 2840–2845, 1999.
46. Hussain, M., Becker, K., von Eiff, C., Schrenzel, J., Peters, G., and Herrmann, M., Identification and characterization of a novel 38.5-kilodalton cell surface protein of *Staphylococcus aureus* with extended-spectrum binding activity for extracellular matrix and plasma proteins. *J. Bacteriol.* 183, 6778–6786, 2001.
47. Lopes, J.D., dos Reis, M., and Brentani, R.R., Presence of laminin receptors in *Staphylococcus aureus*. *Science* 229, 275–277, 1985.
48. Flock, J.I., Froman, G., Jonsson, K., et al., Cloning and expression of the gene for a fibronectin-binding protein from *Staphylococcus aureus*. *EMBO J.* 6, 2351–2357, 1987.
49. Patti, J.M., Bremell, T., Krajewska-Pietrasik, D., et al., The *Staphylococcus aureus* collagen adhesin is a virulence determinant in experimental septic arthritis. *Infect. Immun.* 62, 152–161, 1994.
50. Moreillon, P., Entenza, J.M., Francioli, P., et al., Role of *Staphylococcus aureus* coagulase and clumping factor in pathogenesis of experimental endocarditis. *Infect. Immun.* 63, 4738–4743, 1995.
51. Raja, R.H., Raucci, G., and Hook, M., Peptide analogs to a fibronectin receptor inhibit attachment of *Staphylococcus aureus* to fibronectin-containing substrates. *Infect. Immun.* 58, 2593–2598, 1990.
52. Mohamed, N., Teeters, M.A., Patti, J.M., Hook, M., and Ross, J.M., Inhibition of *Staphylococcus aureus* adherence to collagen under dynamic conditions. *Infect. Immun.* 67, 589–594, 1999.
53. Mohamed, N., Rainier, T.R. Jr., and Ross, J.M., Novel experimental study of receptor-mediated bacterial adhesion under the influence of fluid shear. *Biotechnol. Bioeng.* 68, 628–636, 2000.
54. Cramton, S.E., Gerke, C., and Gotz, F., *In vitro* methods to study staphylococcal biofilm formation. *Methods Enzymol.* 336, 239–255, 2001.
55. McKenney, D., Pouliot, K.L., Wang, Y., et al., Broadly protective vaccine for *Staphylococcus aureus* based on an *in vivo*-expressed antigen. *Science* 284, 1523–1527, 1999.
56. Jefferson, K.K., Cramton, S.E., Gotz, F., and Pier, G.B., Identification of a 5-nucleotide sequence that controls expression of the *ica* locus in *Staphylococcus aureus* and characterization of the DNA-binding properties of IcaR. *Mol. Microbiol.* 48, 889–899, 2003.
57. Heilmann, C., Schweitzer, O., Gerke, C., Vanittanakom, N., Mack, D., and Gotz, F., Molecular basis of intercellular adhesion in the biofilm-forming *Staphylococcus epidermidis*. *Mol. Microbiol.* 20, 1083–1091, 1996.

58. Cramton, S.E., Ulrich, M., Gotz, F., and Doring, G., Anaerobic conditions induce expression of polysaccharide intercellular adhesin in *Staphylococcus aureus* and *Staphylococcus epidermidis. Infect. Immun.* 69, 4079–4085, 2001.

59. Caiazza, N.C., and O'Toole, G.A., Alpha-toxin is required for biofilm formation by *Staphylococcus aureus. J. Bacteriol.* 185, 3214–3217, 2003.

60. Song, L., Hobaugh, M.R., Shustak, C., Cheley, S., Bayley, H., and Gouaux, J.E., Structure of staphylococcal alpha-hemolysin, a heptameric transmembrane pore. *Science* 274, 1859–1866, 1996.

61. Goerke, C., Fluckiger, U., Steinhuber, A., Zimmerli, W., and Wolz, C., Impact of the regulatory loci *agr*, *sarA* and *sae* of *Staphylococcus aureus* on the induction of alpha-toxin during device-related infection resolved by direct quantitative transcript analysis. *Mol. Microbiol.* 40, 1439–1447, 2001.

62. Peacock, S.J., Moore, C.E., Justice, A., et al., Virulent combinations of adhesin and toxin genes in natural populations of *Staphylococcus aureus. Infect. Immun.* 70, 4987–4996, 2002.

63. Cucarella, C., Solano, C., Valle, J., Amorena, B., Lasa, I., and Penades, J.R., Bap, a *Staphylococcus aureus* surface protein involved in biofilm formation. *J. Bacteriol.* 183, 2888–2896, 2001.

64. Toledo-Arana, A., Valle, J., Solano, C., et al., The enterococcal surface protein, Esp, is involved in *Enterococcus faecalis* biofilm formation. *Appl. Environ. Microbiol.* 67, 4538–4545, 2001.

65. Cucarella, C., Tormo, M.A., Knecht, E., et al., Expression of the biofilm-associated protein interferes with host protein receptors of *Staphylococcus aureus* and alters the infective process. *Infect. Immun.* 70, 3180–3186, 2002.

66. Ubeda, C., Tormo, M.A., Cucarella, C., et al., Sip, an integrase protein with excision, circularization and integration activities, defines a new family of mobile *Staphylococcus aureus* pathogenicity islands. *Mol. Microbiol.* 49, 193–210, 2003.

67. Shankar, N., Baghdayan, A.S., and Gilmore, M.S., Modulation of virulence within a pathogenicity island in vancomycin-resistant *Enterococcus faecalis. Nature* 417, 746–750, 2002.

68. Savolainen, K., Paulin, L., Westerlund-Wikstrom, B., Foster, T.J., Korhonen, T.K., and Kuusela, P., Expression of *pls*, a gene closely associated with the *mecA* gene of methicillin-resistant *Staphylococcus aureus*, prevents bacterial adhesion *in vitro. Infect. Immun.* 69, 3013–3020, 2001.

69. Novick, R.P., Autoinduction and signal transduction in the regulation of staphylococcal virulence. *Mol. Microbiol.* 48, 1429–1449, 2003.

70. Said-Salim, B., Dunman, P.M., McAleese, F.M., et al., Global regulation of *Staphylococcus aureus* genes by Rot. *J. Bacteriol.* 185, 610–619, 2003.

71. Pratten, J., Foster, S.J., Chan, P.F., Wilson, M., and Nair, S.P., *Staphylococcus aureus* accessory regulators: expression within biofilms and effect on adhesion. *Microbes Infect.* 3, 633–637, 2001.

72. Wolz, C., Goerke, C., Landmann, R., Zimmerli, W., and Fluckiger, U., Transcription of clumping factor A in attached and unattached *Staphylococcus aureus in vitro* and during device-related infection. *Infect. Immun.* 70, 2758–2762, 2002.

73. Novick, R.P., Ross, H.F., Projan, S.J., Kornblum, J., Kreiswirth, B., and Moghazeh, S., Synthesis of staphylococcal virulence factors is controlled by a regulatory RNA molecule. *EMBO. J.* 12, 3967–3975, 1993.

74. Heinrichs, J.H., Bayer, M.G., and Cheung, A.L., Characterization of the *sar* locus and its interaction with *agr* in *Staphylococcus aureus. J. Bacteriol.* 178, 418–423, 1996.

75. Tegmark, K., Karlsson, A., and Arvidson, S., Identification and characterization of SarH1, a new global regulator of virulence gene expression in *Staphylococcus aureus*. *Mol. Microbiol.* 37, 398–409, 2000.

76. Schmidt, K.A., Manna, A.C., Gill, S., and Cheung, A.L., SarT, a repressor of alpha-hemolysin in *Staphylococcus aureus*. *Infect. Immun.* 69, 4749–4758, 2001.

77. Giraudo, A.T., Mansilla, C., Chan, A., Raspanti, C., and Nagel, R., Studies on the expression of regulatory locus *sae* in *Staphylococcus aureus*. *Curr. Microbiol.* 46, 246–250, 2003.

78. Fournier, B., Klier, A., and Rapoport, G., The two-component system ArlS–ArlR is a regulator of virulence gene expression in *Staphylococcus aureus*. *Mol. Microbiol.* 41, 247–261, 2001.

79. Kullik, I., Giachino, P., and Fuchs, T., Deletion of the alternative sigma factor *sigmaB* in *Staphylococcus aureus* reveals its function as a global regulator of virulence genes. *J. Bacteriol.* 180, 4814–4820, 1998.

80. Bischoff, M., Entenza, J.M., and Giachino, P., Influence of a functional *sigB* operon on the global regulators sar and *agr* in *Staphylococcus aureus*. *J. Bacteriol.* 183, 5171–5179, 2001.

81. McNamara, P.J., Milligan-Monroe, K.C., Khalili, S., and Proctor, R.A., Identification, cloning, and initial characterization of *rot*, a locus encoding a regulator of virulence factor expression in *Staphylococcus aureus*. *J. Bacteriol.* 182, 3197–3203, 2000.

82. Luong, T.T., Newell, S.W., and Lee, C.Y., *Mgr*, a novel global regulator in *Staphylococcus aureus*. *J. Bacteriol.* 185, 3703–3710, 2003.

83. Blevins, J.S., Gillaspy, A.F., Rechtin, T.M., Hurlburt, B.K., and Smeltzer, M.S., The Staphylococcal accessory regulator (*sar*) represses transcription of the *Staphylococcus aureus* collagen adhesin gene (*cna*) in an *agr*-independent manner. *Mol. Microbiol.* 33, 317–326, 1999.

84. Nicholas, R.O., Li, T., McDevitt, D., et al., Isolation and characterization of a *sigB* deletion mutant of *Staphylococcus aureus*. *Infect. Immun.* 67, 3667–3669, 1999.

85. Wolz, C., McDevitt, D., Foster, T.J., and Cheung, A.L., Influence of *agr* on fibrinogen binding in *Staphylococcus aureus* Newman. *Infect. Immun.* 64, 3142–3147, 1996.

86. Valle, J., Toledo-Arana, A., Berasain, C., et al., SarA and not sigmaB is essential for biofilm development by *Staphylococcus aureus*. *Mol. Microbiol.* 48, 1075–1087, 2003.

87. Giraudo, A.T., Cheung, A.L., and Nagel, R., The *sae* locus of *Staphylococcus aureus* controls exoprotein synthesis at the transcriptional level. *Arch. Microbiol.* 168, 53–58, 1997.

88. Cheung, A.L., and Projan, S.J., Cloning and sequencing of *sarA* of *Staphylococcus aureus*, a gene required for the expression of *agr*. *J. Bacteriol.* 176, 4168–4172, 1994.

89. Cheung, A.L., and Zhang, G., Global regulation of virulence determinants in *Staphylococcus aureus* by the SarA protein family. *Front. Biosci.* 7, d1825–d1842, 2002.

90. Lyon, G.J., Wright, J.S., Muir, T.W., and Novick, R.P., Key determinants of receptor activation in the *agr* autoinducing peptides of *Staphylococcus aureus*. *Biochemistry* 41, 10095–10104, 2002.

91. Beenken, K.E., Blevins, J.S., and Smeltzer, M.S., Mutation of *sarA* in *Staphylococcus aureus* limits biofilm formation. *Infect. Immun.* 71, 4206–4211, 2003.

92. Vuong, C., Saenz, H.L., Gotz, F., and Otto, M., Impact of the *agr* quorum–sensing system on adherence to polystyrene in *Staphylococcus aureus*. *J. Infect. Dis.* 182, 1688–1693, 2000.

93. Blevins, J.S., Elasri, M.O., Allmendinger, S.D., et al., Role of *sarA* in the pathogenesis of *Staphylococcus aureus* musculoskeletal infection. *Infect. Immun.* 71, 516–523, 2003.

94. Bayer, M.G., Heinrichs, J.H., and Cheung, A.L., The molecular architecture of the *sar* locus in *Staphylococcus aureus*. *J. Bacteriol.* 178, 4563–4570, 1996.

95. Chien, Y., Manna, A.C., Projan, S.J., and Cheung, A.L., SarA, a global regulator of virulence determinants in *Staphylococcus aureus*, binds to a conserved motif essential for sar-dependent gene regulation. *J. Biol. Chem.* 274, 37169–37176, 1999.

96. Chan, P.F., and Foster, S.J., Role of SarA in virulence determinant production and environmental signal transduction in *Staphylococcus aureus*. *J. Bacteriol.* 180, 6232–6241, 1998.

97. Flock, J.I., Hienz, S.A., Heimdahl, A., and Schennings, T., Reconsideration of the role of fibronectin binding in endocarditis caused by *Staphylococcus aureus*. *Infect. Immun.* 64, 1876–1878, 1996.

98. Kuypers, J.M., and Proctor, R.A., Reduced adherence to traumatized rat heart valves by a low-fibronectin-binding mutant of *Staphylococcus aureus*. *Infect. Immun.* 57, 2306–2312, 1989.

99. Que, Y.A., Francois, P., Haefliger, J.A., Entenza, J.M., Vaudaux, P., and Moreillon, P., Reassessing the role of *Staphylococcus aureus* clumping factor and fibronectin-binding protein by expression in Lactococcus lactis. *Infect. Immun.* 69, 6296–6302, 2001.

100. Hienz, S.A., Schennings, T., Heimdahl, A., and Flock, J.I., Collagen binding of *Staphylococcus aureus* is a virulence factor in experimental endocarditis. *J. Infect. Dis.* 174, 83–88, 1996.

101. Palmqvist, N., Foster, T., Tarkowski, A., and Josefsson, E., Protein A is a virulence factor in *Staphylococcus aureus* arthritis and septic death. *Microb. Pathog.* 33, 239–249, 2002.

102. Gemmell, C.G., Goutcher, S.C., Reid, R., and Sturrock, R.D., Role of certain virulence factors in a murine model of *Staphylococcus aureus* arthritis. *J. Med. Microbiol.* 46, 208–213, 1997.

103. Kielian, T., Cheung, A., and Hickey, W.F., Diminished virulence of an alpha-toxin mutant of *Staphylococcus aureus* in experimental brain abscesses. *Infect. Immun.* 69, 6902–6911, 2001.

104. Francois, P., Tu Quoc, P.H., Bisognano, C., et al., Lack of biofilm contribution to bacterial colonisation in an experimental model of foreign body infection by *Staphylococcus aureus* and *Staphylococcus epidermidis*. *FEMS. Immunol. Med. Microbiol.* 35, 135–140, 2003.

105. Shiro, H., Meluleni, G., Groll, A., et al., The pathogenic role of *Staphylococcus epidermidis* capsular polysaccharide/adhesin in a low-inoculum rabbit model of prosthetic valve endocarditis. *Circulation* 92, 2715–2722, 1995.

106. Abdelnour, A., Arvidson, S., Bremell, T., Ryden, C., and Tarkowski, A., The accessory gene regulator (*agr*) controls *Staphylococcus aureus* virulence in a murine arthritis model. *Infect. Immun.* 61, 3879–3885, 1993.

107. Cheung, A.L., Yeaman, M.R., Sullam, P.M., Witt, M.D., and Bayer, A.S., Role of the *sar* locus of *Staphylococcus aureus* in induction of endocarditis in rabbits. *Infect. Immun.* 62, 1719–1725, 1994.

108. Kupferwasser, L.I., Yeaman, M.R., Nast, C.C., et al., Salicylic acid attenuates virulence in endovascular infections by targeting global regulatory pathways in *Staphylococcus aureus*. *J. Clin. Invest.* 112, 222–233, 2003.

109. Peacock, S.J., Day, N.P., Thomas, M.G., Berendt, A.R., and Foster, T.J., Clinical isolates of *Staphylococcus aureus* exhibit diversity in *fnb* genes and adhesion to human fibronectin. *J. Infect.* 41, 23–31, 2000.
110. Slobodnikova, L., Kotulova, D., and Zahradnikova, I., *Staphylococcus aureus* in chronic and recurrent infections. *Folia Microbiol. (Praha),* 40, 655–658, 1995.
111. Montanaro, L., Arciola, C.R., Baldassarri, L., and Borsetti, E., Presence and expression of collagen adhesin gene (*cna*) and slime production in *Staphylococcus aureus* strains from orthopaedic prosthesis infections. *Biomaterials* 20, 1945–1949, 1999.
112. Montanaro, L., Arciola, C.R., Borsetti, E., Collamati, S., and Baldassarri, L., Detection of fibronectin-binding protein genes in staphylococcal strains from peri-prosthesis infections. *New Microbiol.* 22, 331–336, 1999.
113. Hussain, M., Becker, K., von Eiff, C., Peters, G., and Herrmann, M., Analogs of Eap protein are conserved and prevalent in clinical *Staphylococcus aureus* isolates. *Clin. Diagn. Lab. Immunol.* 8, 1271–1276, 2001.
114. Arciola, C.R., Baldassarri, L., and Montanaro, L., Presence of *icaA* and *icaD* genes and slime production in a collection of staphylococcal strains from catheter-associated infections. *J. Clin. Microbiol.* 39, 2151–2156, 2001.
115. Arciola, C.R., Collamati, S., Donati, E., and Montanaro, L., A rapid PCR method for the detection of slime-producing strains of *Staphylococcus epidermidis* and *S. aureus* in periprosthesis infections. *Diagn. Mol. Pathol.* 10, 130–137, 2001.
116. Knobloch, J.K., Horstkotte, M.A., Rohde, H., and Mack, D., Evaluation of different detection methods of biofilm formation in *Staphylococcus aureus. Med. Microbiol. Immunol. (Berl.)*, 191, 101–106, 2002.
117. Ammendolia, M.G., Di Rosa, R., Montanaro, L., Arciola, C.R., and Baldassarri, L., Slime production and expression of the slime-associated antigen by staphylococcal clinical isolates. *J. Clin. Microbiol.* 37, 3235–3238, 1999.
118. Proctor, R.A., van Langevelde, P., Kristjansson, M., Maslow, J.N., and Arbeit, R.D., Persistent and relapsing infections associated with small-colony variants of *Staphylococcus aureus. Clin. Infect. Dis.* 20, 95–102, 1995.
119. McNamara, P.J., and Proctor, R.A., *Staphylococcus aureus* small colony variants, electron transport and persistent infections. *Int. J. Antimicrob. Agents* 14, 117–122, 2000.
120. Chuard, C., Vaudaux, P.E., Proctor, R.A., and Lew, D.P., Decreased susceptibility to antibiotic killing of a stable small colony variant of *Staphylococcus aureus* in fluid phase and on fibronectin–coated surfaces. *J. Antimicrob. Chemother.* 39, 603–608, 1997.
121. von Eiff, C., Heilmann, C., Proctor, R.A., Woltz, C., Peters, G., and Gotz, F., A site-directed *Staphylococcus aureus hemB* mutant is a small-colony variant which persists intracellularly. *J. Bacteriol.* 179, 4706–4712, 1997.
122. Lewis K., Programmed death in bacteria. *Microbiol. Mol. Biol. Rev.* 64, 503–514, 2000.
123. McConnell, S.A., Gubbins, P.O., and Anaissie, E.J., Do antimicrobial-impregnated central venous catheters prevent catheter-related bloodstream infection? *Clin. Infect. Dis.* 37, 65–72, 2003.
124. Hentzer, M., Wu, H., Andersen, J.B., et al., Attenuation of *Pseudomonas aeruginosa* virulence by quorum sensing inhibitors. *EMBO J.* 22, 3803–3815, 2003.
125. Gov, Y., Bitler, A., Dell'Acqua, G., Torres, J.V., and Balaban, N., RNA III inhibiting peptide (RIP), a global inhibitor of *Staphylococcus aureus* pathogenesis: structure and function analysis. *Peptides* 22, 1609–1620, 2001.

126. Giacometti, A., Cirioni, O., Gov, Y., et al., RNA III inhibiting peptide inhibits *in vivo* biofilm formation by drug-resistant *Staphylococcus aureus*. *Antimicrob. Agents Chemother.* 47, 1979–1983, 2003.

127. Balaban, N., Gov, Y., Bitler, A., and Boelaert, J.R., Prevention of *Staphylococcus aureus* biofilm on dialysis catheters and adherence to human cells. *Kidney Int.* 63, 340–345, 2003.

128. Schennings, T., Heimdahl, A., Coster, K., and Flock, J.I., Immunization with fibronectin binding protein from *Staphylococcus aureus* protects against experimental endocarditis in rats. *Microb. Pathog.* 15, 227–236, 1993.

129. Brouillette, E., Lacasse, P., Shkreta, L., et al., DNA immunization against the clumping factor A (ClfA) of *Staphylococcus aureus*. *Vaccine* 20, 2348–2357, 2002.

130. Nilsson, I.M., Patti, J.M., Bremell, T., Hook, M., and Tarkowski, A., Vaccination with a recombinant fragment of collagen adhesin provides protection against *Staphylococcus aureus*-mediated septic death. *J. Clin. Invest.* 101, 2640–2649, 1998.

131. Pankey, J.W., Boddie, N.T., Watts, J.L., and Nickerson, S.C., Evaluation of protein A and a commercial bacterin as vaccines against *Staphylococcus aureus* mastitis by experimental challenge. *J. Dairy Sci.* 68, 726–731, 1985.

132. Kernodle, D.S., Voladri, R.K., Menzies, B.E., Hager, C.C., and Edwards, K.M., Expression of an antisense *hla* fragment in *Staphylococcus aureus* reduces alpha-toxin production *in vitro* and attenuates lethal activity in a murine model. *Infect. Immun.* 65, 179–184, 1997.

133. Shinefield, H., Black, S., Fattom, A., et al., Use of a *Staphylococcus aureus* conjugate vaccine in patients receiving hemodialysis. *N. Engl. J. Med.* 346, 491–496, 2002.

134. Ziebuhr, W., *Staphylococcus aureus* and *Staphylococcus epidermidis*: emerging pathogens in nosocomial infections. *Contrib. Microbiol.* 8, 102–107, 2001.

7 Coagulase-Negative Staphylococci

Dietrich Mack, Matthias A. Horstkotte,
Holger Rohde, and Johannes K.-M. Knobloch

CONTENTS

7.1   INTRODUCTION

Biomaterial-associated infections, most frequently caused by coagulase-negative staphylococci, are of increasing importance in modern medicine. Regularly, antimicrobial therapy fails without removal of the implanted device. The most important factor in the pathogenesis of biomaterial-associated staphylococcal infections is the formation of adherent, multilayered bacterial biofilms. In this chapter, recent progress regarding the taxonomy and clinical relevance of coagulase-negative staphylococci, factors functional in biofilm formation, their role in pathogenesis and antibiotic susceptibility, and the regulation of their expression is presented.

7.2   TAXONOMY OF THE GENUS *STAPHYLOCOCCUS*

Staphylococci are Gram-positive cocci with a diameter of 0.5 to 1.5 µm, that grow in pairs, tetrads, and small clusters. They usually produce the enzyme catalase (with the exception of *Staphylococcus aureus* subsp. *anaerobius*), are nonspore forming, and with the exception of *S. sciuri*-group species cytochrome-oxidase negative. Almost all staphylococci produce acid from glucose under anaerobic conditions. They have a peptidoglycan containing interpeptide bridges with a high glycine content rendering them susceptible to lysostaphin cleavage and contain teichoic acids in their cell walls (1). The G/C-content of staphylococcal DNA characteristically is between 30 to 39 mol% (2–4).

Staphylococci are separated by their ability to clot plasma into coagulase-positive staphylococci, in specimens from humans almost exclusively *S. aureus*, and coagulase-negative staphylococci. With the recognition of coagulase-negative staphylococci as frequent and significant causes of a vast array of different infections significant interest has been generated in these organisms leading to date to the description of 39 staphylococcal species, which reside within a large variety of animal species and the environment (5–9, Euzeby: http://www.bacterio.cict.fr/index.html). These contain 10 species which are further separated into 21 additional subspecies (Table 7.1).

Whereas *S. aureus* is the almost exclusive coagulase-positive staphylococcal species encountered in humans five other coagulase-positive staphylococcal species including *S. hyicus* (10), *S. intermedius* (11), *S. lutrae* (12), *S. delphini* (13), and *S. schleiferi* subsp. *coagulans* (14) have been described in animal hosts including pigs and cattle, dogs, otters, and dolphins. In contrast, all other staphylococcal species and subspecies are coagulase-negative. Within the coagulase-negative staphylococcal species these may be separated according to their regular appearance in specimens from humans and sometimes animals as opposed to those species which to date occur almost exclusively in specimens from animals or environmental sources (Table 7.1). Additionally, susceptibility to novobiocin is a characteristic used to group coagulase-negative staphylococci (Table 7.1).

Starting from the pioneering studies of Kloos and Schleifer leading to the description of coagulase-negative staphylococci colonizing the human skin including *S. epidermidis*, *S. saprophyticus*, *S. cohnii*, *S. haemolyticus*, *S. xylosus*, *S. warneri*, *S. capitis*, *S. hominis*, *S. simulans*, and *S. auricularis* (15–17) significant additional differentiation has been reached within these species leading to the description of further subspecies of *S. saprophyticus*, *S. hominis*, *S. cohnii*, and *S. capitis* (18–21). Additional coagulase-negative staphylococcal species occur in humans including *S. lugdunensis*, *S. caprae*, *S. pasteuri*, *S. pettenkoferi*, *S. pulvereri*, *S. vitulinus*, and *S. schleiferi* (22–28). Whether *S. pulvereri* and *S. vitulinus* in fact represent a single species has been debated (29,30). This led to recommendations for the description of new staphylococcal species to include at least five different strains (31).

A variety of coagulase-negative staphylococcal species have been described to date which appear to be closely adapted to certain animal species including *S. arlettae* (32), *S. chromogenes* (33), *S. felis* (34), *S. muscae* (35), *S. equorum* subsp. *equorum* (32), *S. gallinarum* (22), *S. kloosii* (32), and *S. nepalensis* (36). Other coagulase-negative staphylococcal species have been described from environmental sources including processed food. These include *S. carnosus* (37,38), *S. condimenti* (37), *S. fleurettii* (39), *S. piscifermentans* (40), and *S. succinus* (41,42).

Identification of coagulase-negative staphylococci in the clinical laboratory is based on morphological, physiological, biochemical, and molecular biological parameters

TABLE 7.1

Taxonomy of the Genus *Staphylococcus*

Source	Coagulase		
	Positive	Negative	
		Novobiocin	
		Susceptible	Resistant
Human	*S. aureus*	*S. auricularis*	*S. saprophyticus*
	S. a. subsp. *aureus*		*S. s.* subsp. *saprophyticus*
	S. a. subsp. *anaerobius*		*S. s.* subsp. *bovis*
		S. capitis	*S. cohnii*
		S. c. subsp. *capitis*	*S. c.* subsp. *cohnii*
		S. c. subsp. *urealyticus*	*S. c.* subsp. *urealyticus*
		S. caprae	*S. pulvereri*
		S. epidermidis	*S. vitulinus*
		S. haemolyticus	*S. xylosus*
		S. hominis	*S. hominis*
		S. h. subsp. *hominis*	*S. h.* subsp. *novobiosepticus*
		S. lugdunensis	
		S. pasteuri	
		S. pettenkoferi	
		S. saccharolyticus	
	S. schleiferi	*S. schleiferi*	
	S. s. subsp.	*S. s.* subsp. *schleiferi*	
	coagulans	*S. simulans*	
		S. warneri	
Animal and	*S. delphini*	*S. carnosus*	*S. arlettae*
environment		*S. c.* subsp. *carnosus*	
		S. c. subsp. *utilis*	
	S. hyicus	*S. caseolyticus*	*S. equorum*
			S. e. subsp. *equorum*
			S. e. subsp. *linens*
	S. intermedius	*S. chromogenes*	*S. fleurettii*
	S. lutrae	*S. condimenti*	*S. gallinarum*
		S. felis	*S. kloosii*
		S. muscae	*S. lentus*
		S. piscifermentans	*S. nepalensis*
		S. succinus	*S. sciuri*
		S. s. subsp. *casei*	*S. s.* subsp. *carnatus*
		S. s. subsp. *succinus*	*S. s.* subsp. *rodentium*
			S. s. subsp. *sciuri*

using modifications of the original scheme described by Kloos and Schleifer (9,43,44). This has been integrated into commercially available manual identification kits such as ID32Staph (bioMerieux), and (semi)-automated instruments for strain identification and susceptibility testing including Vitek II (bioMerieux), Phoenix (BD Becton-Dickinson), and MicroScan Walkaway (Dade-Behring) (45–49). However, it must be emphasized that the databases used for identification with these systems may contain only a subset of the presently known staphylococcal species and therefore identification of species occurring only rarely in human specimens and/or having been newly described will not always be possible using these commercial methods.

The genome sequences of *S. epidermidis* ATCC 12228 (4) and *S. epidermidis* RP62A (TIGR, http://www.tigr.org/tdb/mdb/mdbinprogress.html) have been completely determined. Additionally, the genome sequences of coagulase-negative staphylococcal species *S. haemolyticus* and *S. carnosus* are in progress. The genome sequences of multiple strains of *S. aureus* including MRSA N315, the prototype GISA strain Mu50, epidemic MRSA EMRSA-16, and invasive community acquired MSSA476 and MRSA MW2 (3,50–52) have been published, whereas work on several other strains of *S. aureus* is still in progress (http://www.genomesonline.org/index.cgi?want=Prokaryotic+Ongoing+Genomes).

7.3 CLINICAL RELEVANCE OF COAGULASE-NEGATIVE STAPHYLOCOCCI

Many of the coagulase-negative staphylococcal species described above belong to the normal flora colonizing the skin and mucous membranes of humans, or colonize these habitats transiently (53). Therefore, isolation of coagulase-negative staphylococci from clinical specimens always requires differentiation of contamination from clinical infection. Whereas in earlier decades isolation of coagulase-negative staphylococci had been regularly dismissed as contamination, today these organisms have been recognized as major human pathogens (54,55).

The vast majority of infections due to coagulase-negative staphylococci are of nosocomial origin and according to data of the Centers for Disease Control and the National Nosocomial Infections Surveillance System (NNIS) in fact coagulase-negative staphylococci today belong to the five most frequent causes of nosocomial infections (56,57). There has been a tremendous increase in the incidence of nosocomial bacteremia in the 1980s (58,59). This resulted mainly from the increase of cases caused by coagulase-negative staphylococci and to a more modest degree by *S. aureus* (60). By now coagulase-negative staphylococci account for about 40% of causes of nosocomial bacteremia and are the second most frequent cause of surgical site infection on the ICU (57,61–63). Nosocomial pneumonia or urinary tract infections caused by coagulase-negative staphylococci are less common (56,62).

The successes of modern medicine are closely linked to the ever increasing use of implanted biomedical devices for the intermittent or permanent substitution of failing organs or for management of vital functions of critically ill patients on the intensive care unit. The major risk factor for infection with coagulase-negative staphylococci includes presence of implanted biomedical devices like central venous catheters, prosthetic joints, fracture fixation devices, cardiac pacemakers and heart valves, artificial lenses, vascular grafts, mammary implants, and CSF-shunts (54,64). In Germany alone more than 2.5 million of these biomedical devices are used annually (Table 7.2). A major complication of their use is infection affecting up to 100,000 patients in Germany each year. Similar figures were reported for other industrialized countries like the United States, indicating that millions of patients are at risk worldwide (64).

TABLE 7.2
Estimated Annual Cases of Biomaterial Related Infections in Germany

Device	Device Use/Year[a]	Infection Rates[b]	Infections/Year
Central catheters	~1,750,000	1–5 %	17,500–87,000
Hip prostheses	>200,000	<2%	~4,000
Knee prostheses	60,000	<1–6 %	<600–3,600
Heart valves	18,000	0.8–5.7 %	150–1,000
Pacemakers	70,000	<1–3 %	700–2,100
Artificial lenses	~300,000	<0.1–0.3 %	<300–900
CSF-Shunts	10,000	2–20 %	200–2,000

[a] Numbers were obtained from either manufacturers or experts involved with the use of the respective devices.
[b] Data were taken from (6).

Additional patient related risk factors for infection with coagulase-negative staphylococci are malignancy, chemotherapy, leukopenia, premature birth, care on an ICU, bone marrow transplantation, and immunosuppression due to multiple reason including polytrauma, HIV infection, and transplantation (54,65).

Of great concern is the ever increasing antibiotic resistance of coagulase-negative staphylococci. Up to 90% of coagulase-negative staphylococci causing nosocomial bacteremia are methicillin resistant (57,66,67). This is aggravated by the occurrence of coagulase-negative staphylococci with decreased susceptibility to glycopeptide antibiotics (68,69). Fortunately, to date no clinical coagulase-negative staphylococcal isolate has been described containing the vancomycin resistance determinants *vanA* or *vanB* of enterococci as has been described in *S. aureus* (70). Recent isolates of coagulase-negative staphylococci from nosocomial bacteremia are still uniformly susceptible to linezolid (66).

The most common coagulase-negative staphylococcal species encountered in human infection is *S. epidermidis,* accounting for 60 to 90% of isolates (66,71–78). *S. saprophyticus, S. hominis* and *S. haemolyticus* predominate in occurrence among the other coagulase-negative staphylococci species, while only occasional cases of infection due to *S. capitis, S. warneri, S. caprae, S. lugdunensis, S. cohnii, S. simulans, S. saprophyticus, S. schleiferi,* and *S. auricularis* occur (66,72,79).

S. haemolyticus is of increasing importance as a multiresistant opportunistic pathogen colonizing axilla, perineum and groin but also relatively dry skin areas (53,80). *S. haemolyticus* has been observed as a pathogen in urinary tract and wound infection, osteomyelitis and prosthetic joint infections, peritonitis, sepsis, and endocarditis (81). *S. haemolyticus* is the coagulase-negative staphylococcal species where glycopeptide resistance has been most frequently detected (69,82–85). Glycopeptide resistance was associated with expression of additional membrane proteins and alterations of cell wall composition and morphology (86–88).

S. hominis is, together with *S. epidermidis* and *S. haemolyticus*, the most frequent coagulase-negative staphylococcal species colonizing human skin and mucous membranes (53). However, based on results from different animal infection models it was concluded that *S. hominis* is less virulent than *S. epidermidis, S. haemolyticus*, or *S. saprophyticus* (89–91). As *S. hominis* is isolated quite frequently from clinical specimens care has to be taken in the decision of relevance for this organism in the individual patient.

S. lugdunensis strains may express clumping factor-like activity and a weak thermostable nuclease and therefore can be misidentified as *S. aureus* in the clinical laboratory. *S. lugdunensis* is more virulent in animal models than other coagulase-negative staphylococci (92). *S. lugdunensis* is now recognized as a coagulase-negative staphylococcal species with an exceptionally high pathogenic potential in human infection causing severe forms of destructive endocarditis, which may have a course as devastating as *S. aureus* endocarditis (93–97). *S. lugdunensis* has been observed in other forms of significant infections including bacteremia, pacemaker infection, osteomyelitis and prosthetic joint infection, endophthalmitis, and CSF-shunt infection (73,74,98–102). Soft tissue infections and urinary tract infections caused by *S. lugdunensis* were also reported (103–107). Due to the exceptionally high pathogenic potential of *S. lugdunensis*, isolation of this species should be sufficient reason for the medical microbiologist as well as the clinician to start a thorough clinical investigation of the underlying cause leading to its isolation. In no case *S. lugdunensis* should be dismissed as nonsignificant coagulase-negative *Staphylococcus* even when only a single blood culture is positive. In contrast to most other coagulase-negative staphylococci causing clinically significant infections, *S. lugdunensis* is still rather susceptible to antibiotics. Only 25% of isolates produce penicillinase (103).

S. caprae was originally isolated from goats and has been associated with animals (22). In the last decade *S. caprae* has been recognized as a human pathogen causing community acquired and nosocomial infections (108). Different types of infections have been observed including bacteremia (109–111), urinary tract infection (23,109), skin abscesses (23), and bone and joint infections (23,109–114). A special association of *S. caprae* infection with traumatic fractures and orthopedic prostheses was noted by Shuttleworth and co-workers (113). Kawamura and colleagues reported on a collection of staphylococcal clinical isolates from humans which were characterized for species distribution using DNA–DNA hybridization techniques, noting that 132 strains of *S. caprae* were misidentified as *S. haemolyticus, S. hominis*, and *S. warneri* due to acid production from fructose and mannitol by these strains, in contrast to the original description of *S. caprae* in *Bergey's Manual* (23). Taking these difficulties into account it might be anticipated that *S. caprae* will be recognized as a human pathogen more frequently, when commercial identification systems will have adapted their database to *S. caprae* (113).

S. schleiferi is an opportunistic, primarily nosocomial pathogen (115,116). While *S. schleiferi* subsp. *schleiferi* was originally isolated from humans, *S. schleiferi* subsp. *coagulans* was isolated from dogs (14,24). However, the latter species has recently been described as a pathogen causing post surgical wound infection in man (117). In a series of 28 patients with *S. schleiferi* infections bacteremia, pacemaker infection, wound and soft tissue infection, abscesses and empyema, osteomyelitis, and meningitis were described (116). Although mortality was low the authors concluded that any *S. schleiferi* isolated from clinical specimens should be considered as a serious pathogen (116). *S. schleiferi* was described in an outbreak of deep wound infections after cardiac surgery, leading to increased length of hospital stay similar in duration as for *S. aureus* infections (118). In other studies relation of *S. schleiferi* colonization of the axilla to consecutive pacemaker related infection was demonstrated (119,120). Additional cases of osteomyelitis and prosthetic valve endocarditis were reported (121,122). *S. schleiferi* is infrequently isolated from urine specimens (123).

S. xylosus is only infrequently isolated from blood cultures (75). Only rare reports of human infections have been published including endocarditis (124), pyelonephritis (125), and intra-abdominal infection (126).

S. simulans is predominantly an animal pathogen (81), but rare cases of human infection including native valve endocarditis (127,128), septicemia (129), and osteomyelitis and joint prosthesis infections (129–131) have been reported.

S. cohnii infection is rare in humans, despite its occasional occurrence in blood cultures. Cases of septicaemia, septic arthritis, meningitis, and community-acquired pneumonia, in an immuno-compromised HIV-infected patient, were reported (132–135). Of interest is the report on association of *S. cohnii* and *Ureaplasma urealyticum* in chorioamnionitis and respiratory distress in a neonate (136,137).

S. sciuri and its subspecies have their habitats in different animal species (81). Rare cases of infections in humans were reported including endocarditis, peritonitis during peritoneal dialysis, skin and soft tissue infections, surgical wound infection, and urinary tract infection (138–144).

S. warneri is a minor component of human skin flora (53). Compared to other coagulase-negative staphylococci like *S. epidermidis, S. lugdunensis,* and *S. schleiferi, S. warneri* was less pathogenic in an experimental foreign-body abscess model (92). This species has been observed as an occasional opportunistic pathogen in bacteremia, osteomyelitis, endocarditis, CSF-shunt infection, subdural empyema, and discitis (77,145–149).

S. capitis colonizes predominantly the human forehead (53). *S. capitis* has been associated with bacteremia, urinary tract infection, soft tissue infection, endocarditis, and CSF-shunt infections (77,81,150–152).

Only a few reports exist about infections with other coagulase-negative staphylococcal species including *S. saccharolyticus, S. auricularis, S. gallinarum* and *S. pettenkoferi* (26,142,153–156).

In contrast *S. saprophyticus* represents the only species of coagulase-negative staphylococci, which is regularly encountered in outpatients as the second most frequent organism of urinary tract infections, mostly cystitis, especially in sexually active, young women (157,158). However, more severe manifestations like pyelonephritis and urosepsis, chronic prostatitis, endocarditis, and rare cases of nosocomial infections including septicemia, and pneumonia have been reported to be caused by *S. saprophyticus* (159–164).

7.4 STAPHYLOCOCCAL BIOFILM FORMATION

The great majority of infections caused by coagulase-negative staphylococci are related to implanted materials. Early on it was recognized that *S. epidermidis* strains causing Spitz–Holter shunt- or intravascular catheter-related infections were able to grow on polymer surfaces in an adherent mode of growth leading to a macroscopically visible mucoid bacterial consortium, which was referred to as slime (165–167). As the term slime is reserved for polysaccharides noncovalently attached to a bacterial cell surface, this special mode of staphylococcal growth is now generally referred to as biofilm formation (55,168,169). Staphylococcal implant-related infections are the prototype of a biofilm infection. Microbial biofilms are defined as complex consortia of adherent microorganisms encased by a polymeric matrix (170). In recent years it has been generally accepted that the biofilm mode of growth is the predominant way of life of microorganisms in their natural habitats and that the planktonic state in suspension, as has been used in many laboratory experiments, is not the general form of life of most microorganisms (170,171).

A general model has been proposed for the development of bacterial biofilms which includes several stages (172). First, planktonic bacteria make contact with and adhere to a surface. In a second phase organisms accumulate in multiple layers. Later a phase of biofilm maturation takes place, where characteristic morphologic changes may occur leading to the development of statues and mushrooms of adherent bacteria on a surface. Subsequently detachment from the biofilm can occur leading to release of planktonic cells into the environment which may then initiate another cycle of biofilm development (Figure 7.1). Significant features of the biofilm development cycle are mechanisms involved in attachment of planktonic bacteria to surfaces, those leading to intercellular adhesion between bacterial cells within the biofilm as most cells have no direct contact to the surface, mechanisms determining the form of the mature biofilm, and those counteracting the mechanisms of adherence leading to release of planktonic bacterial cells during biofilm detachment. This also involves inter-bacterial communication which is referred to as *quorum sensing*.

Staphylococci form biofilms of exceptional thickness on surfaces easily visible by the naked eye and with great rapidity. Therefore, biofilm formation of staphylococci has been studied predominantly using a semi-quantitiative 96-well microtiter plate assay as originally described by Christensen and colleagues (173). Many modifications of this assay have been described (174). In general the strains to be studied are inoculated into the microtiter wells in an appropriate growth medium, mostly trypticase soy broth, and grown overnight. Then the plates are washed with buffer, the adherent biofilm is fixed

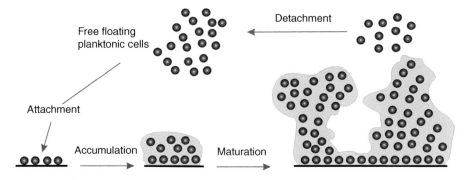

FIGURE 7.1   Model of different phases of biofilm formation.

FIGURE 7.2   Semi-quantitative adherence assay for detection biofilm formation of staphy-lococci showing wells with four different *S. epidermidis* strains with differing degrees of biofilm formation. Second from left: biofilm-negative strain.

and then stained with gentian violet or safranin, and finally the absorbance of light is quantitated in a spectrophotometer (Figure 7.2). Generally, staphylococcal biofilms are so thick that these measurements can be made directly without extracting the absorbed stain from the adherent bacterial cells. The majority of studies related to the mecha-nisms of biofilm formation of staphylococci have used this model biofilm.

In contrast, biofilm formation of Gram-negative bacteria such as *Pseudomonas aeruginosa, Escherichia coli*, and many others are often studied in long term biofilm cultures using continuous flow cells and confocal laser scanning microscopy, where biofilms form on glass or other surfaces under a continuous flow of medium (175–177). However, the microtiter plate assay described above has been used in genetic approaches to the study of biofilm formation by these organisms (178). The modest thickness of biofilms of *P. aeruginosa* or *E. coli* under these conditions largely requires extraction of the stain from cells in the microtiter plate wells for recognition of quantitative differences in biofilm formation (179–181).

Many coagulase-negative staphylococcal species are reported to be able to produce biofilms as detected by the biofilm assay (173,182–185). Epidemiological studies have linked biofilm formation by *S. epidermidis* with pathogenicity (182,186). Therefore, mechanisms of biofilm formation by coagulase-negative staphylococci have been stud-ied most extensively for *S. epidermidis*, which is the predominant coagulase-negative staphylococcal species encountered in human infection.

7.5   PRIMARY ATTACHMENT OF COAGULASE-NEGATIVE STAPHYLOCOCCI TO BIOMATERIALS

After implantation polymer surfaces become rapidly modified by adsorption of host derived plasma proteins, extracellular matrix proteins, and coagulation products, i.e., platelets and thrombi, which is often followed by even more intense integration of the foreign-body implant into host tissue (187,188). Coagulase-negative staphylococci may attach to the implant directly or to the conditioned surface. Detailed study of adhesion indicated that it is a complex process involving a variety of different specific factors.

The ability to attach to polymer surfaces is widespread among coagulase-negative staphylococcal strains. Differences are observed primarily in the quantitative degree of attachment between strains (189–194). Quantitative differences in attachment between individual strains are related to cell surface hydrophobicity and to the respective polymer used (192,195,196). Modification of a polymer by plasma proteins generally decreases attachment of *S. epidermidis* about 100-fold (189,196,197). However, compared to a serum albumin blocked surface, modification with specific extracellular matrix proteins like fibrinogen, fibronectin, vitronectin, and thrombospondin promotes attachment of *S. epidermidis* and other coagulase-negative staphylococci significantly, but adherence is usually below the level of that to an unmodified polymer surface (183,196–204). Similarly, modification of the surface by activated platelets promotes the attachment of *S. epidermidis* RP62A as compared to the plasma protein modified polymer (205).

Specific molecules have been identified which are involved in attachment of *S. epidermidis* to native polystyrene (Figure 7.3). The most prominent is *S. epidermidis* autolysin AtlE identified by transposon mutagenesis in a mutant defective for attachment to polystyrene but still able to attach and form a biofilm on glass (203,206). AtlE encodes a deduced protein of 1335 amino acids with a predicted molecular mass of 148 kDa and lacks the common LPXTG motif responsible for cell surface linkage of many Gram-positive cell wall proteins (203). AtlE contains two bacteriolytic active domains, a 60 kDa amidase and a 52 kDa glucosaminidase, and is located on the cell surface. It is unsettled, whether AtlE mediates attachment directly or by exposing or presenting the actual adhesin, as antiserum against the 60 kDa fragment of AtlE did not inhibit attachment. Interestingly, AtlE has a possible dual role in attachment of *S. epidermidis* as it is able to bind specifically to vitronectin (203). However, promotion of attachment of *S. epidermidis* to vitronectin modified surfaces by AtlE has not yet been reported. Using an isogenic strain pair of *S. epidermidis* 0–47 and mut1 defective in production of AtlE a significant difference in infection rate and recovery of bacteria from infected catheters was observed in a rat central venous catheter infection model (207). Analysis of a 52 kDa vitronectin binding protein by N-terminal aminoacid sequencing and mass spectrometry resulted in identification as AtlE (201).

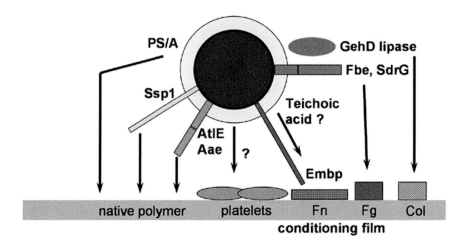

FIGURE 7.3   Mechanisms of *S. epidermidis* primary attachment to biomaterials.

Additional 60 kDa, 21 kDa, and 16 kDa proteins with vitronectin binding activity were identified without homology to known staphylococcal proteins (201).

Interestingly, homologous proteins of AtlE in other coagulase-negative staphylococcal species and *S. aureus* have been characterized (208–210). Aas produced by *S. saprophyticus* and AtlC produced by *S. caprae* both contain the active amidase and glucosaminidase domains (202,208,209). Aas exhibits fibronectin-binding activity and can mediate binding of *S. saprophyticus* to sheep erythrocytes (209). However, whether Aas might be involved in attachment of *S. saprophyticus* to fibronectin coated surfaces or to native polystyrene is not known. AtlC also has binding activity for fibronectin, indicating that matrix protein binding activity might be an universal feature of this class of autolysins from coagulase-negative staphylococci (208). Attachment of *S. caprae* to fibronectin coated surfaces was demonstrated, however, a role of AtlC in this binding could not yet be proven due to lack of an isogenic mutant (183,202). A role of these staphylococcal autolysins in attachment to polystyrene as demonstrated for AtlE has not been reported to date. Nevertheless based on their affinity to matrix proteins these staphylococcal autolysins were proposed as a new class of autolysin/adhesins (204).

Recently, an additional 35 kDa surface-associated autolysin Aae of *S. epidermidis* has been described, which binds to vitronectin (204). Interaction with other matrix proteins, fibronectin, and fibrinogen, was demonstrated using biomolecular interaction analysis. A role in attachment to polystyrene has not been demonstrated.

Concentration dependent inhibition of *S. epidermidis* 354 attachment to polystyrene spheres by a monoclonal antibody (mAb) specifically reacting with a 220 kDa cell wall associated protein was reported (211). The protein identified by the mAb referred to as Ssp1 is organized in a fimbria-like structure on the surface of *S. epidermidis* 354. Proteolytic processing of Ssp1 apparently occurs, rendering a variant of lower molecular weight (Ssp2), which appears to have lower activity regarding attachment of bacterial cells (212). Further molecular characterization of this interesting surface protein has not been reported. Neither is anything known about the prevalence of this protein in *S. epidermidis* strain collections causing biomaterial-related infection.

A capsular polysaccharide adhesin (PS/A) mediating primary attachment to unmodified silastic catheter surfaces was characterized from *S. epidermidis* RP62A (213). PS/A was identified as an activity, which inhibited attachment of *S. epidermidis* strains to silastic catheter tubing. *S. epidermidis* attachment was also inhibited by antiserum raised against purified PS/A. Epidemiological study of *S. epidermidis* strains indicated that PS/A expression alone apparently is not sufficient for biofilm-production of *S. epidermidis* (194). Biofilm-negative transposon mutants of the PS/A-positive *S. epidermidis* M187 displayed decreased attachment to silastic catheter tubing and expressed significantly reduced amounts of PS/A (214). The gene locus inactivated by the transposon has not yet been reported, but is definitely not *icaADBC* (Rohde and Mack unpublished). However, PS/A does not promote attachment to other types of polymers, as no differences in attachment to polyethylene was observed for wild-type and PS/A-negative mutant (215). Later studies revealed that PS/A is structurally closely related if not identical to the polysaccharide intercellular adhesin (PIA) of *S. epidermidis* (216–219, see below).

Genes, referred to as *fbe* or *sdrG* representing variants of the same gene from different strains, encoding fibrinogen-binding proteins were cloned from *S. epidermidis*, which correspond favorably with the property of some *S. epidermidis* strains to bind fibrinogen-coated polymer surfaces (198,220,221). Fbe is a protein with a deduced molecular mass of 119 kDa, which displays homology with *sdrCDE* family of cell surface receptors of *S. aureus* (222). Deletion of the fibrinogen adhesin genes decreased binding to immobilized fibrinogen (221,223). PCR analysis indicated that a large proportion of clinical *S. epidermidis* isolates possessed the gene for Fbe (220,224,225). Despite the presence of the *fbe*-gene a very heterogeneous binding activity to fibrinogen was observed in most *S. epidermidis* strains analyzed. This may reflect different expression levels of Fbe in different strains or it may even indicate that Fbe binds to a ligand different from fibrinogen when expressed on the cell surface (220). Additional genes *sdrF* and *sdrH* encoded in the same locus have not yet been linked to any specific function (226).

Detailed analysis of the domain of fibrinogen involved in binding to the fibrinogen binding proteins of *S. epidermidis* revealed binding to the Bβ-chain of fibrinogen, thereby inhibiting thrombin-induced clotting of fibrinogen due to interference with release of fibrinopeptide B (227,228). The crystal structures of the ligand binding region of Fbe/SdrG as an apoprotein and in complex with a synthetic peptide analogous to its binding site in fibrinogen revealed a dynamic "dock, lock, and latch" mechanism of binding, which represents a general mode of ligand binding for structurally related cell wall-anchored proteins of Gram-positive bacteria (229). It was demonstrated that Fbe-specific antibodies block adherence of *S. epidermidis* to fibrinogen-coated catheters, subcutaneously implanted catheters from rats, and peripheral venous catheters from human patients (230). Antibodies against the fibrinogen binding protein (Fbe) of *S. epidermidis* significantly increased macrophage phagocytosis. The severity of systemic infection of mice with *S. epidermidis* was reduced if the bacteria were preopsonized with anti-Fbe prior to administration, indicating that Fbe might be a vaccine candidate for *S. epidermidis* infections (231).

Using a phage display library Williams and co-workers identified a peptide of *S. epidermidis* which specifically recognized fibronectin (232). The respective reactive peptide Embp32 comprising at least 121 amino acids was part of a very large open reading frame of 30.5 kb encoding a protein, referred to as Embp, of more than 1 MDa present in *S. epidermidis* RP62A and ATCC 12228 (4, TIGR, http://www.tigr.org/tdb/mdb/mdbinprogress.html). A very large protein expressing fibronectin binding activity was also detected in *S. aureus* (233). Embp32 was able to inhibit binding of *S. epidermidis* to surface immobilized fibronectin up to 91% indicating that it represents the major fibronectin-specific adhesin of *S. epidermidis*. As Embp32 did not inhibit binding of *S. aureus* to fibronectin coated surfaces in contrast to the D1–D4 repeat domain of *S. aureus* fibronectin binding protein B (FnbPB), whereas the opposite effect was observed for *S. epidermidis* fibronectin binding, the molecular domains of fibronectin, to which *S. epidermidis* and *S. aureus* adhere, appear to be quite different. Consistent with this hypothesis the fibronectin reactive phages isolated with *S. epidermidis* DNA did not bind to the 30 kDa N-terminal region of fibronectin, which contains the binding domain of *S. aureus* fibronectin binding proteins (232,234). Transcriptional analysis revealed that *embp* appeared to be transcribed throughout the cell cycle (232). Antibodies directed against Embp did not significantly enhance phagocytosis of several *S. epidermidis* isolates (231).

Purified teichoic acids were shown to significantly enhance binding of *S. epidermidis* strains to immobilized fibronectin, where the teichoic acids appeared to have a bridging function between fibronectin and a respective receptor on the bacterial surface (235). Earlier findings of involvement of lipoteichoic acids in binding of *S. epidermidis* to fibrin-platelet clots might relate to these observations (236).

Similar to *S. aureus*, *S. epidermidis* strains were also found to bind thrombospondin (200,237). Immobilized thrombospondin promoted adhesion of *S. epidermidis*, indicating that this matrix protein may be involved in adherence of coagulase-negative staphylococci to biomaterials. No specific *S. epidermidis* receptor has yet been identified to promote this interaction.

GehD, a lipase secreted by *S. epidermidis*, was shown to specifically bind collagen (238). However, lysostaphin cell wall extracts contained GehD, indicating that the lipase may be also cell surface associated. The GehD sequence and structure were radically different from the Cna collagen adhesin of *S. aureus* (238). Recombinant GehD and antibodies specific for GehD are able to inhibit attachment of *S. epidermidis* to immobilized collagen, providing evidence that GehD is a bifunctional molecule acting not only as a lipase but also as a cell surface-associated adhesin (238).

7.6 MECHANISMS OF BIOFILM ACCUMULATION

The polysaccharide intercellular adhesin (PIA) has been identified as the major functional component involved in intercellular adhesion essential for accumulation of multilayered *S. epidermidis* biofilms by a combination of genetic, biochemical and immunochemical evidence (239–241). PIA is synthesized by the gene products of the *icaADBC* locus (242). The *icaADBC* locus has been discovered in other staphylococcal species including *S. aureus* (243,244), and *S. caprae* (208), and homologous DNA-sequences have been detected in a number of other coagulase-negative staphylococci including *S. auricularis, S. capitis, S. condimenti, S. cohnii, S. saprophyticus, S. lugdunensis, S. intermedius,* and *S. piscifermentans* (185,244).

Inactivation of the *icaADBC* locus by Tn917-insertion into *icaA* and *icaC* leads to a biofilm- and PIA-negative phenotype of *S. epidermidis* (206,240,245,246). Similarly, direct evidence for the involvement of *icaADBC* in biofilm accumulation has been obtained for *S. aureus* and *S. caprae* (208,244).

Interestingly, genes with homology to *icaADBC* are not reserved to Gram-positive staphylococci but have been detected in Gram-negative bacteria as *Yersinia pestis* (247) and *E. coli* (248). Apparently, mechanisms of biofilm accumulation depending on synthetic machinery encoded by *icaADBC* are widespread in nature.

Using isogenic transductants of different biofilm-positive *S. epidermidis* wild-type strains containing *icaA*::Tn917 insertions after phage transduction, linkage of abolished PIA synthesis and biofilm formation with abolished hemagglutinating activity of *S. epidermidis* was observed (245). Together with the epidemiological evidence of an association of biofilm formation and hemagglutination of clinical *S. epidermidis* strains these results prove that PIA also functions as the hemagglutinin of *S. epidermidis* (245,249–251).

PIA is a homoglycan composed of β-1,6-linked 2-deoxy-2-amino-D-glucopyra-nosyl residues containing up to 15% de-N-acetylated amino groups and some substitution with ester-linked succinate and phosphate residues introducing simulta-neously positive and negative charges into the polysaccharide (219). The polysac-charide can be separated by Q-Sepharose ion exchange chromatography into two polysaccharide species PIA I and PIA II (219). PIA I (>80%) did not bind to Q-Sepharose and PIA II, a minor polysaccharide (<20%), was moderately anionic. Chain cleavage of PIA I by deamination with HNO_2 revealed a more or less random distribution of the non-N-acetylated glucosaminyl residues, with some prevalence of glucosaminyl-rich sequences. Cation-exchange chromatography separated molecular species of PIA I whose content of non-N-acetylated glucosaminyl residues varied between 2 and 26% (219).

The polysaccharide elutes in the void volume of a Sephadex G200 column (219,240) or a Sephacryl S300 column (C. Fischer and D. Mack, unpublished results) indicating a high apparent molecular weight. However, determination of the ratio of reducing sugar residues at the end of polysaccharide chains and total sugar residues in PIA revealed an average polysaccharide chain length of 130 (219). This indicates that the individual polysaccharide chains form aggregates in solution, which results in a high apparent molecular weight in gel filtration. The joint occurrence of positive and negative charges within the polysaccharide may explain its function of stably linking different cells within the biofilm simply by ionic interactions; however, the presence of a specific proteinaceous receptor specifically binding to PIA is under study.

Reconstitution of PIA synthesis *in vitro* from cloned components in membrane extracts from recombinant *S. carnosus* and UDP-GlcNAc revealed that IcaAD repre-sents a gylcosyltransferase leading to synthesis of oligosaccharides up to a chain length of 20, whereas IcaC appears to be involved in externalization and elongation of the growing polysaccharide as this protein was essential for synthesis of full length, immunoreactive PIA (252). The role of IcaB remains unsettled, but it has been postulated that IcaB may be a deacetylase (253).

Pier and colleagues reported that PS/A of *S. epidermidis* was also produced by the *icaADBC* locus and contained predominantly N-acetylglucosamine (216), which was in contrast to previous compositional analysis revealing primarily galactose as a constituent of PS/A (213). The authors postulated that PS/A is a polysaccharide of high molecular mass composed of β-1,6-linked glucosamine residues with a high degree of substitution with N-linked succinate and acetate (216). In addition, they reported that *S. carnosus*, expressing cloned *icaADBC* from a plasmid pCN27 (219), did produce PS/A but not PIA (216). On analysis of polysaccharide purified from *S. aureus*, referred to as PNSG, the same group reported a similar structure as for PS/A containing β-1,6-linked glucosamine residues with a high degree of substitution with N-linked succinate and acetate and a very high apparent molecular weight (243). In a further report on the structure of PS/A polysaccharide produced by *S. aureus* MN8m, now referred to as PNAG, it was discovered that no N-succinylglucosamine was present in the polysaccharide (218). PNAG was produced with a wide range of molecular masses that could be divided into three major fractions with average molecular masses of 460 kDa (PNAG-I), 100 kDa (PNAG-II), and 21 kDa (PNAG-III) according to gel filtration analysis using Sephacryl S300 (218).

Recently, the structure of polysaccharide purified from *S. aureus* MN8m was analyzed in detail (217), confirming the basic structure of PIA containing β-1,6-linked N-acetylglucosamine, deacetylated amino groups and ester-linked succinate residues (219). *S. aureus* MN8m produced a high-molecular-weight (>300,000 Da) polymer of beta-(1,6)-linked glucosamine containing 45 to 60% N-acetyl, and a small amount of O-succinyl (approx. 10% mole ratio to monosaccharide units). By detailed NMR analyses of polysaccharide preparations it was shown that the previous identification of N-succinyl was an analytical artefact (216,217,243). The exopolysaccharide isolated was active in *in vitro* hemagglutination assays. It was conclude that *S. aureus* strain MN8m produces a polymer that is chemically and biologically closely related to the PIA produced by *S. epidermidis* (217,219).

In *E. coli* evidence has been obtained that the *icaADBC* homologous locus *pgaABCD* is responsible for the synthesis of an unbranched β-1,6-N-acetyl-D-glucosamine polymer previously unknown from the gram-negative bacteria, exhibiting a structure similar to PIA. As predicted, if poly-β-1,6-GlcNAc mediates cohesion, metaperiodate caused biofilm dispersal and the release of intact cells, whereas treatment with protease or other lytic enzymes had no effect. The *pgaABCD* operon exhibits features of a horizontally transferred locus and is present in a variety of eubacteria. Therefore, the polysaccharide serves as an adhesin that stabilizes biofilms of *E. coli* and other bacteria (248).

Interestingly, dispersin B produced by the periodontal pathogen *Actinobacillus actinomycetemcomitans* involved in dispersal of mature biofilms formed by this organism is an N-acetylglucosaminidase (254). Recently, it was reported that dispersin B is able to detach *S. epidermidis* biofilms from polystyrene surfaces (255). Although direct proof is lacking, it is highly likely that the enzyme degrades PIA thereby releasing *S. epidermidis* cells from the established biofilm.

Using a mutant functional in primary attachment but deficient in surface accumulative growth a novel *S. epidermidis* protein relevant for biofilm formation was identified which was subsequently termed "accumulation associated protein" (AAP) (256). While anti-AAP antibodies interfered with biofilm formation in a PIA-positive background, the exact mechanism of this molecule in biofilm formation needs to be elucidated (257).

Although synthesis of PIA is the predominant mechanism of biofilm accumulation of *S. epidermidis*, strains which are biofilm-negative due to lack of *icaADBC* are isolated from proven biomaterial-related infection. This indicates that mechanisms of biofilm accumulation exist which are independent from PIA. Recently, we isolated stable biofilm-positive variants from long-term biofilm cultures of biofilm-negative *S. epidermidis* 5179 causing persistent CSF-shunt infection (219,258). We found that the cell-wall linked Aap alone can specifically mediate intercellular adhesion and biofilm formation in a completely PIA- and polysaccharide independent manner (257). To gain adhesive function, Aap has to be proteolytically processed through staphylococcal proteases (259). Interestingly, exogenously added granulocyte proteases activated Aap, thereby inducing biofilm formation in *S. epidermidis* 5179. It is therefore reasonable to assume that *in vivo* effector mechanisms of innate immunity can directly induce protein dependent *S. epidermidis* cell aggregation and biofilm formation, thereby enabling the pathogen to evade clearance by phagocytes (259).

7.7 ROLE OF BIOFILM FORMATION AND POLYSACCHARIDE INTERCELLULAR ADHESIN IN *S. epidermidis* VIRULENCE

Pathogenicity studies in different infection models have been pursued to confirm the relation of increased pathogenicity and biofilm formation of *S. epidermidis* obtained from epidemiological studies (reviewed in 182). In early studies, clinical isolates were compared in regard to biofilm forming potential. Significant differences in pathogenicity of different coagulase-negative staphylococcal species were observed in a model evaluating decreased weight gain in new-born mice without an implanted foreign body (89). Eight of nine pathogenic *S. epidermidis* strains causing severe or intermediate decrease of weight gain were biofilm-positive, whereas eight less pathogenic strains causing minimal decrease in weight gain were biofilm-negative (89). Deighton and co-workers demonstrated that various biofilm-positive and biofilm-negative clinical *S. epidermidis* isolates induced inflammatory reactions and abscess formation after subcutaneous injection in mice without a foreign body (260). Infection with biofilm-positive *S. epidermidis* strains resulted significantly more frequent in subcutaneous abscesses (260). In a different study, infection of mice after subcutaneous injection could only be induced after implantation of a catheter segment or wounding (261). In this case higher numbers of organisms could be recovered from catheters with biofilm-positive clinical strains, whereas the rate of abscess formation was higher in biofilm-negative *S. epidermidis* strains (261). The same authors reported a significantly reduced ID_{50} for biofilm-positive *S. epidermidis* strains as compared to biofilm-negative clinical isolates in a similar murine model of subcutaneous catheter infection (262). Differences in pathogenic potential of biofilm-positive *S. epidermidis* RP62A and biofilm-negative reference strain SP-2, which was later identified as *S. hominis,* likely resulted from differences in pathogenic potential of these two staphylococcal species (90). In a model of *S. epidermidis* rat endocarditis, the ability to form biofilms was the single parameter that was associated with higher frequency of positive blood cultures (91). However, the interpretation of these results is severely impaired by the fact that clinical isolates, which could differ in their pathogenic potential due to factors independent of their ability for biofilm formation, were compared.

The comparison of biofilm-positive *S. epidermidis* RP62A and several biofilm-negative phase variants of this strain did not lead to clear cut results, but a trend to lower ID_{50} was observed in both the rat endocarditis and the murine subcutaneous catheter infection models (263). With other biofilm-negative phase variants of the same wild-type strain a significant difference in pathogenicity was reported in the rat endocarditis model (264). Phase variation of biofilm formation has been studied intensely in recent years. Additional pleiotropic alterations including colony morphology and antibiotic resistance have been reported, making interpretation of the observed variation of pathogenicity of phase variants difficult (186,264–266).

Similarly, biofilm-negative mutant M7, obtained by chemical mutagenesis from biofilm-positive *S. epidermidis* RP62A, was not significantly less virulent in a rat endocarditis model than the wild-type (256,267). As this mutant is impaired in expression of the AAP but still produces PIA at wild-type levels, the biofilm forming phenotype of this mutant *in vivo* cannot be foreseen.

By using the genetically well-characterized strain pair, biofilm-positive PIA-producing *S. epidermidis* 1457 and its isogenic biofilm-negative *icaA*::Tn917 transductant 1457-M10 where the synthetic gene locus for PIA synthesis was inactivated, conclusive evidence was obtained for the first time that the ability to synthesize PIA and to produce biofilm indeed is a virulence factor of *S. epidermidis* (268–270). This was confirmed by independent analysis of a different biofilm-positive wild-type strain O-47 and its isogenic biofilm-negative *icaADBC*-mutant Mut2 in the rat model (207). The differences in virulence were not related to altered fibronectin binding activity of wild-type and mutant strains (269). Increased phagocytosis and killing by human polymorphonuclear phagocytes was observed with an isogenic icaA::Tn917 insertion mutant as compared to wild-type which may play a role in the observed differences in virulence (268,269,271). Interestingly, the same pairs of isogenic mutants did not show differences in virulence in a guinea-pig tissue cage model indicating that biofilm formation mediated by PIA appears not to be essential in all types of *S. epidermidis* infections (272). Similarly, in the tissue cage model isogenic *S. aureus* variants differing through inactivation of the *icaADBC* locus were not significantly different in their virulence in this model (272,273). These results may be related to the fact that phagocytes are severely impaired in tissue cages, and the advantage of decreased susceptibility to phagocytic killing for PIA-producing wild-type strains is not apparent (271,274). In addition, epidemiological studies which aim to pinpoint the presence of potential *S. epidermidis* virulence genes and the invasive behavior of the strains should clearly take into account that *S. epidermidis* may rely on different virulence determinants in various types of infections.

In a low inoculum rabbit model of endocarditis, biofilm-positive P/SA producing *S. epidermidis* M187 was recovered significantly more often from blood cultures than its biofilm-negative P/SA-negative transposon mutant M187-sn3 (275). In a modification of this model where a catheter colonized by the infecting strain was directly introduced into the left ventricle there was a significant difference in ID_{50} necessary for induction of endocarditis by this strain pair (276). However, endocarditis could be induced by the biofilm-negative mutant M187-sn3 using a sufficiently high inoculum. As it has recently been shown that production of PS/A is dependent on the *icaADBC* locus, and PS/A appears to be structurally almost identical to PIA, these results are further evidence supporting the role of biofilm formation mediated by PIA for pathogenesis of *S. epidermidis* infections (216,217,219, see also discussion above). However, the locus of insertion of the transposon in M187-sn3 is unknown. It is not inserted in *icaADBC* (Rohde and Mack, unpublished results) but may be inserted in a regulatory locus leading to pleiotropic alterations in addition to influencing biofilm formation and exopolysaccharide synthesis.

Vandecasteele and colleagues have used an elegant approach to investigate transcriptional expression of genes related to biofilm formation of *S. epidermidis* in an *in vivo* catheter infection model using real-time RT-PCR (277–279). Exposure to foreign bodies *in vitro* and *in vivo* induced a sharp increase in *icaADBC* expression that was followed by a progressive decrease (279). The *in vivo* expressions of *aap* and *mecA* were high during early but low during late foreign body infections. The *in vivo* expression of *atlE* and *gmk* (a house keeping gene) remained relatively high and stable. In conclusion, biofilm genes encoding for structural elements are mainly expressed during early foreign body infections. The authors concluded that *icaADBC* are associated with initial colonization but not with persistence. The constant expression of *atlE* and *gmk* may indicate a role during the entire course of foreign body infections (279). In addition, after an initial increase in protein synthesis after establishing foreign body infection, a constant decrease of metabolic activity as measured by rRNA content of bacterial cells was detected (280). This decreased metabolic activity during late foreign body infection can explain at least in part the difficulties associated with eradication of such infections with conventional antibiotic treatment.

7.8   USE OF *icaADBC* TO DISCRIMINATE CLINICALLY SIGNIFICANT *S. epidermidis* STRAINS

Currently it is anticipated that invasive *icaADBC*-positive strains are selected at the time of foreign-body implantation from a population of essentially *icaADBC*- and biofilm-negative commensal strains due to their biofilm forming capacity. This model is based on the observation of higher *icaADBC* prevalence in invasive strains compared to commensal strains from healthy individuals (186,281,282). In addition, *mecA*, mediating methicillin resistance, has been proposed as a marker discriminating between invasive and contaminating strains (281). Other factors involved in foreign-body colonization displayed no differential distribution in invasive and commensal *S. epidermidis* (281–284). However, in those studies data from invasive strains were compared against commensal strains from healthy individuals with no contact to the hospital environment. Therefore, as *S. epidermidis* infections typically are of endogenous character, the *S. epidermidis* populations from which invasive strains are recruited, i.e., the commensals of the hospitalized patients, were neglected in most studies.

The distribution of virulence-associated genes, polysaccharide intercellular adhesin (PIA) synthesis and biofilm formation was investigated in *S. epidermidis* strains from independent episodes of catheter related bacteraemia in bone marrow transplant recipients (225). Results were compared with those obtained for commensal *S. epidermidis* from hospitalized patients after bone marrow transplantation and from healthy individuals, respectively. *IcaADBC*, *mecA*, and IS256 were significantly more prevalent in bacteraemia isolates than in commensal isolates from healthy individuals. However, in clonally independent, endogenous commensal strains from bone marrow transplantation patients the prevalence of any of the genes did not differ from bacteraemia strains. Apparently, *icaADBC* and methicillin resistance, factors important for establishment of catheter-related infections, already ensure survival in the physiological habitat in the hospital environment, resulting in a higher contamination probability of indwelling medical devices with virulent *S. epidermidis* strains. Therefore, detection of *icaADBC* and *mecA* appears not to be suitable for discriminating invasive from contaminating *S. epidermidis* strains (225,284).

7.9   REGULATION OF THE EXPRESSION OF BIOFILM FORMATION IN *S. epidermidis*

Phenotypically the amount of biofilm produced by individual *S. epidermidis* strains is highly variable, and is influenced by changing environmental conditions including growth medium, carbohydrate supplementation, oxygen and carbon dioxide content of the atmosphere, iron concentration, sub-inhibitory concentrations of certain antibiotics, and high osmolarity or exposure to ethanol, indicating that expression of PIA and biofilm formation is tightly regulated (166,239,258,286–295).

Quorum sensing of staphylococci is mediated by the *agr* locus encoding four genes transcribed as an operon in RNA II and an additional regulatory RNA III, which is additionally translated to synthesize δ-toxin (296). The system recognizes the concentration of a regulatory octapepetide containing a thiolactone ring structure essential for biologic activity (297). When the *agr* system is activated in *S. aureus* stationary cells, synthesis of cell surface proteins including matrix protein adhesins are down regulated, whereas the synthesis of several secreted effector proteins including α-toxin is upregulated. The regulation of the *agr* quorum sensing system is interconnected with another general staphylococcal regulatory system *sar* (298). Different classes of these auto-inducing peptides have been recognized in *S. aureus* and *S. epidermidis*, which significantly interfere with the activity of the *agr* locus in strains expressing a heterologous auto-inducing peptide (299,300).

Inactivation of the *agr* locus in *S. epidermidis* indicated similar effects on synthesis of cell wall associated proteins and secreted proteins as in *S. aureus*, however, no information was obtained on differential expression of specific gene products in this mutant (301). Recently it was demonstrated that deletion of the *agr* locus in a biofilm-positive *S. epidermidis* strain increased biofilm formation (302). The *agr* mutant additionally displayed increased primary attachment and expression of the autolysin AtlE, but lacked δ-toxin production. Similar effects of *agr*-inactivation on biofilm formation were observed in *S. aureus* (303). In both staphylococcal species the level of polysaccharide intercellular adhesin expression was independent of *agr*-activity (302,303). Delta-toxin appeared to exert an effect on attachment to polystyrene related to its surfactant properties during later stages of biofilm formation (302,303). However, the level of PIA expression was equivalent to the isogenic wild-type strains (302,303). Importantly, addition of cross-inhibiting pheromones mimicked an *agr* mutation and significantly enhanced biofilm formation, which suggests that care should be used when treating *S. epidermidis* infections with cross-inhibiting peptides. *Agr*-specificity group 1 was predominant within a collection of clinical *S. epidermidis* strains (304).

In contrast, a heptapeptide RNA III-inhibiting peptide (RIP), which inhibited *S. aureus* pathogenesis by disrupting quorum-sensing mechanisms, was reported to reduce adherence of *S. epidermidis* to plastic after 3 h incubation to about 25% (305, 306). RIP was added to the bacteria at a concentration of 10 μg/10^7 bacteria indicating that peptide was present at 100 μg/well (306). It is likely that the inhibitory effect observed is related to the non-specific protein blocking effect on *S. epidermidis* attachment to polymers. It should be noted, that the concept of RIP is hotly debated by other groups in the field (307).

Using Tn917 transposon mutagenesis, in the biofilm-positive *S. epidermidis* 1457, three independent unlinked gene loci (class II, III, and IV biofilm-negative mutants) were identified in addition to *icaADBC* (class I biofilm negative mutants), that were essential for expression of biofilm formation and PIA synthesis (Figure 7.4). Expression of *icaADBC in trans* from an independent promoter in these three different biofilm-negative mutant classes restored biofilm formation. Together with transcriptional analysis this revealed a regulatory function of these gene loci on expression

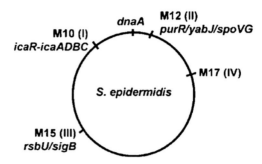

FIGURE 7.4 Relative location of Tn917 insertions in different biofilm-negative mutant classes of *S. epidermidis* (285).

of *icaADBC* at the level of transcription (246). Transfer of the regulatory loci mutations into *mecA*-positive *S. epidermidis* revealed either decreased, more heterogeneous expression of methicillin resistance (class II and III), or increased homogenous expression of methicillin resistance (class IV), indicating that biofilm formation and virulence as well as antibiotic resistance are modulated by the same regulatory mechanisms (308).

In the class III biofilm-negative mutant M15 *rsbU*, the positive regulator of the alternative sigma factor σ^B, was inactivated, indicating that activity of σ^B is essential for transcription of *icaADBC* (309). Using this mutant the complete σ^B-locus of *S. epidermidis* was characterized (309). Interestingly, stress factors acting via *rsbU* in *Bacillus subtilis* like increased osmolarity and ethanol had differential effects on biofilm formation of M15. While ethanol restored biofilm formation and PIA synthesis by the mutant, it remained biofilm-negative in the presence of increased NaCl concentrations. Recently, it was demonstrated by allelic gene replacement that indeed σ^B was essential for biofilm formation of *S. epidermidis* and that the effect of ethanol in mutant M15 was transmitted independent of σ^B (310). The effects of *rsbU* and σ^B inactivation were observed in different *icaADBC*-positive *S. epidermidis* strains, indicating that the essential role for expression of *icaADBC* and PIA is a general phenomenon of *S. epidermidis* in contrast to *S. aureus* (308–311). σ^B acts indirectly on transcription of *icaADBC* by repressing the transcription of *icaR* (Figure 7.5), which has recently been characterized as a negative transcriptional regulator of *icaADBC* (288,310). The role of σ^B in regulation of PIA expression had been questioned by the observation of a biofilm-positive phenotype after expression of *icaADBC* from a plasmid in a $\Delta sigB$ mutant of an *icaADBC*-negative *S. epidermidis* strain (312). However, the plasmid used for *in trans* expression of *icaADBC* lacked the gene for the negative regulator IcaR (242,288,312). Conversely, regulation of biofilm formation and PIA expression is significantly different in *S. aureus* where *sarA* is predominant, whereas σ^B is essential only in special strains (294,311,313).

The Tn917 insertion site of the biofilm-negative mutant M12 (class II) is located in a homologue of the *purR* genes of *S. aureus*, *B. subtilis*, and *Lactococcus lactis* (314). In *B. subtilis purR* encodes a negative regulator of purine synthesis (315,316), whereas in *L. lactis* the regulator PurR has a positive regulatory effect on purine

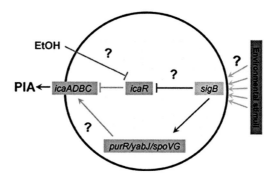

FIGURE 7.5   Regulatory network influencing expression of *icaADBC* and PIA (285).

synthesis (317). In *B. subtilis* a variety of other metabolic genes and operons are regulated by PurR (318). The direct regulation of the *icaADBC* operon by PurR is
unlikely due to a missing putative PurR-binding site. In addition, the gene coding for
the negative regulator IcaR of the *icaADBC* operon is not preceded by a PurR-binding site (288). Therefore, it could be speculated that purine synthesis plays a crucial
role in biofilm formation. However, the T*n917* insertion site is preceded by a putative σ^B dependent promoter, which could explain the biofilm-negative phenotype by
inactivation of genes downstream of *purR*. Interestingly, in *B. subtilis* the σ^B promoter within the *purR* gene seems to be silent (319). However, recently it was
demonstrated in three independent *S. aureus* strains that this promoter is transcribed
in a σ^B-dependent manner (320). In *S. epidermidis* regulation of biofilm expression
by σ^B was predicted, suggesting that these genes could play an additional regulatory
role or could act as the mediator of the σ^B regulation (309). Further experiments to
elucidate the effect of these putative regulatory genes on the regulation of biofilm
expression in *S. epidermidis* are under way.

Besides the regulatory network consisting of σ^B, *icaR*, and the *purR*-locus (Figure
7.5) there exists an additional positive regulator of PIA synthesis represented by
Tn917-mutant M17 (class IV) (246). Inactivation of this locus leads to the biofilm
expression type C1 (reconstitution of biofilm expression by N-acetylglucosamine)
and to a homogeneous expression type of methicillin resistance (258,308). The Tn917
insertion of mutant M17 is located in an open reading frame with unknown function
conserved in many bacterial species (Knobloch et al., unpublished data).

Regulation of the expression of *icaADBC* and PIA synthesis appears to be even
more complex (286). In the presence of glucose, *S. epidermidis* exhibits a PIA- and
biofilm-positive phenotype whereas *icaADBC* transcription is down-regulated in the
post-exponential and stationary phases of growth. Surprisingly, in medium lacking
glucose maximum transcription of *icaADBC* was detectable in the stationary phase of
growth despite the expression of a PIA- and biofilm-negative phenotype. The abundant amount of *icaADBC* mRNA accumulating during growth in medium lacking
glucose is functionally active as shown by *in vitro* PIA synthesis assays. Induction of

biofilm formation by addition of glucose is completely inhibited by chloramphenicol, indicating that translation of an additional, *icaADBC*-independent factor is required for the expression of a biofilm-positive phenotype (286).

Different *S. epidermidis* strains vary widely in the degree of PIA- and biofilm they produce. In several clinical *S. epidermidis* strains the biofilm-forming capacity was analyzed in relation to the amount of *icaADBC* expressed in static biofilm cultures (287). In mid-exponential growth phase no correlation could be detected between the level of *icaADBC* transcription and the biofilm-forming phenotype. When the different strains were grown under conditions leading to a biofilm-negative phenotype, *ica*-expression was still highly upregulated. Sequence analysis demonstrated that the observed differences were not due to major mutations in the *icaADBC* promoter region but apparently to other strain-specific regulators (287). In *S. aureus* strain MN8m a five base pair deletion in the promoter region of *icaA* preventing binding of the transcriptional inhibitor IcaR greatly upregulated transcription of *icaADBC* and synthesis of PIA (321).

Phase variation of biofilm formation has been proposed as an additional mechanism of regulation of *icaADBC* activity (62,186). It appeared that insertion sequences like IS256, which is highly prevalent in clinical *S. epidermidis* isolates, could reversibly insert and excise from integration hotspots within *icaADBC* (62,225,322,323). Insertion of IS256 or IS257 into *icaADBC* has been detected only rarely in clinical isolates (324,325). However, phase variation of biofilm formation occurs also in *S. epidermidis* strains, which are IS256-negative (326). Therefore, additional mechanisms of phase variation exist and remain to be characterized.

Understanding the complex regulatory pathways of PIA synthesis and expression of biofilm formation is of utmost importance as the possibility of targeted modulation of biofilm formation *in vivo* may prove as an efficient means to combat difficult to treat biomaterial-associated infections.

7.10   RELATION OF BIOFILM FORMATION TO ANTIBIOTIC SUSCEPTIBILITY

Biofilm-standing bacteria are suspected to be less susceptible against a variety of antimicrobial substances, depending on the species-drug combination (327–329). Several mechanisms of resistance development are discussed in the literature. The first of these hypotheses suggests that the biofilm glycocalyx prevents the diffusion of antibiotics to the targeted cells within the biofilm. However, for *S. epidermidis* controversial results were obtained by different investigators. Farber et al. detected interference of *S. epidermidis* polysaccharide extracts with the antimicrobial activity of glycopeptides against these organisms (330). In contrast, other investigators showed that a *S. epidermidis* biofilm did not prevent the perfusion of vancomycin or rifampin, however, the biofilm could not be sterilized (331,332). These data indicate that the hypothesis of reduced perfusion of staphylococcal biofilms must be critically reassessed, especially when assuming that bacterial exopolysaccharides contain up to 97% water (333).

The second hypothesis suggests that altered rates of bacterial growth dictate the response to antimicrobial agents. Eng et al. (334) showed that no class of antibiotics was bactericidal against growth limited *S. aureus* and Anwar et al. (335) were able to demonstrate differences of susceptibility in young and aged *S. aureus* biofilms.

For Gram-negative bacteria it could be demonstrated, that only a small part of the cell population within a biofilm is able to persist despite high antibiotic concentrations, leading to the model of single cell persisters during antibiotic therapy (336–338). In *S. epidermidis* recently a similar model of persisting cells within an attached cell population was predicted by our group (339).

The third hypothesis suggests that the microenvironment within a biofilm influences antimicrobial activity and is mainly based on different studies of the biofilm microenvironment compared to standard susceptibility tests under different pH, pCO_2, pO_2, divalent cation concentrations, hydration levels, and pyrimidine concentrations (327).

Many studies focused on new therapeutic regimens to optimize clinical outcome avoiding removal of complex medical devices such as prosthetic heart valves and joint prostheses (340–343). It is therefore highly desirable to establish assay systems which could predict the clinical outcome from susceptibility tests with better correlation than standard susceptibility assays like MIC and MBC determined using planktonic cells or cells growing under optimized conditions on solid media (55,344–348). Key features of such new tests must include reproducibility, as well as ease of performance with respect to their usefulness in a routine laboratory setting, and their correlation with clinical outcome.

Several techniques with differences and difficulties in respect to handling and standardization of *in vivo* grown biofilms have been used to study antibiotic susceptibility of biofilm populations (335,339,340,349–355). Many systems used are effective models of biofilm formation but they are not suitable for rapid susceptibility testing in a clinical laboratory setting. However, two recently developed methods for the investigation of antimicrobial activity of antibiotics against attached bacteria, the MBEC™ device and the MAK assay, could be suitable for susceptibility testing for clinical laboratory use (339,356).

Using the MAK assay two different resistance types could be distinguished. The MAK_{homo} is a homogeneous resistance pattern which represents the resistance of the major population of attached cells. For some strains and antibiotics an additional heterogeneous resistance (MAK_{hetero}) was observed which was up to 4096-fold higher than the MAK_{homo} (339). For isogenic biofilm-negative mutants a lower MAK_{homo} was detected than for the corresponding wild types for several tested antibiotics, which could result from higher bacterial inocula of wild type strains, whereas the MAK_{hetero} was comparable for mutants and wild types to most of the tested antibiotics and strains (339). These data correspond favorably with the model of persistence of single cells (persisters) of Gram-negative bacteria in a biofilm during antibiotic therapy (337,357, and see Chapter 12).

Interestingly, for the MAK_{hetero} of biofilm-positive *S. epidermidis* wild type and biofilm-negative mutant strains no significant differences were observed, indicating that persisters occur with comparable resistance that could lead to therapeutic failure in foreign body-associated infections even with biofilm-negative *S. epidermidis* (339). These persisters may be of relevance especially in foreign body-related infections due to the local granulocyte defects in the area of foreign bodies, whereas persisters in tissue may be easily removed by phagocytes (274,358).

Both, MAK_{hetero} and MAK_{homo} were equal or higher than the respective MICs for all investigated *S. epidermidis* strains and antibiotics, which is similar to the relation of the minimal biofilm eradicating concentration (MBEC) and MICs determined for *E. coli*, *P. aeruginosa*, and *S. aureus* (339,359). Comparison of MAK_{hetero} and MAK_{homo} with the respective MBCs performed under standard conditions and with high bacterial inocula revealed an individual behavior of attached *S. epidermidis* for the different strains. In most cases the MAK_{hetero} correlated with the MBCs observed for high inocula, which indicates that biofilm-standing *S. epidermidis* cells are not significantly less susceptible compared to planktonic cells, and persisters occur in biofilm integrated populations as well as among planktonic cells. Recently, for *P. aeruginosa* it was also demonstrated that cells within a biofilm and planktonic cells have similar resistance against killing by antimicrobials, and persisters could occur in planktonic populations in this species (336,337,357). These data indicate that the observed therapeutic failure of antimicrobial therapy in foreign-body associated infections is mainly caused by the insufficient removal of the persisters by the immune system in an attached bacterial population, whereas changes in the antibiotic susceptibility play only a minor role. For single antibiotics the MAK_{hetero} of *S. epidermidis* was in a range which could possibly lead to a therapeutic success (339), confirming the recently published results, that MIC and MBC determination with high inocula are not suitable to predict response to antibiotic therapy of clinical isolates from medical device-associated infections (340). However, these data indicate that the individual susceptibility testing of infectious isolates in an assay using attached bacteria could be helpful to decide the therapeutic strategy of infected medical devices. Further studies are necessary to evaluate the correlation of assays using attached bacteria to the clinical response of infections to therapy.

ACKNOWLEDGMENTS

The authors gratefully acknowledge permission to reprint Figures 7.4 and 7.5 from the *International Journal of Medical Microbiology*, Vol. 294, Mack et al., Mechanisms of biofilm formation in *Staphylococcus epidermidis* and *Staphylococcus aureus*: functional molecules, regulatory circuits, and adaptive responses, 203-212, 2004. With permission from Elsevier.

REFERENCES

1. Schleifer, K.H., *Micrococcaceae*. In: Sneath, H.A., ed., Vol 2. Baltimore: Williams and Wilkins, 1986:1003.
2. Kloos, W.E., and Schleifer, K.H., Genus IV *Staphylococcus*. In: Sneath, H.A., ed., Baltimore: Williams and Wilkins, 1986:1013–1035.
3. Kuroda, M., Ohta, T., Uchiyama, I., et al., Whole genome sequencing of meticillin-resistant *Staphylococcus aureus*. *Lancet* 357, 1225–1240, 2001.
4. Zhang, Y.Q., Ren, S.X., Li, H.L., et al., Genome-based analysis of virulence genes in a non-biofilm-forming *Staphylococcus epidermidis* strain (ATCC 12228). *Mol. Microbiol.* 49, 1577–1593, 2003.
5. Pulverer, G., and Halswick, R., Coagulase-negative staphylococci (*Staphylococcus albus*) as pathogens. *Dtsch. Med. Wochenschr.* 92, 1141–1145, 1967.
6. Bisno, A.L., and Waldvogel, F.A., Infections Associated with Indwelling Medical Devices. Washington, D.C.: American Society of Microbiology, 1994.

7. Smith, I.M., Beals, P.D., Kingsbury, K.R., and Hasenclever, H.F., Observations on Staphylococcus albus septicemia in mice and men. *AMA Arch. Intern. Med.* 102, 375–388, 1958.

8. Euzeby, J.P., List of bacterial names with standing in nomenclature: a folder available on the Internet. *Int. J. Syst. Bacteriol.* 47, 590–592, 1997.

9. Bannerman, T.L., *Staphylococcus, Micrococcus,* and Other Catalase-Positive Cocci That Grow Aerobically. In: Murray, P.R., Baron, E.J., Jorgensen, J.H., Pfaller, M.A., and Yolken, R.H., eds., 8th. Washington, D.C.: American Society of Microbiology, 2003:384–404.

10. Devriese, L.A., Hajek, V., Oeding, P., Meyer, S.A., and Schleifer, K.H., *Staphylococcus hyicus* (Sompolinski 1953) comb. nov. and *Staphylococcus hyicus* subsp. *chromogenes* subsp. nov. *Int. J. Syst. Bacteriol.* 28, 482–490, 1978.

11. Hajek, V., *Staphylococcus intermedius,* a new species isolated from animals. *Int. J. Syst. Bacteriol.* 26, 401–408, 1976.

12. Foster, G., Ross, H.M., Hutson, R.A., and Collins, M.D., *Staphylococcus lutrae* sp. nov., a new coagulase-positive species isolated from otters. *Int. J. Syst. Bacteriol.* 47, 724–726, 1997.

13. Varaldo, P.E., Kilpper-Bälz, R., Biavasco, F., Satta, G., and Schleifer, K.H., *Staphylococcus delphini* sp. nov., a coagulase-positive species isolated from dolphins. *Int. J. Syst. Bacteriol.* 38, 436–439, 1988.

14. Igimi, S., Takahashi, E., and Mitsuoka, T., *Staphylococcus schleiferi* subsp. *coagulans* subsp. nov., isolated from the external auditory meatus of dogs with external ear otitis. *Int. J. Syst. Bacteriol.* 40, 409–411, 1990.

15. Schleifer, K.H., and Kloos, W.E., Isolation and characterization of staphylococci from human skin. I. Amended descriptions of *Staphylococcus epidermidis* and *Staphylococcus saprophyticus* and description of three new species: *Staphylococcus cohnii, Staphylococcus haemolyticus,* and *Staphylococcus xylosus. Int. J. Syst. Bacteriol.* 25, 50–61, 1975.

16. Kloos, W.E., and Schleifer, K.H., Isolation and characterization of staphylococci from human skin. II. Description of four new species: *Staphylococcus warneri, Staphylococcus capitis, Staphylococcus hominis,* and *Staphylococcus simulans. Int. J. Syst. Bacteriol.* 25, 62–75, 1975.

17. Kloos, W.E., and Schleifer, K.H., *Staphylococcus auricularis* sp. nov.; an inhabitant of the human external ear. *Int. J. Syst. Bacteriol.* 9–14, 1983.

18. Hajek, V., Meugnier, H., Bes M., et al., *Staphylococcus saprophyticus* subsp. *bovis* subsp. nov., isolated from bovine nostrils. *Int. J. Syst. Bacteriol.* 46, 792–796, 1996.

19. Kloos, W.E., George, C.G., Olgiate J.S., et al., *Staphylococcus hominis* subsp. *novobiosepticus* subsp. nov., a novel trehalose- and N-acetyl-D-glucosamine-negative, novobiocin- and multiple-antibiotic-resistant subspecies isolated from human blood cultures. *Int. J. Syst. Bacteriol.* 48(Pt 3), 799–812, 1998.

20. Kloos, W.E., and Wolfshohl, J.F., *Staphylococcus cohnii* subspecies: *Staphylococcus cohnii* subsp. *cohnii* subsp. nov. and *Staphylococcus cohnii* subsp. *urealyticum* subsp. nov. *Int. J. Syst. Bacteriol.* 41, 284–289, 1991.

21. Bannerman, T.L., and Kloos, W.E., *Staphylococcus capitis* subsp. *ureolyticus* subsp. nov. from human skin. *Int. J. Syst. Bacteriol.* 41, 144–147, 1991.

22. Devriese, L.A., Poutrel, B., Kilpper-Bälz, R., and Schleifer, K.H., *Staphylococcus gallinarum* and *Staphylococcus caprae,* two new species from animals. *Int. J. Syst. Bacteriol.* 1983, 480–486.

23. Kawamura, Y., Hou, X.G., Sultana F., et al., Distribution of *Staphylococcus* species among human clinical specimens and emended description of *Staphylococcus caprae. J. Clin. Microbiol.* 36, 2038–2042, 1998.

24. Freney, J., Brun, Y., Bes M., et al., *Staphylococcus lugdunensis* sp. nov. and *Staphylococcus schleiferi* sp. nov., two species from human clinical specimens. *Int. J. Syst. Bacteriol.* 38, 172, 1988.

25. Chesneau, O., Morvan, A., Grimont, F., Labischinski, H., and El Solh N., *Staphylococcus pasteuri* sp. nov., isolated from human, animal, and food specimens. *Int. J. Syst. Bacteriol.* 43, 237–244, 1993.

26. Trulzsch, K., Rinder, H., Trcek, J., Bader, L., Wilhelm, U., and Heesemann, J., *Staphylococcus pettenkoferi*, a novel staphylococcal species isolated from clinical specimens. *Diagn. Microbiol. Infect. Dis.* 43, 175–182, 2002.

27. Zakrzewska-Czerwinska, J., Gaszewska-Mastalarz, A., Lis, B., Gamian, A., and Mordarski, M., *Staphylococcus pulvereri* sp. nov., isolated from human and animal specimens. *Int. J. Syst. Bacteriol.* 45, 169–172, 1995.

28. Webster, J.A., Bannerman, T.L., Hubner R.J., et al., Identification of the *Staphylococcus sciuri* species group with EcoRI fragments containing rRNA sequences and description of *Staphylococcus vitulus* sp. nov. *Int. J. Syst. Bacteriol.* 44, 454–460, 1994.

29. Petras P., *Staphylococcus pulvereri = Staphylococcus vitulus? Int. J. Syst. Bacteriol.* 48(Pt 2), 617–618, 1998.

30. Chesneau, O., Morvan, A., Aubert, S., and El Solh, N., The value of rRNA gene restriction site polymorphism analysis for delineating taxa in the genus *Staphylococcus. Int. J. Syst. Evol. Microbiol.* 50(Pt 2), 689–697, 2000.

31. Freney, J., Kloos, W.E., Hajek V., et al., Recommended minimal standards for description of new staphylococcal species. Subcommittee on the taxonomy of staphylococci and streptococci of the International Committee on Systematic Bacteriology. *Int. J. Syst. Bacteriol.* 49(Pt 2), 489–502, 1999.

32. Schleifer, K.H., Kilpper-Bälz, R., and Devriese, L.A., *Staphylococcus arlettae* sp. nov., *S. equorum* sp. nov., and *S. kloosii* sp. nov.: three new coagulase-negative, novobiocin-resistant species from animals. *Syst. Appl. Microbiol.* 5, 501–509, 1985.

33. Hajek, V., Devriese, L.A., Mordarski, M., Goodfellow, M., Pulverer, G., and Varaldo, P.E., Elevation of *Staphylococcus hyicus* subsp. *chromogenes* (Devriese et al. 1978) to species status: *Staphylococcus chromogenes* (Devriese et al., 1978) comb. nov. *Syst. Appl. Microbiol.* 8, 169–173, 1986.

34. Igimi, S., Kawamura, S., Takahashi, E., Fiedler, F., Etienne, J., and Freney, J., *Staphylococcus felis*, a new species from clinical specimens from cats. *Int. J. Syst. Bacteriol.* 39, 373–377, 1989.

35. Hajek, V., Ludwig, W., Schleifer K.H., et al., *Staphylococcus muscae*, a new species isolated from flies. *Int. J. Syst. Bacteriol.* 42, 97–101, 1992.

36. Spergser, J., Wieser, M., Taubel, M., Rossello-Mora, R.A., Rosengarten, R., and Busse, H.J., *Staphylococcus nepalensis* sp. nov., isolated from goats of the Himalayan region. *Int. J. Syst. Evol. Microbiol.* 53, 2007–2011, 2003.

37. Probst, A.J., Hertel, C., Richter, L., Wassill, L., Ludwig, W., and Hammes, W.P., *Staphylococcus condimenti* sp. nov., from soy sauce mash, and *Staphylococcus carnosus* (Schleifer and Fischer 1982) subsp. *utilis* subsp. nov. *Int. J. Syst. Bacteriol.* 48(Pt 3), 651–658, 1998.

38. Schleifer, K.H., and Fischer U., Description of a new species of the genus *Staphylococcus: Staphylococcus carnosus. Int. J. Syst. Bacteriol.* 32, 153–156, 1982.

39. Vernozy-Rozand, C., Mazuy, C., Meugnier H., et al., *Staphylococcus fleurettii* sp. nov., isolated from goat's milk cheeses. *Int. J. Syst. Evol. Microbiol.* 50(Pt 4), 1521–1527, 2000.

40. Tanasupawat, S., Hashimoto, Y., Ezaki, T., Kozaki, M., and Komagata K., *Staphylococcus piscifermentans* sp. nov., from fermented fish in Thailand. *Int. J. Syst. Bacteriol.* 42, 577–581, 1992.

41. Lambert, L.H., Cox, T., Mitchell K., et al., *Staphylococcus succinus* sp. nov., isolated from Dominican amber. *Int. J. Syst. Bacteriol.* 48(Pt 2), 511–518, 1998.

42. Place, R.B., Hiestand, D., Burri, S., and Teuber M., *Staphylococcus succinus* subsp. *casei* subsp. nov., a dominant isolate from a surface ripened cheese. *Syst. Appl. Microbiol.* 25, 353–359, 2002.

43. Kloos, W.E., and Schleifer, K.H., Simplified scheme for routine identification of human *Staphylococcus* species. *J. Clin. Microbiol.* 1, 82–88, 1975.

44. Martineau, F., Picard, F.J., Ke, D., et al., Development of a PCR assay for identification of staphylococci at genus and species levels. *J. Clin. Microbiol.* 39, 2541–2547, 2001.

45. Spanu, T., Sanguinetti, M., Ciccaglione, D., et al., Use of the VITEK 2 system for rapid identification of clinical isolates of Staphylococci from bloodstream infections. *J. Clin. Microbiol.* 41, 4259–4263, 2003.

46. Ieven, M., Jansens, H., Ursi, D., Verhoeven, J., and Goossens, H., Rapid detection of methicillin resistance in coagulase-negative staphylococci by commercially available fluorescence test. *J. Clin. Microbiol.* 33, 2183–2185, 1995.

47. Weinstein, M.P., Mirrett, S., Van Pelt, L., et al., Clinical importance of identifying coagulase-negative staphylococci isolated from blood cultures: evaluation of MicroScan Rapid and Dried Overnight Gram-Positive panels versus a conventional reference method. *J. Clin. Microbiol.* 36, 2089–2092, 1998.

48. Ligozzi, M., Bernini, C., Bonora, M.G., De Fatima, M., Zuliani, J., and Fontana, R., Evaluation of the VITEK 2 system for identification and antimicrobial susceptibility testing of medically relevant gram-positive cocci. *J. Clin. Microbiol.* 40, 1681–1686, 2002.

49. Chesneau, O., Aubert, S., Morvan, A., Guesdon, J.L., and El Solh, N., Usefulness of the ID32 staph system and a method based on rRNA gene restriction site polymorphism analysis for species and subspecies identification of staphylococcal clinical isolates. *J. Clin. Microbiol.* 30, 2346–2352, 1992.

50. Holden, M.T., Feil, E.J., Lindsay J.A., et al., Complete genomes of two clinical *Staphylococcus aureus* strains: Evidence for the rapid evolution of virulence and drug resistance. *Proc. Natl. Acad. Sci. USA* 2004.

51. Hiramatsu, K., Aritaka, N., Hanaki, H., et al., Dissemination in Japanese hospitals of strains of *Staphylococcus aureus* heterogeneously resistant to vancomycin. *Lancet* 350, 1670–1673, 1997.

52. Baba, T., Takeuchi, F., Kuroda, M., et al., Genome and virulence determinants of high virulence community-acquired MRSA. *Lancet* 359, 1819–1827, 2002.

53. Noble, W.C., *Staphylococcal Carriage and Skin and Soft Tissue Infections.* Crossley, K.B., and Archer G.L., eds. The staphylococci in human disease. New York: Churchill Livingstone, 1997:401–412.

54. Rupp, M.E., and Archer, G.L., Coagulase-negative staphylococci: pathogens associated with medical progress. *Clin. Infect. Dis.* 19, 231–243, 1994.

55. Götz, F., and Peters, G., *Colonization of Medical Devices by Coagulase-Negative Staphylococci.* In: Waldvogel F.A., and Bisno A.L., eds. Infections Associated with Indwelling Medical Devices, 3rd Ed. Washington, D.C.: American Society for Microbiology, 2000:55–88.

56. U.S. Department of Health and Human Services. Public Health Service: National nosocomial infectious surveillance (NNIS) report. Data summary from October 1986–April 1996, issued may 1996. *Am. J. Infect. Control.* 24, 380–388, 1996.

57. Karlowsky, J.A., Jones, M.E., Draghi, D.C., Thornsberry, C., Sahm, D.F., and Volturo, G.A., Prevalence and antimicrobial susceptibilities of bacteria isolated from blood cultures of hospitalized patients in the United States in 2002. *Ann. Clin. Microbiol. Antimicrob.* 3, 7, 2004.

58. Banerjee, S.N., Emori, T.G., Culver D.H., et al., Secular trends in nosocomial primary bloodstream infections in the United States, 1980–1989. National Nosocomial Infections Surveillance System. *Am. J. Med.* 91, 86S–89S, 1991.

59. Thylefors, J.D., Harbarth, S., and Pittet, D., Increasing bacteremia due to coagulase-negative staphylococci: fiction or reality? *Infect. Control. Hosp. Epidemiol.* 19, 581–589, 1998.

60. Emori, T.G., and Gaynes, R.P., An overview of nosocomial infections, including the role of the microbiology laboratory. *Clin. Microbiol. Rev.* 6, 428–442, 1993.

61. Wisplinghoff, H., Seifert, H., Tallent, S.M., Bischoff, T., Wenzel, R.P., and Edmond, M.B., Nosocomial bloodstream infections in pediatric patients in United States hospitals: epidemiology, clinical features and susceptibilities. *Pediatr. Infect. Dis. J.* 22, 686–691, 2003.

62. Ziebuhr, W., Krimmer, V., Rachid, S., Lößner, I., Götz, F., and Hacker, J., A novel mechanism of phase variation of virulence in *Staphylococcus epidermidis*: evidence for control of the polysaccharide intercellular adhesin synthesis by alternating insertion and excision of the insertion sequence element IS256. *Mol. Microbiol.* 32, 345–356, 1999.

63. U.S. Department of Health and Human Services. Public Health Service: National nosocomial infectious surveillance (NNIS) report. Data summary from October 1986-April 1997, issued may 1997. *Am. J. Infect. Control.* 24, 477–487, 1997.

64. Darouiche, R.O., Device-associated infections: a macroproblem that starts with microadherence. *Clin. Infect. Dis.* 33, 1567–1572, 2001.

65. Goldmann, D.A., and Pier, G.B., Pathogenesis of infections related to intravascular catheterization. *Clin. Microbiol. Rev.* 6, 176–192, 1993.

66. Reynolds, R., Potz, N., Colman, M., Williams, A., Livermore, D., and MacGowan, A., Antimicrobial susceptibility of the pathogens of bacteraemia in the UK and Ireland 2001–2002: the BSAC Bacteraemia Resistance Surveillance Programme. *J. Antimicrob. Chemother.* 53, 1018–1032, 2004.

67. National Nosocomial Infections Surveillance (NNIS) System Report, data summary from January 1992 through June 2003, issued August 2003. *Am. J. Infect. Control.* 31, 481–498, 2003.

68. Cercenado, E., Garcia-Leoni, M.E., Diaz, M.D., et al., Emergence of teicoplanin-resistant coagulase-negative staphylococci. *J. Clin. Microbiol.* 34, 1765–1768, 1996.

69. Biavasco, F., Vignaroli, C., and Varaldo, P.E., Glycopeptide resistance in coagulase-negative staphylococci. *Eur. J. Clin. Microbiol. Infect. Dis.* 19, 403–417, 2000.

70. Chang, S., Sievert, D.M., Hageman, J.C., et al., Infection with vancomycin-resistant *Staphylococcus aureus* containing the *vanA* resistance gene. *N. Engl. J. Med.* 348, 1342–1347, 2003.

71. de Silva, G.D., Kantzanou, M., Justice, A., et al., The *ica* operon and biofilm production in coagulase-negative staphylococci associated with carriage and disease in a neonatal intensive care unit. *J. Clin. Microbiol.* 40, 382–388, 2002.

72. Herwaldt, L.A., Geiss, M., Kao, C., and Pfaller, M.A., The positive predictive value of isolating coagulase-negative staphylococci from blood cultures. *Clin. Infect. Dis.* 22, 14–20, 1996.

73. Bannerman, T.L., Rhoden, D.L., McAllister, S.K., Miller, J.M., and Wilson, L.A., The source of coagulase-negative staphylococci in the endophthalmitis vitrectomy study. A comparison of eyelid and intraocular isolates using pulsed-field gel electrophoresis. *Arch. Ophthalmol.* 115, 357–361, 1997.

74. Steinbrink, K., and Frommelt, L., Treatment of periprosthetic infection of the hip using one-stage exchange surgery. *Orthopade* 24, 335–343, 1995.

75. Eng, R.H., Wang, C., Person, A., Kiehn, T.E., and Armstrong, D., Species identification of coagulase-negative staphylococcal isolates from blood cultures. *J. Clin. Microbiol.* 15, 439–442, 1982.

76. Martin, M.A., Pfaller, M.A., and Wenzel, R.P., Coagulase-negative staphylococcal bacteremia. Mortality and hospital stay. *Ann. Intern. Med.* 110, 9–16, 1989.

77. Jarlov, J.O., Hojbjerg, T., Busch-Sorensen, C., et al., Coagulase-negative Staphylococci in Danish blood cultures: species distribution and antibiotic susceptibility. *J. Hosp. Infect.* 32, 217–227, 1996.

78. Bjorkqvist, M., Soderquist, B., Tornqvist, E., et al., Phenotypic and genotypic characterisation of blood isolates of coagulase-negative staphylococci in the newborn. *APMIS* 110, 332–339, 2002.

79. Ruhe, J., Menon, A., Mushatt, D., Dejace, P., and Hasbun, R., Non-epidermidis coagulase-negative staphylococcal bacteremia: clinical predictors of true bacteremia. *Eur. J. Clin. Microbiol. Infect. Dis.* 23, 495–498, 2004.

80. Archer, G.L., and Climo, M.W., Antimicrobial susceptibility of coagulase-negative staphylococci. *Antimicrob. Agen. Chemother.* 38, 2231–2237, 1994.

81. Lina, G., Etienne, J., and Vandenesch, F., Biology and Pathogenicity of Staphy lococci Other Than *Staphylococcus aureus* and *Staphylococcus epidermidis*. In: Fischetti, V., Novick, R., Ferretti, J., Portnoy, D., and Rood, J., eds. Gram positive pathogens. ASM Press, Washington, D.C.: American Society for Microbiology, 2000:450–462.

82. Schwalbe, R.S., Ritz, W.J., Verma, P.R., Barranco, E.A., and Gilligan, P.H., Selection for vancomycin resistance in clinical isolates of *Staphylococcus haemolyticus*. *J. Infect. Dis.* 161, 45–51, 1990.

83. Veach, L.A., Pfaller, M.A., Barrett, M., Koontz, F.P., and Wenzel, R.P., Vancomycin resistance in *Staphylococcus haemolyticus* causing colonization and bloodstream infection. *J. Clin. Microbiol.* 28, 2064–2068, 1990.

84. Aubert, G., Passot, S., Lucht, F., and Dorche, G., Selection of vancomycin- and teicoplanin-resistant *Staphylococcus haemolyticus* during teicoplanin treatment of *S. epidermidis* infection. *J. Antimicrob. Chemother.* 25, 491–493, 1990.

85. Spanik, S., Trupl, J., Studena, M., and Krcmery, V. Jr., Breakthrough nosocomial bacteraemia due to teicoplanin-resistant *Staphylococcus haemolyticus* in five patients with acute leukaemia. *J. Hosp. Infect.* 35, 155–159, 1997.

86. O'Hare, M.D., and Reynolds, P.E., Novel membrane proteins present in teicoplanin-resistant, vancomycin-sensitive, coagulase-negative *Staphylococcus* spp. *J. Antimicrob. Chemother.* 30, 753–768, 1992.

87. Billot-Klein, D., Gutmann, L., Bryant, D., et al., Peptidoglycan synthesis and structure in *Staphylococcus haemolyticus* expressing increasing levels of resistance to glycopeptide antibiotics. *J. Bacteriol.* 178, 4696–4703, 1996.

88. Giovanetti, E., Biavasco, F., Pugnaloni, A., Lupidi, R., Biagini, G., and Varaldo, P.E., An electron microscopic study of clinical and laboratory-derived strains of teicoplanin-resistant *Staphylococcus haemolyticus*. *Microb. Drug Resist.* 2, 239–243, 1996.

89. Gunn, B.A., Comparative virulence of human isolates of coagulase-negative staphylococci tested in an infant mouse weight retardation model. *J. Clin. Microbiol.* 27, 507–511, 1989.

90. Christensen, G.D., Simpson, W.A., Bisno, A.L., and Beachey, E.H., Experimental foreign body infections in mice challenged with slime-producing *Staphylococcus epidermidis*. *Infect. Immun.* 40, 407–410, 1983.

91. Baddour, L.M., Christensen, G.D., Hester, M.G., and Bisno, A.L., Production of experimental endocarditis by coagulase-negative staphylococci: variability in species virulence. *J. Infect. Dis.* 150, 721–727, 1984.

92. Lambe, D.W. Jr., Ferguson, K.P., Keplinger, J.L., Gemmell, C.G., and Kalbfleisch, J.H., Pathogenicity of *Staphylococcus lugdunensis, Staphylococcus schleiferi*, and three other coagulase-negative staphylococci in a mouse model and possible virulence factors. *Can. J. Microbiol.* 36, 455–463, 1990.

93. Etienne, J., Pangon, B., Leport, C., et al., *Staphylococcus lugdunensis* endocarditis. *Lancet* 1, 390, 1989.

94. Vandenesch, F., Etienne, J., Reverdy, M.E., and Eykyn, S.J., Endocarditis due to *Staphylococcus lugdunensis*: report of 11 cases and review. *Clin. Infect. Dis.* 17, 871–876, 1993.

95. Patel, R., Piper, K.E., Rouse, M.S., Uhl, J.R., Cockerill, F.R. III, and Steckelberg, J.M., Frequency of isolation of *Staphylococcus lugdunensis* among staphylococcal isolates causing endocarditis: a 20-year experience. *J. Clin. Microbiol.* 38, 4262–4263, 2000.

96. Petzsch, M., Leber, W., Westphal, B., Crusius, S., and Reisinger, E.C., Progressive *Staphylococcus lugdunensis* endocarditis despite antibiotic treatment. *Wien. Klin. Wochenschr.* 116, 98–101, 2004.

97. Mylonakis, E., and Calderwood, S.B., Infective endocarditis in adults. *N. Engl. J. Med.* 345, 1318–1330, 2001.

98. Ebright, J.R., Penugonda, N., and Brown W., Clinical experience with *Staphylococcus lugdunensis* bacteremia: a retrospective analysis. *Diagn. Microbiol. Infect. Dis.* 48, 17–21, 2004.

99. Bobin, S., Durand-Dubief, A., Bouhour, D., et al., Pacemaker endocarditis due to *Staphylococcus lugdunensis*: report of two cases. *Clin. Infect. Dis.* 28, 404–405, 1999.

100. Murdoch, D.R., Everts, R.J., Chambers, S.T., and Cowan, I.A., Vertebral osteomyelitis due to *Staphylococcus lugdunensis*. *J. Clin. Microbiol.* 34, 993–994, 1996.

101. Greig, J.M., and Wood, M.J., *Staphylococcus lugdunensis* vertebral osteomyelitis. *Clin. Microbiol. Infect.* 9, 1139–1141, 2003.

102. Elliott, S.P., Yogev, R., and Shulman, S.T., *Staphylococcus lugdunensis*: an emerging cause of ventriculoperitoneal shunt infections. *Pediatr. Neurosurg.* 35, 128–130, 2001.

103. Vandenesch, F., Eykyn, S.J., Etienne, J., and Lemozy, J., Skin and post-surgical wound infections due to *Staphylococcus lugdunensis*. *Clin. Microbiol. Infect.* 1, 73–74, 1995.

104. Bellamy, R., and Barkham, T., *Staphylococcus lugdunensis* infection sites: predominance of abscesses in the pelvic girdle region. *Clin. Infect. Dis.* 35, E32–E34, 2002.

105. Mee-Marquet, N., Achard, A., Mereghetti, L., Danton, A., Minier, M., and Quentin, R., *Staphylococcus lugdunensis* infections: high frequency of inguinal area carriage. *J. Clin. Microbiol.* 41, 1404–1409, 2003.

106. Haile, D.T., Hughes, J., Vetter, E., et al., Frequency of isolation of *Staphylococcus lugdunensis* in consecutive urine cultures and relationship to urinary tract infection. *J. Clin. Microbiol.* 40, 654–656, 2002.

107. Casanova-Roman, M., Sanchez-Porto, A., and Casanova-Bellido, M., Urinary tract infection due to *Staphylococcus lugdunensis* in a healthy child. *Scand. J. Infect. Dis.* 36, 149–150, 2004.

108. Kanda, K., Suzuki, E., Hiramatsu, K., et al., Identification of a methicillin-resistant strain of *Staphylococcus caprae* from a human clinical specimen. *Antimicrob. Agents Chemother.* 35, 174–176, 1991.

109. Vandenesch, F., Eykyn, S.J., Bes, M., Meugnier, H., Fleurette, J., and Etienne, J., Identification and ribotypes of *Staphylococcus caprae* isolates isolated as human pathogens and from goat milk. *J. Clin. Microbiol.* 33, 888–892, 1995.

110. Spellerberg, B., Steidel, K., Lutticken, R., and Haase, G., Isolation of *Staphylococcus caprae* from blood cultures of a neonate with congenital heart disease. *Eur. J. Clin. Microbiol. Infect. Dis.* 17, 61–62, 1998.

111. Takemura, K., Takagi, S., Baba, T., Goto, Y., and Nonogi, H., A 72-year-old man with recurrent sepsis due to *Staphylococcus caprae*. *J. Cardiol.* 36, 269–271, 2000.

112. Elsner, H.A., Dahmen, G.P., Laufs, R., and Mack, D., Intra-articular empyema due to *Staphylococcus caprae* following arthroscopic cruciate ligament repair. *J. Infect.* 37, 66–67, 1998.

113. Shuttleworth, R., Behme, R.J., McNabb, A., and Colby, W.D., Human isolates of *Staphylococcus caprae*: association with bone and joint infections. *J. Clin. Microbiol.* 35, 2537–2541, 1997.

114. Blanc, V., Picaud, J., Legros, E., et al., Infection after total hip replacement by *Staphylococcus caprae*. Case report and review of the literature. *Pathol. Biol. (Paris)* 47, 409–413, 1999.

115. Fleurette, J., Bes, M., Brun, Y., et al., Clinical isolates of *Staphylococcus lugdunensis* and *S. schleiferi*: bacteriological characteristics and susceptibility to antimicrobial agents. *Res. Microbiol.* 140, 107–118, 1989.

116. Hernandez, J.L., Calvo, J., Sota, R., Aguero, J., Garcia-Palomo, J.D., and Farinas, M.C., Clinical and microbiological characteristics of 28 patients with *Staphylococcus schleiferi* infection. *Eur. J. Clin. Microbiol. Infect. Dis.* 20, 153–158, 2001.

117. Vandenesch, F., Lebeau, C., Bes, M., et al., Clotting activity in *Staphylococcus schleiferi* subspecies from human patients. *J. Clin. Microbiol.* 32, 388–392, 1994.

118. Kluytmans, J., Berg, H., Steegh, P., Vandenesch, F., Etienne, J., and van Belkum, A., Outbreak of *Staphylococcus schleiferi* wound infections: strain characterization by randomly amplified polymorphic DNA analysis, PCR ribotyping, conventional ribotyping, and pulsed-field gel electrophoresis. *J. Clin. Microbiol.* 36, 2214–2219, 1998.

119. Celard, M., Vandenesch, F., Darbas, H., et al., Pacemaker infection caused by *Staphylococcus schleiferi*, a member of the human preaxillary flora: four case reports. *Clin. Infect. Dis.* 24, 1014–1015, 1997.

120. Da Costa, A., Lelievre, H., Kirkorian, G., et al., Role of the preaxillary flora in pacemaker infections: a prospective study. *Circulation* 97, 1791–1795, 1998.

121. Calvo, J., Hernandez, J.L., Farinas, M.C., Garcia-Palomo, D., and Aguero, J., Osteomyelitis caused by *Staphylococcus schleiferi* and evidence of misidentification of this *Staphylococcus* species by an automated bacterial identification system. *J. Clin. Microbiol.* 38, 3887–3889, 2000.

122. Leung, M.J., Nuttall, N., Mazur, M., Taddei, T.L., McComish, M., and Pearman, J.W., Case of *Staphylococcus schleiferi* endocarditis and a simple scheme to identify clumping factor-positive staphylococci. *J. Clin. Microbiol.* 37, 3353–3356, 1999.

123. Ozturkeri, H., Kocabeyoglu, O., Yergok, Y.Z., Kosan, E., Yenen, O.S., and Keskin, K., Distribution of coagulase-negative staphylococci, including the newly described species *Staphylococcus schleiferi*, in nosocomial and community acquired urinary tract infections. *Eur. J. Clin. Microbiol. Infect. Dis.* 13, 1076–1079, 1994.

124. Conrad, S.A., and West, B.C., Endocarditis caused by *Staphylococcus xylosus* associated with intravenous drug abuse. *J. Infect. Dis.* 149, 826–827, 1984.

125. Tselenis-Kotsowilis, A.D., Koliomichalis, M.P., and Papavassiliou, J.T., Acute pyelonephritis caused by *Staphylococcus xylosus*. *J. Clin. Microbiol.* 16, 593–594, 1982.

126. Mastroianni, A., Coronado, O., Nanetti, A., and Chiodo, F., *Staphylococcus xylosus* isolated from a pancreatic pseudocyst in a patient infected with the human immunodeficiency virus. *Clin. Infect. Dis.* 19, 1173–1174, 1994.

127. Jansen, B., Schumacher-Perdreau, F., Peters, G., Reinhold, G., and Schonemann, J., Native valve endocarditis caused by *Staphylococcus simulans*. *Eur. J. Clin. Microbiol. Infect. Dis.* 11, 268–269, 1992.

128. McCarthy, J.S., Stanley, P.A., and Mayall, B., A case of *Staphylococcus simulans* endocarditis affecting a native heart valve. *J. Infect.* 22, 211–212, 1991.

129. Males, B.M., Bartholomew, W.R., and Amsterdam, D., *Staphylococcus simulans* septicemia in a patient with chronic osteomyelitis and pyarthrosis. *J. Clin. Microbiol.* 21, 255–257, 1985.

130. Razonable, R.R., Lewallen, D.G., Patel, R., and Osmon, D.R., Vertebral osteomyelitis and prosthetic joint infection due to *Staphylococcus simulans*. *Mayo. Clin. Proc.* 76, 1067–1070, 2001.

131. Sturgess, I., Martin, F.C., and Eykyn, S., Pubic osteomyelitis caused by *Staphylococcus simulans*. *Postgrad. Med. J.* 69, 927–929, 1993.

132. Basaglia, G., Moras, L., Bearz, A., Scalone, S., and Paoli, P.D., *Staphylococcus cohnii* septicaemia in a patient with colon cancer. *J. Med. Microbiol.* 52, 101–102, 2003.

133. Mastroianni, A., Coronado, O., Nanetti, A., Manfredi, R., and Chiodo, F., *Staphylococcus cohnii*: an unusual cause of primary septic arthritis in a patient with AIDS. *Clin. Infect. Dis.* 23, 1312–1313, 1996.

134. Okudera, H., Kobayashi, S., Hongo, K., and Mizuno, M., Fatal meningitis due to *Staphylococcus cohnii*. Case report. *Neurosurg. Rev.* 14, 235–236, 1991.

135. Mastroianni, A., Coronado, O., Nanetti, A., Manfredi, R., and Chiodo, F., Community-acquired pneumonia due to *Staphylococcus cohnii* in an HIV-infected patient: case report and review. *Eur. J. Clin. Microbiol. Infect. Dis.* 14, 904–908, 1995.

136. Sorlin, P., Maes, N., Deplano, A., De Ryck, R., and Struelens, M.J., Chorioamnionitis as an apparent source of vertical transmission of *Staphylococcus cohnii* and *Ureaplasma urealyticum* to a neonate. *Eur. J. Clin. Microbiol. Infect. Dis.* 17, 807–808, 1998.

137. Taylor-Robinson, D., *Staphylococcus cohnii* and *Ureaplasma urealyticum* in a neonate. *Eur. J. Clin. Microbiol. Infect. Dis.* 18, 530, 1999.

138. Hedin, G., and Widerstrom, M., Endocarditis due to *Staphylococcus sciuri*. *Eur. J. Clin. Microbiol. Infect. Dis.* 17, 673–675, 1998.

139. Wallet, F., Stuit, L., Boulanger, E., Roussel-Delvallez, M., Dequiedt, P., and Courcol, R.J., Peritonitis due to *Staphylococcus sciuri* in a patient on continuous ambulatory peritoneal dialysis. *Scand. J. Infect. Dis.* 32, 697–698, 2000.

140. Marsou, R., Bes, M., Boudouma, M., et al., Distribution of *Staphylococcus sciuri* subspecies among human clinical specimens, and profile of antibiotic resistance. *Res. Microbiol.* 150, 531–541, 1999.

141. Stepanovic, S., Dakic, I., Djukic, S., Lozuk, B., and Svabic-Vlahovic, M., Surgical wound infection associated with *Staphylococcus sciuri*. *Scand. J. Infect. Dis.* 34, 685–686, 2002.

142. Kolawole D.O., and Shittu, A.O., Unusual recovery of animal staphylococci from septic wounds of hospital patients in Ile-Ife, Nigeria. *Lett. Appl. Microbiol.* 24, 87–90, 1997.

143. Shittu, A., Lin, J., Morrison, D., and Kolawole D., Isolation and molecular characterization of multiresistant *Staphylococcus sciuri* and *Staphylococcus haemolyticus* associated with skin and soft-tissue infections. *J. Med. Microbiol.* 53, 51–55, 2004.

144. Stepanovic, S., Jezek, P., Vukovic, D., Dakic, I., and Petras, P., Isolation of members of the *Staphylococcus sciuri* group from urine and their relationship to urinary tract infections. *J. Clin. Microbiol.* 41, 5262–5264, 2003.

145. Buttery, J.P., Easton, M., Pearson, S.R., and Hogg, G.G., Pediatric bacteremia due to *Staphylococcus warneri*: microbiological, epidemiological, and clinical features. *J. Clin. Microbiol.* 35, 2174–2177, 1997.

146. Torre, D., Ferraro, G., Fiori, G.P., et al., Ventriculoatrial shunt infection caused by *Staphylococcus warneri*: case report and review. *Clin. Infect. Dis.* 14, 49–52, 1992.

147. Announ, N., Mattei, J.P., Jaoua, S., et al., Multifocal discitis caused by *Staphylococcus warneri*. *Joint Bone Spine* 71, 240–242, 2004.

148. Wood, C.A., Sewell, D.L., and Strausbaugh, L.J., Vertebral osteomyelitis and native valve endocarditis caused by *Staphylococcus warneri*. *Diagn. Microbiol. Infect. Dis.* 12, 261–263, 1989.

149. Crichton, P.B., Anderson, L.A., Phillips, G., Davey, P.G., and Rowley, D.I., Subspecies discrimination of staphylococci from revision arthroplasties by ribotyping. *J. Hosp. Infect.* 30, 139–147, 1995.

150. Kamalesh, M., and Aslam, S., Aortic valve endocarditis due to *Staphylococcus capitis*. *Echocardiography* 17, 685–687, 2000.

151. Sandoe, J.A., Kerr, K.G., Reynolds, G.W., and Jain, S., *Staphylococcus capitis* endocarditis: two cases and review of the literature. *Heart* 82, e1, 1999.

152. Latorre, M., Rojo, P.M., Franco, R., and Cisterna, R., Endocarditis due to *Staphylococcus capitis* subspecies ureolyticus. *Clin. Infect. Dis.* 16, 343–344, 1993.

153. Krishnan, S., Haglund, L., Ashfaq, A., Leist, P., and Roat, T., Prosthetic valve endocarditis due to *Staphylococcus saccharolyticus*. *Clin. Infect. Dis.* 22, 722–723, 1996.

154. Steinbrueckner, B., Singh, S., Freney, J., Kuhnert, P., Pelz, K., and Aufenanger, J., Facing a mysterious hospital outbreak of bacteraemia due to *Staphylococcus saccharolyticus*. *J. Hosp. Infect.* 49, 305–307, 2001.

155. Westblom, T.U., Gorse, G.J., Milligan, T.W., and Schindzielorz, A.H., Anaerobic endocarditis caused by *Staphylococcus saccharolyticus*. *J. Clin. Microbiol.* 28, 2818–2819, 1990.

156. Lew, S.Q., Saez, J., Whyte, R., and Stephenson, Y., Peritoneal dialysis-associated peritonitis caused by *Staphylococcus auricularis*. *Perit. Dial. Int.* 24, 195–196, 2004.

157. Gatermann, S.G., Virulence factors of *Staphylococcus saprophyticus*, *Staphylococcus epidermidis*, and enterococci. In: Mobley, H.L.T., and Warren, J.W., eds., 1st. Washington, D.C.: American Society for Microbiology, 1996:313–340.

158. Hovelius, B., and Mardh, P.A., *Staphylococcus saprophyticus* as a common cause of urinary tract infections. *Rev. Infect. Dis.* 6, 328–337, 1984.

159. Olafsen, L.D., and Melby, K., Urinary tract infection with septicaemia due to *Staphylococcus saprophyticus* in a patient with a ureteric calculus. *J. Infect.* 13, 92–93, 1986.

160. Lee, W., Carpenter, R.J., Phillips, L.E., and Faro, S., Pyelonephritis and sepsis due to *Staphylococcus saprophyticus*. *J. Infect. Dis.* 155, 1079–1080, 1987.

161. Bergman, B., Wedren, H., and Holm, S.E., *Staphylococcus saprophyticus* in males with symptoms of chronic prostatitis. *Urology* 34, 241–245, 1989.

162. Singh, V.R., and Raad, I., Fatal *Staphylococcus saprophyticus* native valve endocarditis in an intravenous drug addict. *J. Infect. Dis.*, 162, 783–784, 1990.

163. Nataro, J.P., and St. Geme III, J.W., Septicemia caused by *Staphylococcus saprophyticus* without associated urinary tract infection. *Pediatr. Infect. Dis. J.* 7, 601–602, 1988.

164. Hell, W., Kern, T., and Klouche, M. *Staphylococcus saprophyticus* as an unusual agent of nosocomial pneumonia. *Clin. Infect. Dis.* 29, 685–686, 1999.

165. Bayston, R., and Penny, S.R., Excessive production of mucoid substance in *Staphylococcus* SIIA: a possible factor in colonisation of Holter shunts. *Dev. Med. Child. Neurol. Suppl.* 27, 25–28, 1972.

166. Christensen, G.D., Simpson, W.A., Bisno, A.L., and Beachey, E.H., Adherence of slime-producing strains of *Staphylococcus epidermidis* to smooth surfaces. *Infect. Immun.* 37, 318–326, 1982.

167. Peters, G., Locci, R., and Pulverer, G., Adherence and growth of coagulase-negative staphylococci on surfaces of intravenous catheters. *J. Infect. Dis.* 146, 479–482, 1982.

168. Hussain, M., Wilcox, M.H., and White, P.J., The slime of coagulase-negative staphylococci: biochemistry and relation to adherence. *FEMS Microbiol. Rev.* 10, 191–207, 1993.

169. Mack, D., Bartscht, K., Dobinsky, S., et al., *Staphylococcal Factors involved in Adhesion and Biofilm Formation on Biomaterials.* In: An, Y.H., and Friedman R.J., eds., 1st edition. Totowa, N.J.: Humana Press, 2000:307–330.

170. Hall-Stoodley, L., Costerton, J.W., and Stoodley, P., Bacterial biofilms: from the natural environment to infectious diseases. *Nat. Rev. Microbiol.* 2, 95–108, 2004.

171. O'Toole, G., Kaplan, H.B., and Kolter, R., Biofilm formation as microbial development. *Annu. Rev. Microbiol.* 54, 49–79, 2000.

172. Costerton, J.W., Stewart, P.S., and Greenberg, E.P., Bacterial biofilms: a common cause of persistent infections. *Science* 284, 1318–1322, 1999.

173. Christensen, G.D., Simpson, W.A., Younger, J.J., et al., Adherence of coagulase-negative staphylococci to plastic tissue culture plates: a quantitative model for the adherence of staphylococci to medical devices. *J. Clin. Microbiol.* 22, 996–1006, 1985.

174. Deighton, M.A., Capstick, J., Domalewski, E., and van Nguyen, T., Methods for studying biofilms produced by *Staphylococcus epidermidis*. *Methods Enzymol.* 336, 177–195, 2001.

175. Palmer, R.J. Jr., Microscopy flowcells: perfusion chambers for real-time study of biofilms. *Methods Enzymol.* 310, 160–166, 1999.

176. Lawrence, J.R., and Neu, T.R., Confocal laser scanning microscopy for analysis of microbial biofilms. *Methods Enzymol.* 310, 131–144, 1999.

177. Christensen, B.B., Sternberg, C., Andersen, J.B., et al., Molecular tools for study of biofilm physiology. *Methods Enzymol.* 310, 20–42, 1999.

178. O'Toole, G., Pratt, L.A., Watnick, P.I., Newman, D.K., Weaver, V.B., and Kolter, R., Genetic approaches to study of biofilms. *Methods Enzymol.* 310, 91–109, 1999.

179. O'Toole, G.A., and Kolter, R., Flagellar and twitching motility are necessary for *Pseudomonas aeruginosa* biofilm development. *Mol. Microbiol.* 30, 295–304, 1998.

180. O'Toole, G.A., and Kolter, R., Initiation of biofilm formation in *Pseudomonas fluorescens* WCS365 proceeds via multiple, convergent signalling pathways: a genetic analysis. *Mol. Microbiol.* 28, 449–461, 1998.

181. Pratt, L.A., and Kolter, R., Genetic analysis of *Escherichia coli* biofilm formation: roles of flagella, motility, chemotaxis and type I pili. *Mol. Microbiol.* 30, 285–293, 1998.

182. Christensen, G.D., Baldassarri, L., and Simpson, W.A., Colonization of medical devices by coagulase-negative staphylococci. In: Bisno, A.L., and Waldvogel, F.A., eds. 2nd. Washington D.C.: American Society of Microbiology, 1994:45–78.

183. Allignet, J., Galdbart, J.O., Morvan, A., et al., Tracking adhesion factors in *Staphylococcus caprae* strains responsible for human bone infections following implantation of orthopaedic material. *Microbiology* 145(Pt 8), 2033–2042, 1999.

184. Stepanovic, S., Vukovicc, D., Trajkovic, V., Samardzic, T., Cupic, M., and Svabic-Vlahovic M. Possible virulence factors of *Staphylococcus sciuri*. *FEMS Microbiol. Lett.* 199, 47–53, 2001.

185. Moretro, T., Hermansen, L., Holck, A.L., Sidhu, M.S., Rudi, K., and Langsrud, S., Biofilm formation and the presence of the intercellular adhesion locus *ica* among staphylococci from food and food processing environments. *Appl. Environ. Microbiol.* 69, 5648–5655, 2003.

186. Ziebuhr, W., Heilmann, C., Götz, F., et al., Detection of the intercellular adhesion gene cluster (*ica*) and phase variation in *Staphylococcus epidermidis* blood culture strains and mucosal isolates. *Infect. Immun.* 65, 890–896, 1997.

187. Fuller, R.A., and Rosen, J.J., Materials for medicine. *Sci. Am.* 255, 118–125, 1986.

188. Gristina, A.G., Biomaterial-centered infection: microbial adhesion versus tissue integration. *Science* 237, 1588–1595, 1987.

189. Espersen, F., Wilkinson, B.J., Gahrn-Hansen, B., Thamdrup, R.V., and Clemmensen, I., Attachment of staphylococci to silicone catheters *in vitro*. *APMIS* 98, 471–478, 1990.

190. Hogt, A.H., Dankert, J., and Feijen, J., Adhesion of *Staphylococcus epidermidis* and *Staphylococcus saprophyticus* to a hydrophobic biomaterial. *J. Gen. Microbiol.* 131(Pt 9), 2485–2491, 1985.

191. Hogt, A.H., Dankert, J., and Feijen, J., Adhesion of coagulase-negative staphylococci to methacrylate polymers and copolymers. *J. Biomed. Mater. Res.* 20, 533–545, 1986.

192. Hogt, A.H., Dankert, J., Hulstaert, C.E., and Feijen, J., Cell surface characteristics of coagulase-negative staphylococci and their adherence to fluorinated poly(ethylenepropylene). *Infect. Immun.* 51, 294–301, 1986.

193. Pascual, A., Fleer, A., Westerdaal, N.A., and Verhoef, J., Modulation of adherence of coagulase-negative staphylococci to Teflon catheters *in vitro*. *Eur. J. Clin. Microbiol.* 5, 518–522, 1986.

194. Muller, E., Takeda, S., Shiro, H., Goldmann, D., and Pier, G.B., Occurrence of capsular polysaccharide/adhesin among clinical isolates of coagulase-negative staphylococci. *J. Infect. Dis.* 168, 1211–1218, 1993.

195. Ludwicka, A., Jansen, B., Wadstrom, T., and Pulverer, G., Attachment of staphylococci to various synthetic polymers. *Zentralbl. Bakteriol. Mikrobiol. Hyg. [A]* 256, 479–489, 1984.

196. Mack, D., Bartscht, K., Fischer, C., et al., Genetic and biochemical analysis of *Staphylococcus epidermidis* biofilm accumulation. *Meth. Enzymol.* 336, 215–239, 2001.

197. Muller, E., Takeda, S., Goldmann, D.A., and Pier, G.B., Blood proteins do not promote adherence of coagulase-negative staphylococci to biomaterials. *Infect. Immun.* 59, 3323–3326, 1991.

198. Herrmann, M., Vaudaux, P.E., Pittet, D., et al., Fibronectin, fibrinogen, and laminin act as mediators of adherence of clinical staphylococcal isolates to foreign material. *J. Infect. Dis.* 158, 693–701, 1988.

199. Vaudaux, P., Pittet, D., Haeberli, A., et al., Host factors selectively increase staphylococcal adherence on inserted catheters: a role for fibronectin and fibrinogen or fibrin. *J. Infect. Dis.* 160, 865–875, 1989.

200. Yanagisawa, N., Li, D.Q., and Ljungh, A., The N-terminal of thrombospondin-1 is essential for coagulase-negative staphylococcal binding. *J. Med. Microbiol.* 50, 712–719, 2001.

201. Li, D.Q., Lundberg, F., and Ljungh, A., Characterization of vitronectin-binding proteins of *Staphylococcus epidermidis*. *Curr. Microbiol.* 42, 361–367, 2001.

202. Allignet, J., England, P., Old, I., and El Solh, N., Several regions of the repeat domain of the *Staphylococcus caprae* autolysin, AtlC, are involved in fibronectin binding. *FEMS Microbiol. Lett.* 213, 193–197, 2002.

203. Heilmann, C., Hussain, M., Peters, G., and Götz, F., Evidence for autolysin-mediated primary attachment of *Staphylococcus epidermidis* to a polystyrene surface. *Mol. Microbiol.* 24, 1013–1024, 1997.

204. Heilmann, C., Thumm, G., Chhatwal, G.S., Hartleib, J., Uekötter, A., and Peters, G., Identification and characterization of a novel autolysin (Aae) with adhesive properties from *Staphylococcus epidermidis*. *Microbiology* 149, 2769–2778, 2003.

205. Wang, I.W., Anderson, J.M., and Marchant, R.E., *Staphylococcus epidermidis* adhesion to hydrophobic biomedical polymer is mediated by platelets. *J. Infect. Dis.* 167, 329–336, 1993.

206. Heilmann, C., Gerke, C., Perdreau-Remington, F., and Götz, F., Characterization of Tn917 insertion mutants of *Staphylococcus epidermidis* affected in biofilm formation. *Infect. Immun.* 64, 277–282, 1996.

207. Rupp, M.E., Fey, P.D., Heilmann, C., and Götz, F., Characterization of the importance of *Staphylococcus epidermidis* autolysin and polysaccharide intercellular adhesin in the pathogenesis of intravascular catheter-associated infection in a rat model. *J. Infect. Dis.* 183, 1038–1042, 2001.

208. Allignet, J., Aubert, S., Dyke, K.G., and El Solh, N., *Staphylococcus caprae* strains carry determinants known to be involved in pathogenicity: a gene encoding an autolysin-binding fibronectin and the *ica* operon involved in biofilm formation. *Infect. Immun.* 69, 712–718, 2001.

209. Hell, W., Meyer, H.G., and Gatermann, S.G., Cloning of *aas*, a gene encoding a *Staphylococcus saprophyticus* surface protein with adhesive and autolytic properties. *Mol. Microbiol.* 29, 871–881, 1998.

210. Oshida, T., Sugai, M., Komatsuzawa, H., Hong, Y.M., Suginaka, H., and Tomasz, A., A *Staphylococcus aureus* autolysin that has an N-acetylmuramoyl-L-alanine amidase domain and an endo-beta-N-acetylglucosaminidase domain: cloning, sequence analysis, and characterization. *Proc. Natl. Acad. Sci. USA* 92, 285–289, 1995.

211. Timmerman, C.P., Fleer, A., Besnier, J.M., De Graaf, L., Cremers, F., and Verhoef, J., Characterization of a proteinaceous adhesin of *Staphylococcus epidermidis* which mediates attachment to polystyrene. *Infect. Immun.* 59, 4187–4192, 1991.

212. Veenstra, G.J., Cremers, F.F., van Dijk, H., and Fleer, A., Ultrastructural organization and regulation of a biomaterial adhesin of *Staphylococcus epidermidis*. *J. Bacteriol.* 178, 537–541, 1996.

213. Tojo, M., Yamashita, N., Goldmann, D.A., and Pier, G.B., Isolation and characterization of a capsular polysaccharide adhesin from *Staphylococcus epidermidis* [published erratum appears in J Infect Dis 1988 Jul;158(1):268]. *J. Infect. Dis.* 157, 713–722, 1988.

214. Muller, E., Hübner, J., Gutierrez, N., Takeda, S., Goldmann, D.A., and Pier, G.B., Isolation and characterization of transposon mutants of *Staphylococcus epidermidis* deficient in capsular polysaccharide/adhesin and slime. *Infect. Immun.* 61, 551–558, 1993.

215. Higashi, J.M., Wang, I.W., Shlaes, D.M., Anderson, J.M., and Marchant, R.E., Adhesion of *Staphylococcus epidermidis* and transposon mutant strains to hydrophobic polyethylene. *J. Biomed. Mater. Res.* 39, 341–350, 1998.

216. McKenney, D., Hübner, J., Muller, E., Wang, Y., Goldmann, D.A., and Pier, G.B., The *ica* locus of *Staphylococcus epidermidis* encodes production of the capsular polysaccharide/adhesin. *Infect. Immun.* 66, 4711–4720, 1998.

217. Joyce, J.G., Abeygunawardana, C., Xu, Q., et al., Isolation, structural characterization, and immunological evaluation of a high-molecular-weight exopolysaccharide from *Staphylococcus aureus. Carbohydr. Res.* 338, 903–922, 2003.

218. Maira-Litran, T., Kropec, A., Abeygunawardana, C., et al., Immunochemical properties of the staphylococcal poly-N-acetylglucosamine surface polysaccharide. *Infect. Immun.* 70, 4433–4440, 2002.

219. Mack, D., Fischer, W., Krokotsch, A., et al., The intercellular adhesin involved in biofilm accumulation of *Staphylococcus epidermidis* is a linear beta-1,6-linked glucosaminoglycan: purification and structural analysis. *J. Bacteriol.* 178, 175–183, 1996.

220. Nilsson, M., Frykberg, L., Flock, J.I., Pei, L., Lindberg, M., and Guss, B., A fibrinogen-binding protein of *Staphylococcus epidermidis. Infect. Immun.* 66, 2666–2673, 1998.

221. Hartford, O., O'Brien, L., Schofield, K., Wells, J., and Foster, T.J., The Fbe (SdrG) protein of *Staphylococcus epidermidis* HB promotes bacterial adherence to fibrinogen. *Microbiology* 147, 2545–2552, 2001.

222. Josefsson, E., McCrea, K.W., Ni, E.D., et al., Three new members of the serine-aspartate repeat protein multigene family of *Staphylococcus aureus. Microbiology* 144 (Pt 12), 3387–3395, 1998.

223. Pei, L., and Flock, J.I., Lack of *fbe*, the gene for a fibrinogen-binding protein from *Staphylococcus epidermidis*, reduces its adherence to fibrinogen coated surfaces. *Microb. Pathog.* 31, 185–193, 2001.

224. Arciola, C.R., Campoccia, D., Gamberini, S., Donati, M.E., and Montanaro, L., Presence of fibrinogen-binding adhesin gene in *Staphylococcus epidermidis* isolates from central venous catheters-associated and orthopaedic implant-associated infections. *Biomaterials* 25, 4825–4829, 2004.

225. Rohde, H., Kalitzky, M., Kröger, N., et al., Detection of virulence-associated genes does not discriminate between invasive and commensal *Staphylococcus epidermidis* strains on a bone marrow transplant unit. *J. Clin. Microbiol.* 42, 5614–5619, 2004.

226. McCrea, K.W., Hartford, O., Davis, S., et al., The serine-aspartate repeat (Sdr) protein family in *Staphylococcus epidermidis. Microbiology* 146 (Pt 7), 1535–1546, 2000.

227. Davis, S.L., Gurusiddappa, S., McCrea, K.W., Perkins, S., and Hook, M., SdrG, a fibrinogen-binding bacterial adhesin of the microbial surface components recognizing adhesive matrix molecules subfamily from *Staphylococcus epidermidis*, targets the thrombin cleavage site in the Bbeta chain. *J. Biol. Chem.* 276, 27799–27805, 2001.

228. Pei, L., Palma, M., Nilsson, M., Guss, B., and Flock, J.I., Functional studies of a fibrinogen binding protein from *Staphylococcus epidermidis. Infect. Immun.* 67, 4525–4530, 1999.

229. Ponnuraj, K., Bowden, M.G., Davis, S., et al., A "dock, lock, and latch" structural model for a staphylococcal adhesin binding to fibrinogen. *Cell* 115, 217–228, 2003.

230. Pei, L., and Flock, J.I., Functional study of antibodies against a fibrogenin-binding protein in *Staphylococcus epidermidis* adherence to polyethylene catheters. *J. Infect. Dis.* 184, 52–55, 2001.

231. Rennermalm, A., Nilsson, M., and Flock, J.I., The fibrinogen binding protein of *Staphylococcus epidermidis* is a target for opsonic antibodies. *Infect. Immun.* 72, 3081–3083, 2004.

232. Williams, R.J., Henderson, B., Sharp, L.J., and Nair, S.P., Identification of a fibronectin-binding protein from *Staphylococcus epidermidis. Infect. Immun.* 70, 6805–6810, 2002.

233. Clarke, S.R., Harris, L.G., Richards, R.G., and Foster, S.J., Analysis of Ebh, a 1.1-megadalton cell wall-associated fibronectin-binding protein of *Staphylococcus aureus*. *Infect. Immun.* 70, 6680–6687, 2002.

234. Joh, D., Speziale, P., Gurusiddappa, S., Manor, J., and Höök, M., Multiple specificities of the staphylococcal and streptococcal fibronectin-binding microbial surface components recognizing adhesive matrix molecules. *Eur. J. Biochem.* 258, 897–905, 1998.

235. Hussain, M., Heilmann, C., Peters, G., and Herrmann, M., Teichoic acid enhances adhesion of *Staphylococcus epidermidis* to immobilized fibronectin. *Microb. Pathog.* 31, 261–270, 2001.

236. Chugh, T.D., Burns, G.J., Shuhaiber, H.J., and Bahr, G.M., Adherence of *Staphylococcus epidermidis* to fibrin-platelet clots *in vitro* mediated by lipoteichoic acid. *Infect. Immun.* 58, 315–319, 1990.

237. Herrmann, M., Suchard, S.J., Boxer, L.A., Waldvogel, F.A., and Lew, P.D., Thrombospondin binds to *Staphylococcus aureus* and promotes staphylococcal adherence to surfaces. *Infect. Immun.* 59, 279–288, 1991.

238. Bowden, M.G., Visai, L., Longshaw, C.M., Holland, K.T., Speziale, P., and Höök, M., Is the GehD lipase from *Staphylococcus epidermidis* a collagen binding adhesin? *J. Biol. Chem.* 277, 43017–43023, 2002.

239. Mack, D., Siemssen, N., and Laufs, R., Parallel induction by glucose of adherence and a polysaccharide antigen specific for plastic-adherent *Staphylococcus epidermidis*: evidence for functional relation to intercellular adhesion. *Infect. Immun.* 60, 2048–2057, 1992.

240. Mack, D., Nedelmann, M., Krokotsch, A., Schwarzkopf, A., Heesemann, J., and Laufs, R., Characterization of transposon mutants of biofilm-producing *Staphylococcus epidermidis* impaired in the accumulative phase of biofilm production: genetic identification of a hexosamine-containing polysaccharide intercellular adhesin. *Infect. Immun.* 62, 3244–3253, 1994.

241. Mack, D., Haeder, M., Siemssen, N., and Laufs, R., Association of biofilm production of coagulase-negative staphylococci with expression of a specific polysaccharide intercellular adhesin. *J. Infect. Dis.* 174, 881–884, 1996.

242. Heilmann, C., Schweitzer, O., Gerke, C., Vanittanakom, N., Mack, D., and Götz, F., Molecular basis of intercellular adhesion in the biofilm-forming *Staphylococcus epidermidis*. *Mol. Microbiol.* 20, 1083–1091, 1996.

243. McKenney, D., Pouliot, K.L., Wang, Y., et al., Broadly protective vaccine for *Staphylococcus aureus* based on an *in vivo*-expressed antigen. *Science* 284, 1523–1527, 1999.

244. Cramton, S.E., Gerke, C., Schnell, N.F., Nichols, W.W., and Götz, F., The intercellular adhesion (*ica*) locus is present in *Staphylococcus aureus* and is required for biofilm formation. *Infect. Immun.* 67, 5427–5433, 1999.

245. Mack, D., Riedewald, J., Rohde, H., et al., Essential functional role of the polysaccharide intercellular adhesin of *Staphylococcus epidermidis* in hemagglutination. *Infect. Immun.* 67, 1004–1008, 1999.

246. Mack, D., Rohde, H., Dobinsky, S., et al., Identification of three essential regulatory gene loci governing expression of the *Staphylococcus epidermidis* polysaccharide intercellular adhesin and biofilm formation. *Infect. Immun.* 68, 3799–3807, 2000.

247. Darby, C., Hsu, J.W., Ghori, N., and Falkow, S., *Caenorhabditis elegans:* plague bacteria biofilm blocks food intake. *Nature* 417, 243–244, 2002.

248. Wang, X., Preston, J.F. III, and Romeo, T., The pgaABCD locus of *Escherichia coli* promotes the synthesis of a polysaccharide adhesin required for biofilm formation. *J. Bacteriol.* 186, 2724–2734, 2004.

249. Rupp, M.E., and Archer, G.L., Hemagglutination and adherence to plastic by *Staphylococcus epidermidis*. *Infect. Immun.* 60, 4322–4327, 1992.

250. Rupp, M.E., Sloot, N., Meyer, H.G., Han, J., and Gatermann, S., Characterization of the hemagglutinin of *Staphylococcus epidermidis*. *J. Infect. Dis.* 172, 1509–1518, 1995.

251. Fey, P.D., Ulphani, J.S., Götz, F., Heilmann, C., Mack, D., and Rupp, M.E., Characterization of the Relationship between Polysaccharide Intercellular Adhesin and Hemagglutination in *Staphylococcus epidermidis*. *J. Infect. Dis.* 179, 1561–1564, 1999.

252. Gerke, C., Kraft, A., Süssmuth, R., Schweitzer, O., and Götz, F., Characterization of the N-acetylglucosaminyltransferase activity involved in the biosynthesis of the *Staphylococcus epidermidis* polysaccharide intercellular adhesin. *J. Biol. Chem.* 273, 18586–18593, 1998.

253. Otto, M., Virulence factors of the coagulase-negative staphylococci. *Front. Biosci.* 9, 841–863, 2004.

254. Kaplan, J.B., Ragunath, C., Ramasubbu, N., and Fine, D.H., Detachment of *Actinobacillus actinomycetemcomitans* biofilm cells by an endogenous beta-hexosaminidase activity. *J. Bacteriol.* 185, 4693–4698, 2003.

255. Kaplan, J.B., Ragunath, C., Velliyagounder, K., Fine, D.H., and Ramasubbu N., Enzymatic detachment of *Staphylococcus epidermidis* biofilms. *Antimicrob. Agents Chemother.* 48, 2633–2636, 2004.

256. Schumacher-Perdreau, F., Heilmann, C., Peters, G., Götz, F., and Pulverer, G., Comparative analysis of a biofilm-forming *Staphylococcus epidermidis* strain and its adhesion-positive, accumulation-negative mutant M7. *FEMS Microbiol. Lett.* 117, 71–78, 1994.

257. Hussain, M., Herrmann, M., von Eiff, C., Perdreau-Remington, F., and Peters, G., A 140-kilodalton extracellular protein is essential for the accumulation of *Staphylococcus epidermidis* strains on surfaces. *Infect. Immun.* 65, 519–524, 1997.

258. Rohde, H., Knobloch J.K.M., Horstkotte, M.A., and Mack, D., Correlation of biofilm expression types of *Staphylococcus epidermidis* with polysaccharide intercellular adhesin synthesis: evidence for involvement of *icaADBC* genotype-independent factors. *Med. Microbiol. Immunol. (Berl.)* 190, 105–112, 2001.

259. Rohde, H., Burdelski, C., Bartscht, K., Hussain, M., Buck, F., Horstkotte, M.A., Knobloch, J.K.-M., Heilmann, C., Herrmann, M., and Mack, D., Induction of *Staphylococcus epidermidis* biofilm formation via proteolytic processing of the accumulation-associated protein by staphylococcal and host proteases. *Mol. Microbiol.* 55, 1883–1895, 2005.

260. Deighton, M.A., Borland, R., and Capstick, J.A., Virulence of *Staphylococcus epidermidis* in a mouse model: significance of extracellular slime. *Epidemiol. Infect.* 117, 267–280, 1996.

261. Patrick, C.C., Plaunt, M.R., Hetherington, S.V., and May, S.M., Role of the *Staphylococcus epidermidis* slime layer in experimental tunnel tract infections. *Infect. Immun.* 60, 1363–1367, 1992.

262. Patrick, C.C., Hetherington, S.V., Roberson, P.K., Henwick, S., and Sloas, M.M., Comparative virulence of *Staphylococcus epidermidis* isolates in a murine catheter model. *Pediatr. Res.* 37, 70–74, 1995.

263. Christensen, G.D., Baddour, L.M., and Simpson, W.A., Phenotypic variation of *Staphylococcus epidermidis* slime production *in vitro* and *in vivo*. *Infect. Immun.* 55, 2870–2877, 1987.

264. Christensen, G.D., Baddour, L.M., Madison, B.M., et al., Colonial morphology of staphylococci on Memphis agar: phase variation of slime production, resistance to beta-lactam antibiotics, and virulence. *J. Infect. Dis.* 161, 1153–1169, 1990.

265. Mempel, M., Müller, E., Hoffmann, R., Feucht, H., Laufs, R., and Grüter, L., Variable degree of slime production is linked to different levels of beta-lactam susceptibility in *Staphylococcus epidermidis* phase variants. *Med. Microbiol. Immunol. (Berl.)* 184, 109–113, 1995.

266. Mempel, M., Feucht, H., Ziebuhr, W., Endres, M., Laufs, R., and Grüter, L., Lack of *mecA* transcription in slime-negative phase variants of methicillin-resistant *Staphylococcus epidermidis*. *Antimicrob. Agents. Chemother.* 38, 1251–1255, 1994.

267. Perdreau-Remington, F., Sande, M.A., Peters, G., and Chambers, H.F., The abilities of a *Staphylococcus epidermidis* wild-type strain and its slime-negative mutant to induce endocarditis in rabbits are comparable. *Infect. Immun.* 66, 2778–2781, 1998.

268. Rupp, M.E., Ulphani, J.S., Fey, P.D., and Mack, D., Characterization of *Staphylococcus epidermidis* polysaccharide intercellular adhesin/hemagglutinin in the pathogenesis of intravascular catheter-associated infection in a rat model. *Infect. Immun.* 67, 2656–2659, 1999.

269. Rupp, M.E., Ulphani, J.S., Fey, P.D., Bartscht, K., and Mack, D., Characterization of the importance of polysaccharide intercellular adhesin/hemagglutinin of *Staphylococcus epidermidis* in the pathogenesis of biomaterial-based infection in a mouse foreign body infection model. *Infect. Immun.* 67, 2627–2632, 1999.

270. Rupp, M.E., and Fey, P.D., *In vivo* models to evaluate adhesion and biofilm formation by *Staphylococcus epidermidis*. *Methods Enzymol.* 336, 206–215, 2001.

271. Vuong, C., Voyich, J.M., Fischer, E.R., et al., Polysaccharide intercellular adhesin (PIA) protects *Staphylococcus epidermidis* against major components of the human innate immune system. *Cell. Microbiol.* 6, 269–275, 2004.

272. Francois, P., Tu Quoc, P.H., Bisognano, C., et al., Lack of biofilm contribution to bacterial colonisation in an experimental model of foreign body infection by *Staphylococcus aureus* and *Staphylococcus epidermidis*. *FEMS Immunol. Med. Microbiol.* 35, 135–140, 2003.

273. Kristian, S.A., Golda, T., Ferracin, F., et al., The ability of biofilm formation does not influence virulence of *Staphylococcus aureus* and host response in a mouse tissue cage infection model. *Microb. Pathog.* 36, 237–245, 2004.

274. Zimmerli, W., Lew, P.D., and Waldvogel, F.A., Pathogenesis of foreign body infection. Evidence for a local granulocyte defect. *J. Clin. Invest.* 73, 1191–1200, 1984.

275. Shiro, H., Muller, E., Gutierrez, N., et al., Transposon mutants of *Staphylococcus epidermidis* deficient in elaboration of capsular polysaccharide/adhesin and slime are avirulent in a rabbit model of endocarditis. *J. Infect. Dis.* 169, 1042–1049, 1994.

276. Shiro, H., Meluleni, G., Groll, A., et al., The pathogenic role of *Staphylococcus epidermidis* capsular polysaccharide/adhesin in a low-inoculum rabbit model of prosthetic valve endocarditis. *Circulation* 92, 2715–2722, 1995.

277. Vandecasteele, S.J., Peetermans, W.E., Merckx, R., and Van Eldere, J., Quantification of expression of *Staphylococcus epidermidis* housekeeping genes with Taqman quantitative PCR during *in vitro* growth and under different conditions. *J. Bacteriol.* 183, 7094–7101, 2001.

278. Vandecasteele, S.J., Peetermans, W.E., Merckx, R., Van Ranst, M., and Van Eldere, J., Use of gDNA as internal standard for gene expression in staphylococci *in vitro* and *in vivo*. *Biochem. Biophys. Res. Commun.* 291, 528–534, 2002.

279. Vandecasteele, S.J., Peetermans, W.E., Merckx, R., and Van Eldere, J., Expression of biofilm-associated genes in *Staphylococcus epidermidis* during *in vitro* and *in vivo* foreign body infections. *J. Infect. Dis.* 188, 730–737, 2003.

280. Vandecasteele, S.J., Peetermans, W.E., Carbonez, A., and Van Eldere, J., Metabolic activity of *Staphylococcus epidermidis* is high during initial and low during late experimental foreign-body infection. *J. Bacteriol.* 186, 2236–2239, 2004.

281. Frebourg, N.B., Lefebvre, S., Baert, S., and Lemeland, J.F., PCR-Based assay for discrimination between invasive and contaminating *Staphylococcus epidermidis* strains. *J. Clin. Microbiol.* 38, 877–880, 2000.

282. Galdbart, J.O., Allignet, J., Tung, H.S., Ryden, C., and El Solh, N., Screening for *Staphylococcus epidermidis* markers discriminating between skin-flora strains and those responsible for infections of joint prostheses. *J. Infect. Dis.* 182, 351–355, 2000.

283. Klug, D., Wallet, F., Kacet, S., and Courcol, R.J., Involvement of adherence and adhesion *Staphylococcus epidermidis* genes in pacemaker lead-associated infections. *J. Clin. Microbiol.* 41, 3348–3350, 2003.

284. Vandecasteele, S.J., Peetermans, W.E., Merckx, R., Rijnders, B.J., and Van Eldere, J., Reliability of the *ica*, *aap* and *atlE* genes in the discrimination between invasive, colonizing and contaminant *Staphylococcus epidermidis* isolates in the diagnosis of catheter-related infections. *Clin. Microbiol. Infect.* 9, 114–119, 2003.

285. Mack, D., Becker, P., Chatterjee, I., Dobinsky, S., Knobloch, J. K. -M., Peters, G., Rohde, H., and Herrmann, M., Mechanisms of biofilm formation in *Staphylococcus epidermidis* and *Staphylococcus aureus*: functional molecules, regulatory circuits, and adaptive responses. *Int. J. Med. Microbiol.* 294, 203-212, 2004.

286. Dobinsky, S., Kiel, K., Rohde, H., et al., Glucose related dissociation between *icaADBC* transcription and biofilm expression by *Staphylococcus epidermidis*: evidence for an additional factor required for polysaccharide intercellular adhesin synthesis. *J. Bacteriol.* 185, 2879–2886, 2003.

287. Dobinsky, S., Rohde, H., Knobloch J.K.M., Horstkotte, M.A., and Mack, D., Transcriptional activity of *icaADBC* is not correlated to the degree of biofilm formation in clinical *ica*-positive *Staphylococcus epidermidis* strains. *Biofilms* 1, 101–106, 2004.

288. Conlon, K.M., Humphreys, H., and O'Gara, J.P., *icaR* encodes a transcriptional repressor involved in environmental regulation of *ica* operon expression and biofilm formation in *Staphylococcus epidermidis*. *J. Bacteriol.* 184, 4400–4408, 2002.

289. Cramton, S.E., Ulrich, M., Götz, F., and Döring, G., Anaerobic conditions induce expression of polysaccharide intercellular adhesin in *Staphylococcus aureus* and *Staphylococcus epidermidis*. *Infect. Immun.* 69, 4079–4085, 2001.

290. Deighton, M., and Borland, R., Regulation of slime production in *Staphylococcus epidermidis* by iron limitation. *Infect. Immun.* 61, 4473–4479, 1993.

291. Hussain, M., Wilcox, M.H., White, P.J., Faulkner, M.K., and Spencer, R.C., Importance of medium and atmosphere type to both slime production and adherence by coagulase-negative staphylococci. *J. Hosp. Infect.* 20, 173–184, 1992.

292. Fitzpatrick, F., Humphreys, H., Smyth, E., Kennedy, C.A., and O'Gara, J.P., Environmental regulation of biofilm formation in intensive care unit isolates of *Staphylococcus epidermidis*. *J. Hosp. Infect.* 52, 212–218, 2002.

293. Knobloch J.K.M., Horstkotte, M.A., Rohde, H., Kaulfers P-M, and Mack, D., Alcoholic ingredients in skin disinfectants increase biofilm expression of *Staphylococcus epidermidis*. *J. Antimicrob. Chemother.* 49, 683–687, 2002.

294. Rachid, S., Ohlsen, K., Wallner, U., Hacker, J., Hecker, M., and Ziebuhr, W., Alternative transcription factor sigma(B) is involved in regulation of biofilm expression in a *Staphylococcus aureus* mucosal isolate. *J. Bacteriol.* 182, 6824–6826, 2000.

295. Rachid, S., Ohlsen, K., Witte, W., Hacker, J., and Ziebuhr, W., Effect of subinhibitory antibiotic concentrations on polysaccharide intercellular adhesin expression in biofilm-forming *Staphylococcus epidermidis*. *Antimicrob. Agents Chemother.* 44, 3357–3363, 2000.

296. Novick, R.P., Pathogenicity factors and their regulation. In: Fischetti, V.A., Novick, R.P., Ferretti, J.J., Portnoy, D.A., and Rood J.I., eds. 1st. Washington, D.C.: American Society for Microbiology, 2000:392–407.

297. Otto, M., *Staphylococcus aureus* and *Staphylococcus epidermidis* peptide pheromones produced by the accessory gene regulator *agr* system. *Peptides* 22, 1603–1608, 2001.

298. Bayer, M.G., Heinrichs, J.H., and Cheung, A.L., The molecular architecture of the *sar* locus in *Staphylococcus aureus*. *J. Bacteriol.* 178, 4563–4570, 1996.

299. Otto, M., Sussmuth, R., Jung, G., and Götz, F., Structure of the pheromone peptide of the *Staphylococcus epidermidis agr* system. *FEBS Lett.* 424, 89–94, 1998.

300 Otto, M., Echner, H., Voelter, W., and Götz, F., Pheromone cross-inhibition between *Staphylococcus aureus* and *Staphylococcus epidermidis*. *Infect. Immun.* 69, 1957–1960, 2001.

301. Vuong, C., Götz, F., and Otto, M., Construction and characterization of an *agr* deletion mutant of *Staphylococcus epidermidis*. *Infect. Immun.* 68, 1048–1053, 2000.

302. Vuong, C., Gerke, C., Somerville, G.A., Fischer, E.R., and Otto, M., Quorum-sensing control of biofilm factors in *Staphylococcus epidermidis*. *J. Infect. Dis.* 188, 706–718, 2003.

303. Vuong, C., Saenz, H.L., Götz, F., and Otto, M., Impact of the *agr* quorum-sensing system on adherence to polystyrene in *Staphylococcus aureus*. *J. Infect. Dis.* 182, 1688–1693, 2000.

304. Carmody, A.B., and Otto, M., Specificity grouping of the accessory gene regulator quorum-sensing system of *Staphylococcus epidermidis* is linked to infection. *Arch. Microbiol.* 181, 250–253, 2004.

305. Balaban, N., Goldkorn, T., Nhan, R.T., et al., Autoinducer of virulence as a target for vaccine and therapy against *Staphylococcus aureus* [see comments]. *Science* 280, 438–440, 1998.

306. Balaban, N., Giacometti, A., Cirioni, O., et al., Use of the quorum-sensing inhibitor RNAIII-inhibiting peptide to prevent biofilm formation *in vivo* by drug-resistant *Staphylococcus epidermidis*. *J. Infect. Dis.* 187, 625–630, 2003.

307. Novick, R.P., Ross, H.F., Figueiredo, A.M., Abramochkin, G., and Muir, T., Activation and inhibition of the staphylococcal *agr* system. *Science* 287, 391a, 2000.

308. Mack, D., Sabottke, A., Dobinsky, S., Rohde, H., Horstkotte, M.A., and Knobloch, J.K.M., Differential expression of methicillin resistance by different biofilm-negative *Staphylococcus epidermidis* transposon mutant classes. *Antimicrob. Agents Chemother.* 46, 178–183, 2002.

309. Knobloch J.K.M., Bartscht, K., Sabottke, A., Rohde, H., Feucht, H.H., and Mack, D., Biofilm formation by *Staphylococcus epidermidis* depends on functional RsbU, an activator of the *sigB* operon: differential activation mechanisms due to ethanol and salt stress. *J. Bacteriol.* 183, 2624–2633, 2001.

310. Knobloch J.K.M., Jäger, S., Horstkotte, M.A., Rohde, H., and Mack, D., RsbU dependent regulation *Staphylococcus epidermidis* biofilm formation is mediated via the alternative sigma factor σ^B by repression of the negative regulator gene *icaR*. *Infect. Immun.* 72, 3838–3848, 2004.

311. Valle, J., Toledo-Arana, A., Berasain, C., et al., SarA and not sigmaB is essential for biofilm development by *Staphylococcus aureus*. *Mol. Microbiol.* 48, 1075–1087, 2003.

312. Kies, S., Otto, M., Vuong, C., and Götz, F., Identification of the *sigB* operon in *Staphylococcus epidermidis*: construction and characterization of a *sigB* deletion mutant. *Infect. Immun.* 69, 7933–7936, 2001.

313. Knobloch J.K.M., Horstkotte, M.A., Rohde, H., and Mack, D., Evaluation of different detection methods for biofilm formation in *Staphylococcus aureus*. *Med. Microbiol. Immunol. (Berl.)* 191, 101–106, 2002.

314. Knobloch J.K.M., Nedelmann, M., Kiel, K., et al., Establishment of an arbitrary PCR for rapid identification of Tn917 insertion sites in *Staphylococcus epidermidis*: characterization of biofilm-negative an non-mucoid mutants. *Appl. Environ. Microbiol.* 69, 5812–5818, 2003.

315. Weng, M., Nagy, P.L., and Zalkin, H., Identification of the *Bacillus subtilis pur* operon repressor. *Proc. Natl. Acad. Sci. USA* 92, 7455–7459, 1995.

316. Weng, M., and Zalkin, H., Mutations in the *Bacillus subtilis* purine repressor that perturb PRPP effector function *in vitro* and *in vivo*. *Curr. Microbiol,* 41, 56–59, 2000.

317. Kilstrup, M., and Martinussen, J., A transcriptional activator, homologous to the *Bacillus subtilis* PurR repressor, is required for expression of purine biosynthetic genes in *Lactococcus lactis*. *J. Bacteriol.* 180, 3907–3916.

318. Saxild, H.H., Brunstedt, K., Nielsen, K.I., Jarmer, H., and Nygaard, P., Definition of the *Bacillus subtilis* PurR operator using genetic and bioinformatic tools and expansion of the PurR regulon with *glyA, guaC, pbuG, xpt-pbuX, yqhZ-folD,* and *pbuO*. *J. Bacteriol.* 183, 6175–6183, 2001.

319. Petersohn, A., Bernhardt, J., Gerth, U., et al., Identification of sigma(B)-dependent genes in *Bacillus subtilis* using a promoter consensus-directed search and oligonucleotide hybridization. *J. Bacteriol.* 181, 5718–5724, 1999.

320. Bischoff, M., Dunman, P., Kormanec, J., et al., Microarray-based analysis of the *Staphylococcus aureus* sigmaB regulon. *J. Bacteriol.* 186, 4085–4099, 2004.

321. Jefferson, K.K., Cramton, S.E., Götz, F., and Pier, G.B., Identification of a 5-nucleotide sequence that controls expression of the *ica* locus in *Staphylococcus aureus* and characterization of the DNA-binding properties of IcaR. *Mol. Microbiol.* 48, 889–899, 2003.

322. Kozitskaya, S., Cho, S.H., Dietrich, K., Marre, R., Naber, K., and Ziebuhr, W., The bacterial insertion sequence element IS256 occurs preferentially in nosocomial *Staphylococcus epidermidis* isolates: association with biofilm formation and resistance to aminoglycosides. *Infect. Immun.* 72, 1210–1215, 2004.

323. Arciola, C.R., Campoccia, D., Gamberini, S., et al., Search for the insertion element IS256 within the *ica* locus of *Staphylococcus epidermidis* clinical isolates collected from biomaterial-associated infections. *Biomaterials* 25, 4117–4125, 2004.

324. Cho, S.H., Naber, K., Hacker, J., and Ziebuhr, W., Detection of the *icaADBC* gene cluster and biofilm formation in *Staphylococcus epidermidis* isolates from catheter-related urinary tract infections. *Int. J. Antimicrob. Agents* 19, 570–575, 2002.

325. Rohde, H., Knobloch, J.K., Horstkotte, M.A., and Mack, D., Correlation of biofilm expression types of *Staphylococcus epidermidis* with polysaccharide intercellular adhesin synthesis: evidence for involvement of *icaADBC* genotype-independent factors. *Med. Microbiol. Immunol. (Berl.)* 190, 105–112, 2001.

326. Handke, L.D., Conlon, K.M., Slater, S.R., et al., Genetic and phenotypic analysis of biofilm phenotypic variation in multiple *Staphylococcus epidermidis* isolates. *J. Med. Microbiol.* 53, 367–374, 2004.

327. Dunne, W.M. Jr., Bacterial adhesion: seen any good biofilms lately? *Clin. Microbiol. Rev.* 15, 155–166, 2002.

328. Donlan, R.M., Role of biofilms in antimicrobial resistance. *ASAIO J.* 46, S47–S52, 2000.

329. Donlan, R.M., Biofilm formation: a clinically relevant microbiological process. *Clin. Infect. Dis.* 33, 1387–1392, 2001.

330. Farber, B.F., Kaplan, M.H., and Clogston, A.G., *Staphylococcus epidermidis* extracted slime inhibits the antimicrobial action of glycopeptide antibiotics. *J. Infect. Dis.* 161, 37–40, 1990.

331. Dunne, W.M. Jr., Mason, E.O. Jr., and Kaplan, S.L., Diffusion of rifampin and vancomycin through a *Staphylococcus epidermidis* biofilm. *Antimicrob. Agents Chemother.* 37, 2522–2526, 1993.

332. Zheng, Z., and Stewart, P.S., Penetration of Rifampin through *Staphylococcus epidermidis* Biofilms. *Antimicrob. Agents Chemother.* 46:900–903, 2002.

333. Zhang, X., Bishop, P.L., and Kupferle, M.J., Measurements of polysaccharides and proteins in biofilm extracellular polymers. *Water Sci. Technol.* 37, 345–348, 1998.

334. Eng, R.H., Padberg, F.T., Smith, S.M., Tan, E.N., and Cherubin, C.E., Bactericidal effects of antibiotics on slowly growing and nongrowing bacteria. *Antimicrob. Agents Chemother.* 35, 1824–1828, 1991.

335. Anwar, H., Strap, J.L., and Costerton, J.W., Kinetic interaction of biofilm cells of *Staphylococcus aureus* with cephalexin and tobramycin in a chemostat system. *Antimicrob. Agents Chemother.* 36, 890–893, 1992.

336. Lewis, K., Riddle of biofilm resistance. *Antimicrob. Agents Chemother.* 45, 999–1007, 2001.

337. Lewis, K., Programmed death in bacteria. *Microbiol. Mol. Biol. Rev.* 64, 503–514, 2000.

338. Stewart, P.S., and Costerton, J.W., Antibiotic resistance of bacteria in biofilms. *Lancet* 358, 135–138, 2001.

339. Knobloch J.K.M., von Osten, H., Rohde, H., Horstkotte, M.A., and Mack, D., Minimal attachment killing concentration (MAK): a versatile method for susceptibility testing of attached biofilm-positive and -negative *Staphylococcus epidermidis*. *Med. Microbiol. Immunol. (Berl.)* 191, 107–114, 2002.

340. König, C., Schwank, S., and Blaser, J., Factors compromising antibiotic activity against biofilms of *Staphylococcus epidermidis*. *Eur. J. Clin. Microbiol. Infect. Dis.* 20, 20–26, 2001.

341. Schierholz, J.M., Beuth, J., König, D., Nurnberger, A., and Pulverer, G., Antimicrobial substances and effects on sessile bacteria. *Zentralbl. Bakteriol.* 289, 165–177, 1999.

342. Zimmerli, W., Widmer, A.F., Blatter, M., Frei, R., and Ochsner, P.E., Role of rifampin for treatment of orthopedic implant-related staphylococcal infections: a randomized controlled trial. Foreign-Body Infection (FBI) Study Group. *JAMA* 279, 1537–1541, 1998.

343. Stein, A., Drancourt, M., and Raoult, D., *Ambulatory management of infected orthopedic implants.* In: Waldvogel, F.A., and Bisno, A.L., eds. Infections Associated with Indwelling Medical Devices, 3rd Ed. Washington D.C.: American Society for Microbiology, 2000:211–230.

344. Espersen, F., Frimodt-Moller, N., Corneliussen, L., Thamdrup, R.V., and Skinhoj, P., Experimental foreign body infection in mice. *J. Antimicrob. Chemother.* 31 Suppl D, 103–111, 1993.

345. Lucet, J.C., Herrmann, M., Rohner, P., Auckenthaler, R., Waldvogel, F.A., and Lew, D.P., Treatment of experimental foreign body infection caused by methicillin-resistant *Staphylococcus aureus*. *Antimicrob. Agents Chemother.* 34, 2312–2317, 1990.

346. Schaad, H.J., Chuard, C., Vaudaux, P., Waldvogel, F.A., and Lew, D.P., Teicoplanin alone or combined with rifampin compared with vancomycin for prophylaxis and treatment of experimental foreign body infection by methicillin-resistant *Staphylococcus aureus. Antimicrob. Agents Chemother.* 38, 1703–1710, 1994.

347. Schaad, H.J., Chuard, C., Vaudaux, P., Rohner, P., Waldvogel, F.A., and Lew, D.P., Comparative efficacies of imipenem, oxacillin and vancomycin for therapy of chronic foreign body infection due to methicillin-susceptible and -resistant *Staphylococcus aureus. J. Antimicrob. Chemother.* 33, 1191–1200, 1994.

348. Cagni, A., Chuard, C., Vaudaux, P.E., Schrenzel, J., and Lew, D.P., Comparison of sparfloxacin, temafloxacin, and ciprofloxacin for prophylaxis and treatment of experimental foreign-body infection by methicillin-resistant *Staphylococcus aureus. Antimicrob. Agents Chemother.* 39, 1655–1660, 1995.

349. Yasuda, H., Ajiki, Y., Koga, T., and Yokota, T., Interaction between clarithromycin and biofilms formed by *Staphylococcus epidermidis. Antimicrob. Agents Chemother.* 38, 138–141, 1994.

350. Berthaud, N., and Desnottes, J.F., In-vitro bactericidal activity of quinupristin/dalfopristin against adherent *Staphylococcus aureus. J. Antimicrob. Chemother.* 39 (Suppl A), 99–102, 1997.

351. Hamilton-Miller, J.M., and Shah, S., Activity of quinupristin/dalfopristin against *Staphylococcus epidermidis* in biofilms: a comparison with ciprofloxacin. *J. Antimicrob. Chemother.* 39 Suppl A, 103–108, 1997.

352. Duguid, I.G., Evans, E., Brown, M.R., and Gilbert, P., Growth-rate-independent killing by ciprofloxacin of biofilm-derived *Staphylococcus epidermidis*: evidence for cell-cycle dependency. *J. Antimicrob. Chemother.* 30, 791–802, 1992.

353. Schwank, S., Rajacic, Z., Zimmerli, W., and Blaser, J., Impact of bacterial biofilm formation on *in vitro* and *in vivo* activities of antibiotics. *Antimicrob. Agents Chemother.* 42, 895–898, 1998.

354. Amorena, B., Gracia, E., Monzon, M., et al., Antibiotic susceptibility assay for *Staphylococcus aureus* in biofilms developed *in vivo. J. Antimicrob. Chemother.* 44, 43–55, 1999.

355. Zelver, N., Hamilton, M., Pitts, B., et al., Measuring antimicrobial effects on biofilm bacteria: from laboratory to field. *Methods Enzymol.* 310, 608–628, 1999.

356. Ceri, H., Olson, M., Morck, D., et al., The MBEC Assay System: multiple equivalent biofilms for antibiotic and biocide susceptibility testing. *Methods Enzymol.* 337, 377–385.

357. Spoering, A.L., and Lewis, K., Biofilms and planktonic cells of *Pseudomonas aeruginosa* have similar resistance to killing by antimicrobials. *J. Bacteriol.* 183, 6746–6751, 2001.

358. Vaudaux, P.E., Zulian, G., Huggler, E., and Waldvogel, F.A., Attachment of *Staphylococcus aureus* to polymethylmethacrylate increases its resistance to phagocytosis in foreign body infection. *Infect. Immun.* 50, 472–477, 1985.

359. Ceri, H., Olson, M.E., Stremick, C., Read, R.R., Morck, D., and Buret, A., The Calgary Biofilm Device: new technology for rapid determination of antibiotic susceptibilities of bacterial biofilms. *J. Clin. Microbiol.* 37, 1771–1776, 1999.

8 Pseudomonas aeruginosa Biofilm Infections in Cystic Fibrosis

Andrea Smiley and Daniel J. Hassett

CONTENTS

8.1 INTRODUCTION

Cystic Fibrosis (CF) is one of the most common genetic disorder in humans, with patients typically reaching an average life expectancy of ~31–34 years. The most common reason for their demise is, in part, due to airway infection by the opportunistic pathogenic bacterium, *Pseudomonas aeruginosa*. This organism has the propensity to form hypoxic or even anaerobic biofilms within the thick mucus layer lining the lung epithelia of CF patients. Although numerous therapeutic, dietary and drug regimens have helped to prolong the lives of these individuals, they can only hope to suppress the infection, admittedly a far cry from the ultimate goal of actually clearing the infection. This mini-review will establish a basic understanding of CF in the context of how *P. aeruginosa* forms and established biofilms in the thick mucus layer of the airways. We will address the unique characteristics of *P. aeruginosa* and the type of biofilm infection it causes. Biofilm formation and fundamental bacterial physiological responses will be explored under both aerobic and anaerobic conditions. The link between CF, *P. aeruginosa*, and biofilms will be explained, as well as the activation cascade to biofilm formation involving the quorum sensing system. Lastly, current therapies will be discussed and new avenues for novel therapies will be explored.

8.2   CYSTIC FIBROSIS

Cystic fibrosis (CF) is the most common autosomal recessive disorder in Caucasians, with a frequency of 1 in 2,500 live births (1). Found on chromosome 7, the mutation(s) in question lie in the gene encoding the CF transmembrane conductance regulator (CFTR), a chloride channel, which, in CF, is either partially or completely dysfunctional (2). While those with normal lung function can transport chloride (Cl^-), sodium (Na^+), and water from the basolateral to the apical surfaces of secretory epithelia, those with a mutated CFTR have little or no detectable CFTR-mediated Cl^- transport. This is due to either a lack of CFTR, mutated CFTR, or a truncated CFTR (2). This disrupts the cell's salt/water balance, resulting in production of thick mucus. The thick mucus clogs bronchial tubes and plugs the exits of the pancreas and intestines, which leads to a progressive loss of function of these organs. The highly inspissated airway mucus is difficult to clear, in part, because the cilia are matted down and beat erratically. Impaired ciliary beating significantly slows the velocity of mucociliary transport. Ultimately, the static, poorly cleared mucus becomes a haven for different infectious bacteria, the major "players" being *P. aeruginosa, Staphylococcus aureus, Haemophilus influenzae,* and *Burkholderia cepacia* (3,4).

Another consequence that contributes to the pathology of CF is the influx of proinflammatory cells into the lung, the primary contributors being neutrophils. This contributes to the severe inflammation and lung deterioration that is commonly observed in the lungs of CF patients. Airway inflammation is already present in infants with CF who are as young as four weeks of age (5). This apparent inflammatory dysregulation is from elevated levels of proinflammatory cytokines that have been found in bronchial lavage fluid sampled from CF patients. However, notably negligible amounts of anti-inflammatory interleukin-10 (IL-10) were also found (6). IL-10 is a regulator of the inflammatory response to endobronchial infectious agents, especially *P. aeruginosa.*

As CF lung disease progresses, the mucus layer of epithelial cells becomes increasingly hypoxic and even anaerobic, due to (i) respiring bacteria residing in the mucus, (ii) neutrophils recruited into the lungs by macrophages secreting IL-8 (a neutrophil chemoattractant), and CF epithelial cells that deplete the oxygen at rates 2-3-fold greater than normal epithelia (7). The chronic progressive infections in the lungs are the major contributors to the morbidity and mortality of CF patients. Thus, again, it is not surprising that their life expectancy is only 31–34 years of age.

8.3   *Pseudomonas aeruginosa*

The major infectious organism that causes the hallmark chronic infections in CF that leads to high morbidity and mortality in CF patients is *P. aeruginosa* (9). A Gramnegative bacillus, it is ubiquitous in the environment. Not surprisingly, it can adapt to a myriad of different niches, which is inherently due to its ability to utilize a wide range of organic molecules as a sole carbon source. *P. aeruginosa* is an opportunistic pathogen most likely found complicating cancer patients, those with severe burns, and in the lungs of CF patients, with the airways of most CF patients being colonized before their 10th birthday. In the context of CF, *P. aeruginosa* is inherently highly refractory to multiple antibiotics, especially when embedded within the thick airway mucus. The organism also produces a wide range of virulence determinants that elicit tissue damage and wreak havoc on the host's immune system. These can be potent proteins or redox-compounds that enter and kill host cells or enzymes that disrupt cell membranes and connective tissue. One redox-active phenazine is called pyocyanin, from *pyocyaneous,* meaning *blue pus* (10). When starved for phosphate and in the presence of ample iron (11), *P. aeruginosa* makes copious amounts of pyocyanin, a redox-active tricyclic compound that kills competing microbes (12) and mammalian cells (13,14) by oxidative damage. The oxygen reduction products include both superoxide (O_2^-) and hydrogen peroxide (H_2O_2), mediated by pyocyanin redox-cycling inactivate human vacuolar ATPase, which may have a negative impact on lung function (10). The cross-phylum susceptibility to pyocyanin suggests that the target for this toxin is evolutionarily conserved (10,13,15).

P. aeruginosa is also known to have the potential for invading host epithelia in a CFTR-dependent fashion (16). Without CFTR, there is a distinct possibility that infection will ensue, usually in two to three stages: (a) attachment and colonization, (b) local invasion, and even (c) a disseminated systemic disease. However, in the context of CF, *P. aeruginosa* bacteremia is rare, as organisms are typically localized only to the airways. With the recent publication of the entire 5.2 Mb *P. aeruginosa* genome to the World Wide Web (www.pseudomonas.com), research in the area defining unknown virulence factors has burgeoned.

8.4 BIOFILMS

Bacteria existing in biofilms are highly organized microbial communities, embedded within a carefully formulated matrix that is specifically designed for long-term survival (17). Water is the predominant constituent in these hydrophilic communities, with only 10–20% of the biofilm being actual bacteria (18). The remainder is a polysaccharide-laden (predominantly glucose (19)) matrix, perforated by tiny water channels in the form of what might be considered a primitive circulatory system. In most biofilms there exist oxygen gradients, with higher concentrations of oxygen expectedly being near the outer edges, especially in flow-type biofilms, and lower concentrations near the base, where organisms can agressively attach to various types of substrata (glass, aluminum, stainless steel, cells). Besides an oxygen gradient, there are various gradients of other factors (nutrients, redox, signaling molecules, etc.) due to limited diffusion within the biofilm matrix. Such gradients and the amount of available nutrients dramatically influence the growth rate of organisms within the biofilm, rates that are often significantly slower than those enjoyed by planktonically grown organisms.

When discussing *P. aeruginosa* biofilms in detail, we wish to focus on two well-known examples. The first type, what we term the *classical* biofilm, involves direct adhesion of organisms to surfaces. The second allows for biofilm formation within a thick matrix without the need of direct surface contact (e.g., CF airway mucus) (17). Bacteria biofilms can be found on many surfaces involved in human infection as well. These include teeth, bone, organs, valves, skin, prostheses, and skin. The metabolic activity of some biofilm-related organisms can corrode the underlying surface of metals (20), teeth (21) and blocked narrow tubes (e.g., catheters) (18). Older biofilms tend to be mixed populations of multiple species of bacteria, often referred to as a consortium. These consortia coordinate activities predominantly through the use of quorum sensing systems that will be discussed later in this chapter.

Biofilm formation in the classical, non-CF airway disease biofilm sense is thought to have five basic phases of development that are shown diagrammatically in Figure. 8.1. In the first phase, the bacterium is capable of swimming freely and is commonly referred to as the planktonic mode. The second phase is attachment of the organism to the substratum. This step is mediated by flagella or type IV pili *in vitro* (22) and indirectly mediated by Crc, which controls type IV pilus production, specifically via transcription of *pilA* (23). Phase three depicts the formation of the matrix after cell division and exopolysaccharide biosynthesis. By the fourth phase, the microcolonies have matured into a thicker biofilm, the structures within which are now termed macrocolonies. It is within the macrocolonies that the aforementioned oxygen gradients can be established (24,25). Anaerobic respiration by bacteria such as *P. aeruginosa* can occur in oxygen-depleted pockets if alternative electron acceptors such as NO_3^- or NO_2^- are present. Studies with microelectrodes have shown that the oxygen is depleted within 30 μm of the biofilm surface (24). The fifth, and what the authors cautiously view as the final phase, is a detachment event that is predicted to occur when oxygen (or other attractant gradients) is sensed, a flagellum is formed and the bacterium migrates via swimming motility toward the attractant or away from a repellant. However, very little is known of the genes involved in biofilm detachment (26,27), yet we predict that most are likely various chemotaxis transducers that sense gradients of attractants/repellants whose genes are peppered throughout the *P. aeruginosa* genome. Again, the surface-attached model is restricted to biofilms that are formed on surfaces (17).

To study biofilms *in vitro*, many laboratories around the world have made use of the commonly used strain PAO1, whose genome has been sequenced

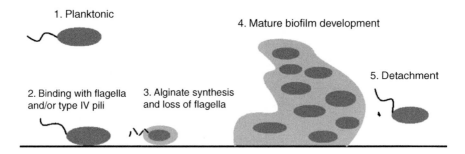

FIGURE 8.1   The five phases of biofilm formation on abiotic surfaces.

(www.pseudomonas.com). The gene products that are necessary to produce biofilms include polyphosphate kinase (28), the *las* quorum sensing circuit (29), the catabolite repressor control protein (23), flagella and type IV pili (22), the global regulator GacA (30), and other unknown factors independent of flagella and twitching motility.

P. aeruginosa also secretes an exopolysaccharide known as alginate, a repeating polymer of mannuronic and glucuronic acid, in response to environmental changes and mutations (for review, see (9)). Its purpose is thought to be in resistance to antibiotics (31), scavenging of oxygen free radicals (32), resistance to macrophage engulfment (33), and for adhesion (34). Many envisioned that attachment would trigger transcription of the alginate biosynthetic genes, similar to those that are known to be upregulated in CF airway mucus (35). However, Wozniak et al. (19) demonstrated that alginate is not a major component of non-mucoid *P. aeruginosa* biofilms. To potentially aid in the detachment process, an enzyme known as alginate lyase, may cleave the sugars into short oligosaccharides that negate anchoring of bacteria (36,37).

8.5   CYSTIC FIBROSIS, *P. aeruginosa* AND BIOFILMS

When grown anaerobically, biofilms of *P. aeruginosa* contain 1.8-fold more live cells than dead ones (38). The gene products that are important for aerobic biofilm formation also are important in anaerobic biofilm formation. Most importantly, type IV pili and the single polar flagellum, by way of the *pilA* and *fliC* genes, are required for optimal biofilms under both conditions. Mutant analyses have shown that the *rhl* quorum sensing circuit (discussed below) is essential for viability of anaerobic *P. aeruginosa* in biofilms. For example, when a *rhlR* mutant was grown as anaerobic biofilms, which are as robust as those of wild-type bacteria, the organisms perished via a metabolic suicide through a by-product of normal anaerobic respiration (38). In fact, the molecular basis for this toxicity was via a dramatic overproduction of respiratory nitric oxide (NO).

Biofilms that are formed in the thick CF airway mucus can grow anaerobically, due to the steep hypoxic gradients that are established within the mucus (39). Generally, the biofilms formed by *P. aeruginosa* in the CF mucus represent ovoid masses or *bacterial rafts*.

The CF mucus is thicker in part, due to the DNA from lysed neutrophils, a decreased water content, and the viscous alginate exopolysaccharide. The mucus is more anaerobic because CF epithelia consume 2-3-fold more oxygen than normal epithelia, the aerobic metabolism of *P. aeruginosa*, and the influx of neutrophils mediated through macrophage IL-8.

Once *P. aeruginosa* infects the mucus layer, it uses its flagella, type IV pili, or mucus turbulence, to finally arrive at the hypoxic zones with an optimum redox status that the organism prefers to grow as biofilms within the mucus. Once in the oxygen-depleted environment, the organism undergoes a rapid transition (typically 7 mins, based upon *in vitro* experiments) from aerobic to anaerobic respiration, especially given that the alternative electron acceptors such as NO_3^- or NO_2^- are amply present in the CF mucus. The subsequent macrocolonies that form are resistant to many defenses, including neutrophils and first tier anti-*Pseudomonas* antibiotics such as ceftazidime, ticarcillin, tobramycin, and ciprofloxacin. Thus, a true chronic infection is established.

During mucoid conversion by *P. aeruginosa*, alginate is overproduced due to mutations in *mucA*, encoding an anti-sigma factor that binds to and, as such, controls AlgT(U). AlgT(U) is a sigma factor that is required for transcriptional activation of genes involved in alginate biosynthesis and gene regulation. Mutants lacking, or those with, MucA have increased expression of a disulfide bond isomerase homolog, and reduced type IV pilus mediated-twitching motility. The reduced expression is co-regulated with the mucoid conversion process and not required for alginate production (40). Mucoid conversion in CF is thought to be an extreme stress response system. This system includes AlgT(U), another sigma factor, and MucABCD, which help to regulate AlgT(U) activity (41). Mutations in this system that activate AlgT(U) aid in the conversion to mucoidy. Loss of MucA, the negative regulator that binds AlgT(U), allows AlgT(U) to freely direct transcription of the alginate biosynthetic and regulatory genes, leading to the classical mucoid colony morphology (42). A second mutation is AlgT(U)-independent, where alginate production and transcription of the *algD* promoter is dependent upon RpoN, an alternative sigma factor that competes with AlgT(U) for binding to the promoter regions upstream of *algD* (42).

In order for *P. aeruginosa* to form robust anaerobic biofilms and survive within such biofilms, the organism requires Rhl quorum sensing system and nitric oxide (NO) reductase. NO reductase is present to modulate or prevent the accumulation of the toxic NO, which is a byproduct of anaerobic respiration (38). *P. aeruginosa* can use NO_3^-, NO_2^-, or N_2O as an alternative electron acceptor for denitrification or arginine for substrate level phosphorylation. In CF sputa, NO_3^- and NO_2^- levels are 383 ± 42 and 125 ± 55 μM respectively, which is more than adequate to support anaerobic respiration (43). *P. aeruginosa* is able to activate anaerobic metabolism, in part, due to the presence of nitrate reductase (NAR) and carbamate kinase. NAR converts NO_3^- to NO_2^-. Carbamate kinase is essential for proper functioning of the arginine deiminase pathway (43).

The fact that NO_2^- and NO_3^- are elevated in sputum of CF patients would suggest that nitric oxide synthase (specifically iNOS2) is active during acute pulmonary exacerbations (44); iNOS2 produces NO in normal and CF airway epithelia. However, it is well established that NO levels per se are lower in CF patients due to a CFTR-dependent reduction in iNOS activity (45–48). The neutrophils that invade the airway to combat *P. aeruginosa* make high concentrations of superoxide (O_2^-) and hydrogen peroxide (H_2O_2). O_2^- and NO combine to make peroxynitrite (HNOO-), which then oxidizes to NO_2^- and finally to NO_3^-. NO_3^- inhibits the conversion of mucoid, alginate producing bacteria to the non-mucoid form (49). So, once the strain is mucoid, with NO_3^- present (especially anaerobically), it cannot revert to the non-mucoid form, is more resistant to phagocytosis, and it is less susceptible to antibiotics. Collectively, these events contribute to the overall clinical demise of CF patients.

As discussed above, we indicated that exhaled NO is low in CF patients, even prior to infection. NO is synthesized by three different forms of NOS: neuronal (n), endothelial (e), and inducible (i). In normal cells, there is an increase in iNOS expression during infection. However, CF cells have reduced levels of NO. Reduced iNOS concentrations are associated with neutrophil sequestration in the lungs. This increases the potential for damage by neutrophil proteases and reactive oxygen species. However, increased NO concentrations may augment the inflammatory process found in acute lung injury from sepsis (50). Although iNOS activity is not the sole source of NO, its reduction in CF clearly plays a significant role in the pathophysiology of the disease. NO is an antimicrobial agent, so reduced amounts would predictably allow for susceptible bacteria to persist and contribute to chronic infections. Furthermore, *P. aeruginosa* has reduced adherence and survival to cells that express iNOS (51). Thus, reduced expression of iNOS in CF epithelia contributes to CF patients' increased susceptibility to infection (48). Loss of NO could also compromise smooth muscle relaxation in the airways and thus contribute to bronchial obstruction (52).

8.6   QUORUM SENSING

The primary quorum sensing (QS) systems of *P. aeruginosa*, *las* and *rhl*, are thought to significantly contribute to the pathogenesis of CF lung infections. QS has been shown to activate expression of multiple virulence factors and differentiation, architecture, and SDS resistance of biofilms. The genes act in tandems; *lasR–lasI* and *rhlR–rhlI*. The transcription of these genes is maximized in the early stationary phase of growth, where cell densities are very high ($\sim 10^8$–10^9 CFU/ml). LasR is a transcriptional activator and LasI catalytically produces the *P. aeruginosa* auto-inducer 3-oxododecanoyl-L-homoserine lactone (PAI-1). PAI-1 combines with LasR to form an active transcriptional complex. However, it remains a mystery as to how LasR-PAI-1 interact with core RNA polymerase. RhlR is also a transcriptional activator while RhlI produces the second auto-inducer N-butyl-L-homoserine lactone (PAI-2). LasR controls *rhlR-rhlI* gene expression, so the *rhl* system is more or less *on hold* until the *las* system is activated. A general skeletal view of most of the QS activation cascade is depicted in Figure 8.2. Due to space constraints, and with all due respect to the hundreds of scientists world-wide that have contributed to this cascade, if players are omitted, it is for this reason. Activation of the *las* system, most specifically the *lasR* gene, requires Vfr, a cAMP receptor protein (CRP) homolog (53). Vfr binds to the CRP-binding consensus sequence to activate transcription at the T1 promoter. As a result, both the *lasR* and downstream *lasI* genes are transcribed. The *lasR* gene was also shown to be under autoregulatory control at palindromic sequences upstream of the *lasR* gene, known as the *lux box* (53). The *las* system proceeds to activate transcription of other genes, such as *lasA, apr, toxA,* hemolysin, *katA, sodA, xpcR, xpcP, lasB, lasI,* and *rhlR*. To initiate *rhlR* transcription, LasR-PAI-1 activates it at the *rhlR* promoter, which will transcribe *rhlR* and *rhlI*. RhlI synthesizes PAI-2, which will combine with RhlR to start transcription of *rhlI, rhlA, rhlB,* chitinase, lipase, pyocyanin, cyanide, *katA,* and *rpoS*. RpoS participates in the activation of alginate gene transcription and biofilm formation (54). Active LasR can inhibit this process by blocking the binding of PAI-2 to RhlR. Other factors and a description of them are described in the legend to Figure 8.2.

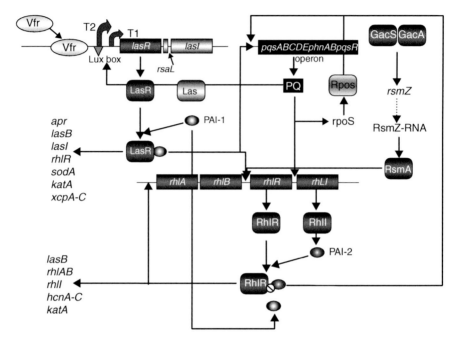

FIGURE 8.2 Abbreviated version of the major players in the quorum sensing regulatory hierarchy in *P. aeruginosa*. First, the master regulator, Vfr, a cyclic AMP receptor-like protein, controls expression of the *lasR* gene at the *lux box* region with two different transcriptional start sites, T1 and T2 (53). Another player in the regulation of the *lasR* gene is the *Pseudomonas* quinolone signal, PQS. PQS is 2-heptyl-3-hydroxy-4-quinolone, and is generated via the collective gene products derived from the *pqsABCDEphnABpqsR* genes (68). This operon requires the *las* quorum sensing circuit, while it is repressed by the *rhl* system (68). The *rsaL* gene encodes an 11 kDa protein that negatively regulates the *lasI* gene (69). The LasR-PAI-1 tandem regulates positive and negatively hundreds of genes, most of which were discovered by two major groups using GeneChip microarrays (70,71). Some classical examples of positively controlled *las* and *rhl* genes are listed. The *las* system regulates the *rhl* system in a positive fashion. However, PAI-1 can dissociate the PAI-2-RhlR dimeric complex, thereby restricting *rhl*-mediated quorum sensing (72). The GacS/GacA rsmZ signal transduction system, in turn, also leads to production of RsmZ regulatory RNA that negatively regulates RsmA. RsmA levels have been shown to be significantly modulated during acute versus chronic infections by *P. aeruginosa* (73). High levels of RsmA cause repression of *rhl* quorum sensing. The RhlR-PAI-2 tandem as well as the stationary phase sigman factor, RpoS, also negatively regulate the *pqsABCDEphnABpqsR* genes.

Transcription of quorum sensing genes also aids in protection of *P. aeruginosa* against the oxygen products, O_2^- and H_2O_2. Mentioned above in one cascade are the genes encoding two superoxide dismutases (Fe-SOD and Mn-SOD) and two haem-containing catalases (KatA and KatB). The Fe-SOD enzyme is active if iron is plentiful, while the Mn-SOD is only present when iron becomes limiting, and in a quorum sensing-dependent fashion (55). KatA is constitutively produced and is elevated when the cell is grown in the presence of high iron, and KatB is only generated when the organisms are exposed to H_2O_2 in an OxyR-dependent fashion (56). Since SOD and catalase are important for resistance to O_2^- and H_2O_2, and genes encoding these important enzymes are controlled by QS, targeting the QS machinery, whether at the auto-inducer, DNA, and/or LasR/RhlR level, indicates that *P. aeruginosa* QS circuitry may make a plausible drug target. QS is also important for resistance of *P. aeruginosa* biofilms to H_2O_2, obviously in a catalase-dependent fashion. In normal epithelia, stimulated inflammatory cells release large amounts of H_2O_2. In CF lung disease, the median level of H_2O_2 is lower in CF patients (57). There is no correlation, however, between H_2O_2 levels and lung function. Biofilm producing infections generally can resist nonopsonic phagocytosis by human neutrophils or macrophages compared to non-mucoid strains. The different surface characteristics render them unrecognizable to this host defense.

Throughout disease progression, mucoid bacteria take on different population characteristics. For example, isolates from the early stages of *P. aeruginosa* colonization are motile, while isolates from late-stage chronic CF patients are non-motile (58). This can be explained by what we have learned; the longer the organism persists in the patient, the more likely it will exist as the mucoid form. *P. aeruginosa* responds to different signals in the mucopurulent airway liquid in CF patients. Quorum sensing is activated and flagellum production is inhibited. The FliC shut-off is rapid and independent of quorum sensing and the known regulatory networks controlling flagellum expression. Since flagellin is immunogenic, specifically inducing a dramatic induction of IL-8, organisms not expressing flagella, such as mucoid organisms (49), would be limited in their ability to activate the host's immune response. Even during the early stages of CF lung disease, the flagellum of *P. aeruginosa* is detected during the early stages of colonization because it is needed for attachment to the epithelial cell's mucins (59). Thus, it is clear that there is little debate that *P. aeruginosa* undergoes both phenotypic and genotypic changes during the course of CF. *P. aeruginosa* has a significantly greater binding affinity to CF epithelia than their normal counterparts (60). Not only are CF cells colonized more efficiently, but the infection had initially been reported to be eradicated because it was thought that the CF airway surface fluid is high in salt, which has been reported to interfere with bactericidal activity (61). Furthermore, biofilm-grown bacteria isolated from adult CF patients show decreased susceptibility to antibiotic combinations than do adherent and planktonically grown bacteria (62). This clearly indicates that there is more to learn about the colonization of CF patients with *P. aeruginosa* in order to help overcome its defenses.

8.7 THERAPIES AND POTENTIAL VACCINES

It has been shown that a chronic infection with transmissible *P. aeruginosa* strains in CF patients raises the issue of cross infection and patient segregation. Such an issue was recently raised and explored with the Liverpool epidemic strain (63). Patients with such strains have a significantly worse prognosis than those with a unique strain and a greater annual loss of lung function (based upon poor forced expiratory volume measurements). Furthermore, the nutritional state of the patient deteriorates to the point of malnutrition. It has been shown therefore that strain identification and segregation has become essential in the treatment of CF chronic infections (63).

There are a few prognosticators for severity of CF pulmonary infection. The sputum amino acid content of a patient with CF is high during infective exacerbations and it correlates with the severity of the disease (64). Also, adult CF patients show decreasing responses to antibiotics, depending on the age and chronicity of infection. The primary antibiotics used during CF airway infection include tobramycin, ticarcillin, ceftazidine, and certain fluoroquinones. The overall efficacy of these antibiotics is dramatically impaired under anaerobic conditions, those in which *P. aeruginosa* biofilms thrive within the thick airway surface mucus in the lungs of the CF patients. Other important components of CF treatment, besides antibiotics, include anti-inflammatory drugs, bronchiodialators, and chest physical therapy. With increasingly futility for treatment, *P. aeruginosa* infections become impossible to eradicate, but growth of the organisms can still be suppressed (65). Strategic combinations of a β-lactam and an aminoglycoside cause a longer clinical remission than a β-lactam alone and a slightly better initial improvement of the patient's clinical status (66). Gene therapy is being explored but there are some obviously frustrating difficulties in trying to target the appropriate cells and toxicity problems with various viral vectors.

Some have explored the possibility of a *P. aeruginosa* vaccine. However, in CF, there is ample antibody generated against multiple *P. aeruginosa* antigens, yet there is no effect at eradicating *P. aeruginosa* from the airways. There is an octavalent *P. aeruginosa* *o*-polysaccharide-toxin A conjugate vaccine (67). This vaccine can induce anti-LPS antibodies that are high affinity and promote opsonophagocytic killing by human peripheral lymphocytes in noncolonized CF patients. However, if it is applied post-infection, the antibodies are low affinity and nonopsonic in nature. There is some apparent efficacy though, after a six year follow up, only 35% of those immunized were infected with *P. aeruginosa* compared to 75% of the control group.

Another potential target for a vaccine or drug could be the major outer membrane porin, OprF. OprF has found to be increased 40-fold under anaerobic conditions and in anaerobic CF sputum (38). It is commonly found in CF mucus and CF patients have antibodies to it. It is also interesting to note that a knockout forms poor anaerobic biofilms (38). A recombinant OprF-OprI vaccine has been shown to raise an antibody response in burn patients and intranasal application induces serum IgA.

Other potential areas for drug development could include a means to restrict the ability of the bacteria to denitrify or undergo arginine substrate-level-phosphorylation. This would limit the ability of *P. aeruginosa* to grow and lead to possible eradication. The novel drug targets would be the enzymes involved in these pathways (17). A homoserine lactone analog would interfere with quorum sensing and possibly cause the buildup of NO, which is toxic. The bacteria would be able to release the quorum sensing signal, but they would not be able to interpret it and they would poison themselves.

8.8 CONCLUSIONS

In this chapter, we have explained what is currently understood about the genetic disorder of CF. We have explored the unfortunate and yet remarkably revealing relationship with the common opportunistic pathogen, *P. aeruginosa*, and how it affects CF patients. Biofilms were discussed in both the classical aerobic formation and anaerobic formation more commonly found in CF. The quorum-sensing activation cascade was explained as well as biofilm formation's link to quorum sensing. The treatments of CF and *P. aeruginosa* infections were discussed and the potential for novel drug targets were explored. With the advent of new technologies and laboratory procedures such as proteomics and microarray analysis, we should see some breakthroughs in the understanding of this disease in years to come.

REFERENCES

1. Ratjen, F., and Doring, G., Cystic fibrosis. *Lancet* 361, 681–689, 2003.
2. Boucher, R.C., An overview of the pathogenesis of cystic fibrosis lung disease. *Adv. Drug. Deliv. Rev.* 54, 1359–1371, 2002.
3. Hassett, D.J., Cuppoletti, J., Trapnell, B., Lymar, S.V., Rowe, J.J., Sun Yoon, S., Hilliard, G.M., Parvatiyar, K., Kamani, M.C., Wozniak, D.J., et al., Anaerobic metabolism and quorum sensing by *Pseudomonas aeruginosa* biofilms in chronically infected cystic fibrosis airways: rethinking antibiotic treatment strategies and drug targets. *Adv. Drug. Deliv. Rev.* 54, 1425–1443, 2002.
4. Hassett, D.J., Lymar, S.V., Rowe, J.J., Schurr, M.J., Passador, L., Herr, A.B., Winsor, G.L., Brinkman, F.S.L., Lau, G.W., Yoon, S.S., et al., Anaerobic metabolism by *Pseudomonas aeruginosa* in cystic fibrosis airway biofilms: role of nitric oxide, quorum sensing and alginate production. In: *Strict and Facultative Anaerobes: Medical and Environmental Aspects* 87–108, 2004.
5. Khan, T.Z., Wagener, J.S., Bost, T., Martinez, J., Accurso, F.J., and Riches, D.W., Early pulmonary inflammation in infants with cystic fibrosis. *Am. J. Respir. Crit. Care Med.* 151, 1075–1082, 1995.
6. Chmiel, J.F., Konstan, M.W., Knesebeck, J.E., Hilliard, J.B., Bonfield, T.L., Dawson, D.V., and Berger, M., IL-10 attenuates excessive inflammation in chronic *Pseudomonas* infection in mice. *Am. J. Respir. Crit. Care Med.* 160, 2040–2047, 1999.
7. Stutts, M.J., Knowles, M.R., Gatzy, J.T., and Boucher, R.C., Oxygen consumption and oubain binding sites in cystic fibrosis nasal epithelium. *Ped. Res.* 20, 1316–1320, 1986.
8. Varlotta, L., Management and care of the newly diagnosed patient with cystic fibrosis. *Curr. Opin. Pulm. Med.* 4, 311–318, 1998.
9. Govan, J.R.W., and Deretic, V., Microbial pathogenesis in cystic fibrosis: mucoid *Pseudomonas aeruginosa* and *Burkholderia cepacia*. *Microbiol. Rev.* 60, 539–574, 1996.
10. Ran, H., Hassett, D.J., and Lau, G.W., Human targets of *Pseudomonas aeruginosa* pyocyanin. *Proc. Natl. Acad. Sci. USA* 100, 14315–14320, 2003.
11. Hassett, D.J., Charniga, L., Bean, K.A., Ohman, D.E., and Cohen, M.S., Antioxidant defense mechanisms in *Pseudomonas aeruginosa:* resistance to the redox-active antibiotic pyocyanin and demonstration of a manganese-cofactored superoxide dismutase. *Infect. Immun.* 60, 328–336, 1992.

12. Hassan, H.M., and Fridovich, I., Mechanism of the antibiotic action of pyocyanine. *J. Bacteriol.* 141, 156–163, 1980.

13. Rahme, L.G., Stevens, E.J., Wolfort, S.F., Shao, J., Tompkins, R.G., and Ausubel, F.M., Common virulence factors for bacterial pathogenicity in plants and animals. *Science* 268, 1899–1902, 1995.

14. Lau, G.W., Ran, H., Kong, F., Hassett, D.J., and Mavrodi, D., *Pseudomonas aeruginosa* pyocyanin is critical for lung infection in mice. *Infect. Immun.* 72, 4275–4278, 2004.

15. Rahme, L.G., Stevens, E., Shao, J., Tompkins, R.G., and Ausubel, F.M., *Pseudomonas aeruginosa* strain that is pathogenic in mice and plants. Symposium on Molecular Genetics of Plant-Microbe Interactions, Rutgers, The State University of New Jersey, 1993.

16. Pier, G.B., Grout, M., Zaidi, T.S., Olsen, J.C., Johnson, L.G., Yankaskas, J.R., and Goldberg, J.B., Role of mutant CFTR in hypersusceptibility of cystic fibrosis patients to lung infections. *Science* 271, 64–67, 1996.

17. Hassett, D.J., Limbach, P.A., Hennigan, R.F., Klose, K.E., Hancock, R.E., Platt, M.D., and Hunt, D.F., Bacterial biofilms of importance to medicine and bioterrorism: proteomic techniques to identify novel vaccine components and drug targets. *Expert Opin. Biol. Ther.* 3, 1201–1207, 2003.

18. Hoiby, N., Krogh Johansen, H., Moser, C., Song, Z., Ciofu, O., and Kharazmi, A., *Pseudomonas aeruginosa* and the *in vitro* and *in vivo* biofilm mode of growth. *Microbes Infect.* 3, 23–35, 2001.

19. Wozniak, D.J., Wyckoff, T.J., Starkey, M., Keyser, R., Azadi, P., O'Toole, G.A., and Parsek, M.R., Alginate is not a significant component of the extracellular polysaccharide matrix of PA14 and PAO1 *Pseudomonas aeruginosa* biofilms. *Proc. Natl. Acad. Sci. USA* 100, 7907–7912, 2003.

20. Beech, I.B., and Sunner, J., Biocorrosion: towards understanding interactions between biofilms and metals. *Curr. Opin. Biotechnol.* 15, 181–186, 2004.

21. Tanzer, J.M., Livingston, J., and Thompson, A.M., The microbiology of primary dental caries in humans. *J. Dent. Educ.* 65, 1028–1037, 2001.

22. O'Toole, G.A., and Kolter, R., Flagellar and twitching motility are necessary for *Pseudomonas aeruginosa* biofilm development. *Mol. Microbiol.* 30, 295–304, 1998.

23. O'Toole, G.A., Gibbs, K.A., Hager, P.W., Phibbs, P.V., Jr., and Kolter, R., The global carbon metabolism regulator Crc is a component of a signal transduction pathway required for biofilm development by *Pseudomonas aeruginosa*. *J. Bacteriol.* 182, 425–431, 2000.

24. Xu, K.D., Stewart, P.S., Xia, F., Huang, C.T., and McFeters, G.A., Spatial physiological heterogeneity in *Pseudomonas aeruginosa* biofilm is determined by oxygen availability. *Appl. Environ. Microbiol.* 64, 4035–4039, 1998.

25. Borriello, G., Werner, E., Roe, F., Kim, A.M., Ehrlich, G.D., and Stewart, P.S., Oxygen limitation contributes to antibiotic tolerance of *Pseudomonas aeruginosa* in biofilms. *Antimicrob. Agents Chemother.* 48, 2659–2664, 2004.

26. Hunt, S.M., Werner, E.M., Huang, B., Hamilton, M.A., and Stewart, P.S., Hypothesis for the role of nutrient starvation in biofilm detachment. *Appl. Environ. Microbiol.* 70, 7418–7425, 2004.

27. Wilson, S., Hamilton, M.A., Hamilton, G.C., Schumann, M.R., and Stoodley, P., Statistical quantification of detachment rates and size distributions of cell clumps from wild-type (PAO1) and cell signaling mutant (JP1) *Pseudomonas aeruginosa* biofilms. *Appl. Environ. Microbiol.* 70, 5847–5852, 2004.

28. Rashid, M.H., and Kornberg, A., Inorganic polyphosphate is needed for swimming, swarming, and twitching motilities of *Pseudomonas aeruginosa*. *Proc. Natl. Acad. Sci. USA* 97, 4885–4890, 2000.

29. Davies, D.G., Parsek, M.R., Pearson, J.P., Iglewski, B.H., Costerton, J.W., and Greenberg, E.P., The involvement of cell-to-cell signals in the development of a bacterial biofilm. *Science* 280, 295–298, 1998.

30. Parkins, M.D., Ceri, H., and Storey, D.G., *Pseudomonas aeruginosa* GacA, a factor in multihost virulence, is also essential for biofilm formation. *Mol. Microbiol.* 40, 1215–1226, 2001.

31. Irvin, R.T., Govan, J.W., Fyfe, J.A., and Costerton, J.W., Heterogeneity of antibiotic resistance in mucoid isolates of *Pseudomonas aeruginosa* obtained from cystic fibrosis patients: role of outer membrane proteins. *Antimicrob. Agents Chemother.* 19, 1056–1063, 1981.

32. Simpson, J.A., Smith, S.E., and Dean, R.T., Scavenging by alginate of free radicals released by macrophages. *Free Rad. Biol. Med.* 6, 347–353, 1989.

33. Simpson, J.A., Smith, S.E., and Dean, R.T., Alginate inhibition of the uptake of *Pseudomonas aeruginosa* by macrophages. *J. Gen. Microbiol.* 134, 29–36, 1988.

34. Baker, N.R., and Svanborg-Eden, C., Role of alginate in the adherence of *Pseudomonas aeruginosa*. *Antibiot. Chemother.* 42, 72–79, 1989.

35. Storey, D.G., Ujack, E.E., Mitchell, I., and Rabin, H.R., Positive correlation of algD transcription to lasB and lasA transcription by populations of *Pseudomonas aeruginosa* in the lungs of patients with cystic fibrosis. *Infect. Immun.* 65, 4061–4067, 1997.

36. Boyd, A., and Chakrabarty, A.M., *Pseudomonas aeruginosa* biofilms: role of the alginate exopolysaccharide. *J. Ind. Microbiol.* 15, 162–168, 1995.

37. Boyd, A., and Chakrabarty, A.M., Role of alginate lyase in cell detachment of *Pseudomonas aeruginosa*. *Appl. Environ. Microbiol.* 60, 2355–2359, 1994.

38. Yoon, S.S., Hennigan, R.F., Hilliard, G.M., Ochsner, U.A., Parvatiyar, K., Kamani, M.C., Allen, H.L., DeKievit, T.R., Gardner, P.R., Schwab, U., et al., *Pseudomonas aeruginosa* anaerobic respiration in biofilms: Relationships to cystic fibrosis pathogenesis. *Dev. Cell.* 3, 593–603, 2002.

39. Worlitzsch, D., Tarran, R., Ulrich, M., Schwab, U., Cekici, A., Meyer, K.C., Birrer, P., Bellon, G., Berger, J., Wei, T., et al., Reduced oxygen concentrations in airway mucus contribute to the early and late pathogenesis of *Pseudomonas aeruginosa* cystic fibrosis airway infection. *J. Clin. Invest.* 109, 317–325, 2002.

40. Malhotra, S., Silo-Suh, L.A., Mathee, K., and Ohman, D.E., Proteome analysis of the effect of mucoid conversion on global protein expression in *Pseudomonas aeruginosa* strain PAO1 shows induction of the disulfide bond isomerase, *dsbA*. *J. Bacteriol.* 182, 6999–7006, 2000.

41. Rowen, D.W., and Deretic, V., Membrane-to-cytosol redistribution of ECF sigma factor AlgU and conversion to mucoidy in *Pseudomonas aeruginosa* isolates from cystic fibrosis patients. *Mol. Microbiol.* 36, 314–327, 2000.

42. Schurr, M.J., and Deretic, V., Microbial pathogenesis in cystic fibrosis: co-ordinate regulation of heat-shock response and conversion to mucoidy in *Pseudomonas aeruginosa*. *Mol. Microbiol.* 24, 411–420, 1997.

43. Hassett, D.J., Proteomics of anaerobic *Pseudomonas aeruginosa* in CF and in biofilms. *Pediatr. Pulmonol.* Supp. 22, 149–150, 2001.

44. Linnane, S.J., Keatings, V.M., Costello, C.M., Moynihan, J.B., O'Connor, C.M., Fitzgerald, M.X., and McLoughlin, P., Total sputum nitrate plus nitrite is raised during acute pulmonary infection in cystic fibrosis. *Am. J. Respir. Crit. Care Med.* 158, 207–212, 1998.

45. Thomas, S.R., Kharitonov, S.A., Scott, S.F., Hodson, M.E., and Barnes, P.J., Nasal and exhaled nitric oxide is reduced in adult patients with cystic fibrosis and does not correlate with cystic fibrosis genotype. *Chest* 117, 1085–1089, 2000.

46. Yu, H., Nasr, S.Z., and Deretic, V., Innate lung defenses and compromised *Pseudomonas aeruginosa* clearance in the malnourished mouse model of respiratory infections in cystic fibrosis. *Infect. Immun.* 68, 2142–2147, 2000.

47. Kelley, T.J., and Drumm, M.L., Inducible nitric oxide synthase expression is reduced in cystic fibrosis murine and human airway epithelial cells. *J. Clin. Invest.* 102, 1200–1207, 1998.

48. Meng, Q.H., Springall, D.R., Bishop, A.E., Morgan, K., Evans, T.J., Habib, S., Gruenert, D.C., Gyi, K.M., Hodson, M.E., Yacoub, M.H., et al., Lack of inducible nitric oxide synthase in bronchial epithelium: a possible mechanism of susceptibility to infection in cystic fibrosis. *J. Pathol.* 184, 323–331, 1998.

49. Wyckoff, T.J., Thomas, B., Hassett, D.J., and Wozniak, D.J., Static growth of mucoid *Pseudomonas aeruginosa* selects for non-mucoid variants that have acquired flagellum-dependent motility. *Microbiology* 148, 3423–3430, 2002.

50. Elphick, H.E., Demoncheaux, E.A., Ritson, S., Higenbottam, T.W., and Everard, M.L., Exhaled nitric oxide is reduced in infants with cystic fibrosis. *Thorax* 56, 151–152, 2001.

51. Darling, K.E., and Evans, T.J., Effects of nitric oxide on *Pseudomonas aeruginosa* infection of epithelial cells from a human respiratory cell line derived from a patient with cystic fibrosis. *Infect. Immun.* 71, 2341–2349, 2003.

52. Mhanna, M.J., Ferkol, T., Martin, R.J., Dreshaj, I.A., van Heeckeren, A.M., Kelley, T.J., and Haxhiu, M.A., Nitric oxide deficiency contributes to impairment of airway relaxation in cystic fibrosis mice. *Am. J. Respir. Cell. Mol. Biol.* 24, 621–626, 2001.

53. Albus, A.M., Pesci, E.C., Runyen-Janecky, L., West, S.E.H., and Iglewski, B.H., Vfr controls quorum sensing in *Pseudomonas aeruginosa*. *J. Bacteriol.* 179, 3928–3935, 1997.

54. Suh, S.J., Silo-Suh, L., Woods, D.E., Hassett, D.J., West, S.E., and Ohman, D.E., Effect of *rpoS* mutation on the stress response and expression of virulence factors in *Pseudomonas aeruginosa*. *J. Bacteriol.* 181, 3890–3897, 1999.

55. Hassett, D.J., Ma, J.-F., Elkins, J.G., McDermott, T.R., Ochsner, U.A., West, S.E.H., Huang, C.-T., Fredericks, J., Burnett, S., Stewart, P.S., et al., Quorum sensing in *Pseudomonas aeruginosa* controls expression of catalase and superoxide dismutase genes and mediates biofilm susceptibility to hydrogen peroxide. *Mol. Microbiol.* 34, 1082–1093, 1999.

56. Ochsner, U.A., Vasil, M.L., Alsabbagh, E., Parvatiyar, K., and Hassett, D.J., Role of the *Pseudomonas aeruginosa oxyR-recG* operon in oxidative stress defense and DNA repair: OxyR-dependent regulation of *katB*, *ahpB*, and *ahpCF*. *J. Bacteriol.* 182, 4533–4544, 2000.

57. Ho, L.P., Faccenda, J., Innes, J.A., and Greening, A.P., Expired hydrogen peroxide in breath condensate of cystic fibrosis patients. *Eur. Respir. J.* 13, 103–106, 1999.

58. Mahenthiralingam, E., Campbell, M.E., and Speert, D.P., Nonmotility and phagocytic resistance of *Pseudomonas aeruginosa* isolates from chronically colonized patients with cystic fibrosis. *Infect. Immun.* 62, 596–605, 1994.

59. Lillehoj, E.P., Kim, B.T., and Kim, K.C., Identification of *Pseudomonas aeruginosa* flagellin as an adhesin for Muc1 mucin. *Am. J. Physiol. Lung Cell. Mol. Physiol.* 282:L751–756, 2002.

60. Imundo, L., Barasch, J., Prince, A., and Al-Awqati, Q., Cystic fibrosis epithelial cells have a receptor for pathogenic bacteria on their apical surface. *Proc. Natl. Acad. Sci. USA* 92, 3019–3023, 1995.

61. Smith, J.J., Travis, S.M., Greenberg, E.P., and Welsh, M.J., Cystic fibrosis airway epithelia fail to kill bacteria because of abnormal airway surface fluid. *Cell* 85, 229–236, 1996.
62. Aaron, S.D., Ferris, W., Ramotar, K., Vandemheen, K., Chan, F., and Saginur, R., Single and combination antibiotic susceptibilities of planktonic, adherent, and biofilm-grown *Pseudomonas aeruginosa* isolates cultured from sputa of adults with cystic fibrosis. *J. Clin. Microbiol.* 40, 4172–4179, 2002.
63. Al-Aloul, M., Crawley, J., Winstanley, C., Hart, C.A., Ledson, M.J., and Walshaw, M.J., Increased morbidity associated with chronic infection by an epidemic *Pseudomonas aeruginosa* strain in CF patients. *Thorax* 59, 334–336, 2004.
64. Thomas, S.R., Ray, A., Hodson, M.E., and Pitt, T.L., Increased sputum amino acid concentrations and auxotrophy of *Pseudomonas aeruginosa* in severe cystic fibrosis lung disease. *Thorax* 55, 795–797, 2000.
65. Bagge, N., Schuster, M., Hentzer, M., Ciofu, O., Givskov, M., Greenberg, E.P., and Hoiby, N., *Pseudomonas aeruginosa* biofilms exposed to imipenem exhibit changes in global gene expression and beta-lactamase and alginate production. *Antimicrob. Agents Chemother.* 48, 1175–1187, 2004.
66. Smith, A.L., Doershuk, C., Goldmann, D., Gore, E., Hilman, B., Marks, M., Moss, R., Ramsey, B., Redding, G., Rubio, T., et al., Comparison of a beta-lactam alone versus beta-lactam and an aminoglycoside for pulmonary exacerbation in cystic fibrosis. *J. Pediatr.* 134, 413–421, 1999.
67. Cryz, S.J., Jr., Lang, A., Rudeberg, A., Wedgwood, J., Que, J.U., Furer, E., and Schaad, U., Immunization of cystic fibrosis patients with a *Pseudomonas aeruginosa* O-polysaccharide-toxin A conjugate vaccine. *Behring Inst. Mitt.* 345–349, 1997.
68. McGrath, S., Wade, D.S., and Pesci, E.C., Dueling quorum sensing systems in *Pseudomonas aeruginosa* control the production of the Pseudomonas quinolone signal (PQS). *FEMS Microbiol. Lett.* 230, 27–34, 2004.
69. de Kievit, T., Seed, P.C., Nezezon, J., Passador, L., and Iglewski, B.H., RsaL, a novel repressor of virulence gene expression in *Pseudomonas aeruginosa*. *J. Bacteriol.* 181, 2175–2184, 1999.
70. Wagner, V.E., Bushnell, D., Passador, L., Brooks, A.I., and Iglewski, B.H., Microarray analysis of *Pseudomonas aeruginosa* quorum-sensing regulons: effects of growth phase and environment. *J. Bacteriol.* 185, 2080–2095, 2003.
71. Schuster, M., Lostroh, C.P., Ogi, T., and Greenberg, E.P., Identification, timing, and signal specificity of *Pseudomonas aeruginosa* quorum-controlled genes: a transcriptome analysis. *J. Bacteriol.* 185, 2066–2079, 2003.
72. Ventre, I., Ledgham, F., Prima, V., Lazdunski, A., Foglino, M., and Sturgis, J.N., Dimerization of the quorum sensing regulator RhlR: development of a method using EGFP fluorescence anisotropy. *Mol. Microbiol.* 48, 187–198, 2003.
73. Goodman, A.L., Kulasekara, B., Rietsch, A., Boyd, D., Smith, R.S., and Lory, S., A signaling network reciprocally regulates genes associated with acute infection and chronic persistence in *Pseudomonas aeruginosa*. *Dev. Cell.* 7, 745–754, 2004.

9 Candida

Stephen Hawser and Khalid Islam

CONTENTS

9.1 INTRODUCTION

Biofilms represent the most prevalent type of microbial growth in nature and are crucial to the development of many infections in humans (1–4). Recently it has been documented that some 65% of all human infections involve biofilms (5). Biofilm microbes involved in human disease range from the Gram-positive pathogens *Staphylococcus epidermidis* and *Staphylococcus aureus* to the Gram-negative pathogens including *Pseudomonas aeruginosa, Escherichia coli* amongst others and to different members of the *Candida* genus, particularly *Candida albicans. Candida* is the fourth most common cause of nosocomial infections, a finding that was recently documented by the United States National Nosocomial Infections Surveillance system (2). Moreover, mortality and morbitity due to *Candida* infections are alarmingly high (5). Furthermore, as many as 40% of catheterized patients have an underlying fungemia attributed to the presence of *Candida* biofilms residing on the catheter surface or within the lumen of the catheter (6,7).

While *C. albicans* is still the most commonly isolated fungal species, other species are being isolated with increasing frequency (8,9). For example, *Candida parapsilosis* has now become the second most commonly detected fungal pathogen in human disease. *C. parapsilosis* is a particular danger in critically-ill neonates, where it is clearly associated with parenteral nutrition and central lines (10–12). Candidiasis associated with central lines and prosthetic devices is especially problematic, since such devices act as surface substrates that facilitate attachment and growth of the yeasts in the form of structured biofilms (1,2). Antifungal therapy aimed at removal of the biofilm on such surfaces is inadequate with removal of the infected device by far the most common approach. However, this approach becomes highly problematic when the clinician is required to remove infected heart valves and joint prostheses through additional surgical procedures.

This chapter aims to describe some of the characteristics of *Candida* biofilms and their association with human disease, difficulties encountered in the treatment of in situ *Candida* biofilms with empiric antifungal therapy and new visions towards future treatment options for caregivers.

9.2 *CANDIDA* BIOFILMS AND INFECTIONS

A relatively small number of *Candida* species cause disease in humans. Typically, those that cause human disease do so in an opportunistic fashion, immunocompromised patients or others with various underlying diseases being the most liable to such opportunism. *Candida* is capable of causing human diseases of different types, namely superficial and deep-seated infections. The major causative pathogen of the *Candida* genus is *C. albicans,* well known for its ability to exist in two distinctly different morphological forms. This organism can proliferate either in the budding or yeast form or as septated hyphae. Although much speculation considered only the hyphal form to be of major virulence in infections, both the yeast and hyphal forms are normally detected in infected human tissues (2,5). The emergence of *Candida* spp., currently representing the fourth most common causative pathogen group of nosocomial infections, is closely related to specific risk factors associated with medical procedures. Examples of such risk factors include the use of immunosuppressive and cytotoxic drugs, the suppression of the normal bacterial flora following antibiotic therapies and the use of implanted medical devices. By nature any implanted medical device, including cannulae, venous catheters, central lines, cardiac replacements, endotracheal tubes, prosthetic heart valves, pacemakers or urinary catheters, automatically provides an inert surface to which human proteins and fluids can adhere to. In a similar fashion, *Candida* spp. known to cause human disease are very capable in adhering to and growing as biofilms on such surfaces in situ (13).

A major distinguishing feature of *C. albicans* biofilm is the presence of both yeast and morphological forms (see Figure 9.1) This has been observed both in different *in vitro* model biofilm systems and also from microscopic examinations of the surface of implanted devices in situ. In terms of how *C. albicans* construct their biofilms, clearly the initial step in the process involves the attachment of the yeast form to the surface of the device material which, in turn, is followed by a period of germ tube formation along the device surface 3 to 7 hours post yeast attachment (14). The earliest report in the literature from Hawser and Douglas (14) effectively demonstrated this in that in an *in vitro* model system based upon the use of small discs of catheter material that the fully matured biofilm consisted in a dense network of all *C. albicans* morphological forms including yeast, hyphal, and pseudohyphal forms. It is worthwhile to note that although the initial step in biofilm formation by *Candida* involves yeast attachment followed by germ tube formation, studies with germ-tube minus/hyphal minus mutants revealed that such mutants are far less capable in forming biofilms on device related surfaces (15). The finding suggests that morphogenesis in *C. albicans* biofilms may effectively be triggered via contact with

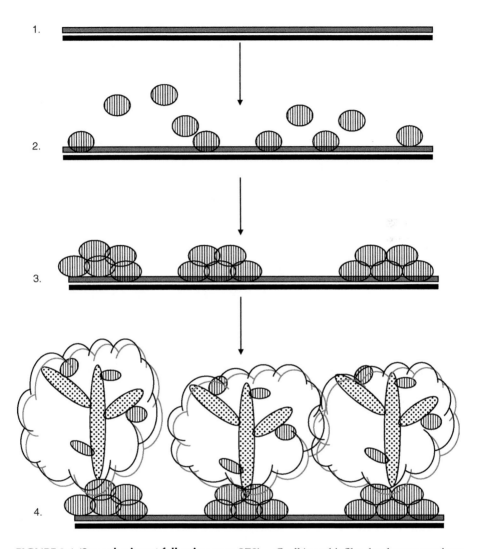

FIGURE 9.1 (See color insert following page 270) *C. albicans* biofilm development on inert surfaces. (1) Inert surface coated with a conditioning film (grey) consisting largely of host proteins. (2) Early attachment and colonization by *C. albicans* yeast-phase cells (striped). (3) Microcolony yeast basal layer formation, involving stacking of yeasts in the formation of the microcolonies. (4) Expansion of the biofilm architecture through the development of a hyphal/pseudohyphal layer (spotted) that protrudes from the inner yeast layer to the outer reaches of the biofilm. The hyphal layer development occurs simultaneously with the development of the thick layer of matrix material that has engulfed both the hyphal and yeast biofilm layers.

the device surface, in some way triggering a cascade of events that lead to the formation of germ tubes (16). Moreover, the recent study by Lewis et al. (15) has shown that *C. albicans* cells that lack the EFG1 gene are poor biofilm formers suggesting the possible involvement of the Efg1p signaling pathway in *C. albicans* biofilm development.

FIGURE 9.2 *C. albicans* biofilm on the surface of a catheter. The biofilm consists of a matrix of yeasts, hyphae and an extensive network of extracellular polymeric material.

In addition to the mixed morphological network in *C. albicans* biofilms, development of such biofilms are almost always associated with the generation of matrix, the majority of which appears to be extracellular material (17). Microscopy studies from various groups strongly suggests that the extracellular material is mainly composed of cell-wall like polysaccharides containing mannose and glucose residues, based mainly upon the use of certain dyes that bind specifically to such carbohydrates (16,18,19). Mature *C. albicans* biofilms have a highly heterogenous architecture in terms of distribution of fungal cells and extracellular materials (19). Interestingly, compared to biofilms grown on the irregular surface of polymethylmethacrylate, those grown on flat hydrophobic surfaces such as silicone elastomer have a distinct biphasic structure composed of an adherent yeast layer covered by a sparser layer of hyphal elements. What gives the biofilm its striking appearance is the vast layer of extracellular matrix that covers almost entirely the *Candida* biofilm. It appears that the matrix is in contact with the hyphal layer only. The precise role that matrix plays is somewhat unclear but many studies have suggested that the matrix assists in some way the biofilm to resist treatment with antifungal therapy. This and other characteristics regarding the recalcitrance of *Candida* biofilms to antifungal drugs will be described in some detail later in this chapter.

The importance of both the yeast and hyphal layers in *C. albicans* biofilms still remains rather unclear. However, studies employing both yeast-minus and hyphal-minus mutants have shown that, at least in *in vitro* biofilm models, that both forms can produce biofilms. For example, the hyphal-minus mutant formed only the yeast layer, whereas the yeast-minus isolate formed a thicker, hyphal biofilm that was similar in structure to the outer layer of wild-type biofilms (2). It is important to note at this stage that considering both forms, albeit to different efficiencies, can form biofilms that dimorphism per se is not an absolute prerequisite for biofilm formation (2,5). This may at least in part explain as to why non-dimorphic *Candida* spp. are also efficient in the formation of biofilms.

The surgically implanted device most frequently colonized by *Candida* is the central venous catheter. This device is frequently employed in order to administer nutrients and fluids and also to administer cytotoxic drugs commonly used in oncology patients. The origin of the infection foci may vary including the infusion fluid itself, the catheter hub, or more frequently the organisms may also be introduced either from the patient's own skin or a finding that appears to be becoming a more frequent origin, from the nursing staff themselves. In certain cases the distal tip of the catheter may become contaminated at the time of insertion. On the other hand, a noninfected catheter already in situ can become infected with *Candida* spp. due to migration of the organisms from the external non-implanted surface towards the distal tip itself (20,21). An additional "source" of *Candida* infection can also come from within the patients. For example, *Candida* is a common organism of the gastrointestinal tract and under certain circumstances, usually whereby an underlying disease has required surgery can cause infection through invasion of the intestinal mucosa and rapid seeding of the organism into the bloodstream. Such "planktonic" *Candida* cells are highly efficient in attaching to and colonizing implanted devices (5,21). The most commonly affected patient population affected by such endogenous "seeding" is that of oncology patients. Such patients normally receive chemotherapeutic treatments, many of which are known to either damage or render fragile the gastrointestinal mucosa (22).

Costerton et al. (23) provided a partial listing of medical devices that have been documented as supporting biofilm formation by various bacteria and by *Candida*. The following is a description of the types of devices that are frequently associated with *Candida* biofilms. Two major groups of prosthetic heart valves are currently used clinically, namely mechanical valves and bioprostheses (also known as tissue valves) (24). Both types appear to be equally susceptible to biofilm formers leading to similar rates for prosthetic valve endocarditis (PVE) or microbial infection of the valve surrounding tissues of the heart. There are several estimates concerning the rate of, PVE that range from 0.5% (25) to 1 and 4% (26). The major risk factor associated with PVE is the fact that device insertion inevitably causes some degree of tissue damage. Damaged tissues are highly susceptible to the accumulation of fibrin, platelets, and also organisms including yeasts (26). In terms of causative pathogens in, PVE, coagulase-negative staphylococci tend to be the most rapid colonizers probably due to their presence on the skin and being very close or at the insertion site. Late stage, PVE, a serious condition, is usually characterized by the presence of multiple organisms ranging from streptococci, coagulase negative staphylococci, enterococci, *S. aureus,* Gram-negative cocci and *C. albicans* (27). Inevitably, late stage, PVE does not respond to either antibiotic or antifungal therapies or combinations of both. The consequence of this is further surgery to remove the device followed by insertion of a replacement device with concomitant empiric antibiotic and antifungal prophylaxis.

Central venous catheters (CVC) have been documented as representing the highest risk associated with device-related infections. The study by Maki effectively demonstrated that CVCs are associated with infection rates of between 3 and 5% (28). These devices are in direct contact with the patients bloodstream. As a consequence the surface of a CVC is rapidly coated with platelets, plasma, and various proteins including laminin, fibronectin, fibrinogen, and albumin (29). Colonization of the CVC and subsequent biofilm formation is relatively rapid, usually occurring within three days post-implantation (30). Initial colonization and biofilm formation tend to occur on the external surface of such CVCs with colonization of and biofilm formation on the internal lumen an event that occurs later, usually within the first 30 days post-implantation (31). As with PVE related biofilms, the biofilm populations on CVCs are rarely if ever mono-microbial. For example, CVCs are typically co-colonized with coagulase negative staphylococci, *P. aeruginosa, K. pneumoniae, E. faecalis* and *C. albicans* (29,32). Patients with CVCs represent a potentially high risk population, a population that has undergone surgery and is receiving parenteral nutrition, medications, and undergoing hemodynamic monitoring. Patients in this group with such mixed biofilm populations are highly susceptible to subsequent septicaemia and candidemia as with time cells are sloughed from the biofilm surface and rapidly "seed" themselves into the bloodstream (1). Although such bloodstream microbes can be treated to a certain degree with empirical antibiotic and antifungal therapies, such therapy does not reduce the extent of the biofilm nor does it reduce the proliferation of the biofilm. Furthermore, despite antibiotic and antifungal treatment regimens of various types and lengths, the majority of seriously infected CVCs are required to be removed surgically to be replaced with a new CVC device using surgical procedures.

Non-device related infections can also involve biofilms. A classical example of one is that of *Candida* endocarditis (2). The primary lesion in *Candida* endocarditis is usually a thrombus, largely made up of platelets and fibrin, which typically develops on the surface of the heart valve itself. It is the thrombotic lesions that then become colonized by *Candida* cells which in its turn often provokes the production of emboli (1). An additional non-device related site where *Candida* biofilms can also occur is represented through the development of *Candida* biofilms upon epithelial surfaces. The most well known example of this is vaginal candidiasis whereby the *Candida* yeasts co-exist in biofilms with the vaginal bacterial flora. This co-existence has been suggested as a potential source for recurrent vaginitis, a disease that is difficult to keep under control and very problematic in curing with today's available antifungal agents.

The most common of superficial *Candida* biofilms is that encountered with the colonization of gum tissue. Although not a life-threatening disease, such denture stomatitis or infection of the oral mucosa by *Candida* is usually promoted through a close fitting upper denture. Biofilms form readily on the acrylic surfaces of dentures and typically consist of large numbers of bacteria, especially streptococci, that co-exist and proliferate in harmony with several of the *Candida* spp. (33). A second example of *Candida*-associated superficial infections involving biofilm formation is in laryngectomized patients fitted with silicone rubber voice prostheses. Biofilms on this type of surface are classically polymicrobial and frequently contain *C. albicans*. The failure of such devices is frequently associated with biofilms on the devices and usually occurs within 2 to 4 months following surgery, the cause of failure of the device being malfunction of the valve mechanism itself due to the valve being blocked by the biofilm (34).

9.3 SUCEPTIBILITY OF *CANDIDA* BIOFILMS TO ANTIFUNGAL AGENTS

Many investigators have studied the susceptibilities of *Candida* biofilms to different antifungal drugs. Drugs used in such investigations have included amphotericin B deoxycholate, lipid-based amphotericin B complexes, azole agents, fluorocytosine, and more recently the echinocandin class of antifungal agents. Additionally, the effects of non-antifungal based drugs including farnesol, chlorhexidine digluconate, aspirin, and other nonsteroidal anti-inflammatory drugs have also been documented.

The majority of such studies have reported that *Candida* biofilm cells are typically much less susceptible as compared to their planktonic counterparts, the earliest report of which came from the group of Douglas (35), who reported that biofilms formed by different *Candida* spp. were typically resistant to amphotericin B deoxycholate, the azoles fluconazole and ketoconazole and to the DNA inhibitor 5-fluorocytosine. Since the publication of that report, a plethora of studies have been additionally reported. For example, it has been reported that *Candida* biofilms respond poorly to amphotericin B deoxycholate, fluconazole, and the most recent of the azoles to be marketed, namely voriconazole. Additionally, this has been shown for amphotericin B deoxycholate, voriconazole and the Phase III drug ravuconazole (36).

Interestingly, several studies have shown that the lipid based amphotericin B formulations, as well as the new echinocandins caspofungin and micafungin, do possess some anti-*Candida* biofilm activity. These findings, previously reported by Kuhn et al. (36), have since been confirmed by studies reported by Bachmann et al. (37), who reported some activity of caspofungin against *Candida* biofilms, and Ramage et al. (38), who reported a 99% killing of *Candida* biofilms at therapeutic concentrations *in vitro*. Notably, although they also observed an effect with amphotericin B deoxycholate, the effect was only observed at concentrations vastly exceeding those attainable therapeutically (39). By contrast to the poor activity of amphotericin B deoxycholate, Kuhn et al. (36) also reported that liposomal amphotericin B and amphotericin B lipid complex (ABLC) to be effective in the *in vitro* treatment of *C. albicans* and *C. parapsilosis* biofilms. The intriguing characteristics of lipid-based formulations of amphotericin B and those of the echinocandin class may prove to be useful in future treatment regimens for *Candida* biofilms. However, these data require to be translated into efficacy *in vivo* in animal models of biofilms. Demonstration of their efficacy in such animal models could prove to be a breakthrough in the treatment of invasive systemic *Candida* infections. In the future, one or a combination of these drugs may allow for the retention of affected intravascular devices or obviate for the need for valve surgery in the case of *Candida* endocarditis.

Several new non-antifungal based compounds have also been investigated for their potential activities against *Candida* biofilms *in vitro*. These include studies with aspirin and other nonsteroidal agents and farnesol, a quorum sensing molecule.

Farnesol, which is associated with *Candida* hyphal development, has been suggested as an important regulatory or quorum-sensing molecule in *C. albicans* biofilm formation. The study by Ramage et al. (39) showed that the addition of farnesol could affect *Candida* biofilm formation and that such effects were dependent on the concentration of the compound used and the initial adherence time. Their study showed that at a concentration of 300 µM, farnesol completely inhibited biofilm formation in their *in vitro* models (38). The importance of farnesol was further highlighted through the report by Hornby et al. (40), who demonstrated that zaragozic acid treatment of *Candida* biofilms leads to accumulation of farnesol in the biofilm, resulting in a toxic effect and biofilm decay and death. The use of farnesol or other quorum sensing molecules in therapy is unclear and further studies in this area are warranted.

Prostaglandins are small lipid molecules that have diverse biological activities, including the modulation of host immune responses (41). They are also known to be produced by *C. albicans, Cryptococcus neoformans* and mammalian cells (42–44). In mammalian systems, arachidonic acid, formed via cleavage of phospholipids, is converted to prostaglandin H2 (PGH2) by the cyclooxygenase (COX) isoenzymes. Interestingly, treatment of *C. albicans* or *C. neoformans* infections with the COX inhibitor indomethacin was shown to significantly reduce the viabilities of the organisms and the subsequent production of prostaglandins, suggesting that an essential COX enzyme may be responsible for fungal prostaglandin synthesis (43). PGE2 is known to enhance the formation of *C. albicans* germ tubes. In the recent study by Alem and Douglas, the effects of nonsteroidal anti-inflammatory drugs that were all known COX inhibitors on *Candida* biofilm formation were investigated (41). They showed that seven of nine inhibitors tested affected biofilm formation. The most notable effects were observed for aspirin, etodolac, and diclofenac with aspirin causing up to 95 % inhibition. Celecoxib, nimesulide, ibuprofen, and meloxicam also affected biofilm formation though to a lesser extent. The potential application of such COX inhibitors in the therapy of human fungal disease is not clear. Further studies in this area may be worthwhile.

9.4 MECHANISMS OF RESISTANCE OF *CANDIDA* BIOFILMS TO ANTIFUNGAL THERAPY

Like their bacterial counterparts, biofilm-grown *Candida* cells are highly resistant to antimicrobials, including antifungal drugs and antiseptics. Bacteria within biofilms are typically 50 to 1,000 times less sensitive to antibiotics as compared with the sensitivities of planktonic bacteria (1). For example, a study by Williams et al. (45) showed that although wild-type *S. aureus* cells were sensitive to vancomycin with an MIC of 1 µg/ml, the *S. aureus* biofilm responded very poorly to vancomycin with a corresponding MIC of 20 µg/ml. This has also been observed by Ceri et al. (46) for *P. aeruginosa* biofilms that did not respond to Imipenem with a biofilm MIC of >1024 µg/ml, while the same cells in liquid culture were highly sensitive to the drug with an MIC of 1 µg/ml. By analogy, corresponding resistance of *Candida* biofilms to antifungal agents was first reported by Hawser and Douglas (35) whereby biofilm populations of *C. albicans* and other *Candida* biofilms were highly resistant to different drugs, including amphotericin B, the azoles and fluorocytosine. MICs for these drugs against biofilms were 30 to 2,000 times higher than the corresponding MICs for planktonic cells (35).

The precise mechanisms that lead to resistance or recalcitrance of *Candida* biofilms are not as yet fully elucidated. Early studies examining the relationship between reduced growth rates in biofilms as compared to growth rates of the cells in liquid culture provided a hint that reduced growth rate may, at least in part, help to explain the recalcitrance of such biofilms to drug therapy. In a similar fashion, many earlier studies with bacterial biofilms demonstrated that biofilms containing *S. aureus, S. epidermidis, K. pneumoniae, P. aeruginosa,* and *E. coli* all possessed greatly reduced growth rates (47–49). In addition, bacterial growth phase and its influence on biofilm susceptibility to antibiotics has also been studied, though did not yield conclusive evidence, suggesting a strong relationship between susceptibility and bacterial growth phase. However, it is evident both in terms of bacterial and fungal biofilms that the biofilm populations tend to exist at reduced growth rates. However, a recent study by Baillie and Douglas (50) showed that with *in vitro* biofilm models artificially increasing the growth rate of *C. albicans* had no strict relationship between antifungal susceptibility and growth rates.

In addition to the possible relationship between growth rates and biofilm susceptibilities, many other studies have focused upon the contributions of the matrix material that in itself may be responsible for a reduction in the perfusion of a drug into the biofilm (1,51), differential gene expression patterns by sessile organisms (52,53) and the link between biofilm recalcitrance and cell membrane efflux pumps (38,39,54). Efflux pumps have been extensively studied in various bacteria and in recent years in *C. albicans* and other *Candida* spp. Efflux pumps are, therefore, critically involved in the antimicrobial resistance of planktonically grown bacteria, however their role in biofilm-related resistance is still not perfectly understood. For example, efflux pumps have been implicated in the resistance of *P. aeruginosa* biofilms to low doses of ofloxacin but not to ciprofloxacin, chloramphenicol, and tobramycin (54,55). The study by Maira-Litran et al. (52) revealed the involvement of an efflux pump in resistance to low concentrations of ciprofloxacin (0.004 µg/ml) though not to high levels of the drug (0.1 µg/ml) in *E. coli* biofilms. Overall, such studies suggest that efflux pumps play both drug-dose and drug-dependent roles in antimicrobial resistance in bacterial biofilms. Concerning the role of efflux in the resistance of *Candida* biofilms to antifungal agents, antifungal resistance of planktonically grown *C. albicans* has been strongly linked to the expression of efflux pumps such as Cdr1p, Cdr2p, and Mdr1p (56–60). A recent study by Mukherjee et al. (61) has deepened our current knowledge in terms of the link between efflux and biofilm resistance in *Candida*. In their study, the investigators employed mutants carrying single, double, or triple mutations of the CDR and MDR1 genes. Their results showed that, at the early phase of development, biofilms formed by the mutant strains were more susceptible to fluconazole than those formed by the wild-type strain. Importantly, among the mutants the triple-knockout strain was by far the most susceptible to fluconazole (MIC of 16 µg/ml), indicating the involvement of efflux pumps in the azole resistance of early-phase biofilms. Interestingly, at later development phases of 12 to 48 hours, biofilms formed by the mutants were totally resistant to fluconazole with MIC values of ≥256 µg/ml. These values were shown to be similar for late stage biofilms of the wild type. In essence, their data would suggest that efflux pumps contribute to azole resistance in the early phase of biofilm formation but not in later phases. In addition, the investigators also studied levels of CDR and MDR1 mRNA showing that expression of CDR genes to be temporally regulated during biofilm formation, with the higher levels of gene transcripts detected in early and intermediate phase biofilms (61). It is also worthwhile noting that CDR and MDR1 were expressed throughout the development phases of these biofilms.

Variations in membrane sterol composition have also been suggested as playing a role in the resistance of *Candida* biofilms. Ghannoum and Rice (62) and Kontoyiannis (63) have shown that sterol variations leading to changes in fluidity and asymmetry of the membrane have significant influence on the sensitivity and resistance of *C. albicans* cells to antifungals. Furthermore, Hitchcock et al. (64,65) showed that *C. albicans* strains resistant to both polyenes and azoles had a larger lipid content and lower polar lipid-to-neutral lipid ratio than other strains. They demonstrated that the major difference between their strain and wild-type strains to be the absence of ergosterol which was replaced by methylated sterols, mainly lanosterol, 24-methylene-24,25-dihydrolanosterol, and 4-methylergostadiene-3-ol. More recent reports have suggested that changes in the status of the membrane lipid phase and asymmetry could contribute to azole resistance in *C. albicans*. Mukherjee et al. (61) recently demonstrated that in early *Candida* biofilms the levels of ergosterol were similar to those detected in planktonic cells. However, in later stage biofilms the level of ergosterol was significantly lower than the level of their planktonic colunterparts (61). They also observed altered levels of other intermediate sterols including zymosterol, 4,14-dimethylzymosterol and obtusifoliol.

9.5 FUTURE OUTLOOK IN THE TREATMENT OF *CANDIDA* BIOFILMS

The plethora of studies that have investigated the architectural properties of *Candida* biofilms, the physiology of *Candida* biofilms and their resistance to antifungal agents have brought our level of knowledge closer to what is known in the heavily exploited area of bacterial biofilms.

Much more is now known as to why *Candida* biofilms do not respond to antifungal therapies, with reduced growth rates, efflux pumps, and the presence of matrix material all possible culprits in the resistance of these biofilms to drugs. It is clear that overall resistance cannot be due to only one of these factors and that drug resistance in *Candida* biofilms is clearly a multifactorial process.

The recent additions to the antifungal armamentarium, including lipid based complexes of amphotericin B and the echinocandin antifungal agent caspofungin may prove to be useful, not necessarily in the complete eradication of *Candida* biofilms in situ but may prolong the life-times of implanted devices, thus reducing mortality and morbidity in patients requiring implanted devices.

REFERENCES

1. Donlon, R.M., and Costerton, J.W., Biofilms: survival mechanisms of clinically relevant microorganisms. *Clin. Microbiol. Rev.* 15, 167–193, 2002.
2. Douglas, L.J., *Candida* biofilms and their role in infection. *TRENDS Microbiol.* 11, 30–36, 2003.
3. Davey, M.E., and O'Toole, G.A., Microbial biofilms: from ecology to molecular genetics. *Microbiol. Mol. Biol. Rev.* 64, 847–867, 2000.
4. O'Toole, G., Kaplan, H.B., and Kolter, R., Biofilm formation as microbial development. *Annu. Rev. Microbiol.* 54, 49–79, 2000.
5. Calderone, R.A., Introduction and historical perspectives. In: Calderone, R.A., ed., Candida *and Candidiasis.* ASM Press, 2002:3–13.
6. Anaissie, E.J., Rex, J.H., Uzun, O., and Vartivarian, S., Predictors of adverse outcome in cancer patients with candidaemia. *Am. J. Med.* 104, 238–245, 1988.
7. Nguyen, M.H., Peacock, J.E., Tanner, D.C., Morris, A.J., Nguyen, M.L., Snydman, D.R., Wagener, M.M., and Lu, V.L., Therapeutic approaches in patients with candidemia: evaluation in a multicenter prospective observational study. *Arch. Intern. Med.* 155, 2429–2435, 1995.
8. Kremery, V. Jr, and Kovacicova, G., Longitudinal 10-year prospective survey of fungaemia in Slovak Republic: trends in etiology in 310 episodes. *Diagn. Microbial. Infect. Dis.* 36, 7–11, 2000.
9. Pagano, L., Antinori, A., Ammassari, A., Mele, L., Nosari, A., Melillo, L., Martino, B., Sanguinetti, M., Equitani, F., Nobile, F., Carotenuto, M., Morra, E., Morace, G., and Leone, G., Retrospective study of candidaemia in patients with hematological malignancies. Clinical features, risk factors and outcome of 76 episodes. *Eur. J. Haematol.* 63, 77–85, 1999.
10. Saiman, L., Ludington, E., Pfaller, M., Rangel-Frausto, S., Wiblin, R.T., Dawson, J., Blumberg, H.M., Patterson, J.E., Rinaldi, M., Edwards, J.E., Wenzel, R.P., and Jarvis, W., Risk factors for candidaemia in neonatal intensive care unit patients. *Pediatr. Infect. Dis.* 19, 319–324, 2000.
11. Sanchez, V., Vazquez, J.A., Barth-Jones, D., Dembry, L., Sobel, J.D., and Zervos, M.J., Nosocomial acquisition of *Candida parapsilosis*: an epidemiologic study. *Am. J. Med.* 94, 577–582, 1993.
12. Weems, J.J. Jr, Chamberland, M.E., Ward, J., Willy, M., Padhye, A.A., and Solomon, S.L., *Candida parapsilosis* fungaemia associated with parenteral nutrition and contaminated blood pressure transducers. *J. Clin. Microbiol.* 25, 1029–1032, 1987.
13. Costerton, J.W., Lewandowski, Z., Caldwell, D.E., Korber, D.R., and Lappin-Scott, H.M., Microbial biofilms. *Annu. Rev. Microbiol.* 49, 711–745, 1995.
14. Hawser, S.P., and Douglas, L.J., Biofilm formation by *Candida* species on the surface of catheter materials *in vitro*. *Infect. Immun.* 39, 915–921, 1994.
15. Lewis, R.E., Antifungal activity of amphotericin B, fluconazole, and voriconazole in an invitro model of *Candida* catheter-related bloodstream infection. *Antimicrob. Agents Chemother.* 46, 3499–3505, 2002.
16. Baillie, G.S., and Douglas, L.J., Role of dimorphism in the development of *Candida albicans* biofilms. *J. Med. Microbiol.* 48, 671–679, 1999.
17. Hawser, S.P., Baillie, G.S., and Douglas, L.J., Production of extracellular matrix by *Candida albicans* biofilms. *J. Med. Microbiol.* 47, 253–256, 1998.
18. Baillie, G.S., and Douglas, L.J., Matrix polymers of *Candida* biofilms and their possible role in biofilm resistance to antifungal agents. *J. Antimicrob. Chemother.* 46, 397–403, 2000.

19. Chandra, J, Kuhn, D.M., Mukherjee, P.K., Hoyer, L.L., McCormick, T., and Ghannoum, M.A., Biofilm formation by the fungal pathogen *Candida albicans*: development, architecture, and drug resistance. *J. Bacteriol.* 183, 5385–5394, 2001.

20. Goldmann, D.A., and Pier, G.B., Pathogenesis of infections related to intravascular catheterization. *Clin. Microbiol. Rev.* 6, 176–192, 1993.

21. Sherertz, R.J., Pathogenesis of vascular catheter infection. In: Waldvogel, F.A., and Bisno, A.L., eds., *Infections Associated with Indwelling Medical Devices*. ASM Press, 2000:111–125.

22. Kullberg, B.J., and Filler, S.G., Candidaemia. In: Calderone, R.A., ed., Candida *and Candidiasis*. ASM Press, 2002:327–340.

23. Costerton, J.W., Cheng, K.J., Geesey, G.G., Ladd, T.I., Nickel, J.C., Dasgupta, M., and Marrie, T.J., Bacterial biofilms in nature and disease. *Annu. Rev. Microbiol.* 41, 435–464, 1987.

24. Braunwald, E., Valvular heart disease. In: Braunwald, E., ed., *Heart Disease*. Philadelphia: W.B. Saunders Co., 1987:1007–1076.

25. Hancock, E.W., Artificial valve disease. In: Schlant, R.C., Alexander, R.W., O'Rourke, R.A., Roberts, R., and Sonnenblick, E.H., eds., *The Heart Arteries and Veins*. New York: McGraw-Hill, 1994:1539–1545.

26. Douglas, J.L., and Cobbs, C.G., Prosthetic valve endocarditis. In: Kaye D., ed., *Infective Endocarditis*. New York: Raven Press Ltd, 1992:375–396.

27. Kaye, D., and Hessen, M.T., Infections associated with foreign bodies in the urinary tract. In: Bisno, A.L., and Waldvogel, F.A., eds., *Infections Associated With Indwelling Medical Devices*. Washington, D.C.: ASM Press, 1994:291–307.

28. Maki, D.G., Infections caused by intravascular devices used for infusion therapy: pathogenesis, prevention, and management. In: Bisno, A.L., and Waldvogel, F.A., eds., *Infections Associated With Indwelling Medical Devices*. Washington, D.C.: ASM Press, 1994:155–212.

29. Raad, I., Intravascular catheter-related infections. *Lancet* 351, 893–898, 1998.

30. Anaissie, E., Samonis, G., Kontoyiannis, D., Costerton, J., Sabharwal, U., Bodey, G., and Raad, I., Role of catheter colonization and infrequent hematogenous seeding in catheter-related infections. *Eur. J. Clin. Microbiol. Infect. Dis.* 14, 135–137, 1995.

31. Raad, I., Costerton, W., Sabharwal, U., Sacilowski, M., Anaissie, W., and Bodey, G.P., Ultrastructural analysis of indwelling vascular catheters: a quantitative relationship between luminal colonization and duration of placement. *J. Infect. Dis.* 168, 400–407, 1993.

32. Elliott, T.S.J., Moss, H.A., Tebbs, S.E., Wilson, I.C., Bonser, R.S., Graham, T.R., Burke, L.P., and Faroqui, M.H., Novel approach to investigate a source of microbial contamination of central venous catheters. *Eur. J. Clin. Microbiol. Infect. Dis.* 28, 210–213, 1997.

33. Budtz-Jorgensen, E., Histopathology, immunology, and serology of oral yeast infections. Diagnosis of oral candidasis. *Acta. Odontol. Scand.* 55, 37–43, 1990.

34. Van der Mai, H.C., Effect of probiotic bacteria on prevalence of yeasts in oropharyngeal biofilms on silicone rubber voice prostheses *in vitro*. *J. Med. Microbiol.* 49, 713–718, 2000.

35. Hawser, S.P., and Douglas, L.J., Resistance of *Candida albicans* biofilms to antifungal agents *in vitro*. *Antimicrob. Agents Chemother.* 39, 2128–2131, 1995.

36. Khun, D.M., Antifungal susceptibility of *Candida* biofilms: unique efficacy of amphotericin B lipid formulations and echinocandins. *Antimicrob. Agents Chemother.* 46, 1773–1780, 2002.

37. Bachman, S.P., *In vitro* activity of caspofungin against *Candida albicans* biofilms. *Antimicrob. Agents Chemother.* 46, 3591–3596, 2002.

38. Ramage, G., Bachmann, S., Patterson, T.F., Wickes, B.L., and Lopez-Ribot, J.L., Investigation of multidrug efflux pumps in relation to fluconazole resistance in *Candida albicans* biofilms. *J. Antimicrob. Chemother.* 49, 973–980, 2002.

39. Ramage, G., Wickes, B.L., and Lopez-Ribot, J.L., Biofilms of *Candida albicans* and their associated resistance to antifungal agents. *Am. Clin. Lab.* 20, 42–44, 2001.

40. Hornby, J.M., Jensen, E.C., Lisec, A.D., Tasto, J.J., Jahnke, B., Shoemaker, R., Dussault, P., and Nickerson, K.W., Quorum sensing in the dimorphic fungus *Candida albicans* is mediated by farnesol. *Appl. Environ. Microbiol.* 67, 2982–2992, 2001.

41. Alem, M.A.S, and Douglas, L.J., Effects of aspirin and other nonsteroidal anti-inflammatory drugs on biofilms and planktonic cells of *Candida albicans*. *Antimicrob. Agents Chemother.* 48, 41–47, 2004.

42. Noverr, M.C., Erb-Downward, J.R., and Huffnagle, G.B., Production of eicosanoids and other oxylipins by pathogenic eukaryotic microbes. *Clin. Microbiol. Rev.* 16, 517–533, 2003.

43. Noverr, M.C., Phare, S.M., Toews, G.B., Coffey, M.J., and Huffnagle, G.B., Pathogenic yeasts *Cryptococcus neoformans* and *Candida albicans* produce immunomodulatory prostaglandins. *Infect. Immun.* 69, 2957–2963, 2001.

44. Noverr, M.C., Toews, G.B., and Huffnagle, G.B., Production of prostaglandins and leukotrienes by pathogenic fungi. *Infect. Immun.* 70, 400–402, 2002.

45. Williams, I., Venables, W.A., Lloyd, D., Paul, F., and Critchley, I., The effects of adherence to silicone surfaces on antibiotic susceptibility in *Staphylococcus aureus*. *Microbiology* 143, 2407–2413, 1997.

46. Ceri, H., Olson, M.E., Stremick, C., Read, R.R., Morck, D., and Buret, A., The Calgary Biofilm Device: a new technology for rapid determination of antibiotic susceptibilities of bacterial biofilms. *J. Clin. Microbiol.* 37, 1771–1776, 1999.

47. Brown, M.R.W, and Gilbert, P., Sensitivity of biofilms to antimicrobial agents. *J. Appl. Bacterial. Symp. Suppl.* 74, 87S–97S, 1993.

48. Brown, M.R.W, Allison, D.G., and Gilbert, P., Resistance of bacterial biofilms to antibiotics: a growth-rate related effect?. *J. Antimicrob. Chemother.* 22, 777–789, 1988.

49. Duguid, I.G., Evans, E., Brown, M.R.W., and Gilbert, P., Growth-rate-independent killing by ciprofloxacin of biofilm-derived *Staphylococcus epidermidis:* evidence for cell-cycle dependency. *J. Antimicrob. Chemother.* 30, 791–802, 1992.

50. Baillie, G.S., and Douglas, L.J., Effect of growth rate on resistance of *Candida albicans* biofilms to antifungal agents. *Antimicrob. Agents Chemother.* 42, 1900–1905, 1998.

51. Hoyle, B.D., Jass, J.D., and Costerton, J.W., The biofilm glycocalyx as a resistance factor. *J. Antimicrob. Chemother.* 26, 1–5, 1990.

52. Maira-Litran, T., Allison, D.G., and Gilbert, P., Expression of the multiple antibiotic resistance operon (mar) during growth of *Escherichia coli* as a biofilm. *J. Appl. Microbiol.* 88, 243–247, 2000.

53. Lewis, K., Riddle of biofilm resistance. *Antimicrob. Agents Chemother.* 45, 999–1007, 2001.

54. De Kievit, T.R., Parkins, M.D., Gillis, R.J., Srikumar, R., Ceri, H., and Poole, K., Multidrug efflux pumps: expression patterns and contribution to antibiotic resistance in *Pseudomonas aeruginosa* biofilms. *Antimicrob. Agents Chemother.* 45, 1761–1770, 2001.

55. Brooun, A., A dose-response study of antibiotic resistance in *Pseudomonas aeruginosa* biofilms. *Antimicrob. Agents Chemother.* 44, 640–646, 2000.

56. Albertson, G.D., Niimi, M., Cannon, R.D., and Jenkinson, H.F., Multiple efflux mechanisms are involved in *Candida albicans* fluconazole resistance. *Antimicrob. Agents Chemother.* 40, 2835–2841, 1996.

57. Lyons, C.N., White, T.C., Transcriptional analyses of antifungal drug resistance in *Candida albicans*. *Antimicrob. Agents Chemother.* 44, 2296–2303, 2000.

58. Prasad, R., Murthy, S.K., Gupta, V., and Prasad, R., Multiple drug resistance in *Candida albicans*. *Acta. Biochem. Pol.* 42, 497–504, 1995.

59. Sanglard, D., Kuchler, K., Ischer, F., Pagani, J.L., Monod, M., and Bille, J., Mechanisms of resistance to azole antifungal agents in *Candida albicans* isolates from, AIDS patients involve specific multidrug transporters. *Antimicrob. Agents Chemother.* 39, 2378–2386, 1995.

60. Sanglard, D., Ischer, F., Monod, M., and Bille, J., Susceptibilities of *Candida albicans* multidrug transporter mutants to various antifungal agents and other metabolic inhibitors. *Antimicrob. Agents Chemother.* 40, 2300–2305, 1996.

61. Mukherjee, P.K., Chandra, J., Kuhn, D.M., and Ghannoum, M.A., Mechanism of fluconazole resistance in *Candida albicans* biofilms: phase-specific role of efflux pumps and membrane sterols. *Infect. Immun.* 71, 4333–4340, 2003.

62. Ghannoum, M.A., and Rice, L.B., Antifungal agents: mode of action, mechanisms of resistance, and correlation of these mechanisms with bacterial resistance. *Clin. Microbiol. Rev.* 12, 501–517, 1999.

63. Kontoyiannis, D.P., Efflux-mediated resistance to fluconazole could be modulated by sterol homeostasis in *Saccharomyces cerevisiae*. *J. Antimicrob. Chemother.* 46, 199–203, 2000.

64. Hitchcock, C.A., Barrett-Bee, K.J., and Russell, N.J., The lipid composition and permeability to azole of an azole- and polyene-resistant mutant of *Candida albicans*. *J. Med. Vet. Mycol.* 25, 29–37, 1987.

65. Hitchcock, C.A., Russell, N.J., and Barrett-Bee, K.J., Sterols in *Candida albicans* mutants resistant to polyene or azole antifungals, and of a double mutant *C. albicans* 6.4. *Crit. Rev. Microbiol.* 15, 111–115, 1987.

Section III

Emerging Issues, Assays, and Models

10 Current Perspectives on the Regulation of the *ica* Operon and Biofilm Formation in *Staphylococcus epidermidis*

Paul D. Fey and Luke D. Handke

CONTENTS

10.1   INTRODUCTION

Staphylococcus epidermidis is a ubiquitous commensal found on the skin of humans and other mammals. However, due to an increase in invasive medical practices, *S. epidermidis* has become the most common pathogen isolated from blood in intensive care unit patients in the United States (1,2). The noted increase of *S. epidermidis* infections in seriously ill patients can be attributed to the ability of this organism to adhere to and form biofilm on the surface of biomaterials, including catheters. The molecular structure of staphylococcal biofilm and the genetics behind its production and regulation have been an active area of research since the early 1980s. This review will focus on loci that have recently been demonstrated to regulate the *ica* operon, the four gene operon that produces enzymes that synthesize staphylococcal biofilm, or polysaccharide intercellular adhesin (PIA).

10.2   STAPHYLOCOCCAL BIOFILM

The study of staphylococcal biofilm and its clinical significance was championed by Gordon Christensen and colleagues in the 1980s (3–8). These, and other investigators, demonstrated that certain *S. epidermidis* isolates could coat biomaterials (including test tubes and microtiter plates) with a gelatinous substance termed "slime." In particular, staphylococcal slime was easily noted and impressive when viewed through electron microscopy. Earlier attempts to characterize staphylococcal slime/biofilm were fraught with issues related to media contamination and therefore was difficult to define (9). In 1996, however, Baldassari and colleagues demonstrated that staphylococcal slime isolated from *S. epidermidis* RP62A (ATCC 35984) was composed of N-acetylglucosamine (8). This polysaccharide was termed slime associated antigen (SAA). Further biochemical characterization of slime by Mack and colleagues found that the polysaccharide was composed of a major and minor component (10). Polysaccharide I (80%) is a linear homoglycan (n = ~130) composed of β-1,6-linked N-acetylglucosamine; approximately 20% of the residues are deacetylated. In contrast, polysaccharide II has a higher proportion of acetylated residues than polysaccharide I and contains phosphate and ester-linked succinate. This structure was thus named polysaccharide intercellular adhesion (PIA); it is generally accepted that the previously named SAA is indeed PIA (11). A highly related compound, differentiated from PIA mostly by its large molecular weight, has been isolated from *S. aureus* MN8m (12). In addition, this compound, termed PNAG, was shown to be the same antigen that was previously described as polysaccharide adhesin (PS/A) by Tojo and colleagues (12–14). McKenney and colleagues have subsequently demonstrated that PNAG has potential as a vaccine candidate against staphylococcal disease (15,16). Since antibodies against PIA cross-react with PNAG, it is possible that these two compounds are identical and the experimental differences are artifacts (11). To avoid confusion, and since we are addressing *S. epidermidis*, staphylococcal slime will be called PIA in this review.

The pathogenesis of biomaterial-related infections caused by *Staphylococcus epidermidis* is thought to occur through a two step process (17–19). Small numbers of *S. epidermidis* found colonizing the skin are thought to contaminate the device during introduction (i.e., insertion of a catheter). The bacteria then rapidly adhere to the biomaterial through a combination of non-specific interactions (van der Waals forces, hydrophobic interactions, etc.) as well as specific interactions through the action of adhesins (e.g., AtlE; discussed below). After initial adherence, PIA, and possibly other proteins or cell wall constituents, play(s) a major role in the accumulation phase, the second step in the pathogenesis process. Electron microscopy studies have shown multilayered clusters of bacteria embedded in a layer of extracellular amorphous material. As the name suggests, PIA also acts as an intercellular adhesin as well as contributing to hemagglutination, a common property of many *S. epidermidis* strains (20,21).

S. *epidermidis* strains that produce PIA have been shown to be more virulent than *ica* mutants in at least two animal models mimicking foreign body infections. Rupp et al. demonstrated that *S. epidermidis* strains 1457 and O-47 were more virulent than their corresponding *ica* mutants in a rat catheter model (1457 and O-47) and a mouse foreign body infection model (1457) (22–25). However, two other models have provided contradictory evidence, suggesting that PIA production may not play a large role in the establishment of infections of native tissue (rabbit endocarditis model) or in infections of biomaterials (guinea pig tissue cage model) that previously have been coated with host cellular proteins (e.g. fibrinogen, fibronectin, etc.) (26,27). These authors suggested that adherence to the biomaterial (before host matrix proteins become established) and subsequent biofilm production before the development of a strong host response may be crucial to *S. epidermidis*-mediated biomaterial related infection (26).

10.3 FACTORS WHICH CONTRIBUTE TO INITIAL ADHERENCE AND/OR BIOFILM FORMATION

As stated above, accumulation of biofilm is preceded by initial adherence of staphylococci to the biomaterial. This process is mediated by several proteins, or other cellular factors, that are essential for initial binding to biomaterials.

10.3.1 AᴛʟE

A *S. epidermidis* Tn*917* mutant (O-47 mut1) that was unable to form biofilm on polystyrene microtiter plates, but still had the ability to produce PIA and mediate hemagglutination, was isolated by Heilmann and colleagues in 1996 (19,21,28). Upon genetic analysis, it was determined that this mutant had an 8 kb deletion that contained an ORF with significant identity to *atl*, the major autolysin of *S. aureus*. This newly identified gene was called *atlE* (28). AtlE is a highly processed protein; however, only the 60 kDa portion is needed to complement and restore biofilm forming capability to the mutant. It is not clear whether AtlE has a direct role in adherence to plastic and other biomaterials or whether deletion of the locus plays an indirect role in changing the overall ability of the cell to adhere to the plastic. It is known that *atlE* mutants are less hydrophobic than wild type and grow as large complex structures of cells, suggesting a major defect/and or change in cell wall structure (11). On the other hand, it was demonstrated that purified AtlE bound vitronectin, suggesting a role for

AtlE in the adherence to surfaces coated with host proteins (28). Furthermore, Rupp and colleagues have found that the *S. epidermidis* O-47 *atlE* mutant was significantly less virulent than wild-type in the rat catheter model of central venous catheter infection (22). Vandecasteele demonstrated that *atlE* gene expression is high and stable during an *in vivo* foreign body infection model, further suggesting the role of AtlE in the pathogenesis of *S. epidermidis* mediated biomaterial infections (29).

A second autolysin with adhesive properties in *Staphylococcus epidermidis*, Aae, has been recently been described by Heilmann and colleagues (30). Aae is a surface associated protein that has bacteriolytic activity as well as the ability to bind vitronectin, fibronectin, and fibrinogen. Aae is the latest addition to a family of proteins (autolysin/adhesions) within the staphylococci with both autolytic activity as well as binding activity (28,31,32).

10.3.2 TEICHOIC ACID

Teichoic acids, found only in Gram-positive bacteria, are polymers of either ribitol phosphate or glycerol phosphate joined through phosphodiester linkages (33,34). These groups can be replaced by D-alanine and N-acetylglucosamine (35,36). The exact physiological role of teichoic acid is not known, however, loss of TagF, a protein responsible for teichoic acid synthesis in *S. epidermidis* ATCC14990, is essential for viability (33). In the search for mutants that were highly sensitive to the activity of positively charged antimicrobial peptides, Peschel and colleagues isolated a *S. aureus* Tn*917* mutant that was not only highly sensitive to peptides, but was also unable to form a functional biofilm (35,36). Genetic studies demonstrated that the mutations were in the *dltA* operon, which is responsible for the production of enzymes that add alanine to cell wall teichoic acid (35). This mutant was unable to bind to polystyrene or glass surfaces thus affecting its ability to form a biofilm; the production of PIA was unaltered however (36). As this mutant was hypersusceptible to positively charged peptides due to its increased negative charge, it was hypothesized that cells could not adhere to polystyrene or glass to do repulsive electrostatic forces. Further work has demonstrated that *dtlA S. aureus* mutants are more susceptible to human neutrophils and are less virulent in a mouse model of sepsis, demonstrating that one potential role of teichoic acids may be a defense mechanism against the action of antimicrobial peptides produced by human neutrophils (37). In addition, Hussain and colleagues have also shown that teichoic acids significantly increased the ability of *S. epidermidis* to bind fibrinogen in a dose dependent manner; the authors suggest that teichoic acid may act as a "bridging molecule" between the cell and immobilized fibrinogen (38).

10.3.3 BIOFILM ASSOCIATED PROTEIN (BAP)

The biofilm associated protein was found in a small percentage (5%) of strong biofilm producing *S. aureus* isolates of bovine origin (39). Interestingly, the *bap* locus was not found in a series of 75 human *S. aureus* isolates (39). The core region of Bap consists of 13 highly related repeats containing 86 amino acids and is cell wall associated. Although isolates that contain Bap have been shown to cause persistent infections in a mouse model of foreign body infections, expression of Bap has been demonstrated to prevent MSCRAMM (microbial surface components recognizing adhesive matrix molecules) proteins from binding to their targets (fibrinogen, fibronectin, etc.) (40,41). Although the principal function of Bap is not known, *bap* has recently been shown to be associated with a pathogenicity island called SaPIbov2 where *bap* is encoded on a transposon-like element along with an ABC transporter (42). Although its prevalence or relevance is not known to date, a Bap-like protein (Bhp) has been identified in *S. epidermidis* RP62A through a genome search (Accn# AY028618).

10.3.4 OTHER FACTORS

Other characterized factors include SSP-1 and SSP-2 (a probable degradation pro-duction of SSP-1), which are involved in adherence of *S. epidermidis* to polystyrene (43). Although the gene associated with these proteins (SSP-1 is 280 kDa) has not been described to date, ultrastructural analysis suggests that this protein exists as a fimbria-like structure. The accumulation-associated protein (AAP) is another protein that may be involved in the accumulation phase of biofilm production (44). As assessed through Western blot, this 140 kDa antigen is associated with *S. epidermidis* isolates that produce abundant biofilm. The gene associated with AAP, *aap*, has recently been deposited into nucleotide databases (Accn. # AJ249487); it appears that expression of *aap* is highest early during a foreign body infection but is signif-icantly lowered in late stages of the infection process (29). Although *S. aureus* produces an abundance of MSCRAMM proteins, *S. epidermidis* does not encode many proteins that allow it to adhere to proteins from the extracellular matrix. To date, two genes encoding MSCRAMM proteins have been identified in *S. epidermidis*, *fbe* encoding a fibrinogen binding protein and *embp* encoding a fibronectin binding protein (45–47). The role of these MSCRAMM proteins in the formation in biofilm in *S. epidermidis* is currently unknown. Perhaps the most surprising finding related to staphylococcal biofilm and related factors is the recent data from Caiazza and colleagues suggesting a role for alpha-toxin in the development of a mature biofilm (48). These authors demonstrated that specific *hla* mutants were unable to form a biofilm in both static and flow conditions. It was suggested that alpha-toxin may be required for functional cell-to-cell interaction during formation of a mature biofilm.

10.4 *ica* OPERON

Genes responsible (*ica*) for biofilm formation and PIA production were identified by Heilmann and colleagues through generalized transposon (Tn917) mutagenesis in 1996 (18,19). The *ica* operon is composed of four genes *icaA* (1238 bp), *icaD* (305 bp), *icaB* (869 bp), and *icaC* (1067 bp). A divergently transcribed repressor, *icaR* (557 bp), is found just upstream of *ica*. This operon, when mobilized into the biofilm-negative *S. carnosus,* promotes intercellular aggregation as well as biofilm formation on glass (18). Although the *ica* operon is found in only approximately half of *S. epidermidis* isolates (49–51), it has been found in all *S. aureus* (52) isolates tested thus far, as well as multiple other coagulase-negative staphylococci (52). Therefore, the ability to form a biofilm through the production of a PIA-like polysaccharide appears to be a conserved trait of all staphylococci.

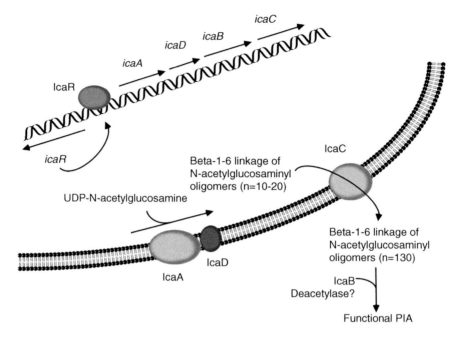

FIGURE 10.1   Model of PIA synthesis as proposed by Götz (11).

IcaA, IcaD and IcaC are all proposed integral membrane proteins whereas IcaB is found in the cytoplasmic fraction (11,53) (Figure 10.1). Through the development of an *in vitro* assay using UDP-N-acetylglucosamine as a substrate, Gerke and colleagues demonstrated that IcaA has N-acetylglucoasminyl-transferase activity when expressed alone; however, the enzyme activity is increased approximately 20-fold when *icaA* is co-expressed with both *icaD* and *icaC* (53). It has been suggested that IcaD may act as a chaperone for IcaA and/or act as a molecular link between IcaA and IcaC (11,53). IcaC is hypothesized to act in translocating polysaccharide synthesized by IcaAD through the cytoplasmic membrane (11,53). IcaB is proposed to act as a deacetylase as it has sequence similarity to the *Rhizobium* NodB, which acts as a chitooligosaccharide deacetylase (11,53). *icaR*, located adjacent to and divergently transcribed from *ica*, appears to be a member of the *tetR* family of transcriptional regulators that mediates transcriptional regulation of *ica* (54). IcaR regulation of *ica* will be discussed below.

10.5   GENETIC FACTORS CONTRIBUTING TO REGULATION OF *ica*

As more research into the regulation of *ica* transcription within the staphylococci has been conducted, it has become increasingly apparent that control of biofilm synthesis is highly complex and involves input from several global regulatory networks. This notion should not be surprising, however, because of the large amount of energy required to synthesize the enzymes and polysaccharide needed to construct PIA. Indeed, approaches utilizing transposon mutagenesis have revealed multiple loci that affect biofilm formation (19,55–57). In this section of the chapter, current studies on regulatory mechanisms affecting biofilm formation will be outlined.

10.5.1 *ICAR*

Located upstream of the *ica* operon is the divergently transcribed *icaR* gene. The role of *icaR* as a repressor of *ica* transcription was initially described by Conlon et al., where *icaR* knockout strains of *S. epidermidis* isolate CSF41498 led to a nearly six-fold upregulation of *ica* transcription (54). Amino acid comparison of IcaR to database proteins by these investigators revealed significant similarity to the *tetR* family of transcriptional regulators. *icaR* expression has been shown to be affected by environmental factors. In CSF41498 and RP62A, transcription of *icaR* was found to be significantly diminished by addition of ethanol to the growth media, thus leading to upregulation of *ica* transcription (54). IcaR and the proposed *ica* promoter sequence have been recently shown to interact directly by Jefferson and colleagues (58). In this work, using a DNase protection assay, recombinant IcaR protein was found to bind to a 42 bp region immediately upstream of *icaA*.

10.5.2 *SIGB*

σ^B is an alternate sigma factor responsible for the stress response within the staphylococci (59–65). The SigB activation pathway involves the action of several gene products: RsbU, RsbV, and RsbW. Under "normal" conditions, SigB is bound to the anti-sigma factor, RsbW. RsbW also uses its kinase activity to phosphorylate the anti-anti-sigma factor, RsbV. Upon sensing of stress, the phosphatase RsbU dephosphorylates RsbV. RsbV can then interact with RsbW, and σ^B is released. At this point, there is contradictory evidence as to what role, if any, SigB has on the expression of biofilm within the staphylococci. Rachid et al. demonstrated that deletion of *sigB* in *S. aureus* isolate MA12 led to a loss of *ica* transcription (66). This effect could be complemented by introduction of *sigB* on the plasmid pSK9. Further, the known *S. aureus rsbU* mutant strain, RN4220, was found to produce a biofilm when pSK9 was inserted. Characterization of one of the three transposon mutant classes of *S. epidermidis* strain 1457 described by Knobloch et al. revealed that Tn*917* had inserted within the phosphatase, RsbU (64). This mutant class exhibited loss of biofilm production when grown in tryptic soy broth (TSB). Unlike the wild-type strain, the mutant was unable to upregulate biofilm synthesis when cultured in TSB containing 4% sodium chloride. Yet, the wild-type and the mutant strains were able to produce biofilms when grown in TSB supplemented with 4% ethanol. These data indicate that σ^B has a central role in the synthesis of biofilms within *S. epidermidis* and *S. aureus*. Conflicting data has been reported by Kies et al., where a *sigB* deletion mutant of *S. epidermidis* strain Tü3298 was found to be unaffected in the cell aggregation that is characteristic of biofilm production (67). In addition, Valle et al. have recently reported that deletion of *sigB* in *S. aureus* isolates 15981 and ISP479c affected neither cellular clumping nor PIA/PNAG production (56). Further experimentation is necessary to determine the role of σ^B in staphylococcal biofilm regulation.

Multiple strains will need to be tested, as differences within the genetic backgrounds of the currently characterized strains may be responsible for the discrepancies in the results. Also unclear is whether σ^B directly or indirectly acts on the *ica* promoter to influence transcription. Several investigators have noted that no σ^B recognition sequence exists within the *ica* promoter (64,66).

10.5.3   *AGR*

The *agr* virulence regulatory system is composed of two divergently transcribed transcripts, RNAII and RNAIII (68–72). RNAII encodes genes responsible for quorum sensing within this system: *agrB*, *agrD*, *agrC*, and *agrA*. AgrC and AgrA are the histidine kinase sensor and response regulator, respectively, of a two-component regulatory system. AgrC senses an octapeptide pheromone that is processed from the AgrD peptide by AgrB. Upon accumulation of sufficient levels of pheromone in the environment, the two-component regulatory system is activated, leading to increased transcription of RNAII and the effector of the *agr* system, RNAIII. By an unknown mechanism, RNAIII mediates the response typical of *agr* activation, where surface proteins are downregulated and secreted proteins are upregulated as the culture reaches post-exponential phase. Otto et al. have recently reported that a *S. epidermidis agr* mutant of strain 1457 has increased levels of biofilm synthesis and adherence to polystyrene (73). These observations also held true for the biofilm-forming *agrC* mutant, *S. epidermidis* strain O-47. These investigators found that transcription of *atlE* was upregulated throughout the growth curve in the *agr* mutant. It was concluded that *agr* acts to repress *atlE* expression thus resulting in decreased initial attachment compared to the *agr* mutant. PIA production, however, was not affected when the 1457 wild-type and *agr* mutant were compared. In contrast to *S. epidermidis*, loss of *agr* appears to have little effect on biofilm synthesis in *S. aureus* (56,74). In addition, as shown by Dunman et al., *agr⁻* strains of *S. aureus* do not have differential expression of *atl* compared to wild type, indicating that expression of *atl* and *atlE* is regulated differently in these two staphylococcal species (75).

10.5.4   *SARA*

The staphylococcal accessory regulator protein, SarA, has been shown by two groups of investigators to play a central role in *S. aureus* biofilm production. Valle et al. showed a significant decrease in biofilm and PIA/PNAG production as well as *icaC* transcription when *sarA* mutants were compared with their isogenic wild-type counterparts (56). Results from Beenken et al. agree with these results, where loss of *sarA* completely diminished biofilm in six strains of *S. aureus* (74). A plasmid containing the *sarA* gene was introduced into two of these strains, and biofilm synthesis was restored both in microtiter plates and in flow cells. It is not known whether SarA interacts directly or indirectly to activate transcription of *ica*.

10.5.5   ADDITIONAL FACTORS

Several other factors have been found to play a role in regulation of biofilm expression within the staphylococci. Within the same collection of transposon mutants as the

S. aureus sarA mutant, Valle et al. have reported that a mutant within the *pnpA* locus, which encodes a polyribonucleotide nucleotidyltransferase, is rendered biofilm-negative (56). Further work is needed to determine the role of this enzyme in the formation of biofilm. In addition, Jefferson and colleagues have recently characterized a *S. aureus* mutant (MN8m) that overproduces PIA/PNAG (58). Upon sequencing of the *ica* promoter region of the mutant strain, these investigators found a 5 bp deletion (TATTT) 104 bp upstream of *icaA*. Using gel shift analysis, a shift was observed when the intact *ica* promoter region was incubated with cell-free lysates. These protein(s) did not bind to promoter fragments containing the 5 bp deletion, suggesting that there are other regulatory proteins, besides IcaR, that regulate *ica* transcription. These investigators also demonstrated that IcaR binds to a region just upstream of *icaA*; binding of IcaR to this region was not affected in promoter fragments containing the 5 bp deletion.

It has been known for some time that *S. epidermidis* does not form a functional biofilm if grown in media that does not contain glucose or certain other polysaccharides such as fructose or sucrose (76). However, it was not known whether the effect that glucose had on the formation of biofilm was due to the lack of a functional substrate for enzymes produced by the *ica* operon or whether glucose induced other protein(s) that are essential for the formation of biofilm. Dobinsky et al. have recently addressed this question by first demonstrating that *S. epidermidis* 1457 growing in media lacking glucose makes abundant, functional *ica* transcript, since induction of cells in stationary phase with glucose led to PIA synthesis (77). As this result could be due to the lack of glucose as a substrate for PIA synthesis or due to a glucose-dependent factor, the protein synthesis inhibitor, chloramphenicol, was added to cells in microtiter plates at time points before, during, and after addition of glucose to the media. Complete inhibition of biofilm formation occurred when the antibiotic was added prior to glucose addition, indicating that a glucose-dependent factor is necessary for biofilm formation. This *icaADBC*-independent factor has not been isolated to date. Associated with this line of work is the recent finding by Lim et al. showing that a gene called *rbf* (regulator of biofilm formation) is associated with the multicellular aggregation stage of biofilm formation (57). Sequence similarity suggests that *rbf* is a member of the AraC/XylS family of regulators; SDS-PAGE analysis of the mutant and wild-type suggest that *rbf* may regulate a glucose and salt responsive 190 kDa protein (57).

10.5.6   ENVIRONMENTAL FACTORS

A number of environmental factors have been reported to affect biofilm synthesis. Anaerobic conditions have been found to increase PIA/PNAG expression by enhanced *ica* transcription in strains of *S. aureus* and *S. epidermidis* (78). *ica* expression in *S. epidermidis* strain 220 was shown in transcriptional fusion and Northern blot studies to be increased in the presence of 4 to 5% sodium chloride or growth at 42°C (79). This study also demonstrated that, while several antibiotics had no inducing effect on biofilm formation, subinhibitory concentrations of tetracycline or streptogrammin antibiotics caused a large increase in *ica* expression. The macrolide, erythromycin, caused a slight increase in biofilm. Addition of ethanol to the growth

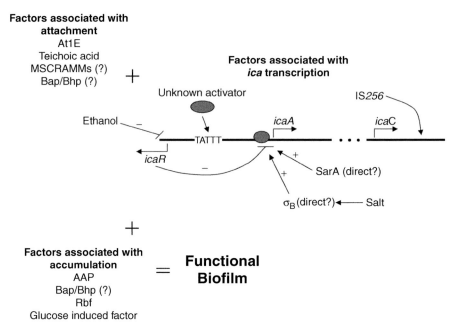

FIGURE 10.2   Factors associated with the production of a functional biofilm in *S. epidermidis.*

media can also induce biofilm synthesis, as shown by Mack et al. (64). An overview of factors that regulate *ica* transcription as well as biofilm formation is shown in Figure 10.2.

10.6 PHENOTYPIC VARIATION

Multiple studies have shown that levels of biofilm production differ consistently among variants of the same strain (80–86). These variants have typically been isolated on Congo Red agar (CRA) or Memphis agar, two media formulations designed to detect differences in ability of individual colonies to produce biofilm (6,87). Colonies on CRA that produce biofilm have been defined as "black and crusty" whereas those that produce little or no biofilm are "red and creamy" or "smooth" (see Figure 10.3). This phenomenon was first well characterized in 1990, when Christensen et al. (86) observed *S. epidermidis* phenotypic variants arise at a frequency of approximately 10^{-5}. These investigators indicated that *S. epidermidis* strain RP62A produced a spectrum of distinct colonial forms on Memphis agar where each member of this series of colony phenotypes (colony phenotypes ρ through ε) was found to have a decreasing level of biofilm synthesis synthesis; ρ had the highest level whereas ε had the lowest level. In addition, all colonies within the RP62A biofilm spectrum were able to give rise to other members of the spectrum with differing biofilm-forming capabilities. These different colony types also had varying levels of phenotypic properties. The biofilm producing ρ colonies were

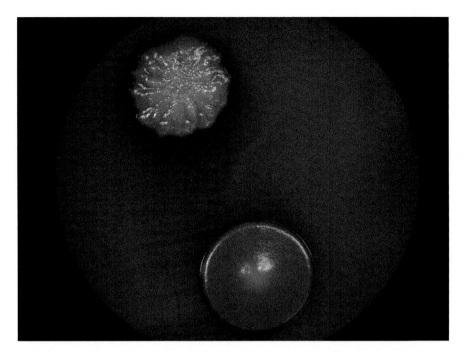

FIGURE 10.3 (See color insert following page 270)   Colonies of biofilm forming ("crusty" uppermost colony) and nonbiofilm forming ("smooth") *S. epidermidis* on Congo red agar.

methicillin-resistant and more virulent in a rabbit endocarditis model than the non-biofilm and methicillin-susceptible ε variants (6).

Mechanisms governing phenotypic variation have recently been studied at the molecular level. Ziebuhr and colleagues demonstrated that, in *S. epidermidis* RP62A, phenotypic variation of biofilm production was due to inactivation of *icaC* by IS*256* in 30% of the variants (82). This process was also shown to be reversible. *icaC* apparently serves as a "hot spot" for insertion of IS*256*, an IS element that appears to excise from DNA in a precise manner and moves to other locations through a circular intermediate (88). Other investigators studying phenotypic variation on IS*256*-negative isolates have found that phenotypic variation is associated with the loss of *ica* transcription (54,81,89). Handke et al. have recently demonstrated that phenotypic variation is a prevailing phenomenon within *S. epidermidis*, including those isolates that do not contain IS*256* (89). Multiple variants with varying levels of biofilm synthesis can be isolated from multiple *S. epidermidis* isolates. These isolates are stable for at least 24 hours of growth in tryptic soy broth, however, most variants revert to wild-type or other "intermediate" forms between 2–5 days incubation at 37°C (89). Two classes of "smooth" phenotypic variants were isolated, those that had decreased amount of *ica* transcription (class I variants) and those that had mutations or deletions within the *ica* operon (class II variants). As expected, biofilm forming revertants (and restoration of *ica* transcription) were isolated from class I phenotypic variants after growth in TSB broth for up to 5 days. Biofilm forming revertants were not isolated from class II phenotypic variants. Although the molecular mechanisms downregulating *ica* transcription and class I phenotypic variation are not known, they may also be involved in the regulation of many more loci than just *ica*.

10.7 CONCLUSIONS

The formation of a biofilm within the staphylococci is dependent upon many factors, both in the initial adherence as well as the accumulation phase. Even when studying just one aspect, regulation of *ica* transcription, there seem to be several layers of regulation. Using techniques (i.e. genomics and proteomics) to analyze global regulatory networks within the staphylococci, investigators will certainly dissect the multiple regulatory pathways that are involved in the regulation of *ica* transcription and biofilm formation in both *S. epidermidis* and *S. aureus*. As production of a biofilm may be related to chronic staphylococcal infections, the development of specific inhibitors of these pathways may lead to novel anti-staphylococcal therapy.

REFERENCES

1. National Nosocomial Infections Surveillance (NNIS) System report, data summary from January 1990–May 1999, issued June 1999. *Am. J. Infect. Control.* 27, 520–532, 1999.
2. Rupp, M.E., and Archer, G.L., Coagulase-negative staphylococci: pathogens associated with medical progress. *Clin. Infect. Dis.* 15, 197–205, 1992.
3. Christensen, G.D., Simpson, W.A., Bisno, A.L., and Beachey, E.H., Adherence of slime-producing strains of *Staphylococcus epidermidis* to smooth surfaces. *Infect. Immun.* 37, 318–326, 1982.
4. Christensen, G.D., Simpson, W.A., Younger, J.J., et al., Adherence of coagulase-negative staphylococci to plastic tissue culture plates: a quantitative model for the adherence of staphylococci to medical devices. *J. Clin. Microbiol.* 22, 996–1006, 1985.
5. Christensen, G.D., Baddour, L.M., and Simpson, W.A., Phenotypic variation of *Staphylococcus epidermidis* slime production *in vitro* and *in vivo*. *Infect. Immun.* 55, 2870–2877, 1987.
6. Christensen, G.D., Baddour, L.M., Madison, B.M., et al., Colonial morphology of staphylococci on Memphis agar: phase variation of slime production, resistance to beta-lactam antibiotics, and virulence. *J. Infect. Dis.* 161, 1153–1169, 1990.
7. Christensen, G.D., Barker, L.P., Mawhinney, T.P., Baddour, L.M., and Simpson, W.A., Identification of an antigenic marker of slime production for *Staphylococcus epidermidis*. *Infect. Immun.* 58, 2906–2911, 1990.
8. Baldassarri, L., Donnelli, G., Gelosia, A., Voglino, M.C., Simpson, A.W., and Christensen, G.D., Purification and characterization of the staphylococcal slime-associated antigen and its occurrence among *Staphylococcus epidermis* clinical isolates. *Infect. Immun.* 64, 3410–3415, 1996.
9. Drewry, D.T., Galbraith, L., Wilkinson, B.J., and Wilkinson, S.G., Staphylococcal slime: a cautionary tale. *J. Clin. Microbiol.* 28, 1292–1296, 1990.
10. Mack, D., Fischer, W., Krokotsch, A., et al., The intercellular adhesin involved in biofilm accumulation of *Staphylococcus epidermidis* is a linear β-1,6-linked glucosaminoglycan. Purification and structural analysis. *J. Bacteriol.* 178, 175–183, 1996.
11. Gotz, F., Staphylococcus and biofilms. *Mol. Microbiol.* 43, 1367–1378, 2002.

12. Maira-Litran, T., Kropec, A., Abeygunawardana, C., et al., Immunochemical properties of the staphylococcal poly-N-acetylglucosamine surface polysaccharide. *Infect. Immun.* 70, 4433–4440, 2002.

13. Tojo, M., Yamashita, N., Goldmann, D.A., and Pier, G.B., Isolation and characterization of a capsular polysaccharide adhesin from *Staphylococcus epidermidis. J. Infect. Dis.* 157, 713–722, 1988.

14. McKenney, D., Hubner, J., Muller, E., Wang, Y., Goldmann, D.A., and Pier, G.B., The *ica* locus of *Staphylococcus epidermidis* encodes production of the capsular polysaccharide/ adhesin. *Infect. Immun.* 66, 4711–4720, 1998.

15. McKenney, D., Pouliot, K.L., Wang, Y., et al., Broadly protective vaccine for *Staphylococcus aureus* based on an *in vivo*-expressed antigen. *Science* 284, 1523–1527, 1999.

16. McKenney, D., Pouliot, K., Wang, Y., et al., Vaccine potential of poly-1-6 beta-D-N-succinylglucosamine, an immunoprotective surface polysaccharide of *Staphylococcus aureus* and *Staphylococcus epidermidis. J. Biotechnol.* 83, 37–44, 2000.

17. Mack, D., Nedelmann, M., Krokotsch, A., Schwarzkopf, A., Heesemann, J., and Laurs, R., Characterization of transposon mutants of biofilm-producing *Staphylococcus epidermidis* impaired in the accumulative phase of biofilm production: genetic identification of hexosamine-containing polysaccharide intercellular adhesin. *Infect. Immun.* 62, 3244–3253, 1994.

18. Heilmann, C., Schweitzer, O., Gerke, C., Vanittanakom, N., Mack, D., and Gotz, F., Molecular basis of intercellular adhesion in the biofilm-forming *Staphylococcus epidermidis. Mol. Microbiol.* 20, 1083–1091, 1996.

19. Heilmann, C., Gerke, C., Perdreau-Remington, F., and Gotz, F., Characterization of *Tn917* insertion mutants of *Staphylococcus epidermidis* affected in biofilm formation. *Infect. Immun.* 64, 277–282, 1996.

20. Mack, D., Riedewald, J., Rohde, H., et al., Essential functional role of the polysaccharide intercellular adhesin of *Staphylococcus epidermidis* in hemagglutination. *Infect. Immun.* 67, 1004–1008, 1999.

21. Fey, P.D., Ulphani, J.S., Gotz, F., Heilmann, C., Mack, D., and Rupp, M.E., Characterization of the relationship between polysaccharide intercellular adhesin (PIA) and hemagglutination (HA) in *Staphylococcus epidermidis. J. Infect. Dis.* 179, 1561–1564, 1999.

22. Rupp, M.E., Fey, P.D., Heilmann, C., and Gotz, F., Characterization of the importance of *Staphylococcus epidermidis* autolysin and polysaccharide intercellular adhesin in the pathogenesis of intravascular catheter-associated infection in a rat model. *J. Infect. Dis.* 183, 1038–1042, 2001.

23. Rupp, M.E., and Fey, P.D., *In vivo* models to evaluate adhesion and biofilm formation by *Staphylococcus epidermidis. Methods Enzymol.* 336, 206–215, 2001.

24. Rupp, M.E., Fey, P.D., Ulphani, J.S., and Mack, D., Characterization of *Staphylococcus epidermidis* polysaccharide intercellular adhesin/hemagglutinin in the pathogenesis of intravascular catheter-associated infection in a rat model. *Infect. Immun.* 67, 2656–2659, 1999.

25. Rupp, M.E., Ulphani, J.S., Fey, P.D., and Mack, D., Characterization of the importance of polysaccharide intercellular adhesin/hemagglutinin of *S. epidermidis* in the pathogenesis of biomaterial based infection in a mouse foreign body model. *Infect. Immun.* 67, 2627–2632, 1999.

26. Francois, P., Tu Quoc, P.H., Bisognano, C., et al., Lack of biofilm contribution to bacterial colonisation in an experimental model of foreign body infection by *Staphylococcus aureus* and *Staphylococcus epidermidis. FEMS Immunol. Med. Microbiol.* 35, 135–140, 2003.

27. Perdreau-Remington, F., Sande, M.A., Peters, G., and Chambers, H.F., The abilities of a *Staphylococcus epidermidis* wild-type strain and its slime-negative mutant to induce endocarditis in rabbits are comparable. *Infect. Immun.* 66, 2778–2781, 1998.

28. Heilmann, C., Hussain, M., Peters, G., and Gotz, F., Evidence for autolysin-mediated primary attachment of *Staphylococcus epidermidis* to a polystyrene surface. *Mol. Microbiol.* 24, 1013–1024, 1997.

29. Vandecasteele, S.J., Peetermans, W.E., Merckx, R., and Van Eldere, J., Expression of biofilm-associated genes in *Staphylococcus epidermidis* during *in vitro* and *in vivo* foreign body infections. *J. Infect. Dis.* 188, 730–737, 2003.

30. Heilmann, C., Thumm, G., Chhatwal, G.S., Hartleib, J., Uekotter, A., and Peters, G., Identification and characterization of a novel autolysin (Aae) with adhesive properties from *Staphylococcus epidermidis*. *Microbiology* 149, 2769–2778, 2003.

31. Allignet, J., England, P., Old, I., and El Solh, N., Several regions of the repeat domain of the *Staphylococcus caprae* autolysin, AtlC, are involved in fibronectin binding. *FEMS Microbiol. Lett.* 213, 193–197, 2002.

32. Hell, W., Meyer, H.G., and Gatermann, S.G., Cloning of aas, a gene encoding a *Staphylococcus saprophyticus* surface protein with adhesive and autolytic properties. *Mol. Microbiol.* 29, 871–881, 1998.

33. Fitzgerald, S.N., and Foster, T.J., Molecular analysis of the tagF gene, encoding CDP-Glycerol:Poly(glycerophosphate) glycerophosphotransferase of *Staphylococcus epidermidis* ATCC 14990. *J. Bacteriol.* 182, 1046–1052, 2000.

34. Moat, A.G., Foster, J.W., M.P.S., Spector, M.P., and Foster, J.W., *Microbial Physiology*. New York, NY: Wiley-Liss, 2002.

35. Peschel, A., Otto, M., Jack, R.W., Kalbacher, H., Jung, G., and Gotz, F., Inactivation of the dlt operon in *Staphylococcus aureus* confers sensitivity to defensins, protegrins, and other antimicrobial peptides. *J. Biol. Chem.* 274, 8405–8410, 1999.

36. Gross, M., Cramton, S.E., Gotz, F., and Peschel, A., Key role of teichoic acid net charge in *Staphylococcus aureus* colonization of artificial surfaces. *Infect. Immun.* 69, 3423–3426, 2001.

37. Collins, L.V., Kristian, S.A., Weidenmaier, C., et al., *Staphylococcus aureus* strains lacking D-alanine modifications of teichoic acids are highly susceptible to human neutrophil killing and are virulence attenuated in mice. *J. Infect. Dis.* 186, 214–219, 2002.

38. Hussain, M., Heilmann, C., Peters, G., and Herrmann, M., Teichoic acid enhances adhesion of *Staphylococcus epidermidis* to immobilized fibronectin. *Microb. Pathog.* 31, 261–270, 2001.

39. Cucarella, C., Solano, C., Valle, J., Amorena, B., Lasa, I., and Penades, J.R., Bap, a *Staphylococcus aureus* surface protein involved in biofilm formation. *J. Bacteriol.* 183, 2888–2896, 2001.

40. Foster, T.J., and Hook, M., Surface protein adhesins of *Staphylococcus aureus*. *Trends. Microbiol.* 6, 484–488, 1998.

41. Cucarella, C., Tormo, M.A., Knecht, E., et al., Expression of the biofilm-associated protein interferes with host protein receptors of *Staphylococcus aureus* and alters the infective process. *Infect. Immun.* 70, 3180–3186, 2002.

42. Ubeda, C., Tormo, M.A., Cucarella, C., et al., Sip, an integrase protein with excision, circularization and integration activities, defines a new family of mobile *Staphylococcus aureus* pathogenicity islands. *Mol. Microbiol.* 49, 193–210, 2003.

43. Veenstra, G.J.C., Cremers, F.F.M., Van Dijk, K.H., and Fleer, A., Ultrastructural organization and regulation of a biomaterial adhesin of *Staphylococcus epidermidis*. *J. Bacteriol.* 178, 537–541, 1996.

44. Hussain, M., Herrmann, M., Von Eiff, C., Perdreau-Remington, F., and Peters, G., A 140-kilodalton extracellular protein is essential for the accumulation of *Staphylococcus epidermidis* strains on surfaces. *Infect. Immun.* 65, 519–524, 1997.

45. Williams, R.J., Henderson, B., Sharp, L.J., and Nair, S.P., Identification of a fibronectin-binding protein from *Staphylococcus epidermidis*. *Infect. Immun.* 70, 6805–6810, 2002.

46. Pei, L., and Flock, J.I., Lack of fbe, the gene for a fibrinogen-binding protein from *Staphylococcus epidermidis*, reduces its adherence to fibrinogen coated surfaces. *Microb. Pathog.* 31, 185–193, 2001.

47. Nilsson, M., Frykberg, L., Flock, J.I., Pei, L., Lindberg, M., and Guss, B., A fibrinogen-binding protein of *Staphylococcus epidermidis*. *Infect. Immun.* 66, 2666–2673, 1998.

48. Caiazza, N.C., and O'Toole, G.A., Alpha-toxin is required for biofilm formation by *Staphylococcus aureus*. *J. Bacteriol.* 185, 3214–3217, 2003.

49. Fitzpatrick, F., Humphreys, H., Smyth, E., Kennedy, C.A., and O'Gara, J.P., Environmental regulation of biofilm formation in intensive care unit isolates of *Staphylococcus epidermidis*. *J. Hosp. Infect.* 52, 212–218, 2002.

50. Arciola, C.R., Campoccia, D., Gamberini, S., Donati, M.E., Baldassarri, L., and Montanaro, L., Occurrence of ica genes for slime synthesis in a collection of *Staphylococcus epidermidis* strains from orthopedic prosthesis infections. *Acta. Orthop. Scand.* 74, 617–621, 2003.

51. Mack, D., Haeder, M., Siemssen, N., and Laufs, R., Association of biofilm production of coagulase-negative staphylococci with expression of a specific polysaccharide intercellular adhesin. *J. Infect. Dis.* 174, 881–884, 1996.

52. Cramton, S.E., Gerke, C., Schnell, N.F., Nichols, W.W., and Gotz, F., The intercellular adhesion (*ica*) locus is present in *Staphylococcus aureus* and is required for biofilm formation. *Infect. Immun.* 67, 5427–5433, 1999.

53. Gerke, C., Kraft, A., Submuth, R., Schweitzer, O., and Gotz, F., Characterization of the N-acetylglucosaminyltransferase activity involved in the biosynthesis of the *Staphylococcus epidermidis* polysaccharide intercellular adhesin. *J. Biol. Chem.* 273, 18586–18593, 1998.

54. Conlon, K.M., Humphreys, H., and O'Gara, J.P., icaR encodes a transcriptional repressor involved in environmental regulation of ica operon expression and biofilm formation in *Staphylococcus epidermidis*. *J. Bacteriol.* 184, 4400–4408, 2002.

55. Mack, D., Rohde, H., Dobinsky, S., et al., Identification of three essential regulatory gene loci governing expression of *Staphylococcus epidermidis* polysaccharide intercellular adhesin and biofilm formation. *Infect. Immun.* 68, 3799–3807, 2000.

56. Valle, J., Toledo-Arana, A., Berasain, C., et al., SarA and not sigmaB is essential for biofilm development by *Staphylococcus aureus*. *Mol. Microbiol.* 48, 1075–1087, 2003.

57. Lim, Y., Jana, M., Luong, T.T., and Lee, C.Y., Control of glucose- and NaCl-induced biofilm formation by *rbf* in *Staphylococcus aureus*. *J. Bacteriol.* 186, 722–729, 2004.

58. Jefferson, K.K., Cramton, S.E., Gotz, F., and Pier, G.B., Identification of a 5-nucleotide sequence that controls expression of the *ica* locus in *Staphylococcus aureus* and characterization of the DNA-binding properties of IcaR. *Mol. Microbiol.* 48, 889–899, 2003.

59. Kullik, I., and Giachino, P., The alternative sigma factor σ^B in *Staphylococcus aureus:* regulation of the *sigB* operon in response to growth phase and heat shock. *Arch. Microbiol.* 167, 151–159, 1997.

60. Deora, R., Tseng, T., and Misra, T.K., Alternative transcription factor $\sigma\beta$ of *Staphylococcus aureus*: characterization and role in transcription of the global regulatory locus *sar. J. Bacteriol.* 179, 6355–6359, 1997.

61. Gertz, S., Engelmann, S., Schmid, R., Ohlsen, K., Hacker, J., and Hecker, M., Regulation of σ^B-dependent transcription of *sigB* and *asp23* in two different *Staphylococcus aureus* strains. *Mol. Gen. Genet.* 261, 558–566, 1999.
62. Miyazaki, E., Chen, J.M., Ko, C., and Bishai, W.R., The *Staphylococcus aureus rsbW* (*orf159*) gene encodes an anti-sigma factor of SigB. *J. Bacteriol.* 181, 2846–2851, 1999.
63. Gertz, S., Engelmann, S., Schmid, R., et al., Characterization of the σ^B regulon in *Staphylococcus aureus*. *J. Bacteriol.* 182, 6983–6991, 2000.
64. Knobloch, J.K.M., Bartscht, K., Sabottke, A., Rohde, H., Feucht, H.H., and Mack, D., Biofilm formation by *Staphylococcus epidermidis* depends on functional RsbU, an activator of the *sigB* operon: differential activation mechanisms due to ethanol and salt stress. *J. Bacteriol.* 183, 2624–2633, 2001.
65. Palma, M., and Cheung, A.L., sigma(B) activity in *Staphylococcus aureus* is controlled by RsbU and an additional factor(s) during bacterial growth. *Infect. Immun.* 69, 7858–7865, 2001.
66. Rachid, S., Ohlsen, K., Sallner, U., Hacker, J., Hecker, M., and Ziebuhr, W., Alternative transcription factor σ^B is involved in regulation of biofilm expression in a *Staphylococcus aureus* mucosal isolate. *J. Bacteriol.* 182, 6824–6826, 2000.
67. Kies, S., Otto, M., Vuong, C., and Gotz, F., Identification of the *sigB* operon in *Staphylococcus epidermidis*: construction and characterization of a *sigB* deletion mutant. *Infect. Immun.* 69, 7933–7936, 2001.
68. Yarwood, J.M., and Schlievert, P.M., Quorum sensing in *Staphylococcus* infections. *J. Clin. Invest.* 112, 1620–1625, 2003.
69. Novick, R.P., Pathogenicity factors and their regulation. In: Fischetti, V.A., Novick, R.P., Ferretti, J.J., Portnoy, D.A., and Rood, J.I., eds., *Gram-positive Pathogens*. Washington, D.C.: American Society for Microbiology, 2000:392–407.
70. Novick, R.P., Ross, H.F., Projan, S.J., Kornblum, J., Kreiswirth, B., and Moghazeh, S., Synthesis of staphylococcal virulence factors is controlled by a regulatory RNA molecule. *EMBO J.* 12, 3967–3975, 1993.
71. Novick, R.P., Projan, S.J., Kornblum, J., et al., The *agr* P2 operon: an autocatalytic sensory transduction system in *Staphylococcus aureus*. *Mol. Gen. Genet.* 248, 446–458, 1995.
72. Peng, H.L., Novick, R.P., Kreiswirth, B., Kornblum, J., and Schlievert, P., Cloning, characterization, and sequencing of an accessory gene regulator *(agr)* in *Staphylococcus aureus*. *J. Bacteriol.* 170, 4365–4372, 1988.
73. Vuong, C., Gerke, C., Somerville, G.A., Fischer, E.R., and Otto, M., Quorum-sensing control of biofilm factors in *Staphylococcus epidermidis*. *J. Infect. Dis.* 188, 706–718, 2003.
74. Beenken, K.E., Blevins, J.S., and Smeltzer, M.S., Mutation of *sarA* in *Staphylococcus aureus* limits biofilm formation. *Infect. Immun.* 71, 4206–4211, 2003.
75. Dunman, P.M., Murphy, E., Haney, S., et al., Transcription profiling-based identification of *Staphylococcus aureus* genes regulated by the *agr* and/or *sarA* loci. *J. Bacteriol.* 183, 7341–7353, 2001.
76. Mack, D., Siemssen, N., and Laurs, R., Parallel induction by glucose of adherence and a polysaccharide antigen specific for plastic-adherent *Staphylococcus epidermidis*: evidence for functional relation to intercellular adhesion. *Infect. Immun.* 60, 2048–2057, 1992.
77. Dobinsky, S., Kiel, K., Rohde, H., et al., Glucose-related dissociation between *icaADBC* transcription and biofilm expression by *Staphylococcus epidermidis*: evidence for an additional factor required for polysaccharide intercellular adhesin synthesis. *J. Bacteriol.* 185, 2879–2886, 2003.

78. Cramton, S.E., Ulrich, M., Gotz, F., and Doring, G., Anaerobic conditions induce expression of polysaccharide intercellular adhesin in *Staphylococcus aureus* and *Staphylococcus epidermidis*. *Infect. Immun.* 69, 4079–4085, 2001.
79. Rachid, S., Ohlsen, K., Witte, W., Hacker, J., and Ziebuhr, W., Effect of subinhibitory antibiotic concentrations on polysaccharide intercellular adhesin expression in biofilm-forming *Staphylococcus epidermidis*. *Antimicrob. Agents Chemother.* 44, 3357–3363, 2000.
80. Rupp, M.E., Sloot, H.G.W., Meyer, J., Han, J., and Gatermann, S., Characterization of the hemagglutinin of *Staphylococcus epidermidis*. *J. Infect. Dis.* 172, 1509–1518, 1995.
81. Ziebuhr, W., Heilmann, C., Gotz, F., et al., Detection of the intercellular adhesion gene cluster (*ica*) and phase variation in *Staphylococcus epidermidis* blood culture strains and mucosal isolates. *Infect. Immun.* 65, 890–896, 1997.
82. Ziebuhr, W., Krimmer, V., Rachid, S., Lobner, I., Gotz, F., and Hacker, J., A novel mechanism of phase variation of virulence in *Staphylococcus epidermidis:* evidence for control of the polysaccharide intercellular adhesin synthesis by alternating insertion and excision of the insertion sequence element IS*256*. *Mol. Microbiol.* 32, 345–356, 1999.
83. Deighton, M.A., Capstick, J., and Borland, R., A study of phenotypic variation of *Staphylococcus epidermidis* using congo red agar. *Epidemiol. Infect.* 109, 423–432, 1992.
84. Deighton, M.A., Pearson, S., Capstick, J., Spelman, D., and Borland, R., Phenotypic variation of *Staphylococcus epidermidis* isolated from a patient with native valve endocarditis. *J. Clin. Microbiol.* 30, 2385–2390, 1992.
85. Christensen, G.D., Baddour, L.M., and Simpson, W.A., Phenotypic variation of *Staphylococcus epidermidis* slime production *in vitro* and *in vivo*. *Infect. Immun.* 55, 2870–2877, 1987.
86. Christensen, G.D., Baddour, L.M., Madison, B.M., et al., Colonial morphology of staphylococci on memphis agar: phase variation of slime production, resistance to beta-lactam antibiotics, and virulence. *J. Infect. Dis.* 161, 1153–1169, 1990.
87. Freeman, D.J., Falkiner, F.R., and Keane, C.T., New method for detection of slime production by coagulase-negative staphylococci. *J. Clin. Pathol.* 42, 872–874, 1989.
88. Loessner, I., Dietrich, K., Dittrich, D., Hacker, J., and Ziebuhr, W., Transposase-dependent formation of circular IS256 derivatives in *Staphylococcus epidermidis* and *Staphylococcus aureus*. *J. Bacteriol.* 184, 4709–4714, 2002.
89. Handke, L.D., Conlon, K.M., Slater, S.R., et al., Genetic and phenotypic analysis of biofilm phenotypic variation in multiple *Staphylococcus epidermidis* isolates. *J. Med. Micro.* 53, 367–374, 2004.

11 Cell-to-Cell Communication in Bacteria

Kenneth D. Tucker and Luciano Passador

CONTENTS

11.1 INTRODUCTION

Historically, bacteria have been studied as isolated organisms. It was appreciated that these organisms could detect and respond to their environment, such as responding with motility towards nutrients. Yet, the observation of various responses to stress, such as fruiting body formation in the myxobacteria, and the swarming motility seen in *Proteus* spp., suggested that bacteria were capable of coordinating their response and reacting as a population to a given environmental stimulus (1,2). Implicit in this belief was the presence of a mechanism for these organisms to communicate with each other. However, only recently has the level of complexity of these interactions become apparent. We now appreciate that simple single-celled organisms can express a complex array of different physiologies in response to their environment. Furthermore, it has become apparent that the single-celled organisms were not acting in isolation, but communicated with organisms of the same species as well as different species of bacteria. The hallmark of these communication systems is the evolution of a mechanism to transfer information from one cell to another using diffusible molecules.

To date various communication systems have been identified. While each system works by a unique mechanism, all share the principle that a signal molecule must be present at a significant concentration to elicit a response. Achieving a threshold amount of the signal requires that a significant population of the bacteria produce the same signal. These systems serve to communicate the presence of a significant number of similar cells, or a quorum. This is referred to as quorum sensing. The term quorum sensing was introduced to connote the idea that the individual cells were coordinating their physiology to act as a collective group. In all of the cases, the bacteria use the communication to adjust their physiology, leaving scientists to speculate on the advantage for the cells to coordinate this adjustment. A hint to the reason for these communication systems is that the accumulation of the communication signal reaches the threshold level typically when the cells are present at a high density. Thus, these systems are also referred to as density dependent communication. Many of the physiologies controlled by the communication systems can be interpreted as being more efficient at high cell densities. Without trying to over-interpret the role of these systems in the cells' response to new environments, it is certain the bacteria can communicate with each other at the sub-species level, the species level and the genus level. In many examples the communication system and the affected physiology impacts the ability of organisms to adapt to and infect humans to cause disease.

While specific signals may differ between species of bacteria, several basic types of communication can be identified. For signals unique to Gram-positive bacteria, the signal is typically a peptide chemotype. In Gram-negative bacteria the signal is typically an acylated homoserine lactone (AHL). However, there are examples of peptide signals in Gram-negative bacteria (3) and of homologues of acylated homoserine lactones acting as a signal in Gram-positive bacteria (4). Additionally a furanosyl borate diester has been identified as a signal in both Gram-positive and Gram-negative bacteria.

The peptide signals used by Gram-positive bacteria may be divided into two basic modes of operation: peptides that trigger a cytoplasmic receptor or peptides that trigger an intermembrane receptor. The cytoplasmic peptide-receptors are varied and do not show significant homology between species. However, most of the intermembrane peptide-receptors belong to a related class of proteins and truly represent a related signaling system (5).

In Gram-negative bacteria the chemotype of the major signal produced (acylated homoserine lactones) defines the type of communication. Furthermore, the proteins involved in the production and detection of the chemical signal share significant amino acid similarity, with one exception. Herein, we discuss examples from each of these groups.

Although cell-to-cell communication has been described in numerous bacteria, it does not appear to be a universal trait, nor does it appear to cluster within a specific family of organisms. Within the three major classes of communication, there are various examples each with specific unique variations. This field has made rapid strides in understanding because of the contributions provided by studies in bacterial physiology and symbiotic relationships involving bacteria with plants and animals. Concomitant with this rapid progress is the accumulation of a vast literature on the topic. To cover this information would require several volumes and is beyond the scope of this chapter. Therefore, in this review we'll focus on examples using bacteria of medical importance, where possible, without meaning any slight to the significant contributions to the understanding of the systems that has been provided by studies of the many organisms that are not mentioned.

11.2 GRAM-NEGATIVE AHL SIGNALS

The first quorum-sensing system in the Gram-negative bacteria to be described and characterized was an acyl-homoserine lactone (AHL) communication system. The symbiotic organism *Vibrio fischeri* colonizes the light organ of squid and is actually responsible for the luminescence of the organ. It was noted that *V. fischeri* was luminescent only at high cell densities, and this light production was dependent upon the presence of a diffusible AHL (6–8). It was subsequently demonstrated that the *V. fischeri lux* operon responsible for luminescence was under the control of two regulatory elements, LuxR and LuxI, where LuxI synthesizes the AHL while LuxR functions as a transcriptional regulatory protein (9). In the presence of its cognate AHL-signal, LuxR directly promotes transcription of the luminescence operon and thereby regulates luminescence (9). Both *lux*R and *lux*I are transcribed from promoters regulated by LuxRI (9). Thus *lux*R and *lux*I are autoregulated and the AHL regulating this is referred to as the auto-inducer (8,9). Subsequently, it was found that the opportunistic human pathogen *Pseudomonas aeruginosa* used a homologous regulatory system to control the expression of many of its virulence factors (10–12). This finding opened the door to the possibility that the LuxRI-type communication system could be more prevalent than originally thought and, further, that it might play a role in bacterial pathogenesis. Indeed, subsequent research has led to the description of over 70 LuxRI-type systems expressed in more than thirty different species of Gram-negative-bacteria, and including examples from alpha, beta and gamma proteobacteria (13,14). As these numbers suggest, many species contain more than one LuxRI-type signaling system. While the LuxRI components in these systems share significant amino acid sequence similarity, they are not identical, and many of the LuxI homologues produce signals with unique acyl sidechain length and composition (see Table 11.1 for examples). In terms of a general mechanism, it has been demonstrated that the acyl-homoserine lactone synthase uses S-adenosyl-L-methionine and acylated acyl carrier protein as substrates to produce the specific acyl-homoserine lactone signal (15–17). While the general structure of the AHL molecules is conserved, there can be significant differences in the composition of the acyl side chain with regard to chain length, oxidation status and degree of chain saturation. It is proposed that the acyl side chain provides specificity to each system and, in those systems described to date, it can vary from 4 to 18 carbons in length (e.g., (18,19)). The variety seen in AHL acyl side chain is specific to the cognate AHL synthase. As would be expected, the signal produced by the acyl-homoserine lactone synthase must co-evolve with the receptor LuxR-homolog, and the cognate LuxR-homologue is relatively specific for the acyl-homoserine lactone produced for that specific LuxRI-type system. Even in species with more than one LuxRI-type system, the LuxR homologue is most efficiently activated only by the AHL produced by its cognate LuxI-like protein and not by the other LuxI-like protein in the same cell. While this specific interaction would suggest signaling isolation, different species harboring non-identical LuxI can produce signals that affect LuxRI-type systems in other species (20). This property was used to discover many of the LuxRI-type systems described subsequent to the discovery of the *P. aeruginosa* LuxRI-type system (21). This indicates the AHL communication systems are not isolated communication systems and that species cross-talk is a possibility.

TABLE 11.1
Examples of Quorum Sensing Signals in Gram-Negative Bacteria

Structure	Designation	Organism	Regulated Physiology	References
	C$_4$-HSL	Pseudomonas aeruginosa	Rhamnolipid, biofilm, elastase	25,51,52,53,80,81
		Aeromonas hydrophilia	Extracellular proteases	21
		Aeromonas salmonicida	Extracellular proteases	21
		Serratia liquifaciens	Extracellular proteases & swarming	18,82
	3-oxo-C$_6$-HSL	Vibrio fischeri	Bioluminescence	6,83
		Erwinia carotovora	Extracellular enzymes	84,85,86,87,88
		Erwinia chrysanthemi	Pectinase	89,90
		Erwinia stewartii	Extracellular virulence factors	91
		Enterobacter agglomerans	?	79
		Yersinia enterocolitica	Motility & clumping	92
		Pseudomonas aeruginosa	?	40
	C$_8$-HSL	Burkholderia cepacia	Extracelluar proteases, siderophores	93
		Rhizobium leguminosarum	Transcription of rhiABC	78,94,95
		Ralstonia solanacearum	?	96
		Yersinia pseudotuberculosis	Motility & clumping	92
	3-oxo-C$_{12}$-HSL	Pseudomonas aeruginosa	Extracellular proteases	10,11,12,34,40, 47,49
	3 Hydroxy-7-cis-C$_{14}$-HSL	Rhizobium leguminosarum	Transcription of rhiABC	78,94,95
	AI-2	Vibrio harveyi	Bioluminescence	99,100

A number of LuxRI-type systems have been identified, but few have been as well characterized as the system in *P. aeruginosa*. As a result of its medical significance and the wealth of detail available, we will focus on this organism. In *P. aeruginosa* there are two LuxRI-type systems (LasRI and RhlRI) that use the AHL signals N-(3-oxododecanoyl)-L-homoserine lactone ($3OC_{12}$-HSL) and N-butyryl homoserine lactone (C_4-HSL) respectively, and interact in a hierarchical manner. Collectively, these systems control expression of virulence factors, such as proteases, associated with tissue damage in the lung and dissemination in infected burn-tissue (22,23). Additionally, these systems regulate rhamnolipid production, which is required for the complex architecture of deep biofilms formed by axenic cultures of *P. aeruginosa* (24–27). Recent transcriptosome analyses indicate the AHL system in *Pseudomonas* regulates approximately 6 to 10% of the organism's genome (28,29). The *in vivo* production of the *Pseudomonas* AHL has been demonstrated in excretions from cystic fibrosis patients suggesting that the quorum-sensing mechanism is operational *in vivo* (30–33). The role of quorum sensing in biofilm formation has been well documented. Strains of *P. aeruginosa* with *lasI* disrupted only formed flat and undifferentiated biofilms (34). When exogenous $3OC_{12}$-HSL was added, the strain produced the deep and complex biofilms seen with wild type *Pseudomonas*. Additionally, chemical antagonism of quorum sensing resulted in the sloughing of formed biofilms (35). Subsequently, it was demonstrated that expression of rhamno-lipid, under the control of RhlRI, was required by *Pseudomonas* to form deep biofilms (24). Further, Yoon et al. demonstrated that bacterial membrane proteins uniquely expressed in anaerobic biofilms are also expressed in infection of human lungs, indicating that a biofilm is formed in the human infection (36). While *Pseudomonas* goes through multiple physiological changes during biofilm development, the *las* and *rhl* systems clearly play a pivotal role in biofilm development in this organism, and the AHL signaling system plays a key role in the control of physiology in this organism (34,37–39).

At the first level of the AHL communication system in *P. aeruginosa* are the LasR and LasI proteins. These proteins are encoded from adjacent and *cis* chromosomal genes with separate promoters. LasI is required for the synthesis of $3OC_{12}$-HSL (Table 11.1). Insertion of *lasI* in an *Escherichia coli* background results in the production of biologically active $3OC_{12}$-HSL by the recombinant *E. coli*. In the LasRI system the cognate AHL has a 12 carbon acyl group and a keto group on C3 (40). Increasing or decreasing the length of the hydrocarbon tail by two carbons, reducing the saturation of the tail, or reducing one of the oxo-groups still results in biologically active AHL, but typically with reduced activity (41). Additional changes to the acyl group results in loss of activity of the AHL in the LasRI system. The AHL-synthase (LasI) provides the specificity to produce the appropriate acyl chain length. However, the available pool of acyl groups in the cell may also affect the type of signal produced and LasI does produce AHL with differing chain lengths (20,42,43). A basal level of expression of LasI allows low, but continuous synthesis of $3OC_{12}$-HSL. In many LuxRI-type systems, the AHL freely diffuses out of the cell and equal concentrations are found inside and outside the cell (8,26,44). However, $3OC_{12}$-HSL has a 12 carbon acyl-chain, which appears to hinder movement across the cell membranes presumably due to its hydrophobicity. In this case the *mexAB-oprM* encoded efflux pump actively transports $3OC_{12}$-HSL out of the cell (45). The concentration of $3OC_{12}$-HSL accumulates in the culture and reaches a critical threshold level in late stationary phase in cultures grown on artificial medium. At the critical threshold level of $3OC_{12}$-HSL, the AHL binds to LasR, which is constitutively expressed. This binding of AHL by LasR is believed to promote a conformational change in LasR and the formation of homodimers of LasR that can act as a transcriptional activator (46).

The genes regulated by LasR are not in an isolated operon (10,11,25,47–49), but are distributed throughout the chromosome of *P. aeruginosa*. Thus LasRI may be considered to control a regulon. The promoters of some of the genes regulated by LasRI contain shared sequences common to many LuxRI-type regulated genes and is referred to as a *las box* (similar to the *lux*-box described in the *V. fischeri* system) (50). However, not all of the *las*-responsive promoters appear to interact with the LasR/3OC12-HSL complex equally and there is an identified hierarchy of promotion of transcription by LasR (48). The promoter that is activated by the lowest concentration of $3OC_{12}$-HSL (i.e., most sensitive) regulates the transcription of *lasI*. In an *E. coli* background containing recombinant *lasR* and $3OC_{12}$-HSL supplied exogenously, as little as 0.1 nM of $3OC_{12}$-HSL results in half-maximal expression from the *lasI* promoter whereas 1.0 nM $3OC_{12}$-HSL is required to activate transcription of the gene for elastase (*lasB*), which is also regulated by the LasRI system. Thus, the LasRI system first promotes the increased production of LasI. This autoinduction of LasI results in a rapid increase in the amount of $3OC_{12}$-HSL required to activate additional LasR to promote the transcription of other genes regulated by the LasRI system. These include proteases such as alkaline protease, LasA, elastase (LasB), and proteins involved in type II secretion (Xpr) (10,11,47,49). Additionally, LasRI regulates the activation of the second *P. aeruginosa* AHL system, RhlRI (25,51,52).

RhlRI make up a second LuxRI-type system in *Pseudomonas*. This system controls transcription of *rhlAB* which encodes the enzymes required for synthesis of rhamnolipid (heat-stable hemolysin) (25,53). Additionally, in an *E. coli* background, RhlRI increases expression of RpoS, a sigma factor expressed during stationary phase that controls numerous stress response genes (54). In the RhlRI system, low level expression of RhlI results in the constitutive synthesis of N-butanoyl-L-homoserine lactone (designated C_4-HSL). As C_4-HSL is synthesized, it passively diffuses out of the cell and accumulates inside and outside the cell. This AHL is required to activate the transcriptional regulator RhlR. However, RhlR is not constitutively expressed. While the transcription of *rhlR* can be regulated by other factors (55), the LasRI system plays a major role in the regulation of RhlR activity at the translational and possibly the post-translational level (26,51,56). LasR with $3OC_{12}$-HSL directly promotes transcription of *rhlR*. Until LasR with $3OC_{12}$-HSL promotes transcription of *rhlR*, the protein is not available to interact with the C_4-HSL signal. The expressed RhlR is activated by the C_4-HSL to promote transcription of genes regulated by the RhlRI system, including increased transcription of *rhlI*. However, while $3OC_{12}$-HSL does not activate RhlR, it does bind to RhlR and competes with C_4-HSL for the protein (56). Thus, $3OC_{12}$-HSL prevents activation of RhlR until the concentration of C_4-HSL approximately equals or exceeds the concentration of

$3OC_{12}$-HSL. This appears to provide a delay in the transcription of RhlRI regulated genes, including *rpoS*, which regulates the expression of genes associated with survival in stationary phase cultures.

In the *Pseudomonas* system there are additional layers of complexity that are not yet well understood. For example, how the cascade down-regulates itself is not defined. A potential mechanism for down regulation involves a recently described protein designated QscR (57). This protein shares sequence similarity to LasR and RhlR, although a cognate LasI/RhlI protein was not identified. It represses quorum sensing controlled genes with the proposed mechanism involving the repression of *lasI*. A recent report demonstrates that QscR also directly interacts with LasR and RhlR (58). Additionally, a gene, *rsaL*, whose product also appears to function as a negative regulator of *lasI* expression has been described (59). This gene lies within the *lasR-lasI* intergenic region but is divergently transcribed, and is also controlled by *lasRI*. Expression of RsaL causes a decrease of detectable $3OC_{12}$-HSL in *P. aeruginosa* and a concurrent decrease in elastase production. Thus, this protein clearly plays a role in down regulating *lasRI*. An additional protein, RsmA, represses synthesis of both $3OC_{12}$-HSL and C_4-HSL (60). It appears to be involved in the control of AHL production during exponential growth since deletion of *rsmA*, results in expression of both $3OC_{12}$-HSL and C_4-HSL earlier in the growth phase than normal (60). While these are tantalizing reports, how *qscR*, *rsaL* and *rsmA* fit within the cascade of AHL-responsive genes is undefined.

Intertwined with the AHL communication system of *Pseudomonas* is a second communication system that uses a quinolone signal, specifically, 2-heptyl-3-hydroxy-4-quinolone (61). The significance of this system is indicated by the demonstration of the quinolone signal in lungs of cystic fibrosis patients infected with *P. aeruginosa* (62,63). The expression of the quinolone signal requires active LasR, and causes increased transcription of *rhlI*, *rhlR* and *lasR* (61,64,65). This system increases expression of elastase in a RhlR dependent manner (61). Thus this system is both regulated and regulates the AHL signaling system in *Pseudomonas*. The genes for the synthesis of the quinolone signal have been identified as *pqsABCDH* by mutation analysis, and the synthesis of the signal involves an anthranilate precursor (65,66). While the specific elements used by the quinolone signaling system to affect the AHL signaling system need to be defined, *pqsE* is required for response to the quinolone signal and may serve as the signal effector (65). This interaction between multiple signaling systems in *Pseudomonas* demonstrates the complexity of transcriptional regulation in this organism, and suggests how this bacterium successfully adapts to a number of different environments.

Through this cascade of transcriptional regulators, *Pseudomonas* is able to coordinate its physiology for optimal efficiency to promote survival in stationary phase (54). Using the LasRI system, it regulates the expression of exported proteins, and the mechanism of export (10,11,47,49). By coordinating the expression of these proteases, the population of cells orchestrates a massive production of proteases for maximal effect and does not expend energy on their synthesis until a cell density is achieved, whereby the export of these proteases would be effective and necessary to release additional nutrients for the cells. Further down this cascade, the cells maintain tight regulation of expression of RhlRI, which provides surfactants that may liberate additional nutrients and allows the cells to develop deep biofilms, which is a condition that requires the coordination of population physiology through cell-to-cell communication. In addition, the RhlRI system then provides regulation over *rpoS*, which sets the stage for the cells to survive starvation conditions. Thus, in *Pseudomonas*, the individual cells appear to be using the cell-to-cell communication to coordinate population starvation and survival activity.

The importance of AHL systems in regulating bacterial physiology suggests that this communication system could provide a point of attack in combating bacterial infections. This is supported by the observation that evolution has apparently selected this as a target for competing bacteria. AHL production has been identified in naturally occurring environmental biofilms (67). Many organisms must encounter bacteria that use AHL communication systems. Numerous plant pathogens have been indicated to use AHL systems (e.g., (68)). Interestingly, a seaweed, *Delisea pulchra* produces a halogenated AHL-analog that antagonizes AHL communication (69). Additionally, some microorganisms produce enzymes that degrade the AHL signal (70,71). It is possible that these represent an evolutionary escalation of defense to counter infection or competition with the AHL producing organisms. Along these lines, antagonists of LasR quorum-sensing have been chemically synthesized (72,73). At least one of these antagonists provides protection in the mouse pulmonary infection model with *Pseudomonas*, which indicates that the AHL signaling system may serve as a potential drug target.

It has been proposed that AHL-communication systems provide only an intraspecies communication because of the specificity imposed by the acyl-chain length (74). Thus, one species using an AHL communication system would not be expected to functionally communicate with other species. However, interspecies cross-talk through the AHL signals is certainly a possibility and examples exist of this being a reality. Steidle *et al.* developed an environmental model in tomato plants and clearly demonstrated that environmental bacteria in the rhizosphere are able to communicate via AHL to other bacterial species that colonize and infect the tomato plant (75). They propose that AHL is a universal language of communication within the rhizosphere (75). Additionally, in mixed biofilms containing *P. aeruginosa* and *Burkholderia cepacia*, it has been demonstrated that the AHL communication system of *B. cepacia* responds to the AHLs produced by *P. aeruginosa* although the *P. aeruginosa* did not respond to the AHLs produced by the *B. cepacia* (76). In a separate report, it was indicated that signals in the media from *P. aeruginosa* increase expression of virulence factors by a strain of *B. cepacia* (77) documenting species cross-talk even though *B. cepacia* uses a C_8-HSL and *P. aeruginosa* uses either 3-oxo-C_{12}-HSL or C_4-HSL.

Beyond these limited examples, a larger potential for species cross-talk is obvious. At one level, many species actually use identical AHL in their systems (see Table 11.1 for examples). For example, *P. aeruginosa* uses C_4-HSL in the RhlRI system, and *Aeromonas hydrophila*, *Aeromonas salmonicida* and *Serratia liquifaciens* each also uses C_4-HSL in its respective AHL communication system. In all four organisms the collective AHL-communication system regulates the expression and release of proteases. Thus, if these organisms did coexist in an environment, then species cross-talk through an AHL communication system would not only be possible, but it would allow the individual species to coordinate a uniform physiological response of protease release. This potential for species cross-talk is not limited to species that use the same AHL for signaling. As described before, the AHL synthase is not absolutely specific, but makes additional AHL that does not act upon the system's AHL-binding protein. For example, LasI has been reported to produce AHL with different acyl-chain lengths besides the 12 carbon chain preferred by LasR (40). Additional examples from other species of AHL-synthase producing AHL with variable acyl chains exist (18,19,78). When this is considered along with the fact that AHL-binding proteins can be activated by AHL with acyl structures not identical to its preferred AHL (18,41,78), then the potential for interspecies cross-talk rapidly expands. This potential is demonstrated by the fact that many of the original LuxRI-like systems were identified by their ability to synthesize AHL that could activate the non-cognate LuxR protein of *V. fischeri* in the reporter system (e.g., (79)). Thus species cross talk using AHL is shown and is potentially more widespread than these examples indicate.

Species cross-talk can impact disease in humans, as exemplified by the cross-talk between *Pseudomonas* and *Burkholderia*. However, the significance of this cross talk upon the bacteria is not so obvious. Cross-talk may allow different species to coordinate their activities to act collectively and facilitate infection of a host. At the same time, these organisms are also in competition with other bacteria. It is possible that the species cross-talk may be used to send a competing organism into a physiological change prematurely and thereby provide a survival disadvantage to that organism. This potential to use AHL signals to befuddle a competing bacterium is suggested by the observation that AHL with acyl chains longer than the cognate AHL can actually inhibit the signal provided by the cognate AHL (67,79). Such communication jamming is a possible reason why some bacteria have two-tiered HSL communication systems. By having a requirement for two separate AHL to control specific genes, the organism can impose a greater degree of control and preclude the potential communication and gene regulation that would be possible by another species synthesizing one cross-communicating AHL. Thus, cross-talk with the potential for communicating information or mis-information exists. However, the significance of this cross-talk in relation to the fitness of the bacteria and their potential to cause disease will require more research.

11.3 FURANOSYL BORATE DIESTER SIGNAL

Coincidentally, the discovery of the second type of cell-to-cell communication in Gram-negative bacteria also resulted from the study of a luminescent *Vibrio*. *Vibrio harveyi* luminescence is regulated in a cell density-dependent manner similar to *V. fischeri*. Diffusible molecules control this luminescence. However, *V. harveyi* has presented several enigmas in its signaling systems. First, it does produce an AHL that regulates luminescence (97). Unlike other Gram-negative bacteria, this AHL is not produced by a LasI-like enzyme (98). Further, the AHL does not bind to a LasR-like protein; instead it interacts with a histidine kinase that indirectly controls transcription of the *lux*-operon encoding the genes required for luminescence. Through studies with strains mutated in their AHL-response, it became apparent the luminescence is regulated by a second circuit that also involves a diffusible molecule referred to as AI-2 (99). (AI-2 is based on the original nomenclature used in the field to designate signaling molecules that auto-induce their own production, with AI as an acronym for auto-inducer. However, AI-2 has not been shown to auto-induce its production.) Through deletion studies, the synthesis of this molecule was traced to the *luxS* gene and subsequently identified as the furanosyl borate diester 3A-methyl-5,6-dihydro-furo [2,3-D][1,3,2] dioxaborole-2,2,6,6 A tetraol (Table 11.1) (99,100). The *luxS* gene is widespread in bacteria, and is found in both Gram-negative and Gram-positive bacteria. While the actual structure of AI-2 has only been confirmed in *V. harveyi*, LuxS dependent signals that can activate the *lux* operon of *V. harveyi* have been demonstrated in a number of bacteria that possess *luxS* (101–103). Thus, a common signal is postulated to be involved in the AI-2 systems.

The potential for bacterial interspecies cross-talk in the AI-2 system has generated much excitement and related research; however, efforts to advance this field have been hampered by the lack of biological assays. The fact that AI-2 acts through a signaling cascade makes the identification of genes directly controlled by AI-2 difficult. Even in the first identified system, the *lux*-operon of *V. harveyi*, there are elements in the cascade that are not identified. Additionally, the *V. harveyi* system is co-regulated by an AHL system. This potential for co-regulation further confounds the identification of AI-2 regulated genes. Even today, the major assay for AI-2 activity is the activation of *V. harveyi* luminescence in a strain with the AHL signaling system disrupted. The role of AI-2 in other bacteria typically has been determined using *luxS* deletions. This approach has indicated that LuxS plays a role in toxin production in *E. coli, Clostridium perfringens, Actinobacillus actinomycetemcomitans, Streptococcus pyogenes* and *Vibrio cholera* (104–108); it regulates motility in *E. coli* and *Campylobacter jejuni* (109–111), and biofilms in *Salmonella typhi* and *V. cholerae* (112,113) to name just a few examples. So in different organisms it regulates different functions. These systems are being teased apart but only two systems have been described in detail *(V. harveyi* and *V. cholera)* with little detail available for other AI-2 associated systems. Therefore, the discussion will focus on the *V. harveyi* system.

The biological synthesis of AI-2 has been demonstrated to involve S-adenosyl-L-methionine (114). S-adenosyl-L-methionine acts as a methyl donor in a number of biosynthetic processes and produces S-adenosylhomocysteine. S-adenosylhomocysteine is hydrolyzed by 5′-methylthioadenosine/S-adenosylhomocysteine nucleosidase to form S-ribosylhomocysteine and adenine. LuxS is proposed to cleave S-ribosylhomocysteine to form homocysteine and 4,5-dihydroxy 2,3-pentanedione. The 4,5-dihydroxy 2,3-pentanedione then cyclizes to form the furanosyl borate diester. This is supported by the observation that mixing S-adenosylhomocysteine with 5′-methylthioadenosine/S-adenosylhomocysteine nucleosidase and LuxS *in vitro* produces AI-2 (114). It is proposed that the 4,5-dihydroxy 2,3-pentanedione cyclizes spontaneously and borate is added in this process (100). In all the organisms studied, AI-2 is produced from the same intermediates (114,115). Since the synthesis of AI-2 is connected to active metabolism, Beeston and Surette have proposed that the production of AI-2 may actually signal growth phase of the cells rather than cell density (116). This fits well with the fact that AI-2 is detected maximally during exponential growth and diminishes as cells enter stationary phase (117–119). Even though AI-2 regulates different physiologies in different species of bacteria, it could serve as a common signal of the physiological state of the producing organisms and thereby the potential growth conditions of the environment.

In *V. harveyi*, transcription of the *lux* operon is under control of an unidentified transcriptional repressor. It is proposed that in the presence of either AI-2 or the AHL signals, a cascade of phosphorelay reactions results in the reduced expression of the transcriptional repressor and allows the activation of transcription of the *lux* operon. For the AI-2 pathway, proteins LuxP and LuxQ act as the sensory apparatus. LuxP has sequence similarity to an *E. coli* ribose binding protein, and in *V. harveyi* is believed to be a periplasmic protein that binds AI-2 because AI-2 contains a structural similarity to ribose (99). In combination with AI-2, LuxP can effect LuxQ (99). LuxQ has sequence similarity to two-component signal transduction proteins that have both histidine protein kinase activity and a response regulator domain (120). Based on sequence analysis, LuxQ is proposed to be a transmembrane protein in the cytoplasmic membrane with its histidine kinase and response regulator domains located in the cytoplasm of the bacterium (121). In a mechanism related to other phosphorelay systems (122,123), the following pathway has been proposed (124,125). LuxQ -phosphorylates a histidine in its kinase domain and transfers this phosphate to an aspartate in its regulatory domain when AI-2 is not present. This phosphate is then relayed to a histidine in the cytoplasmic protein LuxU (124). LuxU acts as a phosphorelay protein that then transfers the phosphate to an aspartate in the protein LuxO. Through an indirect mechanism, phospho-LuxO represses transcription of the *lux*-operon (120). It is proposed that phospho-LuxO interacts with sigma-54 to activate transcription of an unknown repressor of the *lux*-operon, and thereby control luminescence (125) (LuxU and LuxO are also part of the phosphorelay controlled by the AHL-signaling pathway used to control the *lux* operon in *V. harveyi*). In the presence of AI-2, the LuxP-AI-2 complex interacts with LuxQ and shifts LuxQ to phosphatase activity (120). The dephosphorylation of LuxQ results in the phosphorelay transferring phosphate from LuxO through LuxU to LuxQ. This results in dephosphorylation of LuxQ, which does not activate transcription of the repressor of the *lux* operon. With the subsequent reduction in repressor, the *lux* operon promoter is available to *lux*R (not structurally related to LasR), which activates transcription of the *lux* operon (126–128).

An analogous system has been identified to regulate virulence in *Vibrio cholerae* (129). AI-2 regulation of genes has been described in other bacteria; however, the mechanism of regulation is not described. AI-2 signaling responses involving response regulators and sensor kinase are being identified (130), yet if the cascade of signaling seen in the *Vibrio* is typical of this, the signaling system remains to be determined.

At least one AI-2 system appears to use a mechanism different from *Vibrio*. *Salmonella typhimurium* contains the *lsr* operon that encodes an ABC transporter associated with uptake of AI-2 in late exponential growth cultures (131). This results in a rapid decrease in the amount of detectable AI-2 in the culture supernatant and an increase in intracellular AI-2. AI-2 activates transcription of the *lsr* operon, apparently by inactivating the transcriptional repressor LsrR. Interestingly, this involves the uptake of AI-2 and the subsequent phosphorylation of AI-2 by the protein LsrK (132).

The phosphorylation of AI-2 results in a loss of biological activity in a *V. harveyi* detection system, but phospho-AI-2 is postulated to be the actual activator in the *lsr* system (132). The significance of this operon to *S. typhimurium* is not known, but it indicates a transcriptional regulation by AI-2 that is distinct from the described *Vibrio* phosphorelay cascade by requiring both the uptake of AI-2 and the modification of the signal for biological activity.

The potential for interspecies cross-talk using AI-2 is obvious since the *V. harveyi* reporter system indicates many species of bacteria produce a biologically active signal. While this particular field of study in bacterial communication is still developing some clear examples have been reported. The potential cross-talk between *V. harveyi* and *V. cholerae* is apparent, although the significance of this is more difficult to interpret. A recent report demonstrates cross-talk between other vibrios with more obvious implications. Coordination of stress adaptation by species cross-talk using AI-2 has been reported for *V. vulnificus* and *V. angustum* and AI-2 produced by other *Vibrio* can regulate this relevant phenotype (133). In a second example, *Streptococcus gordonii* and *Porphyromonas gingivalis* can coexist in dental plaque, which is a biofilm that contains more than 500 different species of bacteria. Both *S. gordonii* and *P. gingivalis* have *luxS* and produce AI-2 (134,135). These bacteria can form mixed biofilms in which *S. gordonii* initially attaches to the substratum and *P. gingivalis* attaches to the streptococcal cells (136). However, if both species have their *luxS* gene functionally deleted, they can not form the mixed biofilms. McNab et al. found that biofilms could be formed when one of the species contained a functional *luxS*, indicating that the AI-2 produced by one species provided the AI-2 signaling required by both species for mixed biofilm formation (135).

These examples support the potential for species cross-talk through the AI-2 signal. It also indicates the complexity of exploring the role of LuxS *in vivo*, as a strain with *luxS* functionally deleted can still act upon AI-2 provided by other bacteria present in an *in vivo* environment.

11.4   GRAM-POSITIVE PEPTIDE SIGNALS

As a general observation the Gram-positive bacteria use peptides as the messenger in their cell-to-cell communication. In the Gram-positive bacteria there are two basic mechanisms by which the peptide signal may act upon the cell (137,138). The peptide can act directly by entering the cells, as is exemplified with regulation of sporulation in *Bacillus* and by sex pheromone plasmid transfer in the *Enterococcus*. Alternately the peptide signal may act indirectly upon the cell by conveying the signal across the cell membrane via a two component transduction system using a histidine kinase, as is exemplified in *Staphylococcus* virulence regulation and competence regulation in both *Streptococcus* and *Bacillus*.

In the enterococcocal sex pheromone plasmid system, a population of enterococci infected with a plasmid (i.e., donor cells) may be induced to transfer that plasmid to other noninfected enterococci (i.e., recipient cells) in the same population (139). This is accomplished by the noninfected cells producing a peptide signal (i.e., pheromone) that is taken up by the infected cells and induces the genes for cell attachment and plasmid transfer. This system allows donor cells to produce the machinery for plasmid transfer only when recipient cells are present to receive the plasmid. Additionally, the peptide signal must be present at a minimal threshold level to induce plasmid transfer, indicating that a critical number of cells producing the pheromone must be present to achieve this level of pheromone in the environment of the bacteria. Thus the system regulates transfer function so that transfer is attempted only when recipient cells are present at significantly high numbers to allow efficient plasmid transfer.

The enterococcccal system is an example of a peptide hormone being released and exerting an effect by entering the target cells. This system is delicately balanced so that only cells that do not have the sex plasmid will signal cells with the sex plasmid. In the presence of the peptide signal the cells with the sex plasmid are induced to produce a protein promoting clumping of the cells and subsequent transfer of the plasmid to the cells without the plasmid. However, the donor cells are only induced to produce the machinery for plasmid transfer if recipient cells are available.

The sex pheromone plasmids of the enterococci are composed of a family of plasmids that may encode drug resistance, hemolysin or bacteriocins. The most extensively studied plasmids in this group include pAD1, pCF10 and pPD1 (139). The plasmids in this family share the common property of allowing the host cell to respond to a particular peptide signal produced by potential recipient cells. This peptide signal induces the donor cell to produce proteins that promote clumping of the enterococcal cells and the subsequent transfer of the sex pheromone plasmid to recipient cells. There are at least six different sex pheromones identified in the enterococci (139–141). While these pheromones have structural and physical similarities (Table 11.2) (142–146), each pheromone is associated with inducing the transfer of a specific sex pheromone plasmid, so the plasmids can be further divided based on their induction by a particular sex pheromone.

The peptide signals in these sex pheromone systems are encoded on the chromosomal DNA of the enterococci (139). The signal itself is only seven or eight amino acids in length, depending on the actual plasmid system (Table 11.2). Based on sequence analysis it was hypothesized the pheromone is derived from the signal sequences of surface lipoproteins (147), and later it was demonstrated that specific lipoproteins were required for specific pheromone production (148–150). The propheromones are enzymatically removed from the preprotein during the process of acetylation. The proteolytic processing of the protein and the secretion of the protein

TABLE 11.2

Examples of Peptide Signals Used by Enterococcus with Sex Pheromone Plasmids

Plasmid	Pheromone	Pheromone Structure	Inhibitor	Inhibitor Structure	Intracellular Target
pAD1	cAD1	LFSLVLAG	iAD1	LFVVTLVG	TraA
pCF10	cCF10	LVTLVFV	iCF10	AITLIFI	PrgX
pPD1	cPD1	FLVMFLSG	iPD1	ALILTLVS	TraA

and signal sequence appear to occur simultaneously. The resulting peptides are about 21 to 22 amino acids in length (147), so the heptapeptide or octapeptide pheromone is still embedded in the remaining signal sequence, and requires additional processing at both the C-terminal and N-terminal ends of the signal sequence peptide. The actual mechanism involved in degrading the signal sequence to form the peptide pheromone is not known. The Eep protein is involved in the processing of the signal sequence to produce the mature pheromone (148,152) and is predicted to be a membrane bound metalloprotease based on its amino acid sequence (151). Following the processing the resulting mature pheromone is external to the cell and can be released into the surrounding medium. However, these peptides are hydrophobic and can associate with hydrophobic cell wall components. Even so, some of the pheromone is released into the media and is present at low levels. The actual amount of pheromone in the medium is low: about 10^{-11} M pheromone for the pCF10 plasmid is reported. However, as little as 2×10^{-12} M can induce conjugation in this system (142).

When significant amounts of pheromone are in the medium to indicate the presence of recipient cells, the pheromone must bind to a receptor on the membrane of the donor cell, and is transported into the cell to induce conjugation. The binding proteins for the pheromones are plasmid encoded homologs to the OppA family of proteins, which are oligopeptide binding proteins that are part of a permease system in bacteria (153). In the pCF10 system this protein is PrgZ and in the pAD1 and pPD1 systems this is the TraC protein (154–157). OppA is a general oligopeptide binding protein that binds peptides that are 2 to 5 amino acids in length with little specificity for amino acid sequence. In contrast the pheromone binding proteins demonstrate a marked selectivity for their cognate pheromones (154,155). *Enterococcus* also produces a chromosomally encoded OppA protein that can bind and transport the pheromone into the cell. However, only the plasmid encoded pheromone binding protein has the high affinity for the pheromone that is required to transport the peptide at physiologically significant concentrations (154). While OppA does not appear to be required for pheromone transport, the Opp ABC transport system is required, along with the plasmid encoded pheromone binding protein, to transport the pheromone into the donor cell (154).

The pheromones' entry into the cell results in the expression of the aggregation substance, which promotes clumping of enterococci, and the expression of genes associated with gene transfer, with the ultimate transfer of the plasmid from the donor cell to the recipient cells. The steps between pheromone entry into the cell and plasmid transfer are complex. Induction of the plasmid transfer functions are controlled at several levels within the cell, and include regulation involving antisense RNA as well as control by the pheromone. Binding studies indicate the cognate pheromone (cAD1 and cPD1) directly binds to TraA in the pAD1 and pPD1, respectively (155,158). The analogous protein in the pCF10 plasmid system, PrgX, has not been shown to bind the pheromone cPF10 (159). These proteins do not share significant similarity, but genetic deletion studies indicate each acts as transcriptional or post-transcriptional repressors.

These systems certainly operate with nuances unique to each, as is indicated by the lack of significant sequence similarity between the two TraA and PrgX, and different genetic elements associated with other levels of regulation. For the sake of example, we'll focus on the pAD1 system, which provides the most resolved system at this level. Deletions of *traA* in this system result in the constitutive expression of plasmid transfer functions, supporting the role of TraA in repression of expression of these functions (160–162). TraA binds to the DNA of the promoter region of *iad* (i.e., the orf for the inhibitor propeptide) on pAD1 (163). This binding of TraA prevents read through from this promoter (164,165). This read through is also prevented by the transcript mD, which is encoded downstream of the *iad* promoter (166). Addition of cAD1 results in the loss of TraA binding to the *iad* promoter DNA with a reduction in mD which this allows read through of *traE1* (158,164,166–168). TraE1 is a positive transcriptional regulator believed to bind multiple promoters, including the *asa1* gene which encodes the aggregation substance for pAD1, and causes the up-regulation of plasmid transfer functions (163). The expression of aggregation substance at the cell surface promotes cell clumping and is the hallmark associated with plasmid transfer in this system.

Since the peptide pheromone is encoded on the chromosome of *Enterococcus*, both donor and recipient can produce it. However, the production of pheromone by the donor cell could lead to autoinduction of the plasmid transfer. To prevent autoinduction, this endogenous production of pheromone by donor cells is countered by two mechanisms in each of the plasmid systems. In the first level, the plasmid encodes a competitive inhibitor of the pheromone specific to that plasmid's sex pheromone, encoded in *prgQ* on pCF10 and in *iad* on pAD1 (144,169,170). The inhibitor peptides show similarity to the cognate pheromone (Table 11.2) (145,169,170). The expression of the inhibitory peptide appears to be coordinated with the endogenous expression of pheromone so that the level of inhibitor is just sufficient to competitively inhibit uptake of the amount of pheromone produced by the plasmid bearing cell (170). Interestingly, the competitive inhibitors are expressed as peptides of 21 to 23 amino acids in length with structural similarities to the pheromone-propeptides, and are believe to be secreted by the same signal sequence-dependant pathway used to secrete the pheromone (148,152). Additionally, like the pheromone-propeptide, the inhibitor-propeptide must be further processed to produce the mature inhibitory peptide, and this process also involves the Eep protein (148,152). The resulting inhibitory peptide competes with the pheromone for binding with the pheromone receptor protein located on the surface of the cell (designated PrgZ in pCF10, and TraC in pAD1 and pPD1 plasmid systems) (155). The competition of inhibitor peptide and pheromone for the PrgZ or TraC receptor is believed to be the only role for the inhibitory peptide, and there is no evidence the inhibitory peptide is taken into the cell (169,171–172).

In the second level used to counter the endogenous production of pheromone, there is a reduction or sequestering of the endogenous pheromone produced. In some plasmid systems (e.g., pAD1 and pPD1), the increased production of inhibitory peptide is associated with reduced production of pheromone. The observation that Eep is involved in the processing of both inhibitor-propeptides to the signal sequence to produce inhibitor peptide and pheromone suggests that the inhibitory mechanism may involve the processing of the peptide precursors, as well as competing for the receptor protein of the pheromone. This potential competition is also suggested by the observation that the conversion of propheromone to pheromone by Eep is limited by the availability of Eep (152). However, enterococci containing either pAD1 or pPD1 still produce pheromones for other sex pheromone plasmids, indicating if any competition occurs for Eep, there must be a level of specificity that is not obvious at this time. In the pAD1 and pPD1 plasmid systems the plasmids encode the protein TraB, which associates with the cell membrane. This protein causes a reduction in the production of the pheromone by plasmid containing cells, so that endogenous production of pheromone is too low to induce the plasmid transfer functions, although the mechanism involved has not been described (172–174). In the pCF10 system the TraB homolog, PrgY, is not associated with the reduction of pheromone (175). In fact the level of pheromone in the medium of plasmid containing cells is the same as for cells not containing the plasmid (170). Even so, PrgY is required to prevent induction of transfer functions by endogenous pheromone production (175). Plasmid containing cells with *prgY* functionally deleted are induced to express plasmid transfer functions by endogenous pheromone (175,176). Interestingly, functional deletions of *traA* are also sensitive to induction by endogenous pheromone, but this induction may be masked by the addition of exogenous inhibitory peptide. However, in the pCF10 system with *prgY* deletions this induction is not blocked by the addition of exogenous inhibitor peptide. This provides additional support that PrgY and TraB function differently.

The actual mechanism by which PrgY works is not known, but it appears to cause a decrease in the amount of pheromone associated with the cell wall of plasmid containing cells (175). A possible scenario for this reduction is that cell bound pheromone on one cell may bind to the pheromone receptor (i.e., PrgZ) on another cell, and could be taken up by the cell. In this way, reducing the cell bound pheromone in a population of donor cells would reduce this presentation of pheromone by cells that contain the plasmid, whereas recipient cells would have relatively high concentrations of cell wall associated pheromone that could bind to PrgZ expressed on donor cells. In this scenario, the inhibitory peptide would block the induction by soluble pheromone so that induction would be dependant upon cell bound pheromone. Thus, the donor cells not only detect the presence of recipient cells, but the recipient cells must be adjacent to the donor cell to present the pheromone.

Although different mechanisms appear to be involved, either PrgY or TraB are responsible for preventing endogenously produced pheromone from inducing plasmid transfer functions in their respective plasmid system. The net effect for a cell infected with a specific sex-pheromone plasmid is that the functional pheromone produced by the plasmid-infected cell is reduced, and the remaining pheromone is masked by the competitive inhibitor so that the cell does not respond to its own pheromone production. However, additional pheromone provided by adjacent recipient cells overrides the inhibition and the plasmid containing cells are induced to express the components for conjugation and transfer of the plasmid to the pheromone producing recipient cells.

While this system effectively prevents autoinduction of plasmid transfer by the infected cell, it does not prevent an infected cell from receiving unrelated pheromone plasmids. Even when a cell is infected with a specific sex pheromone plasmid, it continues to produce the pheromones for other pheromone sex plasmids that have not infected the cell, thus allowing infected cells to communicate with other cells to receive additional sex pheromone plasmids. The occurrence of this is evident by the fact that clinical isolates of strains of *E. faecium* and *E. faecalis* are commonly infected by several different pheromone plasmids at a time (141). Furthermore, pheromones specifically induce plasmid transfer for their cognate plasmid system. This is exemplified by the observation that donor cells containing both the pCF10 and pPD1 plasmid systems will selectively transfer only one plasmid system to plasmid free cells based on the pheromone with which the donor cells are presented (177). When the cells are exposed to cCF10, pCF10 is transferred with high frequency. When the cells are exposed to cPD1, pPD1 is transferred with high frequency, but transfer of pCF10 is not detected. Thus, the specificity of these pheromone transfer systems for their cognate pheromones is high.

Recent studies involving the binding of pheromones to their cognate intracellular target protein have demonstrated a similar specificity, supporting the postulate that the specificity is due at least in part to the binding of the pheromone to its intracellular target. Nakayama et al. have demonstrated that TraA from pPD1 binds its cognate peptides cPD1 and iPD1, and does not bind significant amounts of pheromone or inhibitory peptide from other pheromone plasmid systems (155). They did not relate the binding of pheromone or inhibitor peptide to the effect on DNA binding by TraA. Fujimoto and Clewell demonstrated that the TraA from the pAD1 system bound not only the cognate pheromone and inhibitory protein cAD1 and iAD1, it also bound cPD1 and iPD1 and, to a much lesser degree, cCF10 (158). However, they demonstrated that the binding of TraA to the *ipd* promoter was affected by only cAD1 and iAD1 (158). Further, Fujimoto and Clewell observed that removal of the carboxyl-terminal 5 to 9 amino acids from TraA caused the binding of peptide to become specific for just cAD1 (158). They suggested that the TraA may be post-translationally modified to remove these amino acids to provide the binding specificity that would be suggested by the highly selective induction by only the cognate pheromone.

The pheromone plasmid system may be viewed as a highly developed detection system actually evolved for plasmid survival. Additionally, factors encoded on these plasmids are indicated to increase the virulence of the cells (162,178–183), so the system appears to favor survival of the bacteria as well. The pheromone plasmid system allows the sub-population of plasmid containing cells to detect and respond to a sub-population of plasmid free cells by transferring the plasmid. This allows the minimal burden on the plasmid containing cell by selecting for plasmid transfer only when the number of potential recipient cells are present at a high enough density to provide significant levels of pheromone. Thus, this system may not represent a communication system that coordinates single cells into a multicellular complex, but rather allows two unique populations to communicate and respond in a manner that facilitates species survival.

In light of the highly specific regulation of the sex pheromone plasmids, it is interesting to note that the coagulase-positive strains of staphylococci also produce peptides that can induce sex pheromone plasmid systems pAD1 or pAM373 (184,185). In *Enterococcus*, the sequence of the pheromone for the pAM373 plasmid system is AIFILAS, and *Staphylococcus* produces a very similar peptide with the sequence AIFILAA (186). As in the *Enterococcus*, the *Staphylococcus* peptide originates from a signal sequence, but the protein from which the sequence derives is different than the protein involved in the expression of the pheromone in *Enterococcus* (187). Likewise the pAD1 active peptide from *S. aureus* is encoded on a plasmid and the protein is not related to the enterococcal protein encoding cAD1 (188). The significance of the "pheromones" production by *Staphylococcus* is not known. The peptides have not been implicated in the regulation of physiology in the staphylococci. While the staph-peptides do induce the plasmid transfer system in enterococci, it has not been demonstrated to induce plasmid transfer to staphylococci. Why *Staphylococcus* would process this signal sequence to a functional pheromone is not known, and may just be fortuitous, especially since *Staphylococcus* does produce and secrete a number of biologically active peptides.

The second type of peptide signaling in Gram-positive bacteria involves a two component transduction system in which a peptide produced by the bacteria serves to induce an intermembrane histidine protein kinase. Analogous to the signal transduction seen with the AI-2 communication system in *Vibrio*, the histidine protein kinase propagates the signal across the bacterial cell membrane without the peptide entering the cell. Through a cascade of reactions, this regulates the expression of specific functions in the cell. This type of system regulates competence in *Bacillus* and *Streptococcus*, toxin production in *Clostridium perfringens*, gelatinase production by *Enterococcus*, lantibiotic production in *Lactococcus lactis* and *B. subtilis*, antimicrobial peptides produced by *Carnobacterium* and *Lactobacillus*, and virulence in *Staphylococcus* (137,138). By way of example, we will focus on the system in *Staphylococcus*.

In the staphylococci the accessory gene regulator locus (*agr*) is a global regulator that controls exoprotein expression through a cascade of regulatory elements. Unlike the pheromone system of the enterococci, the *agr* system in *Staphylococcus* is encoded on the chromosome. At the center of this system is a peptide pheromone that is secreted by *Staphylococcus*. As the density of cells increases so does the concentration of pheromone, until it reaches a threshold concentration achieved at nanomolar amounts of the peptide, which occurs as the cells enter into stationary phase of growth (189). This then triggers the cells in the population to express a number of exoproteins involved in virulence. Depending on the strain of *S. aureus* studied, the *agr* locus is involved in the enhanced expression of alpha-toxin (delta-hemolysin), beta-toxin, delta-toxin, serine protease, DNase, fibrinolysin, enterotoxin B, and toxic shock syndrome toxin-1 (190–196). Unlike the pheromone system in the enterococci, the *Staphylococcus* pheromone does not enter the cell, but is detected via a signal transduction system. This two component system includes a membrane spanning histidine kinase (encoded by *agrC*), that binds the pheromone and is postulated to result in the autophosphorylation of the kinase and subsequently acts on the cognate response regulator (encoded in *agrA*). The response regulator is postulated to control transcription of the *agr* locus from its two promoters, designated P2 and P3. The P3 promoter controls the transcription of RNAIII. While RNAIII contains the open reading frame for delta-hemolysin (*hld*), delta-hemolysin does not regulate the expression of other virulence genes (192). Instead, the RNA III transcript enhances expression of the exoproteins through an unknown mechanism that operates at the level of transcription on most genes and also translation on a few genes (192–194).

Through regulation of RNAIII, the *agr* system is connected with at least two additional regulatory systems found in *Staphylococcus*, including the *sar* and *xpr* loci (197–201). The P2 promoter is adjacent to the P3 promoter, but divergently transcribed to produce RNAII, which encodes four genes (*agrA, agrB, agrC,* and *agrD*) (202). These four genes encode the peptide signal (encoded in *agrD*) (189) and the proteins involved in processing the peptide pheromone (encoded in *argB*) (203–205) and detecting the signal (*agrA* and *agrC*) (189,206,207). Thus, this pheromone system is auto-inducing, controlling the genes required for the synthesis of the peptide pheromone and the elements to detect this peptide. This autoinduction produces an additional amplification of signal to enhance gene expression via RNAIII.

In a general sense, signaling and up-regulation of the *agr* locus causes an increase in the expression of exoproteins associated with virulence and a decrease in expression of surface proteins associated with attachment (190,194). This system may be viewed as a response that allows the bacteria to invade the host or break out of a biofilm for additional nutrients (208). The auto-inducing peptide is expressed at basal levels and accumulates outside the cell. The importance of *agr* in the expression of virulence has been demonstrated in a number of animal models. *Agr* increases virulence in the skin abscess, endocarditis and septic arthritis models (209–211). However, *agr* is not expressed in the peritoneal sepsis model, indicating the importance of the pheromone accumulating to a critical concentration to promote the *agr* response (212,213).

In the *agr* system the signal is a peptide encoded by *agrD* as a propeptide and is ribosomally produced. As a propeptide, AgrD must be processed at both the amino and carboxyl ends to produce the biologically active pheromone referred to as the auto-inducing peptide (AIP). The mature AIP is 7 to 9 amino acids long, depending on the strain of *Staphylococcus* producing it (203,214–216). AIP peptide and its receptor (AgrC) have significant sequence variation within the species of *S. aureus*. This sequence variation affects the specificity of the receptor and ligand. So, not unexpectedly, *S. aureus* can be put into different functional groups based on the ability of a specific ligand to enhance the *agr*-associated response.

Four different *agr* sub-groups are found in the *S. aureus* and one subgroup has been identified in *S. epidermidis* (203,214,217). Unique *agr*-systems are also present in other staphylococci (218). In each case the subgroup represents biological isolation in that only the self-AIP enhances expression in the subgroup, with the exception that *S. aureus* subgroups 1 and 4 show cross-enhancement of virulence expression (214). Linear synthetic peptides with the same primary sequence as the AIP are not biologically active (189). Instead, it has been demonstrated that in *Staphylococcus* the active peptide contains a thiolactone ring between the central cysteine and the carboxyl of the C-terminal amino acid. This produces a ring containing 5 amino acids in the carboxyl end of the molecule and a free tail on the N-terminus of the peptide (203). In all cases, except in some strains of *Staphylococcus intermedius*, the AIP must have the thiolactone ring for biological activity (203,215,218). In other genera of bacteria using the two component signal transduction, the peptide inducer is not modified with a ring, with the exception of *Enterococcus*, which is reported to have a lactone ring (219,220). A lactone or lactame ring in place of the thiolactone ring in the *Staphylococcus* AIP does not propagate signal transduction in its cognate system (211,217). All the staphylococci AIP share the five-member thiolactone ring, although the primary sequence of amino acids varies between the different subgroups. Altering the length of the tail results in a peptide that no longer activates its cognate signaling system (221).

Deletion studies using *agr*-null cells and supplying *agr* components *in trans* have demonstrated that the minimum components needed to produce AIP are *agrD* and *agrB*. While AgrB does not demonstrate significant sequence homology to known proteins, it is predicted to be membrane bound and has been indicated to be associated with membrane fractions in *Staphylococcus* (204). It is postulated that AgrB is responsible for the production of the thiolactone ring. Additionally, because of membrane location of AgrB, it is postulated to be involved in the processing and secretion AIP (203–205). However, hyperexpression of recombinant AgrB and AgrD did not result in an increase in expression of AIP in a *S. epidermidis* background. This suggests that factors in addition to *agrB* and *agrD* are involved in AIP production and export (204).

Recently the protein SvrA was shown to be required for transcription from the P2 and P3, but the function of this protein is not known (222). It is possible that SvrA may also play a role in maturation of AgrD (213). Clearly AgrB is involved in AIP maturation. Additionally, AgrB is highly substrate specific as combinations of *agrD* from one subgroup with *agrB* from another subgroup do not result in biologically active peptide, except for the combination of group 1 and 3 (203). The biochemical events required for AIP maturation need to be investigated further.

AgrC has been demonstrated to be the receptor for AIP by pull-down studies and is the only staphylococcal protein capable of binding AIP (189). Subsequently, it has been demonstrated to be a membrane protein that autophosphorylates in the presence of AIP (206). AgrC and AgrA are postulated to make up a two-component signal transduction circuit. The postulated interaction between AgrC and the response regulator AgrA is based on the protein's sequence similarity to other two component systems and that each are required for the *agr*-response (194,202,206,223). However, the phosphorylation of AgrA by AgrC has not been demonstrated (224). Recently it was proposed that AgrA is constitutively phosphorylated and that AgrC may actually dephosphorylate AgrA to activate the protein (208). Additionally, while AgrA is required for transcription from the P2 and P3 promoters in the *agr* loci, it has not been demonstrated to bind either of the promoters (224). The actual mechanism of regulation by AgrA and AgrC still needs to be determined.

AIP is highly group specific, with a single amino acid change resulting in loss of activation within its group. While AIP activates the *agr* response within its group (i.e., *agr*-self), AIP in one group inhibits the *agr* response of another group (*agr*-nonself). In this way the AIP serves as both a pheromone within its *agr* group and an inhibitor of pheromone within another *agr* group (203). In each case the AIP is active in low nanomolar concentrations (203,204). This dual role of AIP is mediated through the receptor AgrC, with inhibiting AIP competing with inducing AIP (225). The five amino acid ring is still required for a peptide to function as an inhibitor of non-cognate *agr* response; however, the thiolactone can be replaced with either a lactone or lactam bond and still functions as an inhibitor of non-cognate *agr* systems (207,211,226). As indicated above, the amino acid tail plays a significant role in recognition of self-AIP, and in the *S. aureus* groups 1 and 2, removal of the tail from either AIP produces universal inhibitors that block not only nonself *agr*-response, but also their self-*agr* response (216,225). The potential of the therapeutic application of this observation has been noticed (211,227). However, the repression of the *agr* response results in increased transcription of genes associated with biofilm formation. Thus, the suppression of *agr* reduces virulence associated with acute infections, but it promotes biofilms which leads to chronic infections (228).

Since the *agr* response is associated with virulence, repression of the *agr* response decreases the success of infection (211). Thus, AIP serves to enhance the survival of a subgroup of *Staphylococcus* by increasing its cognate *agr* response and thereby enhancing its dissemination during infection. Concurrently, the same AIP blocks the *agr* response from competing *Staphylococcus* species in different *agr*-groups. As an anecdotal observation, the AIP from *S. epidermidis* is more efficient at blocking the *agr* response of *S. aureus* than the AIP from *S. aureus* is at blocking the *agr* response in *S. epidermidis* (229). This may explain why *S. epidermidis* predominates on human skin over *S. aureus* (229). This competition would cause evolutionary isolation. This is supported by the observation that in 13 species of staphylococci 25 *agr* groups were determined by sequence variation, and the grouping of the species based on these *agr* groups was very similar to the grouping of the strains based on 16S rRNA sequence (230). In correlation to this observation, the *agr* groups are associated with certain biotypes. For example most vancomycin intermediately-susceptable *S. aureus* (VISA) strains belong to *agr* group 2, most toxic shock syndrome strains belong to *agr* group 3 and most exfoliation-producing strains belong to *agr* group 4 (203,214,226,230,231). Thus, the *agr* communications system provides an interspecies communication that allows a numerically dominant strain to provide positive selection for like-strains and negative selection for dissimilar strains.

From these examples, the peptide signals appear to represent intraspecies communication systems, but the potential exists for species cross-communication as exemplified by the staphylococci expressing a functional enterococcal sex-pheromone. In addition to transmitting information to the selected audience, the Gram-positive bacteria also use peptides for signal jamming of target related but non-identical systems within their species.

We try to interpret this phenomenon of bacterial cell-to-cell communication within the context of human disease. However, we should remember that in many of these systems of communication the organisms exist in environments other than human and the role for this communication may be lost with our limited perspective on the bacteria's physiology. These communication systems may appear to be limited communications within a species or even a sub-population of a species. However, the intricacies may be much more complex and truly represent a multicellular organism, albeit loosely knitted. Minimally, the bacteria not only wage a war of survival with humans, but they must compete for position in this arena with other bacteria. Possibly the "cross communication" that has been observed between species may be interpreted as benign fortuitous cross-talk, coordinated and co-evolutionary cross-talk, or even subversive cross-talk to subvert the competition for a potential environmental niche. In any case it is now apparent that bacteria communicate within and between species, and often this communication is used to coordinate an activity that has a negative impact on human health.

ACKNOWLEDGMENTS

This project has been funded in part with Federal Funds from the National Cancer Institute, National Institutes of Health, under Contract No. N01-CO-12400 (Article H.36 of the Prime Contract). L.P. was supported in part by funding from both the Cystic Fibrosis Foundation (PASSAD9510, and IGLEWS00V0), and the NIH/NIAID (R01—AI33713). We thank Cathy Hartland for her help proof-reading this chapter.

DISCLAIMER

The content of this publication does not necessarily reflect the views or the policies of the Department of Health and Human Services, nor does mention of trade names, commercial products, or organizations imply endorsement by the U.S. Government.

REFERENCES

1. Kaiser, D., Building a multicellular organism. *Annu. Rev. Genet.* 35, 103–123, 2001.
2. Shapiro, J.A., Thinking about bacterial populations as multicellular organisms. *Annu. Rev. Microbiol.* 52, 81–104, 1998.
3. Holden, M.T.G., Chhabra, S.R., de Nys, R., Stead, P., Bainton, N.J., Hill, P.J., Manefield, M., Kumar, N., Labette, M., England, D., Rice, S., Givskov, M., Salmond, G.P.C., Stewart, G.S.A.B., Bycroft, B.W., Kjelleberg, S., and Williams, P., Quorum-sensing cross-talk: isolation and chemical characterization of cyclic dipeptides from *Pseudomonas aeruginosa* and other Gram-negative bacteria. *Mol. Microbiol.* 33, 1254–1266, 1999.
4. Horinouchi, S., and Beppu, T., A-factor as a microbial hormone that controls cellular differentiation and secondary metabolism in *Streptomyces griseus*. *Mol. Microbiol.* 12, 859–864, 1994.
5. Grebe, T.W., and Stock, J.B., The histidine protein kinase superfamily. *Adv. Microb. Physiol.* 41, 139–227, 1999.
6. Eberhard, A., Burlingame, A.L., Eberhard, C., Kenyon, G.L., Nealson, K.H., and Oppenheimer, J., Structural identification of autoinducer of *Photobacterium fischeri* luciferase. *Biochemistry* 20, 2444–2449, 1981.
7. Kempner, E.S., and Hanson, F.E., Aspects of light production by *Phytobacterium fischeri*. *J. Bacteriol.* 95, 975–979, 1968.
8. Nealson, K.H., Autoinduction of bacterial luciferase. Occurrence, mechanism and significance. *Arch. Microbiol.* 112, 73–79, 1977.
9. Engebrecht, J., and Silverman, M., Identification of genes and gene products necessary for bacterial bioluminescence. *Proc. Natl. Acad. Sci. USA* 81, 4154–4158, 1984.
10. Gambello, M.J., and Iglewski, B.H., Cloning and characterization of the *Pseudomonas aeruginosa lasR* gene: a transcriptional activator of elastase expression. *J. Bacteriol.* 173, 3000–3009, 1991.
11. Gambello, M.J., Kaye, S., and Iglewski, B.H., LasR of *Pseudomonas aeruginosa* is a transcriptional activator of the alkaline protease gene (*apr*) and an enhancer of exotoxin A expression. *Infect. Immun.* 61, 1180–1184, 1993.
12. Passador, L., Cook, J.M., Gambello, M.J., Rust, L., and Iglewski, B.H., Expression of *Pseudomonas aeruginosa* virulence genes requires cell-to-cell communication. *Science* 260, 1127–1130, 1993.

13. Gray, K.M., and Garey, J.R., The evolution of bacterial LuxI and LuxR quorum sensing regulators. *Microbiology* 147, 2379–2387, 2001.

14. Watson, W.T., Minogue, T.D., Val, D.L., von Bodman, S.B., and Churchhill, M.E.A., Structural basis and specificity of acyl-homoserine lactone signal production in bacterial quorum sensing. *Molecular Cell* 9, 685–694, 2002.

15. Parsek, M.R., Val, D.L., Hanzelka, B.L., Cronan, J.E.J., and Greenberg, E.P., Acyl homoserine-lactone quorum-sensing signal generation. *Proc. Natl. Acad. Sci. USA* 96, 4360–4365, 1999.

16. Val, D., and Cronan, J.E.J., *In vivo* evidence that S-adenosyl-methionine and fatty acid synthase intermediates are the substrates for the LuxI family of autoindicer synthases. *J. Bacteriol.* 180, 2644–2651, 1998.

17. Hoang, T.T., and Schweizer, H.P., Characterization of the *Pseudomonas aeruginosa* enoyl-acyl carrier protein reductase: a target for triclosan and its role in acylated homoserine lactone synthesis. *J. Bacteriol.* 181, 5489–5497, 1999.

18. Eberl, L., Winson, M.K., Sternberg, C., Stewart, G.S.A.B., Christiansen, G., Chhabra, S.R., Bycroft, B.W., Williams, P., Molin, S., and Givskov, M., Involvement of N-acyl-L-homoserine lactone autoinducers in controlling the multicellular behavior of *Serratia liquifaciens*. *Mol. Microbiol.* 20, 127–136, 1996.

19. Marketon, M.M., Gronquist, M.R., Eberhard, A., and Gonzalez, J.E., Characterization of the *Sinorhizobium meliloti sinR/sinI* locus and the production of novel N-acyl homoserine lactones. *J. Bacteriol.* 184, 5686–5695, 2002.

20. Hoang, T.T., Sullivan, S.A., Cusick, J.K., and Schweizer, H.P., Beta-ketoacyl acyl carrier protein reductase (FabG) activity of the fatty acid biosynthetic pathway is a determining factor of 3-oxo-homoserine lactone acyl chain lengths. *Microbiology* 148, 3849–3856, 2002.

21. Swift, S., Karlyshev, A.V., Durant, E.L., Winson, M.K., Chhabra, S.R., Williams, P., Macintyre, S., and Stewart, G.S.A.B., Quorum sensing in *Aeromonas hydrophila* and *Aeromonas salmonicida*: identification of the LuxRI homologues AhyRI and AsaRI and their cognate signal molecules. *J. Bacteriol.* 179, 5271–5281, 1997.

22. Pearson, J.P., Feldman, M., Iglewski, B.H., and Prince, A., *Pseudomonas aeruginosa* cell to cell signaling is required for virulence in a model of acute pulmonary infection. *Infect. Immun.* 68, 4331–4334, 2000.

23. Rumbaugh, K.P., Griswold, J.A., Iglewski, B.H., and Hamood, A.N., Contribution of quorum sensing to the virulence of *Pseudomonas aeruginosa* in burn wound infections. *Infect. Immun.* 67, 5854–5862, 1999.

24. Davey, M.E., Caiazza, N.C., and O'Toole, G.A., Rhamnolipid surfactant production affects biofilm architecture in *Pseudomonas aeruginosa*. *J. Bacteriol.* 185, 1027–1036, 2003.

25. Ochsner, U.A., and Reiser, J., Autoinducer-mediated regulation of rhamnolipid biosurfactant synthesis in *Pseudomonas aeruginosa*. *Proc. Natl. Acad. Sci. USA* 92, 6424–6428, 1995.

26. Pearson, J.P., Pesci, E.C., and Iglewski, B.H., Role of *Pseudomonas aeruginosa las* and *rhl* quorum-sensing systems in the control of elastase and rhamnolipid biosynthesis genes. *J. Bacteriol.* 179, 5756–5767, 1997.

27. Purevdorj, B., Costerton, J.W., and Stoodley, P., Influence of Hydrodynamics and Cell Signaling on the Structure and Behavior of *Pseudomonas aeruginosa* Biofilms. *Appl. Environ. Microbiol.* 68, 4457–4464, 2002.

28. Schuster, M., Lostroh, C.P., Ogi, T., and Greenberg, E.P., Identification, timing and signal specificity of *Pseudomonas aeruginosa* quorum-controlled genes: a transcriptome analysis. *J. Bacteriol.* 185, 2066–2079, 2003.

29. Wagner, V.E., Bushnell, D., Passador, L., Brooks, A.I., and Iglewski, B.H., Microarray analysis of *Pseudomonas aeruginosa* quorum-sensing regulons: effects of growth phase and environment. *J. Bacteriol.* 185, 2080–2095, 2003.

30. Erickson, D.l., Endersby, R., Kirkham, A., Stuber, K., Vollman, D.D., Rabin, H.R., Mitchell, I., and Storey, D.G., *Pseudomonas aeruginosa* quorum sensing system may control virulence factor expression in the lungs of patients with cystic fibrosis. *Infect. Immun.* 70, 1783–1790, 2002.

31. Favre-Bonte, S., Pache, J.C., Robert, J., Blanc, D., Pechere, J.C., and van Delden, C., Detection of *Pseudomonas aeruginosa* cell-to-cell signals in lung tissue of cystic fibrosis patients. *Microbiol. Pathogenesis* 32, 143–147, 2002.

32. Middleton, B., Rodgers, H.C., Camara, M., Knox, A.J., Williams, P., and Hardman, A., Direct detection of N-acylhomoserine lactones in cystic fibrosis sputum. *FEMS Microbiol. Lett.* 207, 1–7, 2002.

33. Singh, P.K., Schaefer, A.L., Parsek, M.R., Moninger, T.O., Welsh, M.J., and Greenberg, E.P., Quorum-sensing signals indicate that cystic fibrosis lungs are infected with bacterial biofilms. *Nature* 407, 762–764, 2002.

34. Davies, D.G., Parsek, M.R., Pearson, J.P., Iglewski, B.H., Costerton, J.W., and Greenberg, E.P., The involvement of cell-to-cell signals in the development of a bacterial biofilm. *Science* 280, 295–298, 1998.

35. Hentzer, M., Riedel, K., Rasmussen, T.B., Heydorn, A., Andersen, J.B., Parsek, M.R., Rice, S.A., Eberl, L., Molin, S., Høiby, N., Kjelleberg, S., and Givskov, M., Inhibition of quorum sensing in *Pseudomonas aeruginosa* biofilm bacteria by a halogenated furanone compound. *Microbiology* 148, 87–102, 2002.

36. Yoon, S.S., Hennigan, R.F., Hilliard, G.M., Ochsner, U.A., Parvatiyar, K., Kamani, M.C., Allen, H.L., DeKievit, T.R., Gardner, P.R., Schwab, U., Rowe, J.J., Iglewski, B.H., McDermott, T.R., Mason, R.P., Wozniak, D.J., Hancock, R.E.W., Parsek, M.R., Noah, T.L., Boucher, R.C., and Hassett, D.J., *Pseudomonas aeruginosa* anaerobic respiration in biofilms: Relationships to cystic fibrosis pathogenesis. *Dev. Cell.* 3, 593–603, 2002.

37. de Kievit, T.R., Gillis, R., Marx, S., Brown, C., and Iglewski, B.H., Quorum-sensing genes in *Pseudomonas aeruginosa* biofilms: their role and expression patterns. *Appl. Environ. Microbiol.* 67, 1865–1873, 2001.

38. Sauer, K., Camper, A.K., Ehrlich, G.D., Costerton, J.W., and Davies, D.G., *Pseudomonas aeruginosa* displays multiple phenotypes during development as a biofilm. *J. Bacteriol.* 184, 1140–1154, 2002.

39. Shih, P.C., and Huang, C.T., Effects of quorum-sensing deficiency on *Pseudomonas aeruginosa* biofilm formation and antibiotic resistance. *J. Antimicrob. Chemother.* 49, 309–314, 2002.

40. Pearson, J.P., Gray, K.M., Passador, L., Tucker, K.D., Eberhard, A., Iglewski, B.H., and Greenberg, E.P., Structure of the autoinducer required for expression of *Pseudomonas aeruginosa* virulence genes. *Proc. Natl. Acad. Sci. USA* 91, 197–201, 1994.

41. Passador, L., Tucker, K.T., Guertin, K.R., Journet, M.P., Kende, A.S., and Iglewski, B.H., Functional analysis of the *Pseudomonas aeruginosa* autoinducer PAI *J. Bacteriol.* 178, 5995–6000, 1996.

42. Fray, R.G., Throup, J.P., Daykin, M., Wallace, A., Williams, P., Stewart, G.S., and Grierson, D., Plants genetically modified to produce N-acylhomoserine lactones communicate with bacteria. *Nat. Biotechnology* 171, 1017–1020, 1999.

43. Fuqua, C., and Eberhard, A., Signal generation in autoinduction systems: synthesis of acylated homoserine lactones by LuxI-type proteins. In: *Cell–Cell Communication in Bacteria.* Dunny, G.M., and Winans, S.C., eds., Washington, D.C.: AMS Press, 1999:211–230.

44. Kaplan, H.B., and Greenberg, E.P., Diffusion of autoinducer is involved in regulation of the *Vibrio fischeri* luminescence system. *J. Bacteriol.* 163, 1210–1214, 1985.

45. Pearson, J.P., van Delden, C., and Iglewski, B.H., Active efflux and diffusion are involved in transport of *Pseudomonas aeruginosa* cell-to cell signals. *J. Bacteriol.* 181, 1203–1210, 1999.

46. Kiratisin, P., Tucker, K.D., and Passador, L., LasR, a transcriptional activator of *Pseudomonas aeruginosa* virulence genes, functions as a multimer. *J. Bacteriol.* 184(17), 4912–4919, 2002.

47. Chopon-Herve, V., Akrim, M., Latifi, A., Williams, P., Lazdunski, A., and Bally, M., Regulation of the *xcp* secretion pathway by multiple quorum-sensing modulons in *Pseudomonas aeruginosa. Mol. Microbiol.* 24, 1169–1178, 1997.

48. Seed, P.C., Passador, L., and Iglewski, B.H., Activation of the *Pseudomonas aeruginosa lasI* gene by LasR and the *Pseudomonas* autoinducer PAI: an autoinduction regulatory hierarchy. *J. Bacteriol.* 177, 654–659, 1995.

49. Toder, D.S., Gambello, M.J., and Iglewski, B.H., *Pseudomonas aeruginosa* LasA: a second elastase gene under transcriptional control of *las*R. *Mol. Microbiol.* 5, 2003–2010, 1991.

50. Devine, J.H., Shadel, G.S., and Baldwin, T.O., Identification of the operator of the *lux* regulon from *Vibrio fischeri* strain ATCC7744. *Proc. Natl. Acad. Sci. USA* 86, 5688–5692, 1989.

51. Latifi, A., Winson, M.K., Foglino, M., Bycroft, B.W., Stewart, G.S.A.B., Lazdunski, L., and Williams, P., Multiple homologues of LuxR and LuxI control expression of virulence determinants and secondary metabolites through quorum sensing in *Pseudomonas aeruginosa* PAO1. *Mol. Microbiol.* 17, 333–343, 1995.

52. Pearson, J.P., Passador, L., Iglewski, B.H., and Greenberg, E.P., A second N-acyl-homoserine lactone signal produced by *Pseudomonas aeruginosa. Proc. Natl. Acad. Sci. USA* 92, 1490–1494, 1995.

53. Ochsner, U.A., Koch, A.K., Fiechter, A., and Reiser, J., Isolation and characterization of a regulatory gene affecting rhamnolipid biosurfactant synthesis in *Pseudomonas aeruginosa. J. Bacteriol.* 176, 2044–2054, 1994.

54. Latifi, A., Foglino, M., Tanaka, K., Williams, and P., Lazdunski, A., A hierarchical quorum-sensing cascade in *Pseudomonas aeruginosa* links the transcriptional activators LasR and RhlR (VsmR) to expression of the stationary-phase sigma factor RpoS. *Mol. Microbiol.* 21, 1137–1146, 1996.

55. Medina, G., Juarez, K., Diaz, R., and Soberon-Chavez, G., Transcriptional regulation of *Pseudomonas aeruginosa rhlR*, encoding a quorum-sensing regulatory protein. *Microbiology* 149, 3073–3081, 2003.

56. Pesci, E.C., Pearson, J.C., Seed, P.C., and Iglewski, B.H., Regulation of *las* and *rhl* quorum sensing in *Pseudomonas aeruginosa. J. Bacteriol.* 179, 3127–3132, 1997.

57. Chungani, S.A., Whiteley, M., Lee, K.M., D'Argenio, D., Manoil, C., and Greenberg, E.P., QscR, a modulator of quorum-sensing signal synthesis and virulence in *Pseudomonas aeruginosa. Proc. Natl. Acad. Sci. USA* 98, 2752–2757, 2001.

58. Ledgham, F., Ventre, I., Soscia, C., Foglino, A., Sturgis, J.N., and Lazdunski, A., Interactions of the quorum sensing regulator QscR: interaction with itself and the other regulators of *Pseudomonas aeruginosa* LasR and RhlR. *Mol. Microbiol.* 48, 199–210, 2003.

59. de Kievit, T.R., Seed, P.C., Passador, L., Nezezon, J., Iglewski, and B.H., RsaL, a novel repressor of virulence gene expression in *Pseudomonas aeruginosa. J. Bacteriol.* 181, 2175–2184, 1999.

60. Pessi, G., Williams, F., Hindle, Z., Heurlier, K., Holden, M.T.G., Cámara, M., Haas, D., and Williams, P., The global posttranscriptional regulator RsmA modulates production

of virulence determinants and N-acylhomoserine lactones in *Pseudomonas aeruginosa*. *J. Bacteriol.* 183, 6676–6683, 2001.

61. Pesci, E.C., Milbank, J.B.J., Pearson, J.P., McKnight, S., Kende, A.S., Greenberg, E.P., and Iglewski, B.H., Quinolone signaling in the cell-to-cell communication system of *Pseudomonas aeruginosa. Proc. Natl. Acad. Sci. USA* 96, 11229–11234, 1999.

62. Collier, D.N., Anderson, L., McNight, S.L., Noah, T.L., Knowles, M., Boucher, R., Schwab, U., Gilligan, P., and Pesci, E.C., A bacterial cell to cell signal in the lungs of cystic fibrosis patients. *FEMS Microbial. Lett.* 215, 41–46, 2002.

63. Guina, T., Purvine, S.O., Yi, E.C., Eng, J., Goodlett, D.R., Aebeersold, R., and Miller, S.I., Quantitative proteomic analysis indicates increased synthesis of a quinolone by *Pseudomonas aeruginosa* isolates from cystic fibrosis airways. *Proc. Natl. Acad. Sci. USA* 100, 2771–2776, 2003.

64. McKnight, S.L., Iglewski, B.H., and Pesci, E.C., The *Pseudomonas* quinolone signal regulates *rhl* quorum sensing in *Pseudomonas aeruginosa. J. Bacteriol.* 182, 2702–2708, 2000.

65. Gallagher, L.A., McKnight, S.L., Kuznetsova, M.S., Pesci, E.C., and Manoil, C., Functions required for extracellular quinolone signaling by *Pseudomonas aeruginosa. J. Bacteriol.* 184, 6472–6480, 2002.

66. Calfee, M.W., Coleman, J.P., and Pesci, E.C., Interference with *Pseudomonas* quinolone signal sysnthesis inhibits virlulence factor expression by *Pseudomonas aeruginosa. Proc. Natl. Acad. Sci. USA 98*, 11633–11637, 2001.

67. McClean, R.J.C., Whiteley, M., Stickler, D.J., and Fuqua, W.C., Evidence of autoinducer activity in naturally occuring biofilms. *FEMS Microbiology Lett.* 154, 259–263, 1997.

68. de Kievit, T.R., and Iglewski, B.H., Bacterial quorum sensing in pathogenic relationships. *Infect. Immun.* 68, 4839–4849, 2000.

69. Givskov, M., de Nys, R., Manefield, M., Gram, L., Maximilien, R., Eberl, L., Molin, S., Steinberg, P.D., and Kjelleberg, S., Eukaryotic interference with homoserine lactone-mediated prokaryotic signalling. *J. Bacteriol.* 178, 6618–6622, 1996.

70. Lee, S.J., Park, S-Y., Lee, J-J., Yum, D-Y., Koo, B-T., and Lee, J-K., Genes encoding the N-acyl homoserine lactone-degrading enzyme are widespread in many subspecies of *Bacillus thuringiensis. Appl. Environ. Microbiol.* 68, 3919–3924, 2002.

71. Park, S-Y., Lee, S.J., Oh, T-K., Oh, J-W., Koo, B-T., Yum, D-Y., and Lee, J-K., AhlD, an N-acylhomoserine lactonase in *Arthrobacter* sp., and predicted homologues in other bacteria. *Microbiology* 149, 1541–1550, 2003.

72. Hentzer, M., Wu, H., Andersen, J.B., Riedel, K., Rasmussen, T.B., Bagge, N., Kumar, N., Schembri, M.A., Song, Z., Kristoffersen, P., Manefiled, M., Costerton, J.W., Molin, S., Eberl, L., Steinberg, P., Kjelleberg, S., Hoiby, N., and Givskov, M., Attenuation of *Pseudomonas aeruginosa* virulence by quorum sensing inhibitors. *EMBO J.* 22, 3803–3815, 2003.

73. Smith, K.M., Bu, Y., and Suga, H., Induction and inhibition of *Pseudomonas aeruginosa* quorum sensing by synthetic autoindicer analogs. *Chemistry and Biology* 10, 81–89, 2003.

74. Miller, M.B., and Bassler, B.L., Quorum sensing in bacteria. *Annu. Rev. Microbiol.* 55, 165–199, 2001.

75. Steidle, A., Sigl, K., Schuhegger, R., Ihring, A., Schmid, M., Gantner, S., Stoffels, M., Riedel, K., Givskov, M., Hartman, A., Langebartels, C., and Eberl, L., Visualization of N-acylhomoserine lactone-mediated cell-cell communication between bacteria colonizing the tomato rhizosphere. *Appl. Environ. Microbiol.* 67, 5761–5770, 2001.

76. Riedel, K., Hentzer, M., Geisenberger, O., Huber, B., Steidle, A., Wu, H., Hoiby, N., Givskov, M., Molin, S., and Eberl, L., N-acylhomoserine-lactone-mediated

communication between *Pseudomonas aeruginosa* and *Burkholderia cepacia* in mixed biofilms. *Microbiology* 147, 3249–3262, 2001.

77. McKenney, D., Brown, K.E., and Allison, D.G., Influence *of Pseudomonas aeruginosa* exoproducts on virulence factor production in *Burkholderia cepacia*: evidence of interspecies communication. *J. Bacteriol.* 177, 6989–6992, 1995.

78. Rodelas, B., Lithgow, J.K., Wisniewski-Dye, F., Hardman, A., Wilkinson, A., Economou, A., Williams, P., and Downie, J.A., Analysis of quorum-sensing-dependent control of rhizosphere-expressed (*rhi*) genes *in Rhizobium leguminosarum* bv. *viciae*. *J. Bacteriol.* 181, 3816–3823, 1999.

79. Swift, S., Winson, M.K., Chan, P.F., Bainton, N.J., Birdsall, M., Reeves, P.J., Rees, C.E.D., Chhabra, S.R., Hill, P.J., Throup, J.P., Bycroft, B.W., Salmond, G.P.C., Williams, P., and Stewart, G.S.A.B., A novel strategy for the isolation of *luxI* homologues: evidence for the widespread distribution of a LuxR:LuxI superfamily in enteric bacteria. *Mol. Microbiol.* 10, 511–520, 1993.

80. Brint, J.M., and Ohman, D.E., Synthesis of multiple exoproducts in *Pseudomonas aeruginosa* is under the control of RhlR-RhlI, another set of regulators in strain PAO1 with homology to the autoinducer-responsive LuxR-LuxI family. *J. Bacteriol.* 177, 7155–7163, 1995.

81. Winson, M.K., Camara, M., Latifi, A., Foglino, M., Chhabra, S.R., Daykin, M., Bally, M., Chapon, V., Salmond, G.P.C., Bycroft, B.W., Lazdunski, A., Stewart, G.S.A.B., and Williams, P., Multiple N-acyl-L-homoserine lactone signal molecules regulate production of virulence determinants and secondary metabolites in *Pseudomonas aeruginosa*. *Proc. Natl. Acad. Sci. USA* 92, 9427–9431, 1995.

82. Givskov, M., Eberl, L., and Molin, S., Control of exoenzyme production, motility and cell differentiation in *Serratia liquifaciens*. *FEMS Microbiol. Lett.* 148, 115–122, 1997.

83. Engebrecht, J., and Silverman, M., Nucleotide sequence of the regulatory locus controlling expression of bacterial genes for bioluminescence. *Nucleic Acids Res.* 15, 10455–10467, 1987.

84. Bainton, N.J., Stead, P., Chhabra, S.R., Bycroft, B.W., Salmond, G.P.C., Steward, G.S.A.B., and Williams, P., N-(3-Oxohexanoyl)-L-homoserine lactone regulates carbapenem antibiotic production in *Erwinia carotovora*. *Biochem. J.* 288, 997–1004, 1992.

85. Chhabra, S.R., Stead, P., Bainton, N.J., Salmond, G.P.C., Stewart, G.S.A.B., Williams, P., and Bycroft, B.W., Autoregulation of carbapenem biosynthesis in *Erwinia carotovora* ATCC 39048 by analogues of N-3-(oxohexanoyl)-L-homoserine lactone. *J. Antibiot.* 46, 441–454, 1993.

86. Jones, S., Yu, B., Bainton, N.J., Birdsall, M., Bycroft, B.W., Chhabra, S.R., Cox, A.J.R., Golby, P., Reeves, P.J., Stephens, S., Winson, M.K., Salmond, G.P.C., Stewart, G.S.A.B., and Williams, P., The *lux* autoinducer regulates the production of exoenzyme virulence determinants in *Erwinia carotovora* and *Pseudomonas aeruginosa*. *EMBO J.* 12, 2477–2482, 1993.

87. McGowan, S., Sebaihia, M., Jones, S., Yu, S., Bainton, N., Chan, P.F., Bycroft, B.W., Stewart, G.S.A.B., Salmond, G.P.C., and Williams, P., Carbapenem antibiotic production in *Erwinia carotovora* is regulated by CarR, a homologue of the LuxR transcriptional activator. *Microbiology* 141, 541–550, 1995.

88. Pirhonen, M., Flego, D., Heikinheimo, R., and Palva, E.T., A small diffusible signal molecule is responsible for the global control of virulence and exoenzyme production in the plant pathogen *Erwinia carotovora*. *EMBO J.* 12, 2467–2476, 1993.

89. Nasser, W., Bouillant, M.L., Salmond, G., and Reverchon, S., Characterization of the *Erwinia chrysanthemi expI-expR* locus directing the synthesis of two N-acyl-homoserine lactone signal molecules. *Mol. Microbiol.* 29, 1391–1405, 1998.

90. Reverchon, S., Bouillant, M.L., Salmond, G., and Nasser, W., Integration of the quorum-sensing system in the regulatory networks controlling virulence factor synthesis in *Erwinia chrysanthemi*. *Mol. Microbiol.* 29, 1407–1418, 1998.

91. Beck von Bodman, S., and Farrand, S.K., Capsular polysaccharide biosynthesis and pathogenicity in *Erwinia stewartii* require induction by a N-acyl-homoserine lactone autoinducer. *J. Bacteriol.* 177, 5000–5008, 1995.

92. Atkinson, S., Throup, J.P., Stewart, G.S., and Williams, P., A hierarchical quorum-sensing system in *Yersinia pseudotuberculosis* is involved in the regulation of motility and clumping. *Mol. Microbiol.* 33, 1267–1277, 1999.

93. Lewenza, S., Conway, B., Greenberg, E.P., and Sokol, P.A., Quorum sensing in *Burkholderia cepacia*: identification of the LuxRI homologs CepRI. *J. Bacteriol.* 181, 748–756, 1999.

94. Cubo, M.T., Economou, A., Murphy, G., Johnston, A.W.B., and Downie, J.A., Molecular characterization and regulation of the rhizosphere-expressed genes *rhiABCR* that can influence nodulation by *Rhizobium leguminosarum* biovar *viciae*. *J. Bacteriol.* 174, 4026–4035, 1992.

95. Gray, K.M., Pearson, J.P., Downie, J.A., Boboye, B.E.A., and Greenberg, E.P., Cell-to-cell signaling in the symbiotic nitrogen-fixing bacterium *Rhizobium leguminosarum*: autoinduction of stationary phase and rhizosphere-expressed genes. *J. Bacteriol.* 178, 372–376, 1996.

96. Flavier, A.B., Ganova-Raeva, L.M., Schell, M.A., and Denny, T.P., Hierarchical autoinduction in *Ralstonia solanacearum*: control of acyl-homoserine lactone production by a novel autoregulatory system responsive to 3-hydroxypalmitic acid ester. *J. Bacteriol.* 179, 7089–7097, 1997.

97. Cao, J.G., and Meighen, E.A., Purification and structural identification of an autoinducer for the luminescence system of *Vibrio harveyi*. *J. Biol. Chem.* 264, 21670–21676, 1989.

98. Bassler, B.L., Wright, M., Showalter, R.E., and Silverman, M.R., Intercellular signalling in *Vibrio harveyi*: sequence and function of genes regulating expression of luminescence. *Mol. Microbiol.* 9, 773–786, 1993.

99. Bassler, B.L., Wright, M., and Silverman, M.R., Multiple signalling systems controlling expression of luminescence in *Vibrio harveyi*: sequence and function of genes encoding a second sensory pathway. *Mol. Microbiol.* 13, 273–286, 1994.

100. Chen, X., Schauder, S., Potier, N., van Dorsselaer, A., Pelczer, I., Bassler, B.L., and Hughson, F.M., Structural identification of a bacterial quorum-sensing signal containing boron. *Nature* 415, 545–549, 2002.

101. Bassler, B.L., Greenberg, E.P., and Stevens, A.M., Cross-species induction of luminescence in the quorum-sensing bacteria *Vibrio harveyi*. *J. Bacteriol.* 179, 4043–4045, 1997.

102. Basseler, B.L., How bacteria talk to each other: regulation of gene expression by quorum sensing. *Curr. Opin. Microbiol.* 2, 582–587, 1999.

103. Surette, M.G., Miller, M.B., and Bassler, B.L., Quorum sensing in *Escherichia coli, Salmonella typhimurium* and *Vibrio harveyi*: a new family of genes responsible for autoinducer production. *Proc. Natl. Acad. Sci. USA* 96, 1639–1644, 1999.

104. Day, W.A. Jr., and Maurelli, A.T., *Shigella flexneri* LuxS quorum-sensing system modulates *virB* expression but is not essential for virulence. *Infect. Immun.* 69, 15–23, 2001.

105. Fong, K.P., Chung, W.O., Lamont, R.J., and Demuth, D.R., Intra- and interspecies regulation of gene expression by *Actinobacillus actinomycetemcomitans* LuxS. *Infect. Immun.* 69, 7625–7634, 2001.

106. Lyon, W.R., Madden, J.C., Levin, J.C., Stein, J.L., and Caparon, M.G., Mutation of *luxS* affects growth and virulence factor expression in *Streptococcus pyogenes*. *Mol. Microbiol.* 42, 145–157, 2001.

107. Ohtani, K., Hayashi, H., and Shimizu, T., The *luxS* gene is involved in cell–cell signalling for toxin production in *Clostridium perfringens*. *Mol. Microbiol.* 44, 171–179, 2002.

108. Sperandio, V., Li, C.C., and Kaper, J.B., Quorum-sensing *Escherichia coli* regulator A: a regulator of the LysR family involved in the regulation of the locus of enterocyte effacement pathogenicity island in enterohemorrhagic, *E. coli. Infect. Immun.* 70, 3085–3093, 2002.

109. Elvers, K.T., and Park, S.F., Quorum sensing in *Campylobacter jejuni*: detection of a *luxS* encoded signalling molecule. *Microbiology* 148, 1475–1481, 2002.

110. Giron, J.A., Torres, A.G., Freer, E., and Kaper, J.B., The flagella of enteropathogenic *Escherichia coli* mediate adherence to epithelial cells. *Mol. Microbiol.* 44, 361–379, 2002.

111. Sperandio, V., Torres, A.G., and Kaper, J.B., Quorum sensing *Escherichia coli* regulators B and C (QseBC): a novel two-component regulatory system involved in the regulation of flagella and motility by quorum sensing in *E. coli. Mol. Microbiol.* 43, 809–821, 2002.

112. Prouty, A.M., Schwesinger, W.H., and Gunn, J.S., Biofilm formation and interaction with the surfaces of gallstones by *Salmonella* spp. *Infect. Immun.* 70, 2640–2649, 2002.

113. Hammer, B.K., and Bassler, B.L., Quorum sensing controls biofilm formation in *Vibrio cholerae. Mol. Microbiol.* 50, 101–114, 2003.

114. Schauder, S., Shokat, K., Surrette, M.G., and Basseler, B.L., The *luxS* family of bacterial autoinducers: biosynthesis of a novel quorum-sensing signal molecule. *Mol. Microbiol.* 41, 463–476, 2001.

115. Winzer, K., Hardie, K.R., Burgess, N., Doherty, N., Kirke, D., Holden, M.T., Linforth, R., Cornell, K.A., Taylor, A.J., Hill, P.J., and Williams, P., LuxS: its role in central metabolism and the *in vitro* synthesis of 4-hydroxy 5-methyl 3(2H)-furanone. *Microbiology* 148, 909–922, 2002.

116. Beeston, A.L., and Surette, M.G., Pfs-dependent regulation of autoinducer 2 production in *Salmonella enterica* serovar *thyphimurium. J. Bacteriol.* 184, 3450–3456, 2002.

117. Burgess, N.A., Kirke, D.F., Williams, P., Winzer, K., Hardie, K.R., Meyers, N.L., Aduse-Opoku, J., Curtis, M.A., and Camara, M., LuxS-dependent quorum sensing in *Porphyromonas gingivalis* modulates protease and haemagglutinin activities but is not essential for virulence. *Microbiology* 148, 763–772, 2002.

118. deLisa, M.P., Valdes, J.J., and Bentley, W.E., Mapping stress-induced changes in autoinducer AI-2 production in chemostat-cultivated *Escherichia coli* K-12. *J. Bacteriol.* 183, 2918–2928, 2001.

119. Surette, M.G., and Basseler, B.L., Regulation of autoinducer production in *Salmonella typhimurium. Mol. Microbiol.* 31, 585–595, 1999.

120. Freeman, J.A., and Bassler, B.L., A genetic analysis of the function of LuxO, a two-component response regulator involved in quorum sensing in *Vibrio harveyi. Mol. Microbiol.* 31, 665–678, 1999.

121. Bassler, B.L., and Silverman, M.R., Intercellular communication in marine *Vibrio* species: density-dependent regulation of the expression of bioluminescence, In: Hoch, J.A., and Silhavy, T.J., eds., Two-Component Signal Transduction. American Society for Microbiology, Washington, D.C., 1995:431–435.

122. Appleby, J.L., Parkinson, J.S., and Bourret, R.B., Signal transduction via the multi-step phosphorelay: not necessarily a road less traveled. *Cell* 86, 845–848, 1996.

123. Burbulys, D.K., Trach, A., and Hoch, J.A., The initiation of sporulation in *Bacillus subtilis* is controlled by a multicomponent phosphorelay. *Cell* 64, 545–552, 1991.

124. Freeman, J.A., and Bassler, B.L., Sequence and function of LuxU: a two-component phosphorelay protein that regulates quorum sensing in *Vibrio harveyi. J. Bacteriol.* 181, 899–906, 1999.

125. Lilley, B.N., and Bassler, B.L., Regulation of quorum sensing in *Vibrio harveyi* by LuxO and sigma-54. *Mol. Microbiol.* 36, 940–954, 2000.

126. Martin, M., Showalter, R., and Silverman, M.R., Identification of a locus controlling expression of luminescence genes in *Vibrio harveyi. J. Bacteriol.* 171, 2406–2414, 1989.

127. Showalter, R.E., Martin, M.O., and Silverman, M.R., Cloning and nucleotide sequence of *luxR*, a regulatory gene controlling luminescence in *Vibrio harveyi. J. Bacteriol.* 172, 2946–2954, 1990.

128. Swartzman, E., Silverman, M.R., and Meighen, E.A., The *luxR* gene product of *Vibrio harveyi* is a transcriptional activator of the *lux* promoter. *J. Bacteriol.* 174, 7490–7493, 1992.

129. Miller, M.B., Skorupski, K., Lenz, D.H., Taylor, R.K., and Bassler, B.L., Parallel quorum sensing systems converge to regulate virulence in *Vibrio cholerae. Cell* 110, 303–314, 2002.

130. Sperandio, V., Torres, A.G., and Kaper, J.B., Quorum sensing in *Escherichia coli* regulates B and C (QseBC): a novel two-component regulatory system involved in the regulation of flagella and motility by quorum sensing in *E. coli. Mol. Microbiol.* 43, 809–821, 2002.

131. Taga, M.E., Semmelhack, J.L., and Bassler, B.L., The LuxS-dependent autoinducer AI-2 controls the expression of an ABC transporter that functions in AI-2 uptake in *Salmonella typhimurium. Mol. Microbiol.* 42, 777–793, 2001.

132. Taga, M.E., Miller, S.T., and Bassler, B.L., Lsr-mediated transport and processing of AI-2 in *Salmonella typhimurium. Mol. Microbiol.* 50, 1411–1427, 2003.

133. McDougald, D., Srinivasan, S., Rice, S.A., and Kjelleberg, S., Signal-mediated cross-talk regulates stress adaptation in *Vibrio* species. *Microbiology* 149, 1923–1933, 2003.

134. Chung, W.O., Park, Y., Lamont, R.J., McNab, R., Barbieri, B., and Demuth, D.R., Signaling system in *Porphyromonas gingivalis* based on a LuxS protein. *J. Bacteriol.* 183, 3903–3909, 2001.

135. McNab, R., Ford, S.K., El-Sabaeny, A., Barbieri, B., Cook, G.S., and Lamont, R.J., LuxS signaling in *Streptococcus gordonii*: autoinducer 2 controls carbohydrate metabolism and biofilm formation with *Porphyromonas gingivalis. J. Bacteriol.* 185, 274–284, 2003.

136. Cook, G.S., Costerton, J.W., and Lamont, R.J., Biofilm formation by *Porphyromonas gingivalis* and *Streptococcus gordonii. J. Periodontal. Res.* 33, 323–327, 1998.

137. Dunny, G.M., and Leonard, B.A.B., Cell–cell communication in Gram-positive bacteria. *Annu. Rev. Microbiol.* 51, 527–564, 1997.

138. Dunny, G.M., and Winans, S.C., eds., Cell–cell signaling in bacteria. Washington, D.C.: ASM Press, 1999.

139. Dunny, G.M., Craig, R.A., Carron, R.L., and Clewell, D.B., Plasmid transfer in *Streptococcus faecalis:* production of multiple pheromones by recipients. *Plasmid* 2, 454–465, 1979.

140. Clewell, D.B., Sex pheromone systems in enterococci. In: Cell-Cell Communication in Bacteria. Dunny, G.M., and Winans, S.C., eds. Washington, D.C.: AMS Press, 1999:10–20.

141. Wirth, R., The sex pheromone system of *Enterococcus faecalis*: more than just a plasmid-collection mechanism? *Eur. J. Biochem.* 222, 235–246, 1994.

142. Mori, M., Sakagami, Y., Ishii, Y., Isogai, A., Kitada, C., Fujino, M., Adsit, J.C., Dunny, G.M., and Suzuki, A., Structure of cCF10, a peptide sex pheromone which induces conjugative transfer of the *Streptococcus faecalis* tetracycline resistance plasmid, pCF10. *J. Biol. Chem.* 263, 14574–14578, 1988.

143. Mori, M., Sakagami, Y., Narita, M., Isogai, A., Fujino, M., Kitada, C., Craig, R.A., Clewell, D.B., and Suzuki, A., Isolation and structure of the bacterial sex pheromone,

cAD1, that induces plasmid transfer in *Streptococcus faecalis*. *FEBS Lett.* 178, 97–100, 1984.

144. Mori, M., Tanaka, H., Sakagami, Y., Isogai, A., Fujino, M., Kitada, C., White, B.A., An, F.Y., Clewell, D.B., and Suzuki, A., Isolation and structure of the *Streptococcus faecalis* sex pheromone, cAM373. *FEBS Lett.* 206, 69–72, 1986.

145. Nakayama, J., Ono, Y., and Suzuki, A., Isolation and structure of the sex pheromone inhibitor, iAM373, of *Enterococcus faecalis*. *Biosci. Biotech. Biochem.* 59, 1358–1359, 1995.

146. Suzuki, A., Mori, M., Sakagami, Y., Isogai, A., Fujino, M., Kitada, C., Craig, R.A., and Clewell, D.B., Isolation and structure of bacterial sex pheromone, cPD1. *Science* 226, 849–850, 1984.

147. Clewell, D.B., An, F.Y., Flannagan, S.F., Antiporta, M., and Dunny, G.M., Enterococcal sex pheromone precursors are part of signal sequences for surface lipoproteins. *Mol. Microbiol.* 35, 246–247, 2000.

148. An, F.Y., and Clewell, D.B., Identification of the cAD1 sex pheromone precursor in *Enterococcus faecalis*. *J. Bacteriol.* 184, 1880–1887, 2002.

149. Antiporta, M.H., and Dunny, G.M., *ccfA*, the genetic determinant for the cCF10 peptide pheromone in *Enterococcus faecalis* OG1RF. *J. Bacteriol.* 184, 1155–1162, 2002.

150. Flannagan, S.E., and Clewell, D.B., Identification and characterization of genes encoding sex pheromone cAM373 activity in *Enterococcus faecalis* and *Staphylococcus aureus*. *Mol. Microbiol.* 44, 803–817, 2002.

151. Brown, M.S., Ye, J., Rawson, R.B., and Goldstein, J.L., Regulated intramembrane proteolysis: a control mechanism conserved from bacteria to humans. *Cell* 100, 391–398, 2000.

152. An, F.Y., Sulavik, M.C., and Clewell, D.B., Identification and characterization of a determinant (*eep*) on the *Enterococcus faecalis* chromosome that is involved in production of the peptide sex pheromone cAD1. *J. Bacteriol.* 181, 5915–5921, 1999.

153. Tame, J.R.H., Murshudov, G.N., Dodson, E.J., Neil, T.K., Dodson, G.G., Higgins, C.F., and Wilkinson, A.J., The structural basis of sequence-independent peptide binding by OppA protein. *Science* 264, 1578–1581, 1994.

154. Leonard, B.A.B., Podbielski, A., Hedberg, P.J., and Dunny, G.M., *Enterococcus faecalis* pheromone binding protein, PrgZ, recruits a chromosomal oligopeptide permease system to import sex pheromone cCF10 for induction of conjugation. *Proc. Natl. Acad. Sci. USA* 93, 260–264, 1996.

155. Nakayama, J., Takanami, Y., Horii, T., Sakuda, S., and Suzuki, A., Molecular mechanism of peptide-specific pheromone signaling in *Enterococcus faecalis*: functions of pheromone receptor TraA and pheromone-binding protein TraC encoded by plasmid pPD1. *J. Bacteriol.* 180, 449–456, 1998.

156. Ruhfel, R.E., Manias, D.A., and Dunny, G.M., Cloning and characterization of a region of the *Enterococcus faecalis* conjugative plasmid, pCF10, encoding a sex pheromone binding function. *J. Bacteriol.* 175, 5253–5259, 1993.

157. Tanimoto, K., An, F.Y., and Clewell, D.B., Characterization of the *traC* determinant of the *Enterococcus faecalis* hemolysin-bacteriocin plasmid pAD1: binding of sex pheromone. *J. Bacteriol.* 175, 5260–5264, 1993.

158. Fujimoto, S., and Clewell, D.B., Regulation of the pAD1 sex pheromone response of *Enterococcus faecalis* by direct interaction between the cAD1 peptide mating signal and the negatively regulating, DNA-binding TraA protein. *Proc. Natl. Acad. Sci. USA* 95, 6430–6435, 1998.

159. Bae, T., Clerc-Bardin, S., and Dunny, G.M., Analysis of expression of *prgX*, a key negative regulator of the transfer of the *Enterococcus faecalis* pheromone-inducible plasmid pCF10. *J. Mol. Biol.* 297, 861–875, 2000.

160. Weaver, K.E., and Clewell, D.B., Regulation of the pAD1 sex pheromone response in *Enterococcus fecalis*: construction and characterization of *lacZ* transcriptional fusions in a key control region of the plasmid. *J. Bacteriol.* 170, 4343–4352, 1988.
161. Weaver, K.E., and Clewell, D.B., Construction of *Enterococcus faecalis* pAD1 miniplasmids: identification of a minimal pheromone response regulatory region and evaluation of a novel pheromone-dependent growth inhibition. *Plasmid,* 22, 106–119, 1989.
162. Ike, Y., Hashimoto, H., and Clewell, D.B., Hemolysin of *Streptococcus faecalis* subspecies zymogenes contributes to virulence in mice. *Infect. Immun.* 45, 528–530, 1984.
163. Tanimoto, K., and Clewell, D.B., Regulation of the pAD1-encoded sex pheromone response in *Enterococcus faecalis*: expression of the positive regulator TraE1. *J. Bacteriol.* 175, 1008–1018, 1993.
164. Pontius, L.T., and Clewell, D.B., Conjugative transfer of *Enterococcus faecalis* plasmid pAD1: nucleoride sequence and transcriptional fusion analysis of a region involved in positive regulation. *J. Bacteriol.* 174, 3152–3160, 1992.
165. Muscholl, A., Galli, D., Wanner, G., and Wirth, R., Sex pheromone plasmid pAD1-encoded aggregation substance of Enterococcus faecalis is positively regulated in trans by *traE1*. *Eur. J. Biochem.* 214, 333–338, 1993.
166. Tomita, H., and Clewell, D.B., A pAD1-encoded small RNA molecule, mD, negatively regulates *Enterococcus faecalis* pheromone response by enhancing transcription termination. *J. Bacteriol.* 182, 1062–1073, 2000.
167. Galli, D., Friesenegger, A., and Wirth, R., Transcriptional control of sex-pheromone-inducible genes on plasmid pAD1 of *Enterococcus faecalis* and sequence analysis of a third structural gene for (pPD1-encoded) aggregation substance. *Mol. Microbiol.* 6, 1297–1308, 1992.
168. Weaver, K.E., and Clewell, D.B., Regulation of the pAD1 sex pheromone response in *Enterococcus faecalis*: effects of host strain and *traA, traB* and *C* region mutants on expression of an *E* region pheromone-inducible *lacZ* fusion. *J. Bacteriol.* 172, 2633–2641, 1990.
169. Clewell, D.B., Pontius, L.T., An, F.Y., Ike, Y., Suzuki, A., and Nakayama, J., Nucleotide sequence of the sex pheromone inhibitor (iAD1) determinant of *Enterococcus faecalis* conjugative plasmid pAD1. *Plasmid* 24, 156–161, 1990.
170. Nakayama, J., Ruhfel, R.E., Dunny, G.M., Isogai, A., and Suzuki, A., The *prgQ* gene of the *Enterococcus faecalis* tetracycline resistance plasmid, pCF10, encodes a peptide inhibitor, iCF10. *J. Bacteriol.* 176, 2003–2004, 1994.
171. Bensing, B.A., Manias, D.A., and Dunny, G.M., Pheromone cCF10 and plasmid pCF10-encoded regulatory molecules act post-transcriptionally to activate expression of downstream conjugation functions. *Mol. Microbiol.* 24, 295–308, 1997.
172. Nakayama, J., Yoshida, K., Kobayashi, H., Isogai, A., Clewell, D.B., and Suzuki, A., Cloning and characterization of a region of *Enterococcus faecalis* plasmid pPD1 encoding pheromone inhibitor (*ipd*), pheromone sensitivity (*traC*), and pheromone shutdown (*traB*) genes. *J. Bacteriol.* 177, 5567–5573, 1995.
173. An, F.Y., and Clewell, D.B., Characterization of the determinant (*traB*) encoding sex pheromone shutdown by the hemolysin/bacteriocin plasmid, pAD1 in *Enterococcus faecalis. Plasmid.* 31, 215–221, 1994.
174. Nakayama, J., Dunny, G.M., Clewell, D.B., and Suzuki, A., Quantitative analysis for pheromone inhibitor and pheromone shutdown in *Enterococcus faecalis. Dev. Biol. Stand.* 85, 35–38, 1995.
175. Buttaro(Leonard) B.A., Antiporta, M.H., and Dunny, G.M., Cell-associated pheromone peptide (cCF10) production and pheromone inhibition in *Enterococcus faecalis. J. Bacteriol.* 182, 4926–4933, 2000.

176. Hedberg, P.J., Leonard, B.A.B., Ruhfel, R.E., and Dunny, G.M., Identification and characterization of the genes of *Enterococcus faecalis* plasmid pCF10 involved in replication and in negative control of pheromone-inducible conjugation. *Plasmid.* 35, 46–57, 1996.

177. Dunny, G.M., Antiporta, M.H., and Hirt, H., Peptide pheromone-induced transfer of plasmid pCF10 in *Enterococcus faecalis*: probing the genetic and molecular basis for specificity of the pheromone response. *Peptides* 22, 1529–1539, 2001.

178. Chow, J.W., Thal, L.A., Perri, M.B., Vazquez, J.A., Donabedian, S.M., Clewell, D.B., and Zervos, M.J., Plasmid-associated hemolysin and aggregation substance production contribute to virulence in experimental enterococcal endocarditis. *Antimicrob. Agents Chemother.* 37, 2474–2477, 1993.

179. Jett, B.D., Jensen, H.G., Nordquist, R.E., and Gilmore, M.S., Contribution of the pAD1-encoded cytolysin to the severity of experimental *Enterococcus faecalis* endophthalmitis. *Infect. Immun.* 60, 2445–2452, 1992.

180. Kreft, B., Marre, R., Schramm, U., and Wirth, R., Aggregation substance of *Enterococcus faecalis* mediates adhesion to cultured renal tubular cells. *Infect. Immun.* 60, 25–30, 1992.

181. Olmsted, S.B., Dunny, G.M., Erlandsen, S.L., and Wells, C.L., A plasmid-encoded surface protein on *Enterococcus faecalis* augments its internalization by cultured intestinal epithelial cells. *J. Infect. Dis.* 170, 1549–1556, 1994.

182. Rakita, R.M., Vanek, N.N., Jacques-Palaz, K., Mee, M., Mariscalco, M.M., Dunny, G.M., Snuggs, M., van Winkle, W.B., and Simon, S.I., *Enterococcus faecalis* bearing aggregation substance is resistant to killing by human neutrophils despite phago-cytosis and neutrophil activation. *Infect. Immun.* 67, 6067–6075, 1999.

183. Schlievert, P.M., Gahr, P.J., Assimacopoulos, A.P., Dinges, M.M., Stoehr, J.A., Harmala, J.W., Hirt, H., and Dunny, G.M., Aggregation and binding substances enhance pathogenicity in rabbit models of *Entrococcus faecailis* endocarditis. *Infect. Immun.* 66, 218–223, 1998.

184. Clewell, D.B., An, F.Y., White, B.A., and Gawron-Burke, C., *Streptococcus faecalis* sex pheromone (cAM373) also produced by *Staphylococcus aureus* and identification of a conjugative transposon (Tn918). *J. Bacteriol.* 162, 1212–1220, 1985.

185. Firth, N., Fink, P.D., Johnson, L., and Skurray, R.A., A lipoprotein signal peptide encoded by the staphylococcal conjugative plasmid pSK41 exhibits an activity resembling that of *Enterococcus faecalis* pheromone cAD1. *J. Bacteriol.* 176, 5871–5873, 1994.

186. Nakayama, J., Igarashi, S., Nagasawa, H., Clewell, D.B., An, F.Y., and Suzuki, A., Isolation and structure of staph-cAM373 produced by *Staphylococcus aureus* that induces conjugal transfer of *Enterococcus faecalis* plasmid pAM373. *Biosci. Biotech. Biochem.* 60, 1038–1039, 1996.

187. Flannagan, S.E., and Clewell, D.B., Identification and characterization of genes encoding sex pheromone cAM373 activity in *Enterococcus faecalis* and *Staphylococcus aureus*. *Mol. Microbiol.* 44, 803–817, 2002.

188. An, F.Y., and Clewell, D.B., Identification of the cAD1 sex pheromone precursor in *Enterococcus faecalis*. *J. Bacteriol.* 184, 1880–1887, 2002.

189. Ji, G., Beavis, R., and Novick, R., Cell density control of staphylococcal virulence medi-ated by an octapeptide pheromone. *Proc. Natl. Acad. Sci. USA* 92, 12055–12059, 1995.

190. Dunman, P.M., Murphy, E., Haney, S., Palacios, D., Tucker-Kellogg, G., Wu, S., Browm, L., Zagursky, R.J., Shlaes, D., and Projan, S.J., Transcription profiling-based identification of *Staphylococcus aureus* genes regulated by the *agr* and/or *sarA* loci. *J. Bacteriol.* 183, 7341–7353, 2001.

191. Gaskill, M.E., and Khan, S.A., Regulation of the enterotoxin B gene in *Staphylococcus aureus*. *J. Biol. Chem.* 263, 6276–6280, 1988.

192. Janzon, L., and Arvidson, S., The role of the delta-lysin gene (*hld*) in the regulation of virulence genes by the accessory gene regulator (*agr*) in *Staphylococcus aureus*. *EMBO J.* 9, 1391–1399, 1990.

193. Morfeldt, E., Taylor, D., von Gabain, A., and Arvidson, S., Activation of alpha-toxin translation in *Staphylococcus aureus* by the trans-encoded antisense RNA, RNA III. *EMBO J.* 14, 4569–4577, 1995.

194. Novick, R.P., Ross, H.F., Projan, S.J., Kornblum, J., Kreiswirth, B., and Moghazeh, S., Synthesis of staphylococcal virulence factors is controlled by a regulatory RNA molecule. *EMBO J.* 12, 3967–3975, 1993.

195. Peng H-L, Novick, R.P., Kreiswirth, B., Kornblum, J., and Schlievert, P., Cloning, characterization and sequencing of an accessory gene regulator (*agr*) in *Staphylococcus aureus*. *J. Bacteriol.* 179, 4365–4372, 1988.

196. Recsei, P., Kreiswirth, B., O'Reilly, M., Schlievert, P., Gruss, A., and Novick, R., Regulation of exoprotein gene expression by *agr*. *Mol. Gen. Genet.* 202, 58–61, 1986.

197. Cheung, A.L., Heinrichs, J.H., and Bayer, M.G., Characterization of the *sar* locus and its interaction with *agr* in *Staphylococcus aureus*. *J. Bacteriol.* 178, 418–423, 1996.

198. Chien, Y., and Cheung, A.L., Molecular interactions between two global regulators, *sar* and *agr*, in *Staphylococcus aureus*. *J. Biol. Chem.* 273, 2645–2652, 1998.

199. Morfeldt, E., Tegmark, K., and Arvidson, S., Transcriptional control of the *agr*-dependent virulence gene regulator, RNA III. *Staphylococcus aureus*. *Mol. Microbiol.* 21, 1227–1237, 1996.

200. Smeltzer, M.S., Hart, M.E., and Iandolo, J.J., Phenotypic characterization of *xpr*, a global regulator of extracellular virulence factors in *Staphylococcus aureus*. *Infect. Immun.* 61, 919–925, 1993.

201. Wesson, C.A., Liou, L.E., Todd, K.M., Bohach, G.A., Trumble, W.R., and Bayles, K.W., *Staphylococcus aureus* Agr and Sar global regulators influence internalization and induction of apoptosis. *Infect. Immun.* 66, 5238–5243, 1998.

202. Novick, R.P., Projan, S., Kornblum, J., Ross, H., Kreiswirth, B., and Moghazeh, S., The *agr* P-2 operon: an autocatalytic sensory transduction system in *Staphylococcus aureus*. *Mol. Gen. Genet.* 248, 446–458, 1995.

203. Ji, G., Beavis, R., and Novick, R.P., Bacterial interference caused by autoinducing peptide variants. *Science* 276, 2027–2030, 1997.

204. Saenz, H.L., Augsburger, V., Vuong, C., Jack, R.W., Gotz, F., and Otto, M., Inducible expression and cellular location of AgrB, a protein involved in the maturation of the staphylococcal quorum-sensing pheromone. *Arch. Microbiol.* 174, 452–455, 200.

205. Zhang, L., Gray, L., Novick, R.P., and Ji, G., Transmembrane topology of AgrB, the protein involved in the post-translational modification of AgrD in *Staphylococcus aureus*. *J. Biol. Chem.* 277, 34736–34742, 2002.

206. Lina, G., Jarraud, S., Ji, G., Greenland, T., Pedraza, A., Etienne, J., Novick, R.P., and Vandenesch, F., Transmembrane topology and histidine protein kinase activity of AgrC, the *agr* signal receptor in *Staphylococcus aureus*. *Mol. Microbiol.* 28, 655–662, 1998.

207. Lyon, G.J., Mayville, P., Muir, T.W., and Novick, R.P., Rational design of a global inhibitor of the virulence response in *Staphylococcus aureus*, based in part on localization of the site of inhibition to the receptor-histidine kinase, *agrC*. *Proc. Natl. Acad. Sci. USA* 97, 13330–13335, 2000.

208. Novick, R.P., and Muir, T.W., Virulence gene regulation by peptides in staphylococci and other gram-positive bacteria. *Curr. Opin. Microbiol.* 2, 40–45, 1999.

209. Abdelinour, A., Arvidson, S., Bremell, T., Ryden, C., and Tarkowski, A., The accessory gene regulator (*agr*) controls *Staphylococcus aureus* virulence in a murine arthritis model. *Infect. Immun.* 61, 3879–3885, 1993.

210. Cheung, A.L., Eberhardt, K.J., Chung, E., Yeaman, M.R., Sullam, P.M., Ramos, M., and Bayer, A.S., Diminished virulence of s*ar /agr* mutant of *Staphylococcus aureus* in the rabbit model of endocarditis. *J. Clin. Invest.* 94, 1815–1822, 1994.

211. Mayville, P., Ji, G., Beavis, R., Yang H-M, Goger, M., Novick, R.P., and Muir, T.W., Structure-activity analysis of synthetic autoinducing thiolactone peptides from *Staphylococcus aureus* responsible for virulence. *Proc. Natl. Acad. Sci. USA* 96, 1218–1223, 1999.

212. Yu, J., Bellinger-Kawahara, C., Winterberg, P., and Francis, K., Disruption of *Staphylococcus aureus* RNAIII *agr* locus does not cause virulence attenuation in a mouse sepsis model. *Am. Soc. Microbiol. Gen. Meeting. Abstracts* (B322), 2002.

213. Novick, R.P., Autoindiction and signal transduction in the regulation of staphylococcal virulence. *Mol. Microbiol.* 48, 1429–1449, 2003.

214. Jarraud, S., Lyon, G.J., Figueiredo, A.M.S., Gerard, L., Vandenesch, F., Etienne, J., Muir, T.W., and Novick, R.P., Exfoliatin-producing strains define a fourth *agr* specificity group in *Staphylococcus aureus. J. Bacteriol.* 182, 6517–6522, 2000.

215. Kalkum, M., Lyon, G.J., and Chait, B.T., Detection of secreted peptides using hypothesis-driven multistage mass spectrometry. *Proc. Natl. Acad. Sci. USA* 100, 2795–2800, 2003.

216. Lyon, G.J., Wright, J.S., Muir, T.W., and Novick, R.P., Key determinants of receptor activation in the *agr* autoinducing peptides of *Staphylococcus aureus. Biochemistry* 41, 10095–10104, 2002.

217. Otto, M., Sussmuth, R., Vuong, C., Jung, G., and Gotz, F., Inhibition of virulence factor expression in *Staphylococcus aureus* by the *Staphylococcus epidermidis agr* pheromone and derivatives. *FEMS Lett.* 450, 257–262, 1999.

218. Dufour, P., Jarraud, S., Vandenesch, F., Greenland, T., Novick, R.P., Bes, M., Etienne, J., and Lina, G., High genetic variability of the *agr* locus in *Staphylococcus* species. *J. Bacteriol.* 184, 1180–1186, 2002.

219. Nakayama, J., Cao, Y., Horii, T., Sakuda, S., Akkermans, A.D., de Vos, W.M., and Nagasawa, H., Gelatinase biosynthesis-activating pheromone: a peptide lactone that mediates a quorum sensing in *Enterococcus faecalis. Mol. Microbiol.* 41, 145–154, 2001.

220. Qin, X., Singh, K.V., Weinstock, G.M., and Murray, B.E., Characterization of *fsr*, a regulator controlling expression of gelatinase and serine protease in *Enterococcus faecalis* OG1RF. *J. Bacteriol.* 183, 3372–3382, 2001.

221. Otto, M., Süssmuth, R., Jung, G., and Götz, F., Structure of the pheromone peptide of the *Staphylococcus epidermidis agr* system. *FEMS Lett.* 424, 89–94, 1998.

222. Garvis, S., Mei, J.M., Ruiz-Albert, J., and Holden, D.W., *Staphylococcus aureus svrA*: a gene required for virulence and expression of the *agr* locus. *Microbiology* 148, 3235–3243, 2002.

223. Nixon, B.C., Ronson, C.W., and Ausubel, F.M., Two-component regulatory systems responsive to environmental stimuli share strongly conserved domains with the nitrogen assimilation regulatory genes *ntrB* and *ntrC. Proc. Natl. Acad. Sci. USA* 83, 7850–7854, 1986.

224. Morfeldt, E., Panova-Sapundjieva, I., Gustafsson, B., and Arvidson, S., Detection of the response regulator AgrA in the cytosolic fraction of *Staphylococcus aureus* by monoclonal antibodies. *FEMS Microbiol. Lett.* 143, 195–201, 1996.

225. Lyon, G.J., Wright, J.S., Christopoulos, A., Novick, R.P., and Muir, T.W., Reversible and specific extracellular antagonism of receptor-histidine kinase signaling. *J. Biol. Chem.* 277, 6247–6253, 2002.

226. McDowell, P., Affas, Z., Reynolds, C., Holden, M.T.G., Wood, S.J., Saint, S., Cockayne, A., Hill, P.J., Dodd, C.E.R., Bycroft, B.W., Chan, W.C., and Williams, P., Structure, activity and evolution of the group I thiolactone peptide quorum-sensing system of *Staphylococcus aureus. Mol. Microbiol.* 41, 503–512, 2001.

227. Balaban, N., Goldkorn, T., Nhan, R.T., Dang, L.B., Scott, S., Ridgley, R.M., Rasooly, A., Wright, S.C., Larrick, J.W., Rasooly, R., and Carlson, J.R., Autoinducer of virulence as a target for vaccine and therapy against *Staphylococcus aureus. Science* 280, 438–440, 1998.

228. Vuong, C., Saenz, H., Götz, F., and Otto, M., Impact of the *agr* quorum-sensing system on adherence to polystyrene in *Staphylococcus aureus. J. Infect. Dis.* 182, 1688–1693, 2000.

229. Otto, M., Echner, H., Voelter, W., and Götz, F., Pheromone cross-inhibition between *Staphylococcus aureus* and *Staphylococcus epidermidis. Infect. Immun.* 69, 1957–1960, 2001.

230. Jarraud, S., Mougel, C., Thioulouse, J., Lina, G., Meugnier, H., Forey, F., Nesme, X., Etienne, J., and Vandenesch, F., Relationships between *Staphylococcus aureus* genetic background, virulence factors, *agr* groups (alleles), and human disease. *Infect. Immun.* 70, 631–641, 2002.

231. Sakoulas, G., Eliopoulos, G.M., Moellering, R., Wennersten, C., Venkataraman, L., Novick, R.P., and Gold, H.S., Accessory gene regulator (*agr*) locus in geographically diverse *Staphylococcus aureus* isolates with reduced susceptibility to vancomycin. *Antimicrob. Agents Chemother.* 46, 1492–1502, 2002.

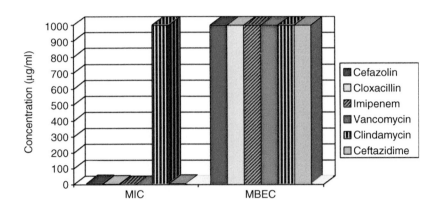

FIGURE 2.1 *Staphylococcus epidermidis* sensitivity.

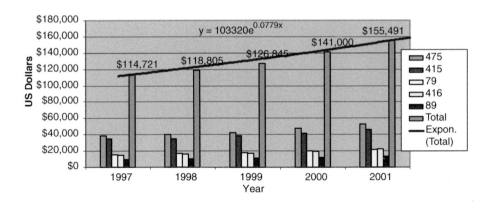

FIGURE 2.2 Mean charge per DRG.

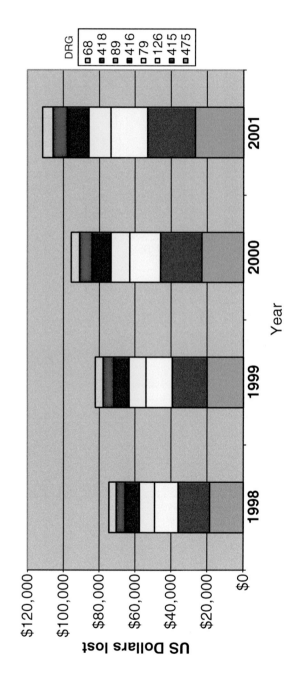

FIGURE 2.3 Loss per incidence for Medicare charges.

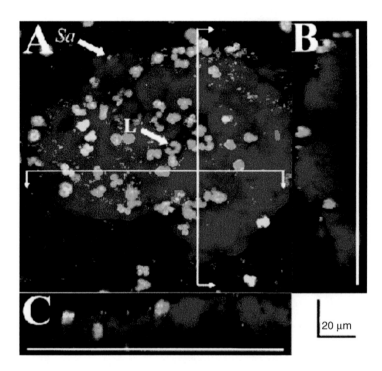

FIGURE 6.1 Scanning laser confocal microscopy image showing leukocytes (L) attached to and within a *S. aureus* biofilm. Note the three-dimensional structure of the biofilm with its red mushroom tower containing bacteria interspersed by channels represented by the black areas within the biofilm. Leukocytes are present within the biofilm. (B) x and (C) y cross-sections showing that the leukocytes have not fully penetrated the biofilm but are lodged within the natural topography of the biofilm. Bar, 30 μm. Reprinted from Leid, J.G., Shirtliff, M.E., Costerton, J.W., and Stoodley, A.P., Human leukocytes adhere to, penetrate, and respond to *Staphylococcus aureus* biofilms. *Infect. Immun.* 70, 6339–6345, 2002. With permission.

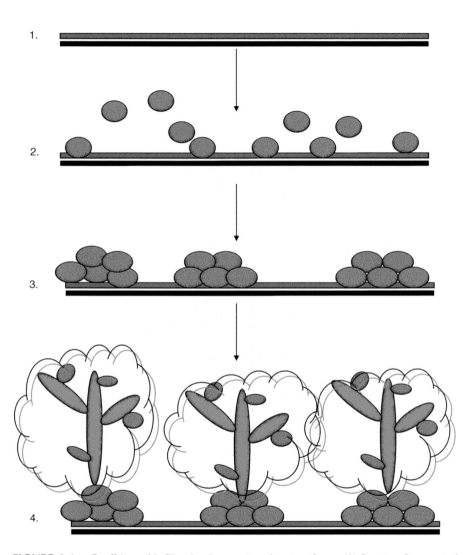

FIGURE 9.1 *C. albicans* biofilm development on inert surfaces. (1) Inert surface coated with a conditioning film (blue) consisting largely of host proteins. (2) Early attachment and colonization by *C. albicans* yeast-phase cells (orange). (3) Microcolony yeast basal layer formation, involving stacking of yeasts in the formation of the microcolonies. (4) Expansion of the biofilm architecture through the development of a hyphal/pseudohyphal layer (turquoise) that protrudes from the inner yeast layer to the outer reaches of the biofilm. The hyphal layer development occurs simultaneously with the development of the thick layer of matrix material that has engulfed both the hyphal and yeast biofilm layers.

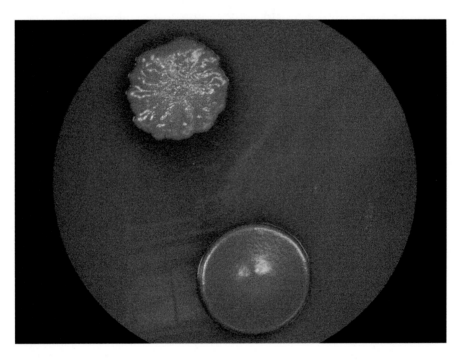

FIGURE 10.3 Colonies of biofilm forming ("crusty" uppermost colony) and nonbiofilm forming ("smooth") *S. epidermidis* on Congo red agar.

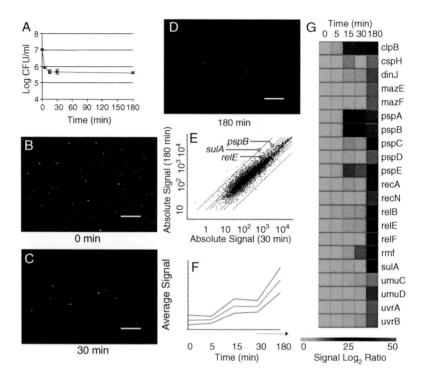

FIGURE 12.5 Gene expression profile of isolated persisters. *E. coli* Hm22 *hip*A7 cells were grown to mid-exponential and treated with 50 μg/ml ampicillin. (A) Samples were taken at indicated times and plated to determine live cells by colony count. (B–D) Samples were stained with a LIVE/DEAD kit and visualized with epifluorescent microscopy. Bar = 50 μm. Green cells are live, red cells (stained by normally impermeant propidium iodide) are dead. Note the extensive red background due to cellular debris, as well as dead intact cells in the 30-min sample. (E) Scatter plot of absolute gene expression at 180 vs. 30 min. Red lines indicate 2-, 3-, and 30-fold changes, respectively. (F) Cluster analysis of persister gene expression profile obtained with Affymetrix Self-Organizing Map (SOM). The profile was obtained by hybridizing labeled mRNA from samples (0, 5, 15, 30, and 180 min) to Affymetrix *E. coli* gene chips. The cluster shown indicates the expression profile of genes specifically upregulated in persisters (180 min). The red line indicates average signal intensity. (G) Heatmap of selected genes from the cluster (F) generated with Spotfire Decisionsite 7.2.

12 Persisters: Specialized Cells Responsible for Biofilm Tolerance to Antimicrobial Agents

Kim Lewis, Amy L. Spoering, Niilo Kaldalu, Iris Keren, and Devang Shah

CONTENTS

12.1 INTRODUCTION

The study of biofilms is a rapidly expanding field of microbiology. Biofilm research encompasses studies of many topics including multispecies communities, cell–cell signaling, virulence, industrial fouling, and bioremediation, to name a few. Yet it is fair to say that the field was propelled to its present prominence due to a singular feature that unites all biofilms—namely their dramatic tolerance to antimicrobial agents. This tolerance is responsible for recalcitrant human infections, accounting for ~60% of all infectious diseases in the West (1). A large number of empirical studies documenting biofilm tolerance have been published in the past two decades. At the same time, there has been a paucity of molecular biology research into the mechanism of tolerance. This may seem surprising, given the prominence of the problem, and our generally good understanding of various mechanisms of antibiotic resistance (2). The reason for this reluctance to tackle the main question of the biofilm field may have stemmed from the suggestion that the problem does not really exist. Biofilms are slow growing, while antibiotics act best against rapidly dividing cells. The action of most β-lactams, for example, depends stringently upon rapid growth (3).

Slow growth alone, or in combination with possible retardation of antibiotic diffusion into the biofilm, could then explain the observed tolerance (4).

Studies reporting binding of aminoglycoside antibiotics to the biofilm matrix seemed to support the idea of limited antibiotic access to cells (5–9). Similarly, it was found that retardation of diffusion combined with active degradation of compounds can effectively protect the biofilm from hydrogen peroxide or a β-lactam antibiotic (10–13). At the same time, fluoroquinolones appeared to diffuse freely into a biofilm (12), and these compounds are able to kill non-growing cells. Other substances, such as antiseptics and disinfectants, can also kill non-growing cells, and show considerably lower effectiveness against biofilms (14,15).

Expression of a possible biofilm-specific resistance mechanism was suggested to contribute to biofilm tolerance (16), and seemed to explain recalcitrance to such compounds as fluoroquinolones or quaternary ammonium antiseptics. This proposition however does not appear realistic. Resistance of biofilms to killing by antimicrobials is dramatic, 100–1000 fold above the MIC, and occurs for all antimicrobials tested and in all species examined. This means that all pathogens potentially harbor an essentially perfect multidrug resistance mechanism, but only "choose" to express it when growing as a biofilm. If such a mechanism were to exist, we should have seen mutants that express it in rapidly growing planktonic cultures as well. This is not the case. A susceptible *Staphylococcus epidermidis*, for example, does not acquire multidrug resistance to "everything" due to a mutation.

But how is it possible that a biofilm can be resistant to killing by all antimicrobials and *not* harbor a resistance mechanism? Herein lies the paradox, and the riddle, of biofilm resistance.

12.2 THE CULPRIT—PERSISTER CELLS

While measuring a dose–response of a *Pseudomonas aeruginosa* biofilm to ofloxacin (17), we noticed that a small fraction of cells was not eliminated even by very high levels of the antibiotic (Figure 12.1). These cells appeared essentially invulnerable. However, the bulk of the biofilm was fairly sensitive. Cells of one of the strains overexpressed a prominent multidrug pump MexAB-OprM, which conferred considerable resistance to the bulk. However, beyond the clinically-achievable concentration, the differences among strains with or without the pump were erased, clearly indicating that some other mechanism was responsible for the remarkable survival of these persister cells. But whatever the mechanism, the culprit appeared in clear view. A small fraction of persisters was imparting survival to the biofilm population. Examination of the biofilm literature showed that persisters were documented in numerous experiments, and ignored (18). It does indeed seem natural to look at the bulk of cells when searching for a property characteristic of the population. However, in the case of a biofilm, the majority of cells appear unremarkable. It is the persistent minority that deserves attention. The biofilm riddle thus shifts to the study of persisters.

Persisters were actually discovered in a rapidly growing, planktonic population. In 1944, Joseph Bigger noticed that the newly-introduced penicillin was unable to "sterilize" a culture of *Staphylococcus* (19). It is interesting to note that the initial focus of that study was to evaluate bactericidal properties of penicillin. Bigger writes: "My results strongly oppose the commonly accepted belief that penicillin is

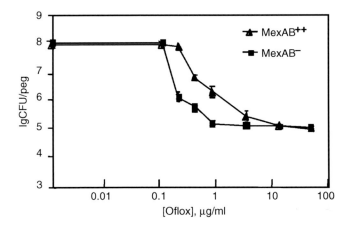

FIGURE 12.1 Role of the MexAB-OprM pump in resistance of biofilms treated with ofloxacin (Oflox).

merely bacteriostatic." It is difficult for us to imagine how the bactericidal property of penicillin might have been missed—the substance rapidly turns a turbid culture into a transparent solution. Bigger plated this transparent-looking medium, and discovered persisters. His experiments indicated that persisters were not mutants, and upon reinoculation produce a population containing a sensitive bulk and new tolerant cells. We repeated this important experiment, quantifying survival of *Escherichia coli* treated with ampicillin (20). The result clearly showed that surviving persisters regenerate the original population, and are therefore phenotypic variants of the wild type.

If persisters are present in planktonic populations, then how unique is the survival capability of a biofilm? In order to address this question, we compared three populations—a logarithmic, stationary, and biofilm culture of *P. aeruginosa* (21). The result was unexpected—the stationary culture produced more persisters, and was more tolerant, than the biofilm (Figure 12.2). In retrospect, this should not be surprising, since a stationary culture has almost no growth, while the biofilm is growing, albeit slowly. We find that persister formation is indeed growth state dependent—an early log culture makes little or no persisters in *E. coli*, *Staphylococcus aureus*, or *P. aeruginosa*, and the level of persisters increases as the population reaches stationary state. It is important to stress that persisters are not simply non-growing cells—in a stationary culture, fluoroquinolones or mitomycin C kill the bulk, leaving 1 to 10% intact persisters (21,22).

The strong dependence of persister formation on growth state appears as a typical quorum-sensing phenomenon. We performed a standard test for quorum sensing, adding spent stationary medium to early log cells and measuring the rate of persisters. Spent medium did not increase persisters in *E. coli* or *P. aeruginosa*, arguing against the involvement of quorum sensing (Kaldalu, Keren and Lewis, unpublished). We also noticed that persisters are rapidly lost if a stationary population is diluted. These observations are consistent with persister formation being inversely dependent upon the level of metabolic activity, rather than on a quorum-sensing type signaling molecule.

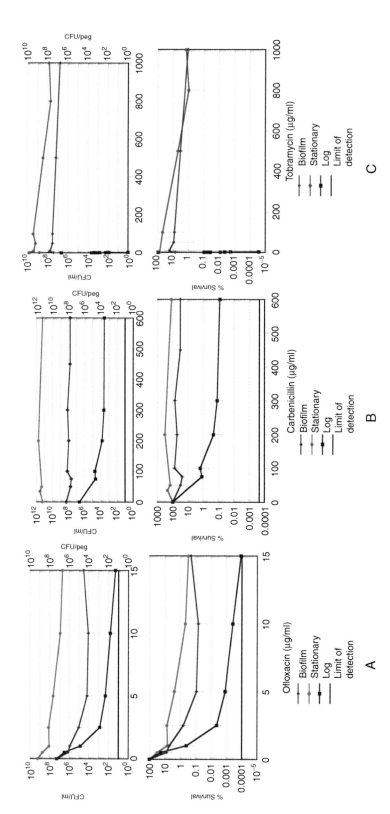

FIGURE 12.2 Killing of logarithmic-phase, stationary-phase, and biofilm cultures by antibiotics. Plotted CFU/ml (upper panel) or as percent survival (lower panel). (A) Ofloxacin, (B) carbenicillin, (C) tobramycin. The limit of detection is indicated by the solid horizontal line.

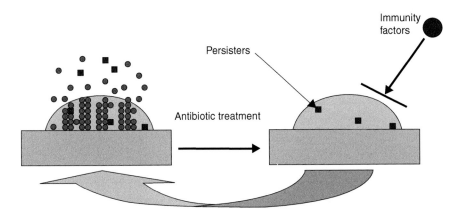

FIGURE 12.3 Model of biofilm resistance based on persister survival. An initial treatment with antibiotic kills planktonic cells and the majority of biofilm cells. The immune system kills planktonic persisters, but the biofilm persister cells are protected from host defenses by the extracellular matrix. After the antibiotic concentration drops, persisters resurrect the biofilm and the infection relapses.

Experiments showing an essential similarity in the tolerance of stationary cells and persisters leads to an obvious question—is drug resistance of biofilms important? The answer would be "no" for a test-tube grown biofilm, and we suggested that tolerance can be much more easily studied with traditional planktonic cultures (21). But there is every reason to believe that biofilms are considerably more tolerant to antibiotics *in vivo*, as compared to planktonic cells, irrespective of their growth state. We argued that *in vivo*, antibiotic treatment will eliminate the bulk of both biofilm and planktonic cells, leaving intact persisters. At this point, the similarity with an *in vitro* experiment probably ends. The immune system will be able to mop up remaining planktonic persisters, just as it eliminates non-growing cells of a population treated with a static antibiotic (Figure 12.3). However, the biofilm matrix protects against immune cells (23–25), and its persisters will survive. After antibiotic concentration drops, persisters will repopulate the biofilm, which will shed off new planktonic cells responsible for disease symptoms. This model explains the relapsing nature of biofilm infections, and suggests that understanding persisters will ultimately allow us to eliminate biofilms.

12.3 PERSISTERS AS SPECIALIZED ALTRUISTIC SURVIVOR CELLS

It was suggested that persisters represent normal cells at a particular stage in the cell cycle (19,26,27). We found that repeated reinoculation of *E. coli* maintaining cells in an early log state leads to a complete loss of persisters (20). Since early log cells undergo a cell cycle but do not produce persisters, we can rule out this particular hypothesis. This simple experiment also suggests that persisters are not formed in response to antibiotic treatment, since early log cells challenged with antibiotics produce no persisters. Neither are persisters cells that temporarily lost their ability to grow due to a reversible defect, such as a stalled replication fork. We would expect defects to occur in early log as well, though no persisters are formed at this stage. What we are left with, then, is an intriguing possibility of persisters representing specialized survivor cells whose production is regulated by the growth stage of the population. Persisters are essentially altruistic cells that forfeit rapid propagation, which ensures survival of the population of kin cells (28) in the presence of lethal factors.

12.4   THE MECHANISM OF TOLERANCE

Realizing that persisters are responsible for biofilm tolerance is helpful, but does not solve the problem. Indeed, how can a cell that does not express specific resistance mechanisms be tolerant to all cidal antibiotics? All resistance mechanisms do essentially the same thing—prevent an antibiotic from binding to the target. We hypothesized that tolerance works in a different way, not by preventing antibiotic binding, but by interfering with the lethal action of cidal compounds (22). Specifically, we proposed that "persister proteins" shut down antibiotic targets (Figure 12.4). It is important to stress that cidal antibiotics kill by corrupting the target function, rather than by merely inhibiting it. For example, erythromycin inhibits protein synthesis and is a static antibiotic. Streptomycin, an aminoglycoside, causes translational misreading, which apparently produces truncated toxic peptides, leading to cell death. Shutting down the ribosome in a persister cell would produce tolerance to aminoglycosides. Fluoroquinolones act by inhibiting DNA gyrase and topoisomerase IV. The action of fluoroquinolones is selective—the ligase activity is inhibited, while the nicking remains intact. As a result, fluoroquinolones force the enzymes to create DNA lesions. Cell-wall acting antibiotics do not merely inhibit peptidoglycan synthesis, but cause death and cell lysis (29,30). By contrast, cells that stop peptidoglycan synthesis due to a lack of nutrients do not usually lyse. Persister proteins could shut down all antibiotic targets, creating a tolerant, dormant persister cell.

12.5   PERSISTER GENES

Finding a persister gene appears straightforward—screening a transposon insertion library should produce clones that die completely, or considerably more so than the

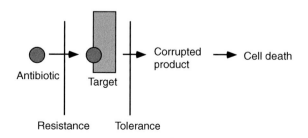

FIGURE 12.4 Antibiotic resistance versus tolerance.

wild type, when challenged with a cidal antibiotic. Our efforts to obtain such a mutant from Tn insertion libraries of *E. coli* and *P. aeruginosa*, however, were not successful, perhaps due to a considerable background variation in the level of persisters among cultures growing in the wells of a microtiter plate (Spoering and Lewis, unpublished). We therefore turned to a gene expression profile approach.

The obvious problem with obtaining an expression profile from persisters is that they first need to be isolated. We reasoned that after treatment with ampicillin, only intact persisters will remain, while the bulk of the cells will lyse. This appeared to be the case, and intact, live persisters were obtained by simple centrifugation. Ampicillin only lyses a rapidly growing, logarithmic culture, where the fraction of persisters is $\sim 10^{-5}$, insufficient material for array work. We therefore took advantage of the *hipA7* strain (26) that produces $\sim 1\%$ persisters (note that a null *hipA* mutant was reported to have no phenotype, see below) surviving ampicillin treatment (Figure 12.5a).

Labeled mRNA from persisters was used to obtain an expression profile with Affymetrix (31) *E. coli* Genechips. A potential problem with analyzing results of this array was that ampicillin treatment produces massive changes in the gene profile of *E. coli* (32), which could obscure persister-specific genes. However, we found that most of the changes accompanying ampicillin treatment occurred between 0 and 30 minute time points. At 30 minutes, the population consisted of mostly dead but intact cells (Figure 12.5b,c). Further incubation produced complete lysis of dead cells, leaving intact persisters after 3 hours with ampicillin (Figure 12.5d). We therefore used cluster analysis to point to genes that specifically increased following the clearing of dead cells and isolation of persisters. In this way, we would be tracking genes that were elevated in persisters prior to ampicillin treatment. We indeed identified such a cluster (Figure 12.5e–g). The cluster contained about 200 genes, including those coding for SOS stress response (*recA*, *sulA*, *uvrBA*, and *umuDC*), phage-shock (*psp*); heat and cold-shock (*cspH*, *htrA*, *ibpAB*, *htpX*, *clpB*). This expression of stress responses is consistent with a survival function of persisters, but does not in itself explain antibiotic tolerance. In practical terms, this set of ~ 200 candidate genes was too large to point to specific persister genes.

Armed with the tolerance hypothesis, we searched for genes that could shut down cellular functions. A number of genes expressed in persisters appeared to fulfill this criterion. RMF inhibits translation by forming ribosome dimers in stationary state (33); UmuDC has been reported to inhibit replication (34); and SulA is an inhibitor of septation (35). Most striking however was the overexpression of well-characterized chromosomal toxin–antitoxin (TA) modules RelBE, MazEF, and putative TA modules PspAB,CD (36) and DinJ/YafQ, homologous to RelBE. Homologs of these genes are found on plasmids where they constitute a maintenance mechanism (37). Typically, the toxin is a protein that inhibits an important cellular function such as translation or replication, and forms an inactive complex with the antitoxin. The toxin is stable, while the antitoxin is degradable. If a daughter cell does not receive a plasmid after segregation, the antitoxin level decreases due to proteolysis, leaving a toxin that either kills the cell or inhibits propagation. TA modules are also commonly found on bacterial chromosomes, but their role is largely unknown. MazEF was proposed to serve as a programmed cell death mechanism (38). However, it was reported recently that MazF and an unrelated toxin RelE do not actually kill cells,

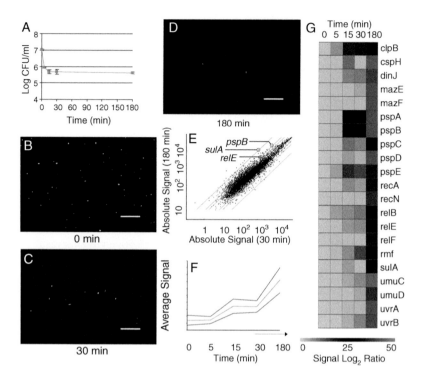

FIGURE 12.5 (See color insert following page 270) Gene expression profile of isolated persisters. *E. coli* HM22 *hip*A7 cells were grown to mid-exponential and treated with 50 μg/ml ampicillin. (A) Samples were taken at indicated times and plated to determine live cells by colony count. (B–D) Samples were stained with a LIVE/DEAD kit and visualized with epifluorescent microscopy. Bar: 50 μm. (E) Scatter plot of absolute gene expression at 180 vs. 30 min. (F) Cluster analysis of persister gene expression profile obtained with Affymetrix Self-Organizing Map (SOM). The profile was obtained by hybridizing labeled mRNA from samples (0, 5, 15, 30, 180 min) to Affymetrix *E. coli* gene chips. The cluster shown indicates the expression profile of genes specifically upregulated in persisters (180 min). (G) Heatmap of selected genes from the cluster (F) generated with Spotfire Decisionsite 7.2.

but induce stasis by inhibiting translation, a condition that can be reversed by expression of corresponding antitoxins (39,40). It was also suggested that MazF and RelE act as attenuators of the stringent response. We reasoned that the ability of "toxin" modules to reversibly block translation makes them excellent candidates for persister genes. By shutting down potential antibiotic targets, toxins will produce tolerant cells.

The best-studied TA module in *E. coli* is RelBE that cleaves mRNA on translating ribosomes, which stalls protein synthesis (41). We tested the ability of RelE to generate persisters tolerant to antibiotics. Cells expressing RelE from an inducible promoter became highly tolerant to ofloxacin, a fluoroquinolone; to cefotaxime,

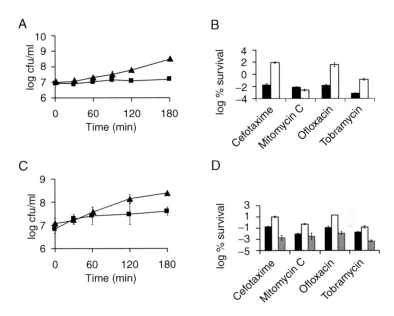

FIGURE 12.6 The effects of toxin overexpression on persister formation. (A) RelE was induced (■) from pkD3035 (pBAD::*relE*) in MG1 (MC1000 Δ*relBE*) at time zero by adding 0.2% arabinose, and MG1 with a blank vector pBAD33 served as the control (▲). Samples were removed at indicated time points and plated for colony counts on LB agar plates containing ampicillin (100 μg/ml), chloramphenicol (50 μg/ml), glucose (0.2%, to suppresses the pBAD promoter), and 1 mM IPTG to induce RelB expression from pKD3033(pA O4/O3::opSD::*relB*). (B) Cells were cultured and RelE was induced as described above. After 3 hours of RelE induction samples were removed and treated with either cefotaxime (100 μg/ml); mitomycin C (10 μg/ml); ofloxacin (5 μg/ml); or tobramycin (25 μg/ml) for 3 hours at 37°C with aeration. Before and after the challenge the cells were plated on media as described in (A) for colony counts. The control (MG1/pBAD; ■) was challenged at a similar cell density to that of the *relE* induced cells (□). (C) HM22 cells (*hipA7*) were cultured as described above and at time 0 were moved 30°C (■) or kept at 37°C (▲). (D) HM22 cells were challenged at 30°C (□) as described in (B). Controls in HM22 (■) and HM21 (K12 *hipA* w.t.; grey bars) cells challenged at 37°C.

a cephalosporin cell wall synthesis inhibitor; and to tobramycin, an aminoglycoside protein synthesis inhibitor (Figure 12.6a,b). This result is consistent with the idea of a toxin blocking antibiotic targets. Indeed, RelE directly blocks the ribosome, which would prevent the lethal action of aminoglycosides. Inhibition of translation in turn blocks DNA and cell wall synthesis, which would diminish the ability of antibiotics acting against DNA gyrase and peptidoglycan synthases to corrupt their targets. Interestingly, RelE did not protect cells from mitomycin C, which forms DNA adducts. This would be consistent with the target blocking idea—RelE does not "block" DNA.

In order to observe persisters, cells were plated on a medium containing an inducer of a recombinant RelB antitoxin. Recovery without inducing RelB was observed as well, but slower. This suggests that a TA module has the ability to both produce and resuscitate persisters.

Given that RelE expression increased persister production prompted us to revisit what was known about the hipBA locus. It was discovered in the first screen for genes specifically affecting persistence that was performed by Moyed and coworkers (26). An EMS-mutagenized population of *E. coli* was enriched with cells surviving ampicillin treatment and then screened for colonies producing higher numbers of persisters. Only mutants whose growth was normally inhibited by ampicillin were examined further. This approach led to the identification of a *hipA7* (high persistence) mutant producing about 1,000-fold more persistent cells as compared to the wild type (26,42,43). This was the first report of a bacterial gene specifically involved in the regulation of persister production and antibiotic tolerance. However, a deletion in either *hipA* or *hipBA* appeared to have no phenotype (44). This suggested that the *hipA7* allele creates a complex pleiotropic artifact, leading to an increase in persisters. The pioneering findings of Moyed's group have been largely ignored.

Interestingly, HipBA is a likely TA module. Indeed, HipB and HipA form a complex (43); overexpression of HipA is "toxic," leading to arrest of cell division (27); a *hipB* mutation could not be obtained due to apparent lethality of free HipA (43); HipB is a repressor of the operon, which is typical for antitoxins; and a homolog of the chromosomal *hipBA* operon is found on the *Rhizobium* symbiotic plasmid pNGR234a where it may play a role in segregation maintenance (45). It was previously reported that mild induction of recombinant HipA increases production of persisters (27). We confirmed this observation, and found that functional expression of HipA strongly increases persisters tolerant to a range of antibiotics (Figure 12.6c,d) (22). The most important remaining unanswered question regarding HipA was whether it had a functional role in creating persisters *in vivo*. We reasoned that the level of persisters is particularly high in stationary state, and that might be the growth phase at which HipA exerts its action.

A *hipBA* deletion strain produced 10 to 100-fold less persisters in stationary state as compared to the wild type (Figure 12.7a,b). This experiment completes the first chapter of the *hip* saga. *hipA* is a persister gene. A biofilm grown from a Δ*hipBA* strain produced considerably less persisters (Figure 12.7c), which makes *hipA* the first validated biofilm tolerance gene as well.

We also tested knockout mutants in other TA modules (*mazEF, dinJ-yafQ, psp*), and in *rmf*. None showed a phenotype in either log or stationary cultures. Given that there are >10 TA modules in *E. coli* (36) which probably are functionally redundant, this is not surprising. What is surprising is that we saw a strong phenotype in the single *hipBA* knockout strain. On the other hand, the lack of an obvious phenotype of a Δ*hipBA* in log state (or in stationary minimal medium, Spoering and Lewis, unpublished) points directly to other players obscuring HipBA under these conditions.

Based on our findings, we proposed the following model of persister production and antibiotic tolerance (Figure 12.8). The ratio of a toxin/antitoxin (such as HipA/HipB) in a population fluctuates, and rare cells will express relatively high levels of a toxin. Cidal antibiotics bind to a target protein and corrupt its function, generating a lethal product (for example, aminoglycosides interrupt translation, resulting in misfolded peptides that damage the cell). A toxin binds to the target and

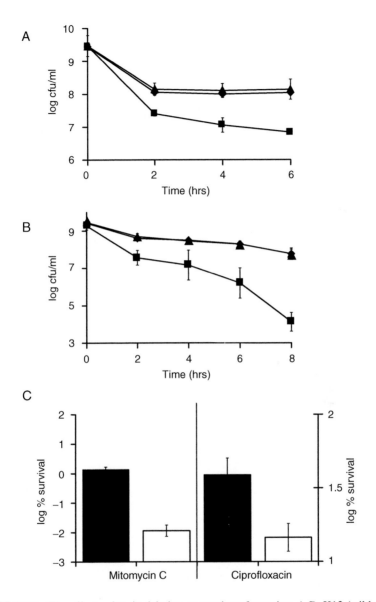

FIGURE 12.7 The effects of toxin deletion on persister formation. A-B. K12 (wild type, ◆)
KL310 (Δ*relBE*; ▲) and KL312 (Δ*hipBA*; ■) were grown to stationary state (16–18 hrs) and
challenged with ofloxacin (5 μg/ml, A) or mitomycin C (10 μg/ml, B). At the designated times
a sample was removed and plated on LB-agar plates for colony count. C. Biofilms (K12
(w.t.; ■) and KL312 (Δ*hipBA*; □) were grown at 37°C for 48 hours, on LB agar to ~2 × 10⁹
CFU/biofilm. They were then exposed to LB agar plates containing 20 mM NaNO₃ with or
without 5 μg/ml mitomycin C (left *y*-axis); and with or without 5 μg/μl ciprofloxacin (right
y-axis). After a 24 hour incubation, biofilms were suspended by vortexing and sonicating in
LB medium, producing a cell suspension that was plated for colony counts.

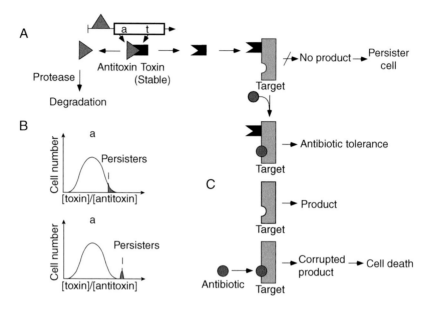

FIGURE 12.8   A model of persister formation and antibiotic tolerance. (A) Chromosomal toxin/antitoxin modules are coded by two-gene operons, with the antitoxin usually acting as a repressor. Once synthesized, the toxin and antitoxin form a stable inactive complex. The antitoxin is labile and can be degraded by proteases. (B) A random fluctuation in relative toxin/antitoxin concentration (a) will produce free active toxin leading to an inhibition of a target function, such as protein synthesis. Suppression of translation will lead to a decrease in the level of this, as well as additional labile antitoxins, producing a distinct population of persister cells with a set of elevated toxin proteins (b). (C) Cidal antibiotics bind to a target protein and corrupt its function. This generates a lethal product (for example, aminoglycosides interrupt translation, resulting in misfolded peptides that damage the cell). A toxin binds to the target and inhibits the function, leading to tolerance. The antibiotic can normally bind to the blocked target, but can no longer corrupt its function. This accounts for antibiotic tolerance of persister cells.

inhibits the function, leading to tolerance. The antibiotic can bind to the blocked target, but can no longer corrupt its function. Inhibition of translation by a toxin such as RelE will further cause a relative increase in the stable toxin (due to antitoxin degradation) of this and other TA modules, which might have an autocatalytic effect on inhibition of translation, leading to a shutdown of other cellular functions, and to dormant, tolerant persister cells. Our observation of several TA modules upregulated in persisters is consistent with this model. Antitoxins act as repressors of TA modules, so a decrease in antitoxin protein level will cause an induction in transcription which we observed in the gene profile of persisters.

Our gene profiling data also support the possibility of persisters being dormant. A cluster of approximately 600 genes involved in metabolism and flagellar synthesis was seen to gradually diminish in cells treated with ampicillin, in accordance with our previous findings (32). Interestingly, the same genes showed a further decrease in isolated persisters, suggesting that these might have been downregulated in persisters prior to the addition of antibiotic. Among these repressed genes were members of the large operons involved in oxidative posphorylation—NADH dehydrogenase, ATP synthase, and cytochrome O ubiquinole oxidase. The general theme of this gene cluster appears to be a decrease in nonessential genes and a shutdown of metabolism, both consistent with a dormant phenotype.

TA modules are widely spread among bacteria (37) and might represent the essential mechanism responsible for persister formation in different species. Production of these phenotypic variants of the wild type appears to be stochastic, and resembles individual variations in the chemotactic behavior of *E. coli* (46,47). Persistence is a reversible phenomenon. Overexpression of toxins inhibits protein synthesis strongly but incompletely (39), which may allow for the synthesis of a neutralizing antitoxin. An even more effective mechanism for resuscitation of persisters would be through expression of tmRNA which releases ribosomes stalled on mRNA lacking a termination codon. tmRNA has been shown to counter the action of artificially expressed RelE and MazF toxins (40). Importantly, tmRNA synthesis may proceed in cells with inhibited translation and might therefore trigger the resuscitation process. Interestingly, "toxins" appear not as damaging factors, but as their opposite, proteins whose function is to protect the cell from damage.

Identification of persister genes, including those described in this work, will lead to a better understanding of tolerance in general and of biofilm resistance to killing in particular. So far, studies aimed at identifying genes responsible for biofilm resistance were limited to the bulk of the population (48), which is fairly susceptible to therapeutic doses of antibiotics or may express resistance genes that act equally well in exponentially growing cells (17,49). It is the persister sub-population, however, present in all species studied, which is responsible for the dramatic tolerance of biofilms to unrelated antibiotics (18). Identification of persister genes is an important first step in understanding recalcitrance of biofilms to antibiotic therapy, and it is likely to shed light on other related but poorly understood phenomena involving a dormant state, such as latent *Mycobacterium tuberculosis* infection (50,51), "viable but not culturable bacteria" (52,53), and "uncultivable" bacteria (54).

REFERENCES

1. Licking, E., Getting a grip on bacterial slime. *Business Week* 1999: 98–100.
2. Lewis, K., et al., *Bacterial Resistance to Antimicrobials: Mechanisms, Genetics, Medical Practice and Public Health.* New York: Marcel Dekker, 2001.
3. Tuomanen, E., et al., The rate of killing of *Escherichia coli* by beta-lactam antibiotics is strictly proportional to the rate of bacterial growth. *J. Gen. Microbiol.* 132, 1297–1304, 1986.
4. Gilbert, P., Collier, P.J., and Brown M.R., Influence of growth rate on susceptibility to antimicrobial agents: biofilms, cell cycle, dormancy, and stringent response. *Antimicrob. Agents Chemother.* 34, 1865–1868, 1990.
5. Gordon, C.A., Hodges N.A., and Marriott, C., Antibiotic interaction and diffusion through alginate and exopolysaccharide of cystic fibrosis-derived *Pseudomonas aeruginosa. J. Antimicrob. Chemother.* 22, 667–674, 1988.

6. Nichols, W.W., et al., Inhibition of tobramycin diffusion by binding to alginate. *Antimicrob. Agents Chemother.* 32, 518–523, 1988.

7. Hoyle, B.D., Jass, J., and Costerton, J.W., The biofilm glycocalyx as a resistance factor. *J. Antimicrob. Chemother.* 26, 1–5, 1990.

8. Shigeta, M., et al., Permeation of antimicrobial agents through *Pseudomonas aeruginosa* biofilms: a simple method. *Chemotherapy* 43, 340–345, 1997.

9. Ishida, H., et al., *In vitro* and *in vivo* activities of levofloxacin against biofilm-producing *Pseudomonas aeruginosa*. *Antimicrob. Agents Chemother.* 42, 1641–1645, 1998.

10. Hassett, D.J., et al., Quorum sensing in *Pseudomonas aeruginosa* controls expression of catalase and superoxide dismutase genes and mediates biofilm susceptibility to hydrogen peroxide. *Mol. Microbiol.* 34, 1082–1093, 1999.

11. Elkins, J.G., et al., Protective role of catalase in *Pseudomonas aeruginosa* biofilm resistance to hydrogen peroxide. *Appl. Environ. Microbiol.* 65, 4594–4600, 1999.

12. Anderl, J.N., Franklin, M.J., and Stewart, P.S., Role of antibiotic penetration limitation in *Klebsiella pneumoniae* biofilm resistance to ampicillin and ciprofloxacin. *Antimicrob. Agents Chemother.* 44, 1818–1824, 2000.

13. Stewart, P.S., Diffusion in biofilms. *J. Bacteriol.* 185, 1485–1491, 2003.

14. Gilbert, P., et al., Do biofilms present a nidus for the evolution of antibacterial resistance? In *Biofilm Community Development: Chance or Necessity?*, Gilbert, P., et al., eds., Cardiff: Bioline Press, 2001.

15. Gilbert, P., Das, J., and Foley, I., Biofilm susceptibility to antimicrobials. *Adv. Dent. Res.* 11, 160–167, 1997.

16. Costerton, J.W., Stewart, P.S., and Greenberg, E.P., Bacterial biofilms: A common cause of persistent infections. *Science* 284, 1318–1322, 1999.

17. Brooun, A., Liu, S., and Lewis, K., A dose-response study of antibiotic resistance in *Pseudomonas aeruginosa* biofilms. *Antimicrob. Agents Chemother.* 44, 640–646, 2000.

18. Lewis, K., Riddle of biofilm resistance. *Antimicrob. Agents Chemother.* 45, 999–1007, 2001.

19. Bigger, J.W., Treatment of staphylococcal infections with penicillin. *Lancet* ii, 497–500, 1944.

20. Keren, I., et al., Persister cells and tolerance to antimicrobials. *FEMS Microbiol. Lett.* 230, 13–18, 2004.

21. Spoering, A.L., and Lewis, K., Biofilms and planktonic cells of *Pseudomonas aeruginosa* have similar resistance to killing by antimicrobials. *J. Bacteriol.* 183, 6746–6751, 2001.

22. Keren, I., Shah, D., Spoering, A., Kaldalu, N., and Lewis, K., Specialized persister cells and the mechanism of antibiotic tolerance in *Escherichia coli*. *J. Bacteriol.* 186, 8172–8180, 2004.

23. Leid, J.G., et al., Human leukocytes adhere to, penetrate, and respond to *Staphylococcus aureus* biofilms. *Infect. Immun.* 70, 6339–6345, 2002.

24. Jesaitis, A.J., et al., Compromised host defense on *Pseudomonas aeruginosa* biofilms: characterization of neutrophil and biofilm interactions. *J. Immunol.* 171, 4329–4339, 2003.

25. Vuong, C., et al., Polysaccharide intercellular adhesin (PIA) protects *Staphylococcus epidermidis* against major components of the human innate immune system. *Cell Microbiol.* 6, 269–275, 2004.

26. Moyed, H.S., and Bertrand, K.P., hipA, a newly recognized gene of *Escherichia coli* K-12 that affects frequency of persistence after inhibition of murein synthesis. *J. Bacteriol.* 155, 768–775, 1983.

27. Falla, T.J., and Chopra, I., Joint tolerance to beta-lactam and fluoroquinolone antibiotics in *Escherichia coli* results from overexpression of *hipA*. *Antimicrob. Agents Chemother.* 42, 3282–3284, 1998.

28. Lewis, K., Pathogen resistance as the origin of kin altruism. *J. Theor. Biol.* 193, 359–363, 1998.

29. Tomasz, A., Albino, A., and Zanati, E., Multiple antibiotic resistance in a bacterium with suppressed autolytic system. *Nature* 227, 138–140, 1970.

30. Rice, K.C., and Bayles, K.W., Death's toolbox: examining the molecular components of bacterial programmed cell death. *Mol. Microbiol.* 50, 729–738, 2003.

31. Selinger, D.W., et al., RNA expression analysis using a 30 base pair resolution *Escherichia coli* genome array. *Nat. Biotechnol.* 18, 1262–1268, 2000.

32. Kaldalu, N., Mei, R., and Lewis, K., Killing by ampicillin and ofloxacin induces overlapping changes in *Escherichia coli* transcription profile. *Antimicrob. Agents Chemother.* 2004 (in press).

33. Wada, A., Growth phase coupled modulation of *Escherichia coli* ribosomes. *Genes Cells* 3, 203–208, 1998.

34. Opperman, T., et al., A model for a *umuDC*-dependent prokaryotic DNA damage checkpoint. *Proc. Natl. Acad. Sci. USA* 96, 9218–9223, 1999.

35. Walker, G.C., The SOS response of *Escherichia coli*. In: Escherichia coli and Samonella. *Cellular and Molecular Biology,* Neidhardt, F.C., ed. Washington, D.C.: SM Press, 1996:1400–1416.

36. Brown, J.M., and Shaw, K.J., A novel family of *Escherichia coli* toxin–antitoxin gene pairs. *J. Bacteriol.* 185, 6600–6608, 2003.

37. Hayes, F., Toxins–antitoxins: plasmid maintenance, programmed cell death, and cell cycle arrest. *Science* 301, 1496–1499, 2003.

38. Sat, B., et al., Programmed cell death in *Escherichia coli*: some antibiotics can trigger *mazEF* lethality. *J. Bacteriol.* 183, 2041–2045, 2001.

39. Pedersen, K., Christensen, S.K., and Gerdes, K., Rapid induction and reversal of a bacteriostatic condition by controlled expression of toxins and antitoxins. *Mol. Microbiol.* 45, 501–510, 2002.

40. Christensen, S.K., et al., Toxin-antitoxin loci as stress-response-elements: ChpAK/MazF and ChpBK cleave translated RNAs and are counteracted by tmRNA. *J. Mol. Biol.* 332, 809–819, 2003.

41. Pedersen, K., et al., The bacterial toxin RelE displays codon-specific cleavage of mRNAs in the ribosomal A site. *Cell* 112, 131–140, 2003.

42. Moyed, H.S., and Broderick, S.H., Molecular cloning and expression of hipA, a gene of *Escherichia coli* K-12 that affects frequency of persistence after inhibition of murein synthesis. *J. Bacteriol.* 166, 399–403, 1986.

43. Black, D.S., et al., Structure and organization of *hip*, an operon that affects lethality due to inhibition of peptidoglycan or DNA synthesis. *J. Bacteriol.* 173, 5732–5739, 1991.

44. Black, D.S., Irwin, B., and Moyed, H.S., Autoregulation of *hip*, an operon that affects lethality due to inhibition of peptidoglycan or DNA synthesis. *J. Bacteriol.* 176, 4081–4091, 1994.

45. Falla, T.J., and Chopra, I., Stabilization of *Rhizobium* symbiosis plasmids. *Microbiology* 145, 515–516, 1999.

46. Spudich, J.L., and Koshland, D.E. Jr., Non-genetic individuality: chance in the single cell. *Nature* 262, 467–471, 1976.

47. Korobkova, E., et al., From molecular noise to behavioural variability in a single bacterium. *Nature* 428, 574–578, 2004.

48. Mah, T.F., et al., A genetic basis for *Pseudomonas aeruginosa* biofilm antibiotic resistance. *Nature* 426, 306–310, 2003.

49. Drenkard, E., and Ausubel, F.M., *Pseudomonas* biofilm formation and antibiotic resistance are linked to phenotypic variation. *Nature* 416, 740–743, 2002.

50. Mukamolova, G.V., et al., A family of autocrine growth factors in *Mycobacterium tuberculosis*. *Mol. Microbiol.* 46, 623–635, 2002.
51. Tufariello, J.M., Chan, J., and Flynn, J.L., Latent tuberculosis: mechanisms of host and bacillus that contribute to persistent infection. *Lancet Infect. Dis.* 3, 578–590, 2003.
52. Colwell, R.R., and Grimes, D.J., *Nonculturable Microorganisms in the Environment.* Washington, D.C.: American Society for Microbiology, 2000.
53. Bogosian, G., and Bourneuf, E.V., A matter of bacterial life and death. *EMBO Rep.* 2, 770–774, 2001.
54. Kaeberlein, T., Lewis, K., and Epstein, S.S., Isolating "uncultivable" microorganisms in pure culture in a simulated natural environment. *Science* 296, 1127–1129, 2002.

13 Minimal Biofilm Eradication Concentration (MBEC) Assay: Susceptibility Testing for Biofilms

Howard Ceri, Merle E. Olson,
Douglas W. Morck, and Douglas G. Storey

CONTENTS

13.1 BIOFILMS AND THEIR PROPERTIES

Biofilms are commonly associated with recurrent, chronic and device related infections (1–3). According to the web site of the U.S. Food and Drug Administration (FDA) more than 60% of human infections are of this type. Biofilms derived from sites of infection and those found in nature or associated with industrial processes bear striking similarities. Infectious biofilms are comprised of microbial cells within microcolonies adherent onto a surface. The surfaces may be abiotic, biotic, or the secreted components of the host's extracellular matrix, or a complex matrix made up of both bacterial secretions and host components (4,5). Attachment and biofilm formation affords several advantages to the infectious bacterial agent within the host. Attachment allows for the bacteria to remain within the host at a site preferential to the success of the organism, and to resist host clearance mechanisms (6). Existence within biofilms also provides a favorable energetic advantage to bacteria, awaiting nutrients to diffuse locally, rather than expending energy to seek them out. The biofilm also provides a security blanket, in context of the glycocalyx secreted by biofilms to encompass the microcolonies. This serves as a barrier against the activity of the immune system, presenting a structure too large for effective phagocytosis (4).

Infection associated biofilms, like those from other sources are encapsulated within a bacterial-derived exopolymer (5). The surrounding matrix also contains substantial host-derived materials such as cells of the immune system (7). Finally, biofilms possess an inherent lack of susceptibility to antimicrobials. In nature this lack of susceptibility may have provided an advantage to bacterial populations by protecting them from antimicrobials produced by other microbes as part of the competition for space in natural environments. This protective advantage has now carried over to infectious biofilms as it has equipped them to deal with antibiotics, many of which are derived from or are synthetically modified from natural products, now used as treatments for infections. This altered susceptibility of bacterial biofilms has made treatment of biofilm-associated disease extremely difficult, resulting in the perceived need for an alternative means of identifying antimicrobials with efficacy against biofilms.

This chapter will focus on the difficulties biofilms pose to medicine, some of the theories for the lack of biofilm susceptibility, and the need for better ways to assess susceptibility of biofilms, in contrast to the minimal inhibitory concentration (MIC) still used today for the selection of new candidate antibiotics and for the confirmation of susceptibility. The need for a minimal biofilm eradication concentration (MBEC) for measuring biofilm susceptibility will be discussed.

13.2 THE RELATIONSHIP OF BIOFILM FORMATION AND DISEASE

13.2.1 ABIOTIC SURFACES

As discussed above, infectious biofilms may be associated both with abiotic surfaces, such as indwelling medical devices, or with biotic surfaces such as the tissues and/or extracellular matrix of the host. The increased utilization of a variety of medical devices in modern medicine has led to a corresponding increase in the numbers of device associated infections. Statistics provided by the U.S. Centers for Disease Control and Prevention NNIS Systems 2000 imply a direct association between the use of indwelling devices and infection rate. These devices would include but are not limited to orthopedic prosthetic joints, cardiac pacemakers and their leads, voice prostheses and prosthetic heart valves, all serving as spare parts to maintain the body, as well as numerous catheters, stents and shunts, such as central venous catheters, urinary catheters, continuous ambulatory peritoneal dialysis (CAPD) catheters, biliary, renal and cerebrospinal fluid shunts, and endotracheal tubes.

Abiotic surfaces are an easy target for biofilm formation as they lack the innate protective coatings and secretions associated with host mucosal and epithelial tissues. Direct attachment of bacteria to polystyrene surfaces *in vitro* has been demonstrated. For example the SSP-1 and SSP-2 surface proteins of *Staphylococcus* spp. can be seen to form structures like pili to allow them to adhere to plastic surfaces (8) *in vitro*. Adhesion *in vivo*, however, is believed to be mediated through conditioning films laid over the abiotic surface by both the host and the infectious agent. Components of blood plasma or interstitial tissue fluid, for example fibrin or fibronectin, can provide either non-specific or specific receptor-mediated targets, which can promote bacterial adhesion and biofilm formation (9,10).

The relationship between a biofilm formed on an indwelling device and clinical disease is often difficult to establish directly. Often biofilm infections may represent low-grade infections where bacteria may be quiescent and adapted to living in the host for extended periods of time. In some cases patients may already possess significant inflammation and other disease symptoms that may mask those associated with the infection. Donlan (11) has recently reviewed the mechanisms by which biofilms may contribute to human disease. The first of these mechanisms is associated with the concentration of organisms at a specific site of infection. This was demonstrated in urinary catheter infections where biofilms on the catheter were seen to be associated with site of bacterial involvement in the associated tissue (12). A similar model could also be used to explain the pneumonias resulting from the sloughing of large biofilm aggregates from endotracheal tubes of ventilated patients producing sufficient inoculum to initiate lung infections (13). Further, the increased concentration of endotoxin associated with increased bacterial numbers from biofilms is likely responsible for pyrogenic reactions in patients (14). While not totally separate from the issue of increased bacterial numbers in biofilms, the increased resistance of biofilms to the host immune system represents another important mechanism by which biofilms contribute to infection. Biofilms have been shown to interfere with phagocytosis (15) as well as killing of cells within the biofilm (16). Many antibodies raised against *Pseudomonas aeruginosa* prove to be non-opsonic against biofilms. Further, pre-immunization of animals to augment their immune systems is ineffective in clearing the biofilm in both a rabbit model of *P. aeruginosa* peritoneal infection or in a rat model of bacterial prostatitis (17,18). It is clear that the longer a medical device remains within the body the greater the chance of infection, especially when there is a connection of the device form the host tissue to the external environment. This makes the treatment of biofilms on devices critical to the practice of modern medicine with its ever-growing dependence on the use of implanted devices.

13.2.2 BIOTIC SURFACES

The body's surfaces have evolved as part of the innate response to provide a barrier function against pathogens. Many host factors play roles in inhibiting bacterial attachment and hence biofilm formation at these surfaces (19). Thus, any process that compromises the integrity of the mucosal surface increases the chances of inflammatory and infectious diseases (20,21). There have been a number of recent reviews of diseases associated with biofilms at tissue surfaces so this paper will consider only a few examples (1–3).

P. aeruginosa, because of its diversity and its association with biofilms in nature and with contamination problems in industry, has become the model system for the study of biofilms. The association of *P. aeruginosa* biofilms in the lungs of cystic fibrosis patients has made this the model for biotic biofilm disease as well (22,23). The organism in infection exists within microcolonies (22), expresses quorum-sensing molecules (23), and displays a lack of susceptibility toward antibiotics that are effective against planktonic cultures (22,24) just as it does *in vitro* (25).

Recurrent urinary tract infections (UTIs) represent another classical biofilm disease problem. Also associated with biofilms in UTIs, intracellular biofilm-like communities have been reported within superficial umbrella cells in the bladder of mice (27,28). The study of this biofilm population is now allowing for the first studies of bacterial differentiation in real time (28). Prostate infections are often chronic and contribute to recurrent UTIs. In a rat model of bacterial prostatitis *Escherichia coli* exist within a biofilm associated at the mucosal surface of the glandular component of the prostate (2,26). The lack of success in treating prostate infections with antibiotics again demonstrates the lack of susceptibility of biofilms to antibiotic when compared to cells grown as planktonic cultures. While there are many more examples of biofilms involved in disease, these few examples serve to put the problem into perspective.

13.3   BIOFILM SUSCEPTIBILITY

There is now a large body of literature dealing with increased levels of antibiotic and biocide required to treat biofilms or, conversely, with the lack of susceptibility of biofilms to MIC levels of antibiotics. Often this difference in susceptibility to antimicrobials seen between biofilm and planktonic cultures of the same organism is referred to as resistance. However the term resistance has become associated with either a permanent genetic change resulting from mutation or the acquisition of a plasmid that confers a continued advantage to the organism. Therefore it would seem more appropriate to refer to this transient phenotypic change in antibiotic efficacy associated with biofilm growth as a reduced susceptibility rather than as resistance. The basis for the change in susceptibility seen in biofilms appears to be complex, multifactorial, and the subject of great debate. A further level of complexity arises due to apparent antibiotic tolerance for biofilm bacteria that likely mediate their refractory character. Recent reviews on the subject of the difference in biofilm susceptibility to antibiotics have discussed the following mechanisms to explain the change in biofilm response to antibiotic treatment (29–31).

13.3.1   ANTIMICROBIAL PENETRATION OF THE BIOFILM MATRIX

The original model for the difference in susceptibility between biofilm and planktonic cultures was attributed to the protection afforded the biofilm by the complex extracellular matrix that is the integral component of the biofilm and surrounds the individual bacterial cells or microcolonies of the biofilm (32). It was believed that the matrix, due to its charge derived from the carbohydrate structure, would act as an affinity ligand binding up antibiotics and preventing their access to the cells within the biofilm. The process of diffusion certainly impacts solute transport in the biofilm and leads to the formation of concentration gradients of metabolites that result in a physiological gradient within the biofilm; this despite the action of water channels to provide access into all levels of the biofilm (33). The barrier to diffusion of antibiotics into the biofilm is dependent upon the nature of the biofilm matrix, the nature of the antibiotic and the ability of the biofilm isolate to affect the diffusion of a specific antibiotic (34,35). However, despite the influences that diffusion has on entrance of antibiotics into the biofilm, it would appear that the time required for diffusion of antibiotics into or the final levels of antibiotic achieved in the biofilms are not consistent with reduced antimicrobial penetration being the major reason for the lack of biofilms susceptibility.

13.3.2   GROWTH RATE OF BIOFILMS

It has been argued that the slow growth rate seen in biofilms would be responsible for the lack of biofilm's susceptibility to antibiotics, as many antibiotics require rapid growth of cells to be effective (36,37). The contribution growth rates make to antibiotic susceptibility may also be important in terms of metabolic activity and heterogeneity within the biofilm (37), even when antibiotics that are not dependent upon rapid cell growth are considered. Interestingly, it is also true that biofilms are less susceptible to many antibiotics whose actions are not dependent upon the growth rate of the population. The contribution of the diverse metabolic activity and growth rate throughout the biofilm remains an important consideration overall in antibiotic susceptibility of the biofilm in total.

13.3.3   PHENOTYPIC PLASTICITY IN THE BIOFILM

The alteration in susceptibility of biofilms to antibiotics may result form the unique nature of the bacterial phenotype expressed within a biofilm. This hypothesis is made more interesting by the realization that close to 30% of the genetic information of sequenced bacteria falls into the category of genes of unknown function. As we know so little about the physiology of bacteria growing as biofilms it is tempting to suggest that many of these genes play a role in the altered life style of a biofilm. Altered gene expression driven under the influence of quorum sensing factors, an important component in the regulation of biofilm formation, would support the concept of altered gene expression in biofilms (38,39). Changes in the phenotype of biofilms compared to planktonic cells have been shown by both differences in genetic expression using subtractive hybridization and by using gene array technology, and at the protein level by employing proteomic approaches (40–45). In many of these cases the genes that were up regulated under biofilm growth conditions were in fact genes of unknown function. However, not all are convinced that changes in gene expression are directly associated with the biofilm phenotype and antibiotic susceptibility (42,46). The results are not always in keeping with our preconceived ideas of what gene changes may result in the lowered susceptibility of biofilms to antibiotics. For example, multi-drug efflux pumps would appear to offer a mechanism of reduced sensitivity of biofilms to antibiotics; however, instead of being expressed at higher levels in biofilms these virulence factors are sometimes repressed (47). The field of defining biofilm specific genes is in its infancy, but holds the promise of identifying genes essential for biofilm survival that would represent the targets for the next generation of antimicrobials.

13.3.4 PERSISTER CELLS

An alternative view of the lack of susceptibility of biofilms is presented by Lewis (48) based on the presence of persister cells in all bacterial cultures. Persister cells are defined as "specialized survivor cells" expressing a temporary tolerance to antibiotics, which is not true resistance as this temporary phenotype doesn't result in a permanent change in MIC (48–50). One argument is that persister cells may survive by escaping apoptosis and that they are not limited to biofilms, and hence the lack of susceptibility of biofilms to antibiotics is not unique to the biofilm mode of growth. The nature and cause of the changes that lead to persistence are not defined; however, if the theory is correct control of the persister population would then become key to treatment of chronic infections.

13.4 BIOFILM SUSCEPTIBILITY ASSAYS: THE NEED FOR A MINIMAL BIOFILM ERADICATION CONCENTRATION (MBEC) VALUE

A number of methods are available to grow biofilm populations. Some of these systems have been utilized to test for antibiotic susceptibility of biofilms. Reviews on many of these methods have been reported (51). In using these different technologies to compare planktonic and biofilm susceptibility to antibiotics a consistent picture has evolved, biofilms are less susceptible to antibiotics than planktonic counterpart cultures of the same organism. For example studies using the constant-depth film fermentor (52–53), the Modified Robbin's Device (54,55), and the MBEC™ Assay system (25,56–58) have consistently demonstrated that biofilms often require from 10 to1,000-fold the concentration of antibiotic to be eradicated as compared to planktonic bacteria (25,52–58). Often the MIC concentration of antibiotic, which by definition is efficacious against the planktonic cells, was not effective against biofilm cultures of the same organism. Typical of such a study is the data presented in Figure 13.1 where it is clear that in MIC assays (carried out according to NCCLS standards) this organism is readily susceptible to a number of antibiotics; however, when the same organism is grown as a biofilm results are markedly different.

Some antibiotics are effective at MIC level concentrations both *in vitro* and *in vivo* while others are not. This represents the problem faced by a physician dealing with a likely biofilm-associated infection. MIC data presents the clinician with a selection of antibiotics that appear efficacious against the patients isolate; however, history of treatment failure with the current and/or past patients have taught the clinician that there may be a poor correlation between MIC values and outcome when treating chronic, recurrent, or device related infections. There are now several studies, which indicate that when dealing with biofilm diseases such as cystic fibrosis (CF)

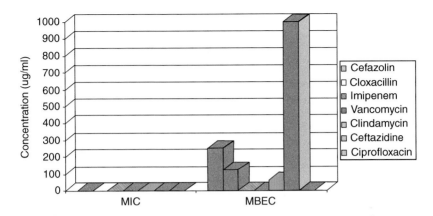

FIGURE 13.1 Antibiotic susceptibility profile of *S. aureus* showing MIC values for a plank-tonic suspension culture carried out by NCCLS protocol compared to the MBEC value obtained for the biofilm culture of the same organism assayed using the MBEC Assay. The isolate is susceptible to all the antibiotics tested in an MIC assay but it is clear that not all antibiotics are effective against the biofilm culture.

and device-related infections that MIC values may not be predictive of outcome (59,60). In the CF study, a retrospective evaluation using spirometry before and after antibiotic treatment with IV tobramycin and ceftazidime (59) the outcome showed no statistical link between susceptibility in an MIC assay with clinical outcome. Similarly, in a retrospective study of CAPD catheter infections by coagulase-negative *Staphylococcus* (CoNS), it was found that MIC values and MBEC values differed considerably and that the international protocol for treatment of these biofilm infections was seldom the best choice of treatment (60). The change in treatment from the defined protocol to that predicted by MBEC values over the next year resulted in a significant improvement represented as a reduction in the loss of catheters due to CoNS infections, and a marked reduction in antibiotic costs for the hospital ward (Sepandj, personal communication).

Resolution of the problem regarding choice of antibiotics to treat biofilm infections could be achieved by utilizing an assay system that can select antibiotics and biocides based on activity against biofilms. The MBEC would represent the concentration of antibiotic or biocide capable of killing a biofilm and would be the equivalent to the MBC used to determine killing efficacy of antibiotics against planktonic populations. The many benefits derived from having such assay capabilities would be equivalent to those now provided by MIC and MBC tests for planktonic susceptibility, including: selecting for new antimicrobials that are effective against biofilms; identifying biofilm targets against which new antimicrobials can be developed; selecting more approrpriate antimicrobials for patient treatment; and tracking the possible future development of antibiotic resistance against biofilm drugs. The criteria for a successful biofilm assay will include the ability to form biofilms of organisms associated with biofilm infection, the formation of multiple equivalent biofilms for susceptibility testing with characteristics consistent to those observed *in vivo*, the ability to carry out assays as simply with the MIC assay and the possibility of automation, and obtaining the data in a time frame mandated by the need to initiate therapy.

As stated above many different methods used to develop biofilms have produced comparable data demonstrating biofilms to be less susceptible to antibiotics. Our laboratories have directly compared data obtained by the Modified Robbins Device to that obtained by the MBEC assay, showing the data to be very reproducible between the two systems (Ceri, unpublished data). The major difference between the two systems was ease of use, as a study that took months to complete on the Modified Robbins Device was completed in less than a week on the MBEC system. This was achieved because all 96 biofilms formed on the MBEC plate can be moved and assayed as a single unit (Figure 13.2) as compared to having to form, wash, treat, wash, and then assay each individual data point in the Modified Robbin's Device and the Constant-depth Film Fermentor.

The next step in the process of defining a MBEC value system will be the collection of data to support the value of the MBEC in patient treatment and anti-microbial development. There are still several protocol issues as to the use of these technologies. For example, what age and thickness of a biofilm should be used to assay antimicrobial susceptibility? The shift to decreased antibiotic susceptibility has been correlated to early stages of biofilm formation shortly following attachment (56,61) and previous to excessive secretion of matrix material, while other studies

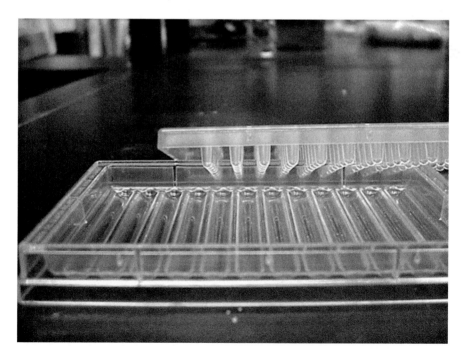

FIGURE 13.2   MBEC™ Device allows for the formation of 96 equivalent biofilms for rapid screening of antibiotic or biocode susceptibility of biofilms.

have defined the aged biofilm as being more resistant to antimicrobial attack (62). More important is the need for additional studies that compare *in vitro* MBEC values to clinical outcomes. We have looked at the activity of fleroxacin, ampicillin, trimethoprim-sulfamethoxazole and gentamicin in the treatment of catheter-associated UTIs in a rabbit model (63). Differences in MIC, MBC, and MBEC for these antibiotics against the *E. coli* strain used in the study, were demonstrated as was a correlation between *in vitro* activity and efficacy in the infection model. Fleroxacin was able to clear the catheter and adjoining tissue of bacteria at MBEC concentrations, while both ampicillin and gentamicin were able to clear the biofilm from the catheter, but not the adjoining tissues. MBECs were determined using the Modified Robbin's Device, but were later reproduced using the MBEC assay system (Ceri, unpublished data) (63). A chronic rat *Staphylococcus aureus* osteomyelitis model also was used to evaluate antibiotic efficacy *in vitro* and *in vivo* (64). Again a correlation for MBEC values with efficacy could be seen. Additionally, MBEC values correlated with data obtained from field studies of a biocide that reduced biofilm and *Legionella pneumophila* in water-cooling towers (65). Most importantly, the successful use of the protocol developed in retrospective studies with the MBEC assay for the treatment of CAPD catheter infections indicates the potential value of such technology (Sepandj, personal communication). While we still have much to learn about the mechanisms of biofilm formation and antibiotic susceptibility, it is important to keep the practical need in mind that an assay that can contribute to the selection of the next generation of antibiotics and biocides with efficacy against biofilms is needed. We believe that the slow response to embrace the concept of biofilms was in part due to the fact that no answer to the problem seemed imminent, and that to keep the field moving forward answers to the issues of biofilms must be forthcoming. There has been an explosion in interest in biofilms, biofilm morphology and biofilm physiology, and with more concentrated efforts by multiple laboratories using methodologies currently being developed solutions to biofilm-mediated infections will be developed. We are at the onset of having the tools in which to practically address the problems of biofilms in infections and infectious diseases of humans and other animals.

REFERENCES

1. Parsek, M.R., and Sing, P.K., Bacterial biofilms: an emerging link to disease pathogens. *Annu. Rev. Microbiol.* 57, 677–701, 2003.
2. Olson, M.E., Ceri, H., and Morck, D.W., Interaction of biofilms with tissues. In: Jass, J., Surman, S., and Walker, J., eds., *Medical Biofilms Detection, Prevention and Control.* Chichester: Wiley, 2003:125–148.
3. Donlan, R.M., and Costerton, J.W., Biofilms: survival mechanisms of clinically relevant microorganisms. *Clin. Microbiol. Rev.* 15, 167–193, 2002.
4. Costerton, J.W., Stewart, P.S., and Greenberg, E.P., Bacterial biofilms: a common cause of persistent infections. *Science* 284, 1318–1322, 1999.
5. Jass, J., Shurman, S., and Walker, J.T., Microbial biofilms in medicine. In: Jass, J., Surman, S., Walker, J., eds., *Medical Biofilms Detection, Prevention and Control.* Chichester: Wiley, 2003:1–28.

6. Ofek, I., and Beachey, E.H., General concepts and principals of bacterial adherence in animals and man. In: Beachey, E.H., ed., *Bacteria Adherence.* London: Chapman and Hall, 1980:1–29.

7. Buret, A., Ward, K.H., Olson, M.E., and Costerton, J.W., An *in vivo* model to study the pathobiology of infectious biofilms on biomaterial surfaces. *J. Biomed. Mater. Res.* 25, 865–874, 1991.

8. Veenstra, G.J., Cremers, F.F., van Dijk, H., and Fleer, A., Ultrastructural organization and regulation of a biomaterial adhesin in Staphylococal epidermidis. *J. Bacteriol.* 178, 537–541, 1996.

9. Herrmann, M., Vaudaux, P.E., Pittet, D., Auckenthaler, R., Lew P.D., Schumacher-Perdeau, F., Peters, G., and Waldvogel, F.A., Fibronectin, fibrinogen and laminin act as mediators of adherence of clinical staphylococcal isolates to foreign material. *J. Infect. Dis.* 158, 693–701, 1988.

10. Cheung, A.I., Projan, S.J., Edelstein, R.E., and Fischetti, V.A., Cloning, expression, and nucleotide sequence of a *Staphylococcus aureus* gene (*fbpA*) encoding a fibrogen-binding protein. *Infect. Immun.* 63, 1914–1920, 1995.

11. Donlan, R.M., Problems of biofilms associated with medical devices and implants. In: Jass, J., Surman, S., and Walker, J., eds., *Medical Biofilms Detection, Prevention and Control.* Chichester: Wiley, 2003:31–49.

12. Rogers, J., Norkett, D.I., Bracegirdle, P., Dowsett, A.B., Walker, J.T., Brooks, T., and Keevil, C.W., Examination of biofilm formation and risk of infection associated with the use of urinary catheters with leg bags. *J. Hosp. Infect.* 32, 105–115, 1996.

13. Adair, C.G., Gorman, S.P., Feron, B.M., Byers, L.M., Jones, D.S., Goldsmith, C.E., Moore, J.E., Kerr, J.R., Curran, M.D., Hogg, G., Webb, C.H., McCarthy, G.J., and Milligan, K.R., Implications of endotracheal tube biofilm for ventilator-associated pneumonia. *Int. Care. Med.* 25, 1072–1076, 1999.

14. Riofolo, C., Devys, C., Meunier, G., Perraud, M., and Goullet, D., Quantitative determination of endotoxins released by biofilms. *J. Hosp. Infect.* 43, 203–209, 1999.

15. Shiau, A-L., and Wu, C-L., The inhibitory effect of *Staphylococcus epidermidis* slime on the phagocytosis of murine peritoneal macrophages is interferon independent. *Microbiol. Immun.* 42, 33–30, 1998.

16. Yasuda, H., Akiki, Y., Aoyama, J., and Yokota, T., Interaction between human polymorphonuclear leucocyte and bacteria released from *in vitro* bacterial biofilm models. *J. Med. Microbiol.* 41, 359–367, 1994.

17. Ward, K.H., Olson, M.E., Lam, K., and Costerton, J.W., Mechanism of persistent infection associated with peritoneal implants. *J. Med. Microbiol.* 36, 406–413, 1992.

18. Ceri, H., Schmidt, S., Olson, M.E., Nickel, J.C., and Benediktsson, H., Specific mucosal immunity in the pathophysiology of bacterial prostatitis in a rat model. *Can. J. Microbiol.* 45, 849–855, 1999.

19. Graham, L.L., Ceri, H., and Costerton, J.W., Lectin-like proteins from uroepithelial cells which inhibit *in vitro* adherence of three urethral bacterial isolates to uroepithelial cells. *Microbial Ecol. in Health and Disease* 5, 77–86, 1992.

20. Lang, M.D., Nickel, J.C., Olson, M.E., Howard, S., and Ceri, H., A rat model of experimentally induced abacterial prostatitis. *Prostate* 45, 201–206, 2000.

21. Rippere-Lampe, K.E., Lang, M.D., Ceri, H., Olson, M.E., Lockman, H.A., and O'Brien, A.D., Cytotoxic Necrotizing Factor Type 1 causes increased inflammation and tissue damage to the prostate in a rat prostatitis model. *Infect. Immun.* 69, 6515–6519, 2001.

22. Lam, J., Chan, R., Lam, K., and Costerton, J.W., Production of mucoid microcolonies by *Pseudomonas aeruginosa* within infected lungs in cystic fibrosis. *Infect. Immun.* 28, 546–556, 1980.

23. Singh, P.K., Schaefer, A.L., Parsek, M.R., Moninger, T.O., Welsh, M.J., and Greenberg, E.P., Quorum-sensing signals indicate the cystic fibrosis lungs are infected with bacterial biofilms. *Nature* 407, 762–764, 2000.

24. Hoiby, N., Krough, J.H., Moser, C., Song, Z., Ciofu, O., and Kharazmi, A., *Pseudomonas aeruginosa* and the *in vitro* and *in vivo* biofilm mode of growth. *Microbes. Infect.* 3, 23–35, 2001.

25. Ceri, H., Olson, M.E., Stremick, C., Read, R.R., Morck, D., and Buret, A., The Calgary Biofilm Device: A new technology for the rapid determination of antibiotic susceptibility of bacterial biofilms. *J. Clin. Microbiol.* 37, 1771–1776, 1999.

26. Ceri, H., Olson, M.E., and Nickel, J.C., Prostatitis: role of the animal model. In: Nickel, J.C., ed., *Textbook of Prostatitis*. Oxford: Isis Medical Media, 1999:109–114.

27. Anderson, G.G., Palermo J.J., Schilling J.D., Roth R., and Heuser J., Hultgren S.J., Intracellular bacterial biofilm-like pods in urinary tract infections. *Science* 301, 105–107, 2003.

28. Justice, S.S., Hung, C., Theriot, J.A., Fletcher, D.A., Anderson, G.G., Footer M.J., and Hultgren, S.J., Differentiation and developmental pathways of uropathogenic *Escherichia coli* in urinary tract pathogenesis. *Proc. Natl. Acad. Sci. USA* 101, 1333–1338, 2004.

29. Stewart, P.S., Mechanisms of antibiotic resistance in bacterial biofilms. *Int. J. Med. Microbiol.* 292, 107–113, 2002.

30. Gilbert, P., Allison, D.G., and McBain, A.J., Biofilms *in vitro* and *in vivo*: do singular mechanisms imply cross-resistance? *J. Appl. Microbiol.* 92, Suppl, 98S–110S, 2002.

31. Lewis, K., Riddle of biofilm resistance. *Antimicrob. Agents Chemother.* 45, 999–1007, 2001.

32. Costerton, J.W., Cheng, K.J., Geesey, G.G., Ladd, T.I., Nickel, J.C., Dasgupta, M., and Marrie, T.J., Bacterial biofilms in nature and disease. *Annu. Rev. Microbiol.* 41, 435–464, 1987.

33. Stewart, P.S., Diffusion in biofilms. *J. Bacteriol.* 185, 1485–1491, 2003.

34. Zheng, Z., and Stewart, P.S., Penetration of rifampin through *Staphylococcus epidermidis* biofilms. *Antimicrob. Agents Chemother.* 46, 900–903, 2002.

35. Anderl, J.N., Franklin, M.J., and Stewart, P.S., Role of antibiotic penetration limitation in *Klebsiella pneumoniae* biofilm resistance to ampicillin and ciprofloxacin. *Antimicrob. Agents Chemother.* 44, 1818–1824, 2000.

36. Brown, M.R., Allison, D.G., and Gilbert, P., Resistance of bacterial biofilms to antibiotics: a growth-rate related effect? *J. Antimicrob. Chemother.* 22, 777–780, 1988.

37. Walters, M.C. 3rd, Roe, F., Bugnicourt, A., Franklin, M.J., and Stewart, P.S., Contributions of antibiotic penetration, oxygen limitation, and low metabolic activity to tolerance of *Pseudomonas aeruginosa* biofilms to ciprofloxacin and tobramycin. *Antimicrob. Agents. Chemother.* 47, 317–323, 2003.

38. Bassler, B.L., How bacteria talk to each other: regulation of gene expression by quorum sensing. *Curr. Opin. Microbiol.* 2, 582–587, 1999.

39. Aguilar, C., Friscina, A., Devescovi, G., Kojic, M., and Venturi, V., Identification of quorum-sensing-regulated genes of Burkholderia cepacia. *J. Bacteriol.* 185, 6456–6462, 2003.

40. Parkins, M.D., Altebaeumer, M., Ceri, H., and Storey, D.G., Subtractive hybridization based identification of genes uniquely or hyperexpressed during biofilm growth. In: Doyle, R.J., ed., *Microbial Growth in Biofilms. Methods Enzymol.* Vol 336, Part A. San Diego: Academic Press, 2001:76–84.

41. Beloin, C., Valle, J., Latour-Lambert, P., Faure, P., Kzreminski, M., Balestrino, D., Haagensen, J.A., Molin, S., Prensier, G., Arbeille, B., and Ghigo, J.M., Global impact

of mature biofilm lifestyle on *Escherichia coli* K-12 gene expression. *Mol. Microbiol.* 51, 659–674, 2004.

42. Whiteley, M., Bangera, M.G., Bumgarner, R.E., Parsek, M.R., Teitzel, G.M., Lory, S., and Greenberg, E.P., Gene expression in *Pseudomonas aeruginosa* biofilms. *Nature* 413, 860–864, 2001.

43. Ren, D., Bedzyk, L.A., Thomas, S.M., Ye, R.W., and Wood, T.K., Gene expression in *Escherichia coli* biofilms. *Appl. Microbiol. Biotechnol.* 64, 515–524, 2004.

44. Arevalo-Ferro, C., Hentzer, M., Reil, G., Gorg, A., Kjelleberg, S., Givskov, M., Riedel, K., and Eberl, L., Identification of quorum-sensing regulated proteins in the opportunistic pathogen *Pseudomonas aeruginosa* by proteomics. *Environ. Microbiol.* 5, 1350–1369, 2003.

45. Sauer, K., The genomics and proteomics of biofilm formation. *Genome Biol.* 4, 219–229, 2003.

46. Ghigo, J.M., Are there biofilm-specific physiological pathways beyond a reasonable doubt? *Res. Microbiol.* 154, 1–8, 2003.

47. De Kievit, T.R., Parkins, M.D., Gillis, R.J., Ceri, H., Poole, K., Iglewski, B.H., and Storey, D.G., The contribution of multidrug efflux pumps to the antibiotic resistance of *Pseudomonas aeruginosa* biofilms. *Antimicrob. Agents Chem.* 45, 1761–1770, 2001.

48. Spoering, A.L., and Lewis, K., Biofilms and planktonic cells of *Pseudomonas aeruginosa* have similar resistance to killing by antimicrobials. *J. Bacteriol.* 183, 6746–6751, 2001.

49 Lewis, K., Riddle of biofilm resistance. *Antimicrob. Agents Chemother.* 45, 999–1007, 2001.

50. Keren, I., Kaldalu, N., Spoering, A., Wang, Y., and Lewis, K., Persister cells and tolerance to antimicrobials. *FEMS Microbiol. Lett.* 230, 13–18, 2004.

51. *Methods Enzymology*, Vol. 336, *Microbial Growth in Biofilms, Part A.*

52. Wilson, M., Susceptibility of oral bacterial biofilms to antimicrobial agents. *J. Med. Microbiol.* 44, 79–87, 1996.

53. Lamfon, H., Porter, S.R., McCullough, M., and Pratten, J., Susceptibility of *Candida albicans* biofilms grown in a constant depth film fermentor to chlorhexidine, fluconazole and miconazole: a longitudinal study. *J. Antimicrob. Chemother.* 53, 383–385, 2004.

54. Raad, I., Darouiche, R., Hachem, R., Sacilowski, M., and Bodey, G.P., Antibiotics and prevention of microbial colonization of catheters. *Antimicrob. Agents Chemother.* 39, 2397–2400, 1995.

55. Kumon, H., Ono, N., Iida, M., and Nickel, J.C., Combination effect of fosfomycin and ofloxacin against Pseudomonas aeruginosa growing in a biofilm. *Antimicrob. Agents Chemother.* 39, 1038–1044, 1995.

56. Ceri, H., Olson, M.E., Morck, D., Storey, D., Read, R.R., Buret, A., and Olson, B., The MBEC™ Assay System: multiple equivalent biofilms for antibiotic and biocide susceptibility testing. In: Doyle R.J., ed., *Methods Enzymol.* Vol. 337, *Microbial Growth in Biofilms*, Part B. San Diego: Academic Press, 2001:377–385.

57. Bardouniotis, E., Ceri, H., and Olson, M.E., Biofilm formation and biocide susceptibility testing of *Mycobacterium fortuitum* and *Mycobacterium marinum*. *Cur. Microbiol.* 46, 28–32, 2003.

58. Olson, M.E., Ceri, H., Morck, D.W., Buret, A.G., and Read, R.R., Biofilm bacteria: formation and comparative susceptibility to antibiotics. *Can. J. Vet. Res.* 66, 86–92, 2002.

59. Smith, A.L., Fiel, S.B., Mayer-Hamblett, N., Ramsey, B., and Burns, J.L., Susceptibility testing of *Pseudomonas aeruginosa* isolates and clinical response to parenteral antibiotic administration Lack of association in cystic fibrosis. *Chest* 123, 1495–1502, 2003.

60. Sepandj, F., Ceri, H., Gibb, A.P., Read, R.R., and Olson, M.E., Minimum inhibitory concentration (MIC) versus minimum biofilm eliminating concentration (MBEC) in evaluation of antibiotic sensitivity of Gram-negative bacilli causing peritonitis. *Peritoneal. Dial. Int.* 24, 1–3, 2004.

61. Konig, C., Schwank, S., and Blaser, J., Factors compromising antibiotic activity against biofilms of *Staphylococcus epidermidis. Eur. J. Clin. Microbiol. Infect. Dis.* 20, 20–26, 2001.

62. Anwar, H., Strap, J.L., and Costerton, J.W., Establishment of aging biofilms: possible mechanism of bacterial resistance to antimicrobial therapy. *Antimicrob. Agents Chemother.* 36, 1347–1351, 1992.

63. Morck, D.W., Lam, K., McKay, S.G., Olson, M.E., Prosser, B., Ellis, B.D., Cleeland R., and Costerton, J.W., Comparative evaluation of fleroxacin, ampicillin, trimethoprim-sulfamethoxazole, and gentamicin as treatments of catheter-associated urinary tract infections in a rabbit model. *Int. J. Antimicrobial. Agents* 2, S21–S27, 1994.

64. Monzon, M., Garcia-Alvarez, F., Lacleriga, A., Gracia, E., Leiva, J., Oteiza, C., and Amorena, B., A simple infection model using pre-colonized implants to reproduce rat chronic Staphylococcus aureus osteomyelitis and study antibiotic treatment. *J. Orthop. Res.*19, 820–826, 2001.

65. Nalepa, C.J., Howarth, J.N., Liimatta, E., Ceri, H., Stremick, C.A., Stout, J.E., and Lin, Y.E., The control of bacteria on surfaces: effectiveness of bromine-based biocides toward microbial biofilms and biofilm-associated *Legionella pneumophila. CTI J.* 24, 12–21, 2003.

14 Environmental Cues Regulate Virulence and Biofilm Formation

John L. Pace and Steven M. Frey

CONTENTS

14.1 INTRODUCTION

Pathogens encounter numerous environmental stimuli and mount coordinated responses to ensure survival and colonization within the host (1–13). Osmolarity, pH, temperature, anaerobiosis or elevated CO_2, and the concentrations of several cations including calcium, magnesium, and iron are important cues (14–21). Host-specific chemical factors play an additional central role in the regulation of virulence and contribute to host and tissue tropism (22–26). Bile acids, porphyrins, and cate-cholamine neurotransmitters among others are implicated as host signaling factors that impact bacterial virulence (27–45). In addition to affecting expression of specific virulence factors, production of biofilms and efflux pumps that affect antibacterial susceptibility appear to be co-regulated (46–60).

A number of environmentally-regulated genetic loci for virulence traits, including those encoding the type III translocation systems, are conserved across diverse human, animal, and plant pathogens (4,5,60–62). The low-calcium-response (LCR) of *Yersinia* spp. encoded by these genes is observed *in vitro* at 37°C, in response to low calcium culture conditions as might be encountered in some host compartments, and mediates production of important factors necessary for *in vivo* growth and virulence (64–66). Expression of homologous genes in *Salmonella* spp. mediate invasion of human epithe-lial cells when cultured to early log phase under low oxygen, and disease upon oral chal-lenge (61–63,67). *Shigella* spp. possessing similar loci, are also invasive when grown at 37°C but not at 30°C, and invasiveness is dependent on *virB*-regulation of the *ipa* operon (8,20,68). *In vitro Shigella* invasiveness is further enhanced when bacteria are cultured under anaerobic conditions as is virulence in the Sereny model, and culture with the bile acid chenodeoxycholic acid similarly increases *in vitro* epithelial cell adherence and invasion (43). The related responses, in addition to the ability to exploit restricted nutri-ents such as iron, render microorganisms successful pathogens (16,43,44,69).

Clearly, environmental conditions impact virulence, physiology, and antibiotic sus-ceptibility (10,37,39,51,53,70). This is significant when considering extrapolation of *in vitro* findings to intervention strategies in the clinic (46–49,51,55,58,59). Simulation of *in vivo* conditions during *in vitro* culture and characterization of pathogens would ensure genuine responses and facilitate identification of appropriate therapies.

14.2   *Campylobacter jejuni*

Campylobacter jejuni is a major cause of bacterial gastroenteritis Worldwide (71). Infections caused by this organism result in a profuse watery to bloody diarrhea, often indistinguishable from that caused by *Shigella* spp. (71,72). The virulence factors of *C. jejuni* are not well defined (103,139). However, epithelial cell adhesion and invasion have been suggested as virulence factors for *C. jejuni*, and diverse strains may be capable of this phenotype (41,42,71,73–76). In fact, severity of the disease correlates with the invasive ability of the infecting *C. jejuni* strain (72,75). Clinical isolates are more invasive than environmental isolates, and *in vivo* invasion has been observed from biopsies of experimentally infected infant macaque mon-keys (71–77). Mutations in regulatory loci result in reduced *in vitro* invasivenss and attack rate in the ferret model (77). Increasing viscosity of *in vitro* culture medium results in bacteria that are better able to invade epithelial cells, but effects of other environmental factors on *C. jejuni* haven't been widely explored (78).

14.2.1   *Campylobacter jejuni* Adhesion and Invasion of Epithelial Cells

Adhesion to and invasion of human intestinal epithelial cells by wild-type *C. jejuni* strains are increased by *in vitro* culture in medium containing bile (38,41,42).

Bile acids are the component of bile responsible for affecting virulence trait of the bacteria. Bile-acid depleted bile is ineffective, while the bile acid deoxycholic acid (DOC) is as effective as bile at enhancing virulence traits of *C. jejuni*. A concentration effect is also observed and is physiologically significant. When concentration of DOC between 0.25 to 2.5 mM were utilized, enhancement was greatest at 2.5 mM, which is the level found in the intestine (38,41,42,79–83). Similarly, *Shigella flexneri* cultured with chenodeoxycholic acid, a bile acid structurally related to deoxycholic acid, are more adherent and secrete greater quantities of Ipa proteins (43). This effect is specific and not due to increased medium viscosity, as effects of bile acids on medium viscosity are insufficient to mediate any effect (43–78).

A co-requisite for enhancement of virulence traits is culture at 37°C in 10% CO_2. No increase in adhesion or invasion is observed when bacteria are cultured at 42°C, or at lower concentrations of atmospheric CO_2 even in the presence of DOC (38,41,42). The co-requisite for growth at 37°C in the presence of increased CO_2 may explain the host tissue tropism of the pathogen in the human intestine (14,23,38,41,42,71,84). *C. jejuni* often colonizes chickens, the suspected source of the pathogen in human disease (14,23,71,84,85). However, *C. jejuni* rarely causes disease in this animal. A possible explanation for this phenomenon is that virulence expression by human-pathogenic strains does not occur at 42°C, near the chicken's higher body temperature (14,23,71,84).

Temporal (growth-phase) regulation of DOC-enhanced *C. jejuni* adhesion and invasion overlap but are not identical (38,41,42). Enhanced adhesion is maintained longer, but the increased adhesion and invasiveness both occur during only a 4 hour time period. This distinction might result from subsequent increased auto-aggregation observed with the bile acid cultured bacteria that could render aggregates too large to be taken up by the induced endocytotic event (38,41,42,86).

Bile acid-induced adherent properties of *C. jejuni* may be due in part to expression of the *flaA* gene product. Flagellin expression is required for virulence, as flagellin-less mutants do not cause disease (85,87–92). It has also been reported that the *flaB* flagellin is environmentally regulated, and that phase variation of *C. jejuni* flagellin during infection may play a role in disease (85,90). Adhesion to or invasion of epithelial cells by *C. jejuni flaA* or *flaAflaB* mutants were not enhanced by bile or DOC (38,41,42). Although lack of invasiveness may be due to a low level of adhesion, lack of motility was compensated for by centrifuging bacteria onto the epithelial cell monolayer. In *C. jejuni*, it would appear that *flbA* regulates both motility and invasiveness (91). *flbA* from *C. jejuni* is homologous to *lcrD*, a gene known to be regulated by environmental factors and involved in the LCR phenotype of *Yersinia* (91). The homologous genes, encoding the Type III protein translocation systems described in *Salmonella* and other bacteria, also share homologies with flagellar assembly genes (5,61–63). *C. jejuni flbA* were non-adherent even when cultured with DOC (38,41,42). The homologies may explain why *C. jejuni flbA* is neither invasive nor expresses flagellin.

14.2.2   CULTURE-INDUCED BROAD AGGLUTINATION
OF UNRELATED LIOR SEROTYPE STRAINS

Following growth in DOC-containing medium cells of several *C. jejuni* Lior serotypes become more immunologically cross-reactive (38,41,42). Unrelated Lior serotype strains grown with DOC, but not cells grown conventionally, strongly agglutinate with antibodies from rabbits immunized with the DOC-grown *C. jejuni* 81–176 (38,41,42). Lior serotype strains, that are moderately cross-reactive with the strain used to raise the antibody when grown conventionally, show even stronger agglutination when cells are grown in DOC-containing medium. This effect is independent of flagellin because similar observations have been made with isogenic *fla* and *flbA* mutant strains (38,41,42,87,91).

Enhanced auto-agglutination is also observed among *C. jejuni* cultured in bile- or DOC-containing medium (38,41,42). Auto-agglutination is a characteristic associated with the expression of some fimbrial types (86). Misawa and Blaser (86) observed that auto-agglutination of *C. jejuni* correlated with adhesion to epithelial cells, and production of flagella. Auto-agglutination was greater at 18 than 24 hour growth, and when cultured at 37°C than 25°C consistent with temporal and temperature-dependent regulation of epithelial cell interactions (38,41,42,86).

14.2.3   PRODUCTION OF SURFACE PROTEINS AND EFFECTS ON ANTIGENICITY

Flagellin protein is an important virulence factor in *C. jejuni* (77,78,85,87–91). In addition to facilitating contact of the pathogen with host epithelial cells, through its role in bacterial movement through the mucus layer, the protein may act as an adhesin (88). Two flagellins, FlaA and FlaB, are produced in *Campylobacter* spp., and their production is regulated in response to environmental conditions, as well as a phase variation effect (85,90). Two remarkable changes are observed with SDS-PAGE in the bacteria cultured with bile or DOC. Flagellin is increased in quantity, as well as two proteins in the size range of 25–28 kDa (38,41,42,93). Western blots confirm enhanced reactivity of convalescent IgA from animals orally infected with *C. jejuni* (81–176) and flagellin, particularly for cells grown in bile or DOC medium, and anti-flagellin antibody reduces colonization of the intestinal tract by *C. jejuni* (38,41,42,92). The 25 to 28 kDa bands observed in bile and DOC-cultured bacteria resemble the Peb1a adhesin (38,41,42,94). Peb1a mutants are 50 to 100-fold less adherent, 15-fold less invasive, and exhibit substantially inferior mouse colonization potential (94). Increased expression of this protein might be expected to contribute the increased adhesiveness observed with the bile and DOC-cultured bacteria.

14.2.4   CONGO RED DYE BINDING

Congo red binding (CRB) has been reported as an indicator of the ability to bind hemin, a source of iron, an essential virulence trait for pathogen survival in the restricted host environment (17,19,69,95). *C. jejuni* cultured on Congo-red agar containing bile or DOC bind greater levels of dye than controls (38,41,42). This effect is observed with wild-type and mutant *C. jejuni* in contrast to effects on the adherent and invasive phenotypes. In contrast to temperature regulation of adherent and invasive traits, increased CRB occurs at both 37°C and 42°C (38,41,42). DOC may regulate CRB through a different mechanism than for other virulence-associated phenotypes.

CRB may also be indicative of biofilm formation by pathogens, and has been utilized to differentiate invasive *Shigella* from noninvasive mutants, so binding of the dye may indicate increased expression of several co-regulated virulence factors (96). *C. jejuni* culture with DOC does enhance levels of cell-associated capsular polysaccharide (38,41,42). CRB, like invasivenss, occurs when *Shigella* are cultured at 37°C but not 30°C or 42°C in contrast to *C. jejuni* (38,41,42,95,97,98). Interestingly, Congo red itself induces increased invasiveness of *Shigella* and Ipa protein secretion which could be detected by convalescent antisera (98). Congo red may mimic some host factor like DOC utilized as a pathogenic cue by the bacterium (38,41,42,98).

14.3   *Vibrio parahaemolyticus*: EARLY STUDIES DEFINE A ROLE FOR A HOST-DEPENDENT CUE

Vibrio parahaemolyticus is an incidental pathogen in much of the World, although it is associated with substantial numbers of cases in many countries where consumption of raw seafoods is prevalent (99–106). Where *V. parahaemolyticus* gastroenteritis does occur, pathogenic strains are only infrequently isolated from the estuarine environment (99–106). This paradox may be due to the fact that bacteria in nonhost environments can enter into a quiescent state that has been termed viable nonculturable rendering the pathogens difficult to detect (107,108). While being metabolically active, the bacteria often will not grow on typical bacteriological media, but viable nonculturable pathogens can cause disease (107,108). Bile components like DOC have been demonstrated to play a role in activating these dormant pathogens and likely are one environmental cue signaling entry into the host and that trigger production of virulence factors (37,39,40,105,107,108).

When disease attributable to *V. parahaemolyticus* was first described, it was noted that the preponderance of clinical isolates from patients with gastroenteritis were hemolytic while most environmental isolates were not (99–104,110). This original correlation of the Kanagawa Phenomenon, hemolysis on a special medium, was attributed to the thermostable direct hemolysin (TDH) (104,109). Subsequently, a Kanagawa Phenomenon-negative (KP-) strain was identified as a cause of gastroenteritis, and found to produce a hemolysin closely related to TDH, the TDH-related hemolysin (TRH) (102,110,111). These hemolysins are important virulence factors producing increases in short-circuit current of intestinal tissue, fluid secretion, cardiotoxicity, and isogenic mutants exhibit reduced colonization potential and virulence (110–112). In *V. parahaemolyticus* cultured with bile salts (acids), TDH production is substantially enhanced (up to 32-fold), further suggesting that bacterial response to bile components plays an important role in disease (36,113). Other *Vibrio* spp. also contain related TDH genes, and TDH production is regulated by a homolog of the *V. cholerae toxRS* operon (36,113).

14.3.1   BILE EFFECTS ON OTHER *Vibrio parahaemolyticus* VIRULENCE FACTORS

Adhesion and invasion of human intestinal cells have also been suggested as virulence factors in addition to production of hemolysins for *V. parahaemolyticus* (37,39,40,99, 115–117). Several reports have described adherence of *V. parahaemolyticus* to human epithelial cells, and in one report the bacteria were also shown to be invasive.

When *V. parahaemolyticus* is cultured with bile or DOC, epithelial-cell adherence is substantially increased (37,39,40,94). Invasiveness, however, could not be assayed because bacteria cultured with the host factors were so cytotoxic to the human cells as to preclude completion of the assay (37,99). This finding is consistent with increased production of hemolysin described earlier (36,113).

14.3.2 IMPACT OF BILE ON LIPOPOLYSACCHARIDE STRUCTURE, OUTER-MEMBRANE PERMEATION, AND VIRULENCE

Bile and in particular bile acids are inhibitory for many bacteria due to their detergent effects and act as a barrier to colonization (118–123). As would be expected for an important environmental cue, steroid molecules, metabolites of cholesterol like bile acids, can penetrate the bacterial outer membrane (119,120). Thus there is a trade-off between growth-limiting and beneficial effects for the pathogen (i.e., induction of virulence gene expression) that result from interaction of the pathogen with the bile acid (118–123).

Two responses that mediate bile-resistance and contribute to pathogenicity are changes in lipopolysaccharide (LPS) composition and porin production (124–134). *V. parahaemolyticus* cultured with bile or DOC produce a longer O-antigen moiety as demonstrated by SDS-PAGE (37,39,40,128,129). This is consistent with roles for LPS in colonization and bile resistance. *Vibrio cholerae* O139 mutants incapable of producing O-antigen side chain exhibit a 30-fold reduction in ability to colonize the infant mouse small intesting (124,130). The *galU V. cholerae* El Tor and rough-LPS mutants are more sensitive to bile acids and other detergents, hydrophobic antibiotics, short chain organic acids, cationic antimicrobial peptides and complement (124,130). Core-oligosaccharide mutants are also 50 to 100-fold attenuated in colonization. Similar observations have been made for *Salmonella* spp., and while *S. flexneri galU* and *rfe* mutants still invade HeLa cells, they fail to induce signs of intercellular migration, and are less virulent *in vivo* (131–133). *V. parahaemolyticus* cultured in bile-containing medium are more virulent as quantified by time-to-death in a CD-1 mouse peritoneal-challenge model (39). This is expected because virulence traits have been induced by *in vitro* simulation of *in vivo* environmental conditions prior to animal challenge.

V. cholerae OmpU porin production is increased by culture with DOC (126,127). This event is ToxR-dependent, and the *toxR* mutants exhibit lower minimal bactericidal concentration for bile, and reduced CD-1 suckling mouse intestinal colonization (134). These effects are due to the lower permeability of bile acids through OmpU porin than the OmpT porin (126,127). Cephaloridine, utilized as a probe, permeates more slowly through outer membranes containing predominantly OmpU, and bile acids compete with cephaloridine for diffusion through OmpT (126,127). Similar results are observed with other *Vibrio* spp. including *V. parahaemolyticus* (134,135).

14.3.3 ENVIRONMENTAL REGULATION OF CAPSULE FORMATION AND DISEASE

Vibrio spp. form biofilms that share both common and unique architecture with other bacteria (136–138). Extracellular polysaccharides, both capsular and loosely-associated, are a major component of these biofilms (139–144). *V. parahaemolyticus* possesses greater potential for capsular polysaccharide (CPS) production than has been typically observed and it is environmentally regulated (142–145).

Opaque colony formation is indicative of the ability for the increased CPS formation and adherence (139–144). The phenotype of capsule formation in *V. parahaemolyticus* is regulated by the *Vibrio harveyi* LuxR-type homolog OpaR, and is a quorum sensing dependent event in a number of bacteria (144–148). ToxR plays a central role in the multilevel signaling pathway where luxO mutants cannot form biofilms and are colonization defective (144–146).

Culture of *V. parahaemolyticus* with bile increases CPS production (37,39,40,147). Concomitant with this effect are formation of microcolonies at the liquid–air interface, encapsulated cell aggregates, and increased epithelial-cell adhesion (39,40). In addition to the effect on *in vitro* biofilm formation and adherence, CPS mediates ability to colonize the small intestine (147). Further, antibody to the CPS reduces the bacterial adherence consistent with the importance of the biofilm formation (147).

14.4 EFFECT OF BIOFILM FORMATION ON ANTIBIOTIC ACTIVITY

Heterogeneous phenomena working in conjunction likely mediate the apparent antibacterial resistance for biofilm bacteria (149–165). These include: physico-chemical barrier effects of capsular and other extracellular materials reducing access of some drugs to the pathogen; efflux of toxic host factors and antibiotics; effect of quorum sensing-regulated responses enhanced by increased localized auto-inducer concentration; quiescent metabolic status or activation of a stress response mediating tolerance or a reduced bactericidal effect; and alternatively emergence of biofilm-specific phenotypes such as persistors or adaptive mutation that impact antibacterial efficacy (149–165).

Efflux pumps play a substantial role in antibacterial resistance as well as in colonization of the host (122,159,161,162). Interestingly, efflux pumps are quite apparent in biofilm bacteria although levels may vary (46,48–50,55,159,161–165). Efflux pumps remove toxic substrates from the bacterial cell (159,161,162). This includes host-specific agents like bile acids. Everted membrane vesicles from wild-type bacteria accumulate bile acids, chenodeoxycholic acid accumulates in bacteria when *acrAB* and *emrAB* are mutated, and overproduction of *baeR* in an *acrAB*-null mutant confers bile acid resistance through enhanced MdtABC production (159,161,162). Production of the broad substrate efflux pump AcrAB is regulated by XylS/AraC, MarA, SoxS, and Rob (162). Lipophilic bile acids induce expression of AcrAB in a Rob-dependent manner. Structure–activity relationships of this bile acid-dependent induction are identical to that of regulation for other virulence traits, including adhesion/invasion of epithelial cells and biofilm formation (162). *vceAB* and *tolc* in *V. cholerae*, and *cmeABC* are also important for bile efflux and resistance (53,54). All of these efflux pumps also mediate resistance to a wide variety of antibiotics (159,161,162).

Susceptibility to some antibacterials of *C. jejuni* and *V. parahaemolyticus* cultured with bile or DOC has been reported (37–39,41,42). Unexpectedly susceptibility of

C. jejuni to metronidazole was greatly enhanced by inclusion of bile or DOC in the assay medium (38,41,42). In addition to other physiological changes in these bacteria, levels of ferredoxin reductase were elevated possibly explaining the increased susceptibility (38,41,42). With bile- and DOC-cultured *V. parahaemolyticus*, susceptibility was largely equivalent with results determined utilizing traditional medium (37,39). This finding is similar to many others for susceptibility of biofilm ensconced bacteria, where remarkable changes in antibacterial action are observed primarily as reduced bactericidal effects (46,48,49,58,158). However, alterations in antibacterial susceptibility with bile-cultured bacteria has been reported in some cases (163).

14.5   CONCLUSIONS

Utilizing culture conditions that optimize expression of virulence phenotypes can lead to clarification of the potentially significant role for physiological traits necessary for both growth in the host (disease) and antibacterial susceptibility. Previously it has been suggested that current antibacterial susceptibility assay methods are limited due to *in vitro* and *in vivo* differences in pathogen susceptibility (164–166). Susceptibility determined utilizing culture conditions more closely mimicking *in vivo* conditions might result in a better correlation between *in vitro* findings and clinical efficacy (164–166). *In vitro* susceptibility assay of bacteria in biofilms may be particularly relevant and *in vivo* simulation may facilitate modeling of bacteria in biofilms (164–167).

Further issues become evident when considering these findings. Pathogenicity, defined broadly as physiology, specific virulence traits, and other factors such as efflux pumps required for resistance to host innate defenses, may also mediate reduced susceptibility to antibacterials (1,2,12,26,39,41,46,48,49,53–56,58, 59,70,122,131,134, 149–152,159,161,162,164–169). Co-evolution of pathogen and host is evident, and increasing evidence of linkage between virulence and antibacterial resistance needs to be considered in light of strategies that restrict the use of antibacterials (70,168–170). Obviously, if the two factors have become linked then the condition of being present in the host may provide sufficient selective pressure to maintain antibiotic resistance, even in the absence of the antibiotic. Continued identification of host molecules that regulate expression of traits, including biofilm formation, and their use for *in vivo* simulation, may help to clarify the relationship between *in vitro* susceptibility and clinical efficacy, leading to improved correlation.

REFERENCES

1. Cotter, P.A., and Miller, J.F., *In vivo* and *ex vivo* regulation of bacterial virulence gene expression. *Curr. Opin. Microbiol.* 1, 17–26, 1998.
2. Deretic, V., Schurr, M.J., Boucher, J.C., and Martin, D.W., Conversion of *Pseudomonas aeruginosa* to mucoidy in cystic fibrosis: Environmental stress and regulation of bacterial virulence by alternative sigma factors. *J. Bacteriol.* 176, 2773–2780, 1994.
3. Finlay, B.B., and Falkow, S., Common themes in microbial pathogenicity revisited. *Microbiol. Mol. Biol. Rev.* 61, 136–139, 1997.

4. Forsberg, A., and Rosqvist, R., *In vivo* expression of virulence genes of *Y. pseudotuberculosis*. *Infect Agents Dis.* 2, 275–278, 1994.
5. Galan, J.E., Alternative strategies for becoming an insider: Lessons from the bacterial World. *Cell* 103, 363–366, 2000.
6. Gross, R., Signal transduction and virulence regulation in human and animal pathogens. *FEMS Microbiol. Lett.* 104, 301–326, 1993.
7. Hughes, K.T., Gillen, K.L., Semon, M.J., and Karlinsey, J.E., Sensing structural intermediates in bacterial flagellar assembly by export of a negative regulator. *Science* 262, 1277–1280, 1993.
8. Maurelli, A.T., Hromockyi, A.E., and Bernardini, M.L., Environmental regulation of *Shigella* virulence. *Curr. Top. Microbiol. Immunol.* 180, 95–116, 1992.
9. Mekalanos, J.J., Environmental signals controlling expression of virulence determinants in bacteria. *J. Bacteriol.* 174, 1–7, 1992.
10. Miller, J.F., Mekalanos, J.J., and Falkow, S., Coordinate regulation and sensory transduction in the control of bacterial virulence. *Science* 243, 916–922, 1989.
11. Sen, A., Leon, M.A., and Palchaudhuri, S., Environmental signals induce major changes in virulence of *Shigella* spp. *FEMS Microbiol. Lett.* 84, 231–236, 1991.
12. Terry, J.M., Pina, S.E., and Mattingly, S.J., Environmental conditions which influence mucoid conversion in *Pseudomonas aeruginosa* PAO1. *Infect. Immun.* 59, 471–477, 1991.
13. Bolin, I., Portnoy, D.A., and Wolf-Watz, H., Expression of the temperature-inducible outer membrane proteins of Yersiniae. *Infect. Immun.* 48, 234–240, 1985.
14. Bras, A.M., Chatterjee, S., Wren, B.W., Newell, D.G., and Ketley, J.M., A novel *Campylobacter jejuni* two-component regulatory system important for temperature-dependent growth and colonization. *J. Bacteriol.* 181, 3298–3302, 1999.
15. Brubaker, R.R., The Vwa+ virulence factor of Yersiniae: The molecular basis of the attendant nutritional requirement for Ca^{++}. *Rev. Infect. Dis.* 5, S748–S758, 1983.
16. Garcia-Vescovi, E., Soncini, F.C., Groisman, E.A., Mg^{2+} as an extracellular signal: environmental regulation of *Salmonella* virulence. *Cell* 84, 165–174, 1996.
17. Lawlor, K.M., Daskaleros, P.A., Robinson, R.E., and Payne, S.M., Virulence of iron transport mutants of *Shigella flexneri* and utilization of host iron compounds. *Infect. Immun.* 55, 594–599, 1987.
18. Makino, S., Sasakawa, C., Uchida, I., Terakado, N., and Yoshikawa, M., Cloning and CO_2-dependent expression of the genetic region for encapsulation from *Bacillus anthracis. Mol. Microbiol.* 2, 371–376, 1988.
19. Payne, S.M., and Finkelstein, R.A., Detection and differentiation of iron-responsive avirulent mutants on Congo red agar. *Infect. Immun.* 18, 94–98, 1977.
20. Pepe, J.C., Badger, J.L., and Miller, V.L., Growth phase and low pH affect the thermal regulation of the *Yersinia pseudotuberculosis inv* gene. *Mol. Microbiol.* 11, 123–135, 1994.
21. Tobe, T., Nagai, S., Okada, N., Adler, B., Yoshikawa, M., and Sasakawa, C., Temperature-regulated expression of invasion genes in *Shigella flexneri* is controlled through the transcriptional activation of the *virB* gene on the large plasmid. *Mol. Microbiol.* 5, 887–893, 1991.
22. Cornelis, G., Contact with eukaryotic cells: a new signal triggering bacterial gene expression. *Trends Microbiol.* 5, 43–44, 1997.
23. Konkel, M.E., Mead, D.J., and Cieplak, W. Jr., Kinetics and antigenic characterization of altered protein synthesis by *Campylobacter jejuni* during cultivation with human epithelial cells. *J. Infect. Dis.* 168, 948–954, 1993.

24. Kreig, D.P., Dass, J.A., and Mattingly, S.J., Phosphorylcholine stimulates capsule formation of phosphate-limited mucoid *Pseudomonas aeruginosa. Infect. Immun.* 56, 864–873, 1988.

25. Sperandio, V., Torres, A.G., Jarvis, B., Nataro, J.P., and Kaper, J.B., Bacteria-host communication: The language of hormones. *PNAS* 100, 8951–8956, 2003.

26. Zielinski, N.A., Maharaj, R., Roychoudhury, S., Danganan, C.E., Hendrickson, W., and Chakrabarty, A.M., Alginate synthesis in *Pseudomonas aeruginosa*: environmental regulation of the *algC* promoter. *J. Bacteriol.* 174, 7680–7688, 1992.

27. Coulanges, V., Andre, P., Ziegler, O., Buchheit, L., and Vidon, D.J., Utilization of iron-catecholamine complexes involving ferric reductase activity in *Listeria monocytogenes. Infect. Immun.* 65, 2778–2785, 1997.

28. D'Mello, A., and Yotis, W.W., The action of sodium deoxycholate on *Escherichia coli. Appl. Environ. Microbiol.* 53, 1944–1946, 1987.

29. Freestone, P.P.E., Haigh, R.D., Williams, P.H., and Lyte, M., Stimulation of bacterial growth by heat-stable norepinephrine-induced autoinducers. *FEMS Microbiol. Lett.* 172, 53–60, 1999.

30. Gunn, J.S., Mechanisms of bacterial resistance and response to bile. *Microbes Infection* 2, 907–913, 2000.

31. Gupta, S., and Chowdhury, R., Bile affects production of virulence fators and motility of *Vibrio cholerae. Infect. Immun.* 65, 1131–1134, 1997.

32. Kinney, K.S., Austin, C.E., Morton, D.S., and Sonnenfeld, G., Catecholamine enhancement of *Aeromonas hydrophila* growth. *Microb. Pathogen.* 26, 85–91, 1999.

33. Lyte, M., Arulanandam, B., Nguyen, K., Frank, C., Erickson, A., and Francis, D., Norepinephrine induced growth and expression of virulence associated factors in enterotoxigenic and enterohemorrhagic strains of *Escherichia coli. Adv. Exp. Med. Biol.* 412, 331–339, 1997.

34. Lyte, M., and Ernst, S., Alpha and beta adrenergic receptor involvement in catecholamine-induced growth of gram-negative bacteria. *Biochem. Biophys. Res. Commun.* 190, 447–452, 1993.

35. Lyte, M., Erickson, A.K., Arulanandam, B., Frank, C.D., Crawford, M.A., and Francis, D.H., Norepinephrine-induced expression of the K99 pilus adhesion of enterotoxigenic *Escherichia coli. Biochem. Biophys. Res. Commun.* 232, 682–682, 1997.

36. Osawa, R., and Yamai, S., Production of thermostable direct hemolysin by *Vibrio parahaemolyticus* enhanced by conjugated bile acids. *Appl. Environ. Microbiol.* 62, 3023–3025, 1996.

37. Pace, J.L., Growth effects on the cell envelope composition of food-poisoning *Vibrio parahaemolyticus.* Ph.D. dissertation, University of Maryland, College Park, MD, 1989.

38. Pace, J., *Antigen-Enhanced Bacterial Vaccines. Vaccines: New Technologies and Applications*, Cambridge Healthtech Institute Conference, Tysons Corner, V.A., 1996.

39. Pace, J.L., Vibrionaceae and in-vivo simulation of enhanced virulence and antibiotic susceptibility. Amer. Soc. Microbiol. Annual Meeting, Salt Lake City, UT, May, 2002.

40. Pace, J.L., Chai, T., Rossi, H.A., and Jiang, X., Effect of bile on *Vibrio parahaemolyticus. Appl. Environ. Microbiol.* 63, 2372–2377, 1997.

41. Pace, J.L., Frey, S., Walker, R., Burr, D., Johnson, W., Rollins, D. Yang, H., and Lee, L., Culture conditions affect *Campylobacter jejuni* virulence and composition. *Cold Spring Harbor Laboratory Conf. Molecular Approaches to the Control of Infectious Diseases*, Cold Spring Harbor, N.Y., October, 1994.

42. Pace, J.L., Walker, R.I., and Frey, S.M., Methods for producing enhanced antigenic *Campylobacter* bacteria and vaccines. US Pat No 5,679,564, 1997.

43. Pope, L.M., Reed, K.E., and Payne, S.M., Increased protein secretion and adherence to HeLa cells by *Shigela* sp, following growth in the presence of bile salts. *Infect. Immun.* 63, 3642–3648, 1995.

44. Schmitt, M.P., Identification of a two-component signal transduction system from *Corynebacterium diphtheriae* that activates gene expression in response to the presence of heme and hemoglobin. *J. Bacteriol.* 181, 5330–5340, 1999.

45. Wandersman, C., and Stojiljkovic, I., Bacterial heme sources: the role of heme, hemo-protein receptor and hemophores. *Curr. Opin. Microbiol.* 3, 215–220, 2000.

46. Costerton, J.W., Cheng, K.J., Geesey, G.G., Ladd, T.I., Nickel, J.C., Dasgupta, M., and Marrie, T.J., Bacterial biofilms in nature and disease. *Annu. Rev. Microbiol.* 41, 435–464, 1987.

47. Costerton, J.W., Geesey, G.G., and Cheng, G.K., How bacteria stick. *Sci. Am.* 238, 86–95, 1978.

48. Davey, M.E., and O'Toole, G.A., Microbial biofilms: from ecology to molecular genetics. *Microbiol. Molecul. Biol. Rev.* 64, 847–867, 2000.

49. Donlan, R.M., and Costerton, J.W., Biofilms: survival mechanisms of clinically relevant microorganisms. *Clin. Microbiol. Rev.* 15, 167–193, 2002.

50. Dunne, W.M. Jr., Bacterial adhesion: Seen any good biofilms lately? *Clin. Microbiol. Rev.* 15, 155–166, 2002.

51. Fitzpatrick, F., Humphreys, H., Smyth, E., Kennedy, C.A., and O'Gara, J.P., Environmental regulation of biofilm formation in intensive care unit isolates of *Staphylococcus epidermidis*. *J. Hosp. Infect.* 42, 212–218, 2002.

52. Gotz, F., *Staphylococcus* and biofilms. *Mol. Microbiol.* 43, 1367–1378, 2002.

53. Lin, J., Overbye Michel, L., and Zhang, Q., CmeABC functions as a multidrug efflux system in *Campylobacter jejuni. Antimicrob. Agents Chemother.* 46, 2124–2131, 2002.

54. Lin, J., Sahin, O., Overbye Michel, L., and Zhang, Q., Critical role of multidrug efflux pump CmeABC in bile resistance and *in vivo* colonization of *Campylobacter jejuni. Infect. Immun.* 71, 4250–4259, 2003.

55. Mah, T.C., and O'Toole, G.A., Mechanisms of biofilm resistance to antimicrobial agents. *Trends. Microbiol.* 9, 34–38, 2001.

56. Prouty, A.M., Schwesinger, W.H., and Gunn, J.S., Biofilm formation and interaction with the surfaces of gallstones by *Salmonella* spp. *Infect. Immun.* 70, 2640–2649, 2002.

57. Rashid, M.H., Rumbaugh, K., Passador, L., Davies, D.G., Hamood, A.N., Iglewski, B.H., and Kornberg, A., Polyphosphate kinase is essential for biofilm development, quorum sensing, and virulence of *Pseudomonas aeruginosa. PNAS* 97, 9636–9641, 2000.

58. Schwank, S., Rajacic, Z., Zimmerli, W., and Blaser, J., Impact of bacterial biofilm formation on an *in vitro* and *in vivo* activities of antibiotics. *Antimicrob. Agents Chemother.* 42, 895–898, 1998.

59. Widmer, A.F., Frei, R., Rajacic, Z., and Zimmerli, W., Correlation between *in vivo* and *in vitro* efficacy of antimicrobial agents against foreign body infections. *J. Infect. Dis.* 162, 96–102, 1990.

60. Yasuda, H., Bacterial biofilms and infectious diseases. *Trends Glycoscicence Glycotechnol.* 8, 409–417, 1996.

61. Collazo, C.M., and Galan, J.E., Requirement for exported proteins in secretion through the invasion-associated type III system of *Salmonella typhimurium. Infect. Immun.* 64, 3254–3531, 1996.

62. Galan, J.E., and Collmer, A., Type III secretion machines: Bacterial devices for protein delivery into host cells. *Science* 284, 1322–1328, 1999.

63. Ginocchio, C.C., and Galan, J.E., Functional conversation among members of the *Salmonella typhimurium* or InvA family of proteins. *Infect. Immun.* 63, 729–732, 1995.

64. Price, S.B., Cowan, C., Perry, R.D., and Straley, S.C., The *Yersinia pestis* V antigen is a regulatory protein necessary for Ca^{2+}-dependent growth and maximal expression of low- Ca^{2+} response virulence genes. *J. Bacteriol.* 173, 2649–2657, 1991.

65. Straley, S.C., Plano, G.V., Skrzypek, E., Haddix, L., and Fields, K.A., Regulation by Ca^{2+} in the *Yersinia* low- Ca^{2+} response. *Mol. Microbiol.* 8, 1005–1010, 1993.

66. Galan, J.E., and Curtiss, III., Cloning and molecular characterization of genes whose products allow *Salmonella typhimurium* to penetrate tissue culture cells. *Proc. Natl. Acad. Sci. USA* 86, 6383–6387, 1989.

67. Schiemann, D.A., and Shope, S.R., Anaerobic growth of *Salmonella typhimurium* results in increased uptake by Henle. 407 epithelial and mouse peritoneal cells *in vitro* and repression of a major outer membrane protein. *Infect. Immun.* 59, 437–440, 1991.

68. High, N., Mounier, J., Prevost, M.C., and Sansonetti, P., IpaB of *Shigella flexneri* causes entry into epithelial cells and escape from the phagocytic vacuole. *EMBO J.* 11, 1991–1999, 1992.

69. Daskaleros, P.A., and Payne, S.M., Congo red binding phenotype is associated with hemin binding and increased infectivity of *Shigella flexneri* in the HeLa cell model. *Infect. Immun.* 55, 1393–1398, 1987.

70. Martinez, J.L., and Baquero, F., Interactions among strategies associated with bacterial infection: pathogenicity, epidemicity, and antibiotic resistance. *Clin. Microbiol. Rev.* 15, 647–679, 2002.

71. Ruiz-Palacios, G.M., Cervantes, L.E., Newberg, D.S., Lopez-Vidal, Y., and Calva, J.J., *In vitro* study of virulence factors of enteric *Campylobacter* spp. In: Nachamkin, I., Blaser M.J., and Tompkins, L.S., eds., *Campylobacter jejuni*: Current status and future trends. Washington, D.C.: American Society for Microbiology, 1992:168–175.

72. Black, R.E., Levine, M.M., Clements, M.L., Hughes, T.P., and Blaser, M.J., Experimental *Campylobacter jejuni* infection in humans. *J. Infect. Dis.* 157, 472–479, 1988.

73. Newell, D.G., McBride, H., Saunders, F., Dehele, F., and Pearson, A.D., The virulence of clinical and environmental isolates of *Campylobacter jejuni*. *J. Hyg. Camb.* 94, 45–54, 1985.

74. Fauchere, J.L., Rosenau, A., Veron, M., Moyen E.N., Richard, S., and Pfister, A., Association with HeLa cells of *Campylobacter jejuni* and *Campylobacter coli* isolated from human feces. *Infect. Immune.* 54, 283–287, 1986.

75. Klipstein, F.A., Engert, R.F., Short, H.B., and Schenk, E.A., Pathogenic properties of *Campylobacter jejuni*: assay and correlation with clinical manifestation. *Infect. Immun.* 50, 43–49, 1985.

76. Woolridge, K.G., and Ketley, J.M., *Campylobacter* host cell interactions. *Trends Microbiol.* 5, 96–102, 1997.

77. Yao, R., Burr, D.H., Doig, P., Trust, T.J., Haiying, N., and Guerry, P., Isolation of motile and non-motile insertional mutants of *Campylobacter jejuni*: the role of motility in adherence and invasion of eukaryotic cells. *Mol. Microbiol.* 14, 883–893, 1994.

78. Szymanski, C.M., King, M., Haardt, M., and Armstrong, G.D., Campylobacter jejuni motility and invasion of Caco-2 cells. *Infect. Immun.* 63, 4295–4300, 1995.

79. Baker, R.D., and Searle, G.W., Bile salt adsorption at various levels of rat small intestine. *Proc. Soc. Exp. Biol. Med.* 105, 521–523, 1960.

80. Borgstrom, B., Bile salts – their physiological functions in the gastrointestinal tract. *Acta. Med. Scand.* 196, 1–10, 1974.

81. Ganong, W.F., Review of Medical Physiology, 17th ed., p. 781. Appleton and Lange, 1995.

82. Hofmann, A.F., and Mysels, Bile acid solubility and precipitation *in vitro* and *in vivo*: the role of conjugation, pH, and Ca2+ ions. *J. Lipid. Res.* 33, 617–626, 1992.

83. Hofmann, A.F., and Small, D.M., Detergent properties of bile salts: correlation with physiological function. *Annu. Rev. Med.* 18, 333–376, 1967.

84. Walker, R.I., Caldwell, M.B., Lee, E.C., Guerry, P., Trust, T.J., and Ruiz-Palacios, G.M., Pathophysiology of *Campylobacter* enteritis. *Microbiol. Rev.* 50, 81–94, 1986.

85. Wassenaar, T.M., Bleumink-Pluym, N.M.C., and van der Zeijst, B.A.M., Inactivation of *Campylobacter jejuni* flagellis genes by homologous recombination demonstrates that *flaA* but not *flab* is required for invasion. *EMBO J.* 10, 2055–2061, 1991.

86. Misawa, N., and Blaser, M.J., Detection and characterization of autoagglutination activity by *Campylobacter jejuni*. *Infect. Immun.* 68, 6168–6175, 2000.

87. Grant, C.C.R., Konkel, M.E., Cieplak, W., Jr., and Tompkins, L.S., Role of flagella in adherence, internalization, translocation of *Campylobacter jejuni* in nonpolarized and polarized epithelial cell cultures. *Infect Immun.* 61, 1764–1771, 1993.

88. McSweegan, E., and Walker, R.I., Identification and characterization of two *Campylobacter jejuni* adhesions for cellular and mucous substrates. *Infect. Immun.* 53, 141–148, 1986.

89. Pavlovskis, O.R., Rollins, D.M., Harberger, R.L., Jr., Green, A.E., Habash, L., Strocko, S., and Walker, R.I., Significance of flagella in colonization resistance of rabbits immunized with *Campylobacter* spp. *Infect. Immun.* 59, 2259–2264, 1991.

90. Alm, R.A., Guerry, P., and Trust, T.J., The *Campylobacter* σ54 *flaB* flagellin promoter is subject to environmental regulation. *J. Bacteriol.* 175, 4448–4455, 1993.

91. Miller, S., Pesci, E.C., Pickett, C.L., A *Campylobacter jejuni* homolog of the LcrD/FlbF family of proteins is necessary for flagellar biogenesis. *Infect. Immun.* 61, 2930–2936, 1993.

92. Ueki, Y., Umeda, A., Fujimoto, S., Mitsuyama, M., and Amako, K., Protection against *Campylobacter jejuni* infection in suckling mice by anti-flagellar antibody. *Microbiol. Immunol.* 31, 1161–1171, 1987.

93. Kostrzynsak, M., Betts, J.D., Austin, J.W., and Trust, T.J., Identification, characterization, and spatial localization of two flagellin species in *Helicobacter pylori*. *J. Bacteriol.* 173, 937–946, 1991.

94. Pei, Z., Burucoa, C., Grignon, B., Baqar, S., Huang, X., Kopecko, D.J., Bourgeois, A.L., Fauchere, J., and Blaser, M.J., Mutation in the *pebIa* locus of *Campylobacter jejuni* reduces interactions with epithelial cells and intestinal colonization of mice. *Infect. Immun.* 66, 938–943, 1998.

95. Andrews, G.P., and Maurelli, A.T., *mxiA* of *Shigella flexneri* 2a, which facilitates export of invasion plasmid antigens, encodes a homolog of the low-calcium-response protein, LcrD, of *Yersinia pestis*. *Infect. Immun.* 60, 3287–3295, 1992.

96. Karlyshev, A., and Wren, B.W., Detection and initial characterization of novel capsular polysaccharide among diverse *Campylobacter jejuni* strains using alcian blue dye. *J. Clin. Microbiol.* 39, 279–284, 2001.

97. Sankaran, K., Ramachandran, V., Subrahmanyam, Y.V.B.K., Rajarathnam, S., Elango, S., and Roy, R.K., Congo-red-mediated regulation of levels of *Shigella flexneri* 2a membrane proteins. *Infect. Immun.* 57, 2364–2371, 1989.

98. Bahrani, F.K., Sansonetti, P.J., Parsot, C., Secretion of Ipa proteins by *Shigella flexneri*: inducer molecules and kinetics of activation. *Infect. Immun.* 65, 4005–4010, 1997.

99. Chai, T., and Pace, J., *Vibrio parahaemolyticus*. In: Hui, Y.H., Gorham, J.R., Murrell, K.D., and Oliver, D.O., eds., *Compendium of Food Poisoning Organisms*, 2nd Edition. New York: Marcel Dekker, Inc., 2000.

100. Hara-Kudo, Y., Sugiyama, K., Nishibuchi, M., Chowdhury, A., Yatsuyanagai, J., Ohtomo, Y., Saito, A., Nagano, H., Nishina, T., Nakagawa, H., Konuma, H., Miyahara, M., and Kumagai, S., Prevalence of pandemic thermostable direct hemolysin-producing *Vibrio parahaemolyticus* O3:K6 in seafood and the coastal environment in Japan. *Appl. Environ. Microbiol.* 69, 3883–3891, 2003.

101. Beuchat, L.R., *Vibrio parahaemolyticus*: public health significance. *Food Technol.* 36, 80–92, 1982.

102. Nishibuchi, M., and Kaper, J.B., Thermostable direct hemolysin gene of *Vibrio parahaemolyticus*: a virulence gene acquired by a marine bacterium. *Infect. Immun.* 63, 2093–2099, 1995.

103. Shirai, H., Ito, H., Hirajama, T., Nakamoto, Y., Nakabayoshi, N., Kumagai, K., Takeda, Y., and Nishbuchi, M., Molecular epidemiologic evidence for association of thermostable direct hemolysin (TDH) and TDH-related hemolysin of *Vibrio parahaemolyticus* with gastroenteritis. *Infect. Immun.* 58, 3568–3573, 1990.

104. Miyamoto, Y., Kato, T., Obara, Y., Akiyama, S., Takizawa, K., and Yamai, S., *in vitro* hemolytic characteristic of *V. parahaemolyticus*: Its close correlation with human pathogenicity. *J. Bacteriol.* 100, 1147–1149, 1969.

105. Pace, J., Chai, T., Comparison of *Vibrio parahaemolyticus* grown in estuarine water and rich medium. *Appl. Environ. Microbiol.* 55, 1877–1887, 1989.

106. Janda, J.M., Powers, C., Bryant, R.G., and Abbott, S.L., Current perspectives on the epidemiology and pathogenesis of clinically relevant *Vibrio* spp. *Clin. Microbiol. Rev.* 1, 245–267, 1988.

107. Roszak, D.B., and Colwell, R.R., Survival strategies of bacteria in the natural environment. *Microbiol. Rev.* 51, 365–379, 1987.

108. Chaiyanan, S., Chaiyanan, S., Huq, A., Maugel, T., and Colwell, R.R., Viability of the nonculturable *Vibrio cholerae* O1 and O139. *Syst. Appl. Microbiol.* 24, 331–341, 2001.

109. Honda, T., Faga, S., Takeda, T., Hasibuan, M.A., Takeda, Y., and Miwatani, T., Identification of lethal toxin with the thermostable direct hemolysin produced by *Vibrio parahaemolyticus*, and some physicochemical properties of the purified toxin. *Infect. Immun.* 13, 133–139, 1976.

110. Nishibuchi, M., Fasano, A., Russell, R.G., and Kaper, J.B., Enterotoxigenicity of *Vibrio parahaemolyticus* with and without genes encoding thermostable direct hemolysin. *Infect. Immun.* 60, 3539–3545, 1992.

111. Honda, T., Ni, Y., and Miwatani, T., Purification and characterization of a hemolysin produced by a clinical isolate of Kanagawa phenomenon-negative *Vibrio parahaemolyticus* and related to the thermostable direct hemolysin. *Infect. Immun.* 56, 961–965, 1988.

112. Raimondi, F., Kao, J.P.Y., Fiorentini, C., Fabbri, A., Donelli, G., Gasparini, N., Rubino, A., and Fasano, A., Enterotoxicity and cytotoxicity of *Vibrio parahaemolyticus* thermostable direct hemolysin in *in vitro* systems. *Infect. Immun.* 68, 3180–3185, 2000.

113. Osawa, R., Arakawa, E., Okitsu, T., Yamai, S., and Watanabe, H., Levels of thermostable direct hemolysin produced by *Vibrio parahaemolytics* O3:K6 and other serovars grown anaerobically with the presence of bile acid. *Curr. Microbiol.* 44, 302–305, 2002.

114. Lin, Z., Kumagai, K., Baba, K., Mekalanos, J.J., and Nishibuchi, M., *Vibrio parahaemolyticus* has a homolog of the *Vibrio cholerae toxRS* operon that mediates environmentally-induced regulation of the thermostable direct hemolysin gene. *J. Bacteriol.* 175, 3844–3855, 1993.

115. Akeda, Y., Nagayama, K., Yamamoto, K., and Honda, T., Invasive phenotype of *Vibrio parahaemolyticus. J. Infect. Dis.* 176, 822–824, 1997.

116. Gingras, S.P., and Howard, L.V., Adherence of *Vibrio parahaemolyticus. Appl. Environ. Microbiol.* 39, 369–371.

117. Carruthers, M.M., *In vitro* adherence of Kanagawa positive *Vibrio parahaemolyticus* to epithelial cells. *J. Infect. Dis.* 136, 588–592, 1977.

118. Margalith, P.Z., Steroid Microbiology. Springfield, IL, Charles C. Thomas, 1986.

119. Plesiat, P., Aires J.R., Godard, C., and Kohler, T., Use of steroids to monitor alterations in the outer membrane of *Pseudomonas aeruginosa. J. Bacteriol.* 179, 7004–7010, 1997.

120. Plesiat, P., and Nikaido, H., Outer membranes of Gram-negative bacteria are permeable to steroid probes. *Mol. Microbiol.* 6, 1323–1333, 1992.

121. Begley, M., Gahan, C.G.M., and Hill, C., Bile stress response in *Listeria monocytogenes* LO28: adaptation, cross-protection, and identification of a genetic locus involved in bile resistance. *Appl. Environ. Microbiol.* 68, 6005–6012, 2002.

122. Bina, J.E., and Mekalanos, J.J., *Vibrio cholerae tolC* is required for bile resistance and colonization. *Infect. Immun.* 69, 4681–4685, 2001.

123. Van Velkinburgh, J.C., and Gunn, J.S., PhoP-PhoQ-regulated loci are required for enhanced bile resistance in *Salmonella* spp. *Infect. Immun.* 67, 1614–1622, 1999.

124. Nesper, J., Lauriano, C.M., Klose, K.E., Kopfhammer, D., Kraib, A., and Reidl, J., Characterization of *Vibrio cholerae* 01 El Tor galU and gale mutants: influence on lipopolysaccharide structure, colonization, and biofilm formation. *Infect. Immun.* 69, 435–445, 2001.

125. Picken, R.N., and Beacham, I.R., Bacteriophage-resistant mutants of *Escherichia coli* K-12. Location of receptors within the lipopolysaccharide. *J. Gen. Microbiol.* 102, 305–318, 1977.

126. Provenzano, D., and Klose, K.E., Altered expression of the ToxR-regulated porins OmpU and OmpT diminishes *Vibrio cholerae* bile resistance, virulence factor expression, and intestinal colonization. *PNAS* 97, 10220–10224, 2000.

127. Wibbenmeyer, J.A., Provenzano, D., Landry, C.F., Klose, K.E., and Delcour, A.H., *Vibrio cholerae* OmpU and OmpT porins are differentially affected by bile. *Infect. Immun.* 70, 121–126, 2002.

128. Tsai, C., and Frasch, C.E., A sensitive silver stain for detecting lipopolysaccharides in polysaccharide gels. *Anal. Biochem.* 119, 115–119, 1982.

129. Westphal, O., and Jann, K., Bacterial lipopolysaccharide extraction with phenol-water and further applications of the procedure. *Methods Carbohydr. Chem.* 5, 83–91, 1965.

130. Nesper, J., Schild, S., Lauriano C.M., Kraiss, A., Klose, K.E., and Reidl, J., Role of *Vibrio cholerae* O139 surface polysaccharide in intestinal colonization. *Infect. Immun.* 70, 5990–5996, 2002.

131. Prouty, A.M., Van Velkinburgh, J.C., and Gunn, J.S., *Salmonella enterica* serovar *typhimurium* resistance to bile: identification and characterization of the tolQRA cluter. *J. Bacteriol.* 184, 1270–1276, 2002.

132. Ramos-Morales, F., Prieto, A.I., Beuzon, C.R., Holden, D.W., and Casadesus, J., Role for *Salmonella enterica* enterobacterial common antigen in bile resistance and virulence. *J. Bacteriol.* 185, 5328–5332, 2003.

133. Sandlin, R.C., Lampel, K.A., Keasler, S.P., Goldberg, M.B., Stolzer, A.L., and Maurelli, A.T., Avirulence of rough mutants of *Shigella flexneri*: Requirement of O antigen for correct unipolar localization of IcsA in bacterial outer membrane. *Infect. Immun.* 63, 229–237, 1995.

134. Provenzano, D., Schuhmacher D.A., Barker J.L., and Klose, K.E., The virulence regulatory protein ToxR mediates enhanced bile resistance in *Vibrio cholerae* and other pathogenic *Vibrio* species. *Infect. Immun.* 68, 1491–1497, 2000.

135. Wang, S., Lauritz, J., Jass, J., and Milton, D.L., A ToxR homolog from *Vibrio anguillarum* serotype O1 regulates its own production, bile resistance, and biofilm formation. *J. Bacteriol.* 184, 1630–1639, 2002.

136. Belas, M.R., and Colwell, R.R., Adsorption kinetics of laterally and polarly flagellated *Vibrio parahaemolyticus. J. Bacteriol.* 151, 1568–1580, 1982.

137. Lawrence, J.R., Korber, D.R., Hoyle, B.D., Costerton, J.W., and Caldwell, D.E., Optical sectioning of microbial biofilms. *J. Bacteriol.* 173, 6558–6567, 1991.

138. Watnick, P.I., and Kolter, R., Steps in the development of a *Vibrio cholerae* El Tor biofilm. *Mol. Microbiol.* 34, 586–595, 1999.

139. Yildiz, F.A., and Schoolnik, G.K., *Vibrio cholerae* O1 El Tor: identification of a gene cluster required for the rugose colony type, exopolysaccharide production, chlorine resistance, and biofilm formation. *PNAS* 96, 4028–4033, 1999.

140. Pratt, L.A., and Kolter, R., Genetic analyses of bacterial biofilm formation. *Curr. Opin. Microbiol.* 2, 598–603, 1999.

141. Ali, A., Rashid, M.H., and Karolis, K.R., High-frequency rugose exopolysaccharide production by *Vibrio cholerae. Appl. Environ. Microbiol.* 68, 5773–5778, 2002.

142. Enos-Berlage, J.L., and McCarter, L.L., Relation of capsular polysaccharide production and colonial cell organization to colony morphology in *Vibrio parahaemolyticus. J. Bacteriol.* 182, 5513–5520, 2000.

143. Guvener, Z.T., and McCarter, L.L., Multiple regulators control capsular polysaccharide production in *Vibrio parahaemolyticus. J. Bacteriol.* 185, 5431–5441, 2003.

144. McCarter, L., OpaR, a homolog of *Vibrio harveyi* LuxR, controls opacity of *Vibrio parahaemolyticus. J. Bacteriol.* 180, 3166–3173, 1998.

145. Vance, R.E., Zhu, J., and Mekalanos, J.J., A constitutively active variant of the quorum sensing regulator LuxO affects protease production and biofilm formation in *Vibrio cholerae. Infect. Immun.* 71, 2571–2576, 2003.

146. Zhu, J., Miller, M.B., Vance, R.E., Dziyman, M., Bassler, B.L., and Mekalanos, J.J., Quorum-sensing regulators control virulence gene expression in *Vibrio cholerae. PNAS* 99, 3129–3134, 2002.

147. Hsieh, Y., Liang, S., Tsai, W., Chen, Y., Liu, T., and Liang, C., Study of capsular polysaccharide from *Vibrio parahaemolyticus. Infect. Immun.* 71, 3329–3336, 2003.

148. Cronatto, A., Chalker, V.J., Lauritz, J., Jass, J., Hardman, A., Williams, P., Camara, M., and Milton, D.L., VanT, a homologue of *Vibrio harveyi* LuxR, regulates serine, metalloprotease, pigment, and biofilm production in *Vibrio anguillarum. J. Bacteriol.* 184, 1617–1629, 2002.

149. Farber, B.F., Kaplan, M.H., and Clogston, A.G., *Staphylococcus epidermidis* extracted slime inhibits the antimicrobial action of glycopeptide antibiotics. *J. Infect. Dis.* 161, 37–40, 1990.

150. Hoyle, B.D., Alcantara, J., and Costerton, J.W., *Pseudomonas aeruginosa* biofilm as a diffusion barrier to piperacillin. *Antimicrob. Agents Chemother.* 36, 2054–2056, 1992.

151. Nichols, W.W., Dorrington, S.M., Slack, M.P.,E., and Walmsley, H.L., Inhibition of tobramycin diffusion by binding to alginate. *Antimicrob. Agents Chemother.* 32, 518–523, 1988.

152. Nickel, J.C., Ruseka, I., Wright, J.B., and Costerton, J.W., Tobramycin resistance of *Pseudomonas aeruginosa* cells growing as biofilm on urinary catheter material. *Antimicrob. Agents Chemother.* 27, 619–624, 1985.

153. Bassler, B.L., Small talk. Cell-to-cell communication in bacteria. *Cell* 109, 421–424, 2002.

154. Wai, S., Mizunoe, Y., Takade, A., Kawabata, S., and Yoshida, S., *Vibrio cholerae* O1 strain TSI-4 produces the exopolysaccharide materials that determine colony morphology, stress resistance, and biofilm formation. *Appl. Environ. Microbiol.* 64, 3648–3655, 1998.

155. Waar, K., van der Mei, H.C., Harmsen, H.J.M., Degener, J.E., and Busscher, H.J., Adhesion to bile drain materials and physiochemical surface properties of *Enterococcus faecalis* strains grown in the presence of bile. *Appl. Environ. Microbiol.* 68, 3855–3858, 2002.

156. Shapiro, J.A., Thinking about bacterial populations as multicellular organisms. *Annu. Rev. Microbiol.* 52, 81–104, 1998.

157. Mackenzie, G.J., Harris, R.S., Lee, P.L., and Rosenberg, S.M., The SOS response regulates adaptive mutation. *Proc. Natl. Acad. Sci. USA* 97, 6646–6651, 2000.

158. Keren, I., Kaldalu, N., Spoering, A., Wang, Y., and Lewis, K., Persister cells and tolerance to antimicrobials. *FEMS Microbiol. Lett.* 230, 13–18, 2004.

159. Nagakubo, S., Nishino, K., Hirata, T., and Yamaguchi, A., The putative response regulator BaeR stimulates multidrug resistance of *Escherichia coli* via a multidrug exporter system, MdtABC. *J. Bacteriol.* 184, 4161–4167, 2002.

160. Davies, D.G., Parsek, M.R., Pearsen, J.P., Iglewski, B.H., Costerton, J.W., and Greenberg, E.P., The involvement of cell-to-cell signals in the development of a bacterial biofilm. *Science* 280, 295–298, 1998.

161. Thanassi, D.G., Cheng, L.W., and Nikaido, H., Active efflux of bile salts by *Escherichia coli. J. Bacteriol.* 179, 2512–2518, 1997.

162. Rosenberg, E.Y., Bertenthol, D., Nilles, M.L., Bertand, K.P., and Nikaido, H., Bile salts and fatty acids induce the expression of *Escherichia coli* AcrAB multidrug efflux pump through their interaction with Rob regulatory protein. *Mol. Microbiol.* 48, 1609–1619, 2003.

163. Rees, E.N., and Elliott, T.S.J., The influence of bile on antimicrobial activity *in vitro. J. Antimicrob. Chemother.* 41, 659–660, 1998.

164. Anwar, H., Dasgupta, M.K., and Costerton, J.W., Testing the susceptibility of bacteria in biofilms to antibacterial agents. *Antimicrob. Agents Chemother.* 34, 2043–2046, 1990.

165. Anwar, H., Strap, J.L., and Costerton, J.W., Eradication of biofilm cells of *Staphylococcus aureus* with tobramycin and cephalexin. *Can. J. Microbiol.* 38, 618–625, 1992.

166. Stoodley, P., Sauer, K., Davies, D.G., and Costerton, D.G., Biofilms as complex differentiated communities. *Annu. Rev. Microbiol.* 56, 187–209, 2002.

166. Ceri, H., Olsen, M.E., Stremick, C., Read, R.R., Morck, D., and Buret, A., The Calgary Biofilm Device: new technology for rapid determination of antibiotic susceptibilities of bacterial biofilms. *J. Clin. Microbiol.* 37, 1771–1776, 1999.

168. Baquero, F., Negri, M.C., Morosini, M.I., and Blazquez, J., Antibiotic-selective environments. *Clin. Infect. Dis.* 27, S5–S11, 1998.

169. Luck, S.N., Turner, S.A., Rajakumar, K., Sakellaris, H., and Adler, B., Ferric dicitrate transport system (Fec) of *Shigella flexneri* 2a YSH6000 is encoded on a novel pathogenicity island carrying multiple antibiotic resistance genes. *Infect. Immun.* 69, 6012–6021, 2001.

170. Schmidt, H., and Hensel, H., Pathogenicity islands in bacterial pathogenesis. *Clin. Microbiol. Rev.* 17, 14–56, 2004.

15 *In Vivo* Models for the Study of Biomaterial-Associated Infection by Biofilm-Forming Staphylococci

Luke D. Handke and Mark E. Rupp

CONTENTS

15.1   INTRODUCTION

Over billions of years of evolution, bacteria have developed a very successful mode of survival that has recently been characterized as the "biofilm phenotype." In nature, opposed to existing in a free-floating planktonic form, the vast majority of bacteria are found attached to solid strata in structured communities encased in a self-produced polymeric matrix. This preferential, biofilm-associated means of bacterial growth has been recently defined by Donlan and Costerton as "a microbially derived sessile community characterized by cells that are irreversibly attached to a substratum or interface or to each other, are embedded in a matrix of extracellular polymeric substances that they have produced, and exhibit an altered phenotype with respect to growth rate and gene transcription" (1). Bacterial biofilms are believed to have evolved in nature to give organisms a competitive advantage in high shear conditions, to protect microbes from environmental hazards (toxic substances or predatory cells), and to adapt to various nutritional conditions. Similarly, bacteria have exploited the biofilm model of existence in various infectious diseases and biomaterial-associated infections.

The production of biofilms by a variety of bacterial species has recently been associated with a number of diseases. Biofilm-associated infections involving native tissues in humans include infective endocarditis (2), otitis media (3), chronic bacterial prostatitis (4), osteomyelitis (5) pneumonia in cystic fibrosis patients (6), Legionnaires' disease (7), and periodontal disease (8). Biofilms are thought to be important in a variety of medical device infections including prosthetic valve endocarditis (9,10), artificial joint infections (11,12), vascular catheter infections (13), cerebrospinal fluid (CSF) shunt infections (14), urinary catheter infections (15,16), and infections of other implanted devices such as ocular lenses, breast implants, and penile prostheses (1,9). The biofilm is thought to offer organisms an advantage in mediating these diseases by offering a barrier to the diffusion of some antibiotics and other toxic compounds, shielding the organism from the host immune system, and optimization of local nutritional conditions.

In recent years our knowledge of biofilms has been expanded greatly by the use of innovative microscopy techniques such as confocal laser scanning microscopy and epifluorescence microscopy, and a molecular genetics approach. It is evident that bacterial biofilms are complex, heterogeneous structures involving sub-populations of cells exhibiting differing patterns of gene expression (1,17).

15.2   STAPHYLOCOCCAL BIOFILMS

Bayston and Penny are credited with first noting the possible clinical significance of coagulase-negative staphylococcal biofilms in patients with CSF shunt infections (18). Early reports of the ultrastructure of coagulase-negative staphylococcal biofilms and their possible clinical relevance came from the research groups of Christensen (19), Peters (20), and Costerton (21). Shortly thereafter, investigators noted the potential importance of biofilms associated with *Staphylococcus aureus* infections (22). A great deal was learned about staphylococcal biofilms through the application of various *in vitro* models. The effect of the host conditioning film (serum proteins that quickly coat devices shortly after implantation) was studied in a variety of settings, flow and shear effects were mimicked in dynamic flow chambers, and artificial clots and vegetations were devised to study endocarditis. Although great strides were made in understanding staphylococcal adherence and biofilm production under *in vitro* conditions (a more thorough discussion of the clinical significance and molecular genetics of staphylococcal biofilm formation can be found in Chapters 6 and 7), *in vivo* conditions involving flow dynamics, complex mixtures of host proteins and cells in conditioning films, and the interaction of cellular and humoral immunity are difficult to duplicate. Therefore, in order to mimic more accurately the complex interactions that occur between the medical device, the pathogen, and the host environment, many investigators have developed animal models. The purpose of this chapter is to outline *in vivo* models for the characterization of biofilm-related foreign body infection by staphylococcal species and to explore current applications of these models.

15.3   UNIVERSAL CONSIDERATIONS IN THE USE OF *IN VIVO* MODELS TO STUDY STAPHYLOCOCCAL BIOFILM-ASSOCIATED DISEASES

In vivo models offer investigators unparalleled opportunity to understand the pathogenesis and treatment of biofilm-associated infections. Often, because of the complex interactions between the host, the microbe, and the biomaterial, there is no *in vitro* alternative in the study of a biofilm-associated infection. However, development or use of an *in vivo* model should never be done without careful consideration of a variety of issues. First, *in vitro* methods of study should be thoroughly explored and if questions can be adequately addressed through these means, *in vivo* studies should not be employed. Second, the bacterial strains utilized should be fully characterized, preferably through a molecular genetics approach (specific genes or traits should be defined). If an *in vivo* model is necessary, it is incumbent upon the investigator to consider the following issues: the potential value of the study (will utilization of animals be associated with a reasonable expectation that such utilization will contribute to the enhancement of human or animal health), the selection of animal species (the lowest form of life on the phylogenetic scale that will yield adequate results should be chosen), the minimization of animal usage (sound scientific and statistical principles should be used to determine sample size), alternatives to painful procedures and minimization of potential pain, and criteria and method for euthanasia. Procedures should be carefully crafted to ensure the reliability of the data. Items to consider include measures to avoid bacterial contaminants, endpoints for the study, adequacy of assessing disease, and dosing regimens that reflect reasonable achievable human levels of therapeutic agents.

15.4   *EX VIVO* MODELS

Several investigators have utilized innovative methods to partially duplicate certain *in vivo* conditions via models that involve an *in vivo* component. These hybrid models are known as *ex vivo* models.

15.4.1   RAT SUBCUTANEOUS EXUDATE MODEL

In order to include the effect of host inflammatory products and phagocytic cells, investigators developed a means to harvest a sterile inflammatory exudate which is then used in the preparation of a bacterial biofilm on a biomaterial sample (23). Briefly, an air pouch is produced in the subcutaneous tissues of a laboratory rodent. One day later, carboxymethyl cellulose (CMC) is injected into the pouch. CMC incites an inflammatory response and the exudate is harvested by aspiration. The biomaterial sample, bacteria, and exudate are next incubated together to produce a biomaterial-based, biofilm-associated *in vitro* system. This model has been primarily utilized to study interactions between bacterial biofilms and antibiotics (24). The limitations of this model are that it is fairly static and that ongoing host effects are not measured.

15.4.2   EXPLANTED DEVICES

Staphylococcal adherence and biofilm formation have been studied using devices that have been coated with a host-derived conditioning film. These studies typically are directed at understanding the role of blood proteins in staphylococcal adherence and utilize vascular catheters ex-planted from patients or normal volunteers (25,26). Vaudaux and colleagues have also used a canine *ex vivo* arteriovenous shunt model to study the significance of *S. aureus* adhesins (27).

15.5   *IN VIVO* MODELS

In vivo models used to study staphylococcal biofilm-associated diseases are summarized in Table 15.1.

15.5.1   SUBCUTANEOUS FOREIGN BODY INFECTION MODEL

One of the first models used to study coagulase-negative staphylococcal foreign body infection was put forward by Christensen et al. (28). In this model, the skin of anesthetized Swiss Albino mice was sterilized with povidone-iodine and incisions were made in subcutaneous space of the flanks. Catheter segments of 14-gauge were implanted into the subcutaneous spaces which were secured with a silk suture. A sham procedure was used as a control. At these sites, incisions were made as described, but no catheter was inserted. The incisions were allowed to heal for three days and were then injected with 10^9 cfu as 100 µl of a 10^{10} cfu/ml solution of *Staphylococcus epidermidis* at the site of catheter insertion or sham procedure. After 10 days, the animals were sacrificed, the catheters were cultured, and the tissue surrounding the catheter was swabbed and cultured. Isolates from these cultures were shown to be identical to the parental strain by antibiotic susceptibility testing. They were also shown to possess the same biofilm-forming capacity as the parental strain by a tube test. When a biofilm-positive and -negative isolate were compared in this model, the biofilm-producing strain infected 53% of catheters compared with only 17% of catheters by the nonproducer.

Variations on this mouse model have been used by several investigators (29–31). Differences in these investigations are primarily in the method and timing of

TABLE 15.1
In Vivo **Models Utilized to Study Staphylococcal Biofilm-Associated Infections**

Model	Animal Species	Description	Comment
Foreign body Infection	Mouse, rat, rabbit	Vascular catheter segment or other biomaterial sample is implanted in subcutaneous space	Technically easy; does not fully mimic vascular space dynamics or immunity; often requires large inoculum
Tissue cage	Rabbit, guinea pig	Cylindrical or spherical object with access ports implanted in subcutaneous space	Technically easy; allows for serial sampling of fluid that collects inside of tissue cage
Intravascular catheter	Rat, rabbit, dog	Intravascular catheter inserted in central venous system; catheter may be subcutaneously tunneled	Technically difficult; long-term catheter patency difficult to maintain
Endocarditis	Rabbit, rat	Mechanical trauma induces valvular vegetation, infection established by intravascular inoculation	Technically difficult; models infection of native tissue
Osteomyelitis	Rat, rabbit, dog, guinea pig	Orthopedic implant biomaterial placed in medullary canal and infected by direct inoculation	Technically moderately difficult; models chronic condition and animals maintained for 30–90 days
Septic arthritis	Mouse	Intravascular inoculation results in septic arthritis in inbred mouse strain	Technically easy; mimics hematogenous route of infection

inoculation and catheter implantation. For example, Gallimore and colleagues grew organisms under investigation at 37°C for 72 hours in the presence of a perforated catheter with sealed ends (31). These catheters were then implanted in the peritoneal space of mice. Espersen et al. implanted catheters in the abdominal wall of mice (29). Inoculation in this study was performed 10 to 15 minutes after surgery by injection into the lateral abdominal wall.

We have used the mouse foreign body infection model to assess the pathogenic significance of biofilm formation (32,33). In these studies, a genetically characterized strain of *S. epidermidis*, 1457, and its isogenic *ica* transposon mutant, 1457 M-10, were tested for their ability to proliferate on catheter segments and form subcutaneous abscesses in the mice. At each inoculum level tested, the transposon mutant was significantly less able to cause abscess formation (Figure 15.1). In addition, fewer cells of 1457 M-10 were obtained when the catheters were cultured. Several steps were taken to ensure that data obtained in this study were trustworthy. For example, complete removal of adherent cells from catheters was ensured by electron microscopy. Pulsed-field gel electrophoresis was performed on recovered isolates to make certain they were identical to the inoculated strain (34).

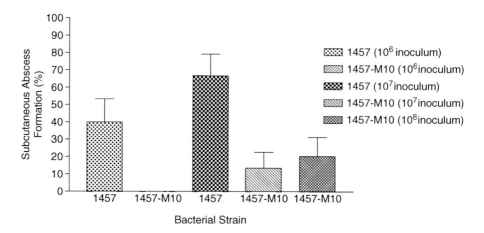

FIGURE 15.1 Subcutaneous abscess formation by *Staphylococcus epidermidis* in the mouse foreign body infection model. Wild type strain 1457 was compared to an isogenic polysaccharide intercellular adhesion (PIA) deficient mutant. Bars represent the mean abscess formation and lines represent standard error of the mean. (From Rupp, M.E., Ulphani, J.S., Fey, P.D., Bartscht, K., and Mack, D., *Infect. Immun.* 67, 2627–2632, 1999. With permission.)

Kadurugamuwa et al. used the mouse foreign body infection model for real-time monitoring of biofilm formation (35). Using a *S. aureus* strain containing the biolu-minescence *lux* operon on a plasmid and biophotonic imaging, these investigators could visualize the progression of biofilm-related infection on catheters in a nonin-vasive manner. In a separate study, this group used this method to follow the effects of antimicrobial therapy on chronic infection (36). The mouse foreign body infec-tion model has also been used recently to explore the virulence of the recently sequenced, biofilm-negative *S. epidermidis* strain, ATCC 12228 (37).

The subcutaneous foreign body model has been adapted for use in other species. Sherertz and colleagues have introduced a rabbit model for evaluation of antibiotic-coated catheters in prevention of device-related subcutaneous *S. aureus* infection (38). In these studies, the back of the rabbit was shaved and depilated chemically. The skin was disinfected by application of povidone iodine. Two incisions, separated by 0.5 to 1 cm, were made for each catheter to be implanted. A bone marrow needle was used to form 3 cm tunnels lateral to each incision. A 6 cm catheter segment that had been sealed at the ends was threaded into these tunnels so that the middle of the catheter was exposed, resting on the back of the animal. Inoculation of *S. aureus* was performed by injection in the subcutaneous space close to the implanted catheter. This model was used to show that catheters coated with dicloxacillin, clindamycin, or fusidic acid were able to significantly reduce numbers of *S. aureus* recovered from explanted catheters when compared to uncoated catheters (38). The topical disinfec-tant chlorhexidine also decreased viable cell numbers. The rabbit model described in this paper is attractive in that up to ten catheters can be implanted per animal, max-imizing data points and allowing several conditions to be tested in the same animal.

Further, as performed by the authors of this study, tissues surrounding catheters can be easily harvested for histological examination (38).

The subcutaneous foreign body model has been further modified in rats to evaluate prophylactic treatment strategies. Van Wijngaerden and colleagues (39) performed experiments in which, prior to implantation, 1 cm catheter segments were incubated in a cell suspension of known density for 2 hours at 0°C. The number of cells adhering to each catheter at the outset of the experiment was determined by plate counts performed on three non-implanted catheters. The backs of rats were shaved and treated with 0.5% chlorhexidine in 70% ethanol. An incision was made, and a tunnel was made in the subcutaneous tissue for implantation of the catheter segment. To evaluate clearance of infection by antibiotic treatment, catheters were explanted after 2 and 3 days and quantitatively cultured. This model has several attractive features: As pointed out by the authors, most foreign body infections occur perioperatively, such as during catheter insertion. This notion is taken into account in the model, where catheter segments are co-incubated with a bacterial suspension as compared to the injection of a bacterial inoculum postoperatively as is done in some models. The dilute suspension also more closely mimics the low inoculum levels encountered in clinical practice; 10^2 to 10^4 cfu/mL of *S. epidermidis* was sufficient to cause disease. The authors further assert that, because catheter contamination may occur at the time of implantation, the biomaterial is not coated with host proteins at the time of infection. The method of inoculation performed in this model mimics the uncoated condition. Vandecasteele and colleagues have recently employed the rat model to analyze the expression of several genes associated with the synthesis of biofilm (40).

15.5.2 INTRAVASCULAR CATHETER INFECTION MODEL

The subcutaneous foreign body infection models outlined above offer simple, quick procedures for the establishment of a foreign body infection *in vivo*. Mice are also an inexpensive means to conduct these studies. However, inoculum sizes in these experiments are often much higher than what is observed in clinical practice. Further, the high-shear environment of the vascular system, influence of humoral immunity, and effects of blood proteins are not fully appreciated in these models.

For these reasons, our laboratory developed an intravascular catheter-associated infection model in the rat (41,42). This model features a lumen-within-lumen catheter, the outer cannula protecting the inner lumen from stress during manipulation of the catheter or from movement of the rat. Briefly, the neck area of anesthetized animals is treated with disinfectant, shaved, and dissected. The catheter is inserted into the right external jugular vein of the animal and is advanced into the superior vena cava. The catheter is held in place using a rodent restraint jacket. This jacket allowed for free motion of the animal following surgery and easy access to the catheter. The catheter is fitted with a heparin lock injection adapter (Figure 15.2). Twenty-four hours after surgery, the animal is inoculated with *S. epidermidis* via the catheter. Eight days after inoculation, the animals are sacrificed, and the catheters are explanted. Quantitative cultures are performed on the catheters, as well as on tissue homogenates obtained from lung, heart, liver, and kidney. Using this model, we were able to show an *ica* transposon mutant was less able to cause catheter-associated

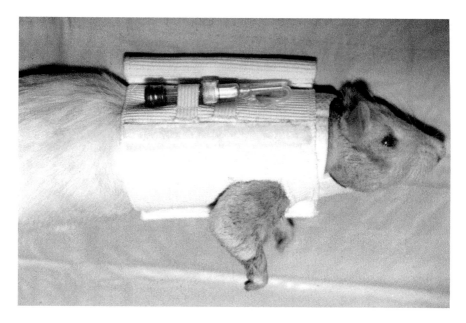

FIGURE 15.2 Rat central venous catheter (CVC) model. Silastic CVC is subcutaneously tunneled and inserted in jugular vein. The rodent restraint jacket allows for unencumbered movement of the animal, protects the catheter, and allows for ready access to the central venous system.

infection than its isogenic parental strain at a 10^4 cfu inoculum (Figure 15.3). The transposon mutant was also less likely to cause metastatic disease. In addition, using this model, we observed that *S. aureus* was able to cause catheter-associated bacteremia and metastatic disease at an inoculum of 100 cfu (41).

15.5.3 TISSUE CAGE MODEL

The guinea pig tissue cage model was first put forth by Zimmerli and colleagues in 1982 (43). In this model, cylinders of polymethacrylate and polytetrafluoroethylene are perforated with 250 holes and sealed at the ends. These tissue cages are implanted into incisions made in the subcutaneous spaces of anaesthetized tricolor guinea pigs. Two weeks following surgery, the animals are inoculated by injection directly into the tissue cage. This model is useful in that bacterial cells can be extracted from the tissue cage fluid to determine expression profiles within the host. Goerke and colleagues have used this model to characterize regulation of α-hemolysin expression *in vivo* (44). For these studies, reverse transcriptase-polymerase chain reaction (RT-PCR) was used directly on tissue cage exudates. More recently, biofilm-positive strains and their isogenic biofilm-negative mutant counterparts in tissue cage fluid were found to behave similarly in this model (45). The guinea pig tissue cage model has been adapted for use in the rat by Yasuda and colleagues (24).

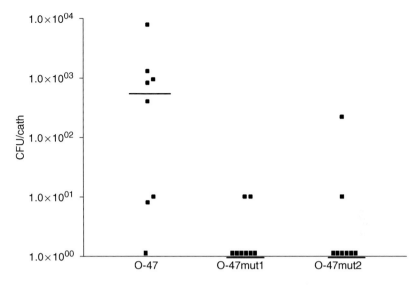

FIGURE 15.3 Recovery of *Staphylococcus epidermidis* from explanted central venous catheters (CVC) in the rat CVC-associated infection model. Wildtype *S. epidermidis* O-47 was compared to an isogenic autolysin deficient mutant (O-47 mut1) and an isogenic poly-saccharide intercellular adhesion deficient mutant (O-47 mut2). (From Rupp, M.E., Fey, P.D., Heilmann, C., and Gotz, F., *J. Infect. Dis.* 183, 1038–42, 2001. With permission.)

15.5.4 ENDOCARDITIS MODELS

Although not a biomaterial-based infection, staphylococcal endocarditis is now regarded as an infection involving a biofilm on native tissues.

15.5.4.1 Rabbit Model

In vivo models to study staphylococcal endocarditis were developed in the 1970s by a variety of investigators (46–48). Briefly, these descriptions all involve induction of mechanical trauma to the valvular structures of the heart through the introduction of a catheter via the right carotid artery into the left ventricle. A sterile vegetation is formed involving the deposition of fibrin, platelets, other serum proteins, and inflammatory cells. Infective endocarditis can then be induced by inoculation of staphylococci into the bloodstream via the catheter or other routes. *S. aureus* and coagulase-negative staphylococci have been studied in this model. The rabbit endo-carditis model has been used to examine the importance of fibronectin binding (49), platelet binding and thrombin-induced platelet microbicidal protein (50,51) and the influence of *S. aureus* regulatory domains, *agr* and *sar* (52). Pier and colleagues used the rabbit model to examine the importance of polysaccharide intercellular adhesion (PIA) (termed polysaccharide adhesion (PS/A)) in *S. epidermidis* infection (53,54). In addition, the rabbit endocarditis model has been used extensively to study therapeutic agents (55–57).

15.5.4.2   Rat Model

The rat endocarditis model was first described by Santoro and Levison (58). It is similar to the rabbit model described above. The rat endocarditis model has been used to examine various anti-staphylococcal therapeutic agents (59–61) and the importance of fibronectin binding protein (62,63), collagen binding protein (64), coagulase (65), autolysin (66), and capsule (67).

15.5.5   ORTHOPEDIC PROSTHETIC DEVICE-ASSOCIATED OSTEOMYELITIS

In vivo models of staphylococcal prosthetic device associated osteomyelitis have been developed in rabbits, rats, and dogs. All of these models involve introduction of a foreign body or substance into the medullary space of long bones, typically the tibia or femur (68–71). Foreign materials used to form the nidus of infection consist of a wide variety of materials found in orthopedic implants, including polymethyl-methacrylate bone cement, stainless steel, cobalt chromium, titanium, silicone elastomer, and polyethylene. Infection with *S. aureus* or *S. epidermidis* is induced by inoculation of 10^4 to 10^9 cfu into the medullary cavity. To mimic the chronic nature of osteomyelitis the animals are typically maintained for 30 to 90 days. These models have been used primarily to study therapeutic modalities and were reviewed by Cremiex and Carbon (72).

15.5.6   SEPTIC ARTHRITIS

Similar to endocarditis, staphylococcal arthritis is now felt to represent a biofilm-associated infection involving native tissues. Early models of septic arthritis involved direct inoculation of staphylococci into the joint capsule. However, this does not mimic the usual hematogenous route of inoculation observed in clinical practice. Bremell and colleagues described a mouse model of hematogenous septic arthritis (73). This model mimics the human condition with regard to histopathologic findings and has been used to assess the importance of a variety of microbial factors including collagen adhesion (74), bone sialoprotein binding (75), *S. aureus* regulatory domains (76) and polysaccharide capsule (77).

5.6   CONCLUSIONS

In conclusion, biofilms are increasingly recognized as being crucial in the patho-genesis of a variety of human diseases. Staphylococci, initially coagulase-negative staphylococci and, more recently *S. aureus*, are well-documented examples of bacteria causing biofilm-associated infections. Because biofilm-associated infections often involve complex interactions between the host, the microbe, and a biomaterial-based prosthetic device, *in vitro* methods may not suffice in the study of these infections. Therefore, a variety of *in vivo* models have been developed. Many factors must be considered in the development of an animal model suitable for the study of biofilm-associated infection. The organism's ability to adhere to the biomaterial to be utilized and the pathogen's capacity for biofilm synthesis should be worked out before beginning any *in vivo* studies. Careful consideration of these and other experimental conditions at the outset will ensure a minimum number of animals will be required for experimentation. The capacity to establish infection reliably and generate reproducible data has been seen by several investigators as drawbacks of currently available animal models. The development of new models will hopefully result in increased reliability while granting insights on means to better treat biofilm-associated infections.

ACKNOWLEDGMENTS

This work was partially supported by the Blanche Widaman predoctoral fellowship from the University of Nebraska Medical Center (LDH).

REFERENCES

1. Costerton, J.W., Stewart, P.S., and Greenberg, E.P., Bacterial biofilms: a common cause of persistent infections. *Science* 284, 1318–1322, 1999.
2. Parsek, M.R., and Singh, P.K., Bacterial biofilms: an emerging link to bacterial pathogenesis. *Annu. Rev. Microbiol.* 57, 677–701, 2003.
3. Post, J.C., Direct evidence of bacterial biofilms in otitis media. *Laryngoscope* 111, 2083–2094, 2001.
4. Nickel, J.C., Olson, M.E., Barabas, A., Benediktsson, H., Dasgupta, M.K., and Costerton, J.W., Pathogenesis of chronic bacterial prostatitis in an animal model. *Br. J. Urol.* 66, 47–54, 1990.
5. Gristina, A.G., Oga, M., Webb, L.X., and Hobgood, C.D., Adherent bacterial colonization in the pathogenesis of osteomyelitis. *Science* 228, 990–993, 1985.
6. Head, N.E., and Yu, H., Cross-sectional analysis of clinical and environmental isolated of *Pseudomonas aeruginosa*: biofilm formation, virulence, and genome diversity. *Infect. Immun.* 72, 133–144, 2004.
7. Murga, R., Forster, T.S., Brown, E., Pruckler, J.M., Fields, B.S., and Donlan, R.M., Role of biofilms in the survival of *Legionella pneumophila* in a model potable-water system. *Microbiology* 147, 3121–3126, 2001.
8. Bernimoulin, J.P., Recent concepts in plaque formation. *J. Clin. Periodontol.* 30(Suppl. 5), 7–9, 2003.
9. Costerton, J.W., Stewart, P.S., and Greenberg, E.P., Bacterial biofilms: a common cause of persistent infections. *Science* 284, 1318–1322, 1999.
10. Ing, M.B., Baddour, L.M., and Bayer, A.S., Bacteremia and infective endocarditis: pathogenesis, diagnosis, and complications. In: Crossley, K.B., and Archer G.L., eds., *The Staphylococci in Human Disease*. Churchill Livingstone, New York, 1997:331–354.
11. Bernard, L., Hoffmeyer, P., Assai, M., Vaudaux, P., Schrenzel, J., and Lew, D., Trends in the treatment of orthopedic prosthetic infections. *J. Antimicrob. Chemoth.* 53, 127–129, 2004.
12. Trampuz, A., Osmon, D.R., Hanssen, A.D., Steckelberg, J.M., and Patel, R., Molecular and antibiofilm approaches to prosthetic joint infection. *Clin. Orthop.* 414, 69–88, 2003.
13. Rupp, M.E., Infection of intravascular catheters and vascular devices., In: Crossley, K.B., and Archer, G.L., eds., *The Staphylococci in Human Disease*. Churchill Livingstone, New York, 1997:379–399.

14. Yogev, R., and Bisno, A.L., Infections of central nervous system shunts. In: Waldvogel, F.A., and Bisno, A.L., eds., *Infections Associated with Indwelling Medical Devices*. ASM Press, Washington D.C., 2000:231–246.

15. Saint, S., and Chenoweth, C.E., Biofilms and catheter-associated urinary tract infections. *Infect. Dis. Clin. North Am.* 17, 411–432, 2003.

16. Tsukamoto, T., Matsukawa, M., Sano, M., Takahashi, S., Hotta, H., Itoh, N., Hirose, T., and Kumamoto, Y., Biofilm in complicated urinary tract infection. *Int. J. Antimicrob. Agents* 11, 233–236, 1999.

17. Dunne, W.M. Jr., Bacterial adhesion: seen any good biofilms lately? *Clin. Microbiol. Rev.* 15, 155–166, 2002.

18. Bayston, R., and Penny, S.R., Excessive production of mucoid substance in staphylococcus SIIA: a possible factor in colonization of Holter shunts. *Dev. Med. Child. Neurol. Suppl.* 27, 25–28, 1972.

19. Christensen, G.D., Simpson, W.A., Bisno, A.L., and Beachey, E.H., Adherence of slime producing strains of *Staphylococcus epidermidis* to smooth surfaces. *Infect. Immun.* 37, 318–326, 1982.

20. Peters, G., Locci, R., and Pulverer, G., Microbial colonization of prosthetic devices: scanning electron microscopy of naturally infected intravenous catheters. *Zentrabl. Bakteriol. Mikrobiol. Hyg. B* 173, 293–299, 1981.

21. Marrie, T.J., Nelligan, J., and Costerton, J.W., A scanning and transmission electron microscopic study of an infected endocardial pacemaker lead. *Circulation* 66, 1339–1341, 1982.

22. Marrie, T.J., and Costerton, J.W. Scanning and transmission electron microscopy of in situ bacterial colonization of intravenous and intrarterial catheters. *J. Clin. Microbiol.* 19, 687–693, 1984.

23. Yasuda, H., Ajiki, Y., Koga, T., Kawada, H., and Yokota, T., Interaction between biofilms formed by *Pseudomonas aeruginosa* and clarithromycin. *Antimicrob. Agents Chemoth.* 37, 1749–1755, 1993.

24. Yasuda, H., Koga, T., and Fukuoka, T., *In vitro* and *in vivo* models of bacterial biofilms. *Methods Enzymol.* 310, 577–595, 1999.

25. Vaudaux, P., Pittet, D., Haeberli, A., Huggler, E., Nydegger, U.E., Lew, D.P., and Waldvogel, F.A., Host factors selectively increase staphylococcal adherence on inserted catheters : a role for fibronectin and fibrinogen or fibrin. *J. Infect. Dis.* 160, 865–875, 1989.

26. Muller, E., Takeda, S., Goldmann, D.A., and Pier, G.B., Blood proteins do not promote adherence of coagulase-negative staphylococci to biomaterials. *Infect. Immun.* 59, 3323–3326, 1991.

27. Vaudaux, P.E., Francios, P., Proctor, R.A., McDevitt, D., Foster, T.J., Albrecht, R.M., Lew, D.P., Wabers, H., and Cooper, S., Use of adhesion-defective mutants of *Staphylococcus aureus* to define the role of specific plasma proteins in promoting bacterial adhesion in canine arteriovenous shunts. *Infect. Immun.* 63, 585–590, 1995.

28. Christensen, G.D., Simpson, W.A., Bisno, A.L., and Beachey, E.H., Experimental foreign body infections in mice challenged with slime-producing *Staphylococcus epidermidis*. *Infect. Immun.* 40, 407–410, 1983.

29. Espersen, F., Frimodt-Moller, N., Corneliussen, L., Thamdrup Rosdahl, V., and Skinhoj, P., Experimental foreign body infection in mice. *J. Antimicrob. Chemother.* 31 Suppl. D, 103–111, 1993.

30. Patrick, C.C., Plaunt, M.R., Hetherington, S.V., and May, S.M., Role of the *Staphylococcus epidermidis* slime layer in experimental tunnel tract infections. *Infect. Immun.* 60, 1363–1367, 1992.

31. Gallimore, B., Gagnon, R.F., Subang, R., and Richards, G.K., Natural history of chronic *Staphylococcus epidermidis* foreign body infection in a mouse model. *J. Infect. Dis.* 164, 1220–1223, 1991.

32. Rupp, M.E., and Fey, P.D., *In vivo* models to evaluate adhesion and biofilm formation by *Staphylococcus epidermidis*. *Methods Enzymol.* 336, 206–215, 2001.

33. Rupp, M.E., Ulphani, J.S., Fey, P.D., Bartscht, K., and Mack, D., Characterization of the importance of polysaccharide intercellular adhesin/hemagglutinin of *Staphylococcus epidermidis* in the pathogenesis of biomaterial-based infection in a mouse foreign body infection model. *Infect. Immun.* 67, 2627–3234, 1999.

34. van Belkum, A., van Leeuwen, W., Kaufmann, M.E., et al., Assessment of resolution and intercenter reproducibility of results of genotyping *Staphylococcus aureus* by pulsed-field gel electrophoresis of *Sma*I macrorestriction fragments: a multicenter study. *J. Clin. Microbiol.* 36, 1653–1659, 1998.

35. Kadurugamuwa, J.L., Sin, L., Albert, E., et al., Direct continuous method for monitoring biofilm infection in a mouse model. *Infect. Immun.* 71, 882–890, 2003.

36. Kadurugamuwa, J.L., Sin, L.V., Yu, J., et al., Rapid direct method for monitoring antibiotics in a mouse model of bacterial biofilm infection. *Antimicrob. Agents Chemother.* 47, 3130–3137, 2003.

37. Zhang, Y.Q., Ren, S.X., Li, H.L., et al., Genome-based analysis of virulence genes in a non-biofilm-forming *Staphylococcus epidermidis* strain (ATCC 12228). *Mol. Microbiol.* 49, 1577–1593, 2003.

38. Sherertz, R.J., Carruth, W.A., Hampton, A.A., Byron, M.P., and Solomon, D.D., Efficacy of antibiotic-coated catheters in preventing subcutaneous *Staphylococcus aureus* infection in rabbits. *J. Infect. Dis.* 167, 98–106, 1993.

39. Van Wijngaerden, E., Peetermans, W.E., Vandersmissen, J., Van Lierde, S., Bobbaers, H., and Van Eldere, J., Foreign body infection: a new rat model for prophylaxis and treatment. *J. Antimicrob. Chemother.* 44, 669–674, 1999.

40. Vandecasteele, S.J., Peetermans, W.E., Merckx, R., and Van Eldere, J., Expression of biofilm-associated genes in *Staphylococcus epidermidis* during *in vitro* and *in vivo* foreign body infections. *J. Infect. Dis.* 188, 730–737, 2003.

41. Ulphani, J.S., and Rupp, M.E., Model of *Staphylococcus aureus* central venous catheter-associated infection in rats. *Lab. Anim. Sci.* 49, 283–287, 1999.

42. Rupp, M.E., Ulphani, J.S., Fey, P.D., and Mack, D., Characterization of *Staphylococcus epidermidis* polysaccharide intercellular adhesin/hemagglutinin in the pathogenesis of intravascular catheter-associated infection in a rat model. *Infect. Immun.* 67, 2656–2655, 1999.

43. Zimmerli, W., Waldvogel, F.A., Vaudaux, P., and Nydegger, U.E., Pathogenesis of foreign body infection: description and characteristics of an animal model. *J. Infect. Dis.* 146, 487–497, 1982.

44. Goerke, C., Fluckiger, U., Steinhuber, A., Zimmerli, W., and Wolz, C., Impact of the regulatory loci agr, sarA and sae of *Staphylococcus aureus* on the induction of alpha-toxin during device-related infection resolved by direct quantitative transcript analysis. *Mol. Microbiol.* 40, 1439–1447, 2001.

45. Francois, P., Tu Quoc, P.H., Bisognano, C., et al., Lack of biofilm contribution to bacterial colonization in an experimental model of foreign body infection by *Staphylococcus aureus* and *Staphylococcus epidermidis*. *FEMS Immunol. Med. Microbiol.* 35, 135–140, 2003.

46. Garrison, P.K., and Freedman, L.R., Experimental endocarditis. Staphylococcal endocarditis in rabbits resulting form placement of a polyethylene catheter in the right side of the heart. *Yale J. Biol. Med.* 42, 394–410, 1970.

47. Durack, A.T., and Beeson, P.B., Experimental bacterial endocarditis., Survival of a bacteria in endocardial vegetations. *Br. J. Exp. Pathol.* 53, 50–53, 1972.

48. Sande, M.A., and Johnson, M.L., Antimicrobial therapy of experimental endocarditis caused by *Staphylococcus aureus. J. Infect. Dis.* 131, 367–375, 1975.

49. Scheld, W.M., Strunk, R.W., Balian, G., and Calderone, R.A., Microbial adhesion to fibronectin *in vitro* correlates with production of endocarditis in rabbits. *Proc. Soc. Exp. Biol. Med.* 180, 474–482, 1985.

50. Dhawan, V.K., Yeaman, M.R., Cheung, A.L., Kim, E., Sullam, P.M., and Bayer, A.S., Phenotypic resistance to thrombin-induced platelet microbicidal protein *in vitro* is correlated with enhanced virulence in experimental endocarditis due to *Staphylococcus aureus. Infect. Immun.* 65, 3293–3299, 1997.

51. Sullam, P.M., Bayer, A.S., Foss, W.M., and Cheung, A.L., Diminished platelet binding *in vitro* by *Staphylococcus aureus* is associated with reduced virulence in a rabbit model of infective endocarditis. *Infect. Immun.* 64, 4915–4921, 1996.

52. Cheung, A.L., Eberhardt, K.J., Chung, E., Yeaman, M.R., Sullam, P.M., Ramos, M., and Bayer, A.S., Diminished virulence of a sar-/agr- mutant of *Staphylococcus aureus* in the rabbit model of endocarditis. *J. Clin. Invest.* 94, 1815–1822, 1994.

53. Takeda, S., Pier, G.B., Kojima, Y., et al., Protection against endocarditis due to *Staphylococcus epidermidis* by immunization with capsular polysaccharide/adhesin. *Circulation* 84, 2539–2546, 1991.

54. Shiro, H., Meluleni, G., Groll, A., Muller, E., Tosteson, T.D., Goldmann, D.A., and Pier, G.B., The pathogenic role of *Staphylococcus epidermidis* capsular polysaccharide/adhesion in a low-inoculum rabbit model of prosthetic valve endocarditis. *Circulation* 92, 2715–2722, 1995.

55. Lowy, F.D., Wexler, M.A., and Steigbigel, N.H., Therapy of methicillin-resistant *Staphylococcus epidermidis* experimental endocarditis. *J. Lab. Clin. Med.* 100, 94–104, 1982.

56. Sande, M.A., Evaluation of antimicrobial agents in the rabbit model of endocarditis. *Rev. Infect. Dis.* 3 Suppl, S240–S249, 1981.

57. Wheat, L.J., Smith, J.W., Reynolds, J., Bemis, A.T., Treger, T., and Norton, J.A., Comparison of cefazolin, cefmandole, vancomycin, and LY146032 for prophylaxis of experimental *Staphylococcus epidermidis* endocarditis. *Antimicrob. Agents Chemoth.* 32, 63–67, 1988.

58. Santoro, J., and Levinson, M.E., Rat model of experimental endocarditis. *Infect. Immun.* 19, 915–918, 1978.

59. Andes, D.R., and Craig, W.A., Pharmacodynamics of fluoroquinolones in experimental models of endocarditis. *Clin. Infect. Dis.* 27, 47–50, 1998.

60. Boucher, H.W., Thauvin-Eliopoulos, C., Loebenberg, D., and Eliopoulos, G.M., *In vivo* activity of evernimicin against methicillin-resistant *Staphylococcus aureus* in experimental infective endocarditis. *Antimicrob. Agents Chemoth.* 45, 208–211, 2001.

61. Baumgartner, J.D., and Glauser, M.P., Comparative imipenem treatment of *Staphylococcus aureus* endocarditis in the rat. *J. Antimicrob. Chemoth.* 12, Suppl. D, 79–87, 1983.

62. Rennermalm, A., Li, Y.H., Bohaufs, L., Jarstrand, C., Brauner, A., Brennan, F.R., and Flock, J.I., Antibodies against a truncated *Staphylococcus aureus* fibronectin-binding protein protect against disseminated infection in the rat. *Vaccine* 19, 3376–3383, 2001.

63. Greene, C., Vaudaux, P.E., Francois, P., Proctor, R.A., McDevitt, D., and Foster, T.J., A low fibronectin-binding mutant of *Staphylococcus aureus* 879R4S has Tn918 inserted into its single fnb gene. *Microbiology* 142, 2153–2160, 1996.

64. Hienz, S.A., Schennings, T., Heimdahl, A., and Flock, J.I., Collagen binding of *Staphylococcus aureus* is a virulence factor in experimental endocarditis. *J. Infect. Dis.* 174, 83–88, 1996.
65. Moreillon, P., Entenza, J.M., Francioli, P., McDevitt, D., Foster, T.J., Francois, P., and Vaudaux, P., Role of *Staphylococcus aureus* coagulase and clumping factor in pathogenesis of experimental endocarditis. *Infect. Immun.* 63, 4738–4743, 1995.
66. Mani, N., Baddour, L.M., Offutt, D.Q., Vijaranskul, U., Nadakavukaren, M.J., and Jayaswai, R.K., Autolysis-defective mutant of *Staphylococcus aureus*: pathological considerations, genetic mapping, and electron microscopic studies. *Infect. Immun.* 62, 1406–1409, 1994.
67. Baddour, L.M., Lowrance, C., Albus, A., Lowrance, J.H., Anderson, S.K., and Lee, J.C., *Staphylococcus aureus* microcapsule expression attenuates bacterial virulence in a rat model of experimental endocarditis. *J. Infect. Dis.* 165, 749–753, 1992.
68. Fitzgerald, R.H., Experimental osteomyelitis: description of a canine model and the role of depot administration of antibiotics in the prevention and treatment of sepsis. *J. Bone Joint. Surg. Br.* 65, 1983 .
69. Gerhart, T.N., Roux, R.D., Hanff, P.A., Horowitz, G.L., Renshaw, A.A., and Hayes, W.C., Antibiotic-loaded biodegradable bone cement for prophylaxis and treatment of experimental osteomyelitis in rats. *J. Orthop. Res.* 11, 250–255, 1993.
70. Petty, W., Spanier, S., Shuster, J.J., and Silverthorne, C., The influence of skeletal implants on incidence of infection: experiments in a canine model. *J. Bone Joint Surg. Am.* 37, 1236–1244, 1985.
71. Eerenberg, J.P., Patka, P., Haarman, H.J.T.H.M., and Dwars, B.J., A new model for posttraumatic osteomyelitis in rabbits. *J. Invest. Surg.* 7, 453–465, 1994.
72. Cremieux, A.C., and Carbon, C., Experimental models of bone and prosthetic joint infections. *Clin. Infect. Dis.* 25, 1295–1302, 1997.
73. Bremell, T., Abdelnour, A., and Tarkowski, A., Histopathologic and serological progression of experimental *Staphylococcus aureus* arthritis. *Infect. Immun.* 60, 2976–2985, 1992.
74. Patti, J.M., Bremell, T., Krajewska-Piertrasik, D., Abelnour, A., Tarkowski, A., Ryden, C., and Hook, M., The *Staphylococcus aureus* collagen adhesion is a virulence determinant in experimental septic arthritis. *Infect. Immun.* 62, 152–161, 1994.
75. Bremell, T., Lange, S., Yacoub, A., Ryden, C., and Tarkowski, S., Experimental *Staphylococcus aureus* arthritis in mice. *Infect. Immun.* 59, 2615–2623, 1991.
76. Nilsson, I.M., Bremell, T., Ryden, C., Cheung, A.L., and Tarkowski, A., Role of the Staphylococcal accessory gene regulator (*sar*) in septic arthritis. *Infect. Immun.* 64, 4438–4443, 1996.
77. Nilsson, I.M., Lee, J.C., Bremell, T., Ryden, C., and Tarkowski, A., The role of staphylococcal polysaccharide microcapsule expression in septicemia and septic arthritis. *Infect. Immun.* 65, 4216–4221, 1997.

16 Host Response to Biofilms

Susan Meier-Davis

CONTENTS

16.1 INTRODUCTION

Biofilms are the physical result of an adaptive achievement by many bacterial species. Indeed, these stratified, multi-cellular bacterial structures are reminiscent of higher eucaryotic organisms comprised of analogous component systems. Their outer polysaccharide integument encloses cells in various stages of growth. Further, this integument or slime protects the cells from the onslaught of host defenses or environmental stress. The cells nutritional needs are met via primitive channels transporting nutrients in and metabolic wastes out of the biofilm structure. Cell propagation occurs through gene induction with consequent dispersion and attachment at distant locations. The biofilm's well-organized structure and function significantly contribute not only to its survival but also its pathogenesis and persistence.

Biofilms are a serious human health problem. Chronic diseases such as periodontitis, otitis media, biliary tract infection, and endocarditis most likely have a biofilm component (1). Additionally, medical implants made from a variety of bio-compatible materials serve as substrates for biofilm formation (2). The persistence of these complex bacterial structures, relative to its free-floating, planktonic congener, contributes to the biofilm's ability to withstand both the host immune system and antibiotic chemotherapy.

How then does the host respond to biofilms? In biofilm-induced disease progression, the host reaction can be either destructive or beneficial. This antithetical response depends upon the organism, the infectious stage and the organ system infected. This chapter will explore the host response relative to biofilms in an effort to understand the relationship between disease and potential intervention.

16.2   HOST IMMUNITY

The host immune system is aggressive. Over the millennia, higher organisms have developed specific defensive mechanisms to combat single-celled and simple organisms. How ironic that these single-celled organisms have formed multi-cellular, organized structures as mechanisms to evade the host's defenses. To comprehend how biofilms evade the immune system, the innate and acquired immune reactions against bacterial infections are key knowledge.

16.3   INNATE IMMUNITY

When a microbe enters the systemic circulation, host recognition of this invading pathogen must occur before the defensive response can ensue. Bacteria, segregated into the two general classifications of Gram-negative and -positive, have on their surface distinct recognition sequences, termed pathogen-associated microbial patterns (3). The microbial patterns are predominantly lipoteichoic acid in Gram-positive bacteria and lipopolysaccharide in Gram-negative organisms. Once recognized by the host, the response cascade commences. First, complement with pentraxins and collectins, found in the plasma, opsonize or coat the microbe. The opsonized microbe is then bound by opsonic receptors on host phagocytes. The predominant cells responsible for this first-line phagocytic defense are the polymorphonuclear neutrophil (PMN) and macrophage. Circulating at a concentration of roughly ~1 to 2000 cells/µl in the plasma, the PMN or macrophage binds the complement covering the microbe. Additional phagocytic cell recognition and binding is mediated directly through membrane-bound C-type lectins, leucine-rich proteins, scavenger receptors and integrins (4–6).

Once PMNs or macrophages recognize and bind to the offending organism, the phagocytic process begins. These phagocytes, measuring approximately 15 to 20 µm in diameter, are capable of engulfing particles >0.5 µm in diameter. Once the receptor is bound to its microbial ligand, the receptors cluster and internalization is initiated (4). The phagosome plasma membrane forms a pocket around the particle referred to as the phagocytic cup. Through a signal transduction cascade involving a complex interaction of tyrosine kinases, phosphoinositidyl-inositol 3-kinase (PI3-kinase), Rac1 and ARF6, among others, the actin encircling the phagocytic cup is assembled and polymerized, resulting in the cup's closure (4,5,7–9). The phagosome, containing the internalized bacterial particle, goes through a maturation process that involves the following steps: (a) depolymerization of F-actin from the phagosome and fusion with early endosomes, requiring Rab5; (b) phagosome sorting wherein the Rab5 marker is lost and Rab7 becomes the predominant marker of the late endosome; (c) acidification of the late endosome, induced by enzymes such as cathepsin D and proton pump ATPase; and (d) completion of lysosomal fusion finalizes phagosome maturation (5,10). The lysosome is a low pH organelle found within host cells with fully functional degradative enzymes. Once the organism is subjected to this compartment, the majority are degraded and destroyed. Thus, the innate cellular host defense against planktonic bacteria is complete.

There are, however, other important components to host microbial killing within the phagocyte, including limitation of nutrients (e.g., iron) and the respiratory burst. Iron is a critical nutrient for bacterial virulence. To restrict this crucial element from pathogens, the host iron receptor, transferrin, is progressively downregulated as the phagosome matures and binds to the lysosome (11). Active removal of divalent cations, including iron, is thought to occur due to the natural resistance-associated macrophage protein 1 (Nramp-1) (12–14). First described relative to host susceptibility to tuberculosis, Nramp-1 homologs have been recognized in response to other bacterial species (15).

The respiratory burst is another microbial killing mechanism employed by host phagocytes. Once inside the cell, the membrane-associated NADPH oxidase generates superoxide anion that produces effective toxicants including hydroxyl radicals and hydrogen peroxide. Within the granules of the phagocyte is myeloperoxidase, an enzyme that catalyzes the conversion of hydrogen peroxide to hypochlorous acid (16,17). Although hydrogen peroxide is an effective antibacterial agent, hypochlorous acid is much more potent.

The difference in bacterial formation has a profound effect on the neutrophil function. Biofilms decrease complement activation relative to the planktonic forms (18). In *Pseudomonas aeruginosa* grown in biofilms, a reduced chemiluminescence response was elicited from neutrophils, a sign of decreased respiratory burst activity (19). The superoxide production and myeloperoxidase activity can be sufficiently stimulated by biofilm bacteria, but both are markedly decreased compared to that of the planktonic form. Mechanical disruption, however, of nonopsonized biofilm structures, increases the extent of chemiluminescence and oxidative burst activity (20,21). These findings suggest that the major goal of the organized biofilm structure is to encase the bacteria against exposure to host defenses. In other words, the bacteria themselves, may not be intrinsically different, but they are protected from host defenses within their slime integument. Further evidence supporting this theory is that enzymatic disruption of the alginate capsule markedly increases mucoid *P. aeruginosa* phagocytosis by macrophages (22,23).

Adherence is a critical factor for biofilm formation. When adherence is established, the biofilm microorganisms secrete their protective polysaccharide slime layer and the biofilm is established. Many bactericidal proteins are found in host secretions. One of these, lactoferrin, is an iron-binding protein that has various protective functions. Lactoferrin chelates iron, a critical element for bacterial virulence, prevents adherence to the host tissues and disrupts bacterial membranes by binding to the lipopolysaccharide (24,25). Although a range of lactoferrin concentrations does not affect the growth rate of planktonic *P. aeruginosa*, adherence is inhibited by concentrations less than the normal levels found in host secretions. Similar results were obtained with other iron chelators, conalbumin and deferoxamine. Adherence is the significant step for lactoferrin antagonism because once micro-colonies form, lactoferrin is ineffective (24). Recently, the FDA announced the approval of marketed lactoferrin as an antibacterial preventive for beef to diminish the gastrointestinal disease potential caused by *Escherichia coli*. Lactoferrin, as announced by the FDA, is a generally regarded as safe (GRAS) substance (26). By preventing the biofilm formation, potential human disease may be prevented similarly.

16.4 ACQUIRED IMMUNITY

The ultimate goal of acquired immunity is the recognition of foreign antigens. By no means is acquired immunity a separate entity from innate immunity. In fact, acquired immunity development, the processes and type involved are inextricably linked to innate immunity. The acquired immune response is divided into two categories: the T-cell, or cell-mediated response, and the B-cell, or humoral response.

The T-cell response has two main functions, that of (a) induction or antigen recognition and T-cell activation and (b) effectors, which includes cytokine production and cell killing. Once pathogenic bacteria enter the host, foreign peptides processed by phagocytic cells (i.e., macrophages or dendritic cells) stimulate increased expression of the major histocompatability complex (MHC) class II molecules on the phagocytic cell surface or antigen presenting cell (APC). Macrophages, which normally express low levels of MHC Class II on their cell surfaces predominantly depend upon gamma interferon (IFN-γ) to become fully functional as antigen-presenting cells (27). The foreign antigen in combination with the MHC Class II molecule is transferred to the cell surface (28). Once on the cell surface, the antigen/MHC II complex is recognized by CD4 T lymphocytes. The CD4 lymphocytes then differentiate into T helper cells. T helper cells are divided into two effector cell categories: Th1 and Th2; each is differentiated based upon the local cytokine levels, the level of antigen on the APC surface and the MHC density. Th1 effector cells depend upon local concentrations of IL-12 and IL-18, which are secreted by macrophages during antigen processing. Bacterial cell wall components, including lipoteichoic acid and lipopolysaccharide, are potent stimulators of IL-12 (and some IL-18) from the macrophages (29). Therefore, bacterial invasion, preferentially stimulates the Th1 subclass differentiation. IL-4 stimulates Th2 differentiation, although the initiation process is less understood. Both Th1 and Th2 are required for plasma cell differentiation and antibody production.

Th1 through IFN-γ production directs two processes. IFN-γ further activates macrophage's phagocytic activity by producing nitric oxide and oxygen radicals that are important for microbiocidal activity. The second process mediated by IFN-γ, demonstrated in mice, is plasma cell activation and subsequent production of the opsonizing antibodies, namely; subtypes IgG2a and IgG3. Th2 produces both IL-4 and IL-5, which are important for B cell maturation into plasma cells. Further, these cytokines control the antibody subtype switching (27).

T cells also serve another function. Acting through the MHC class 1 molecules, T lymphocytes are differentiated from CD8 cells into cytotoxic T effectors. This differentiation process, also depends upon the antigen presentation on the cell surface and IL-2 and IFN-γ produced by CD4 cells. Once cytotoxic T cells are differentiated, they have two major microbiocidal activities: (a) secretion of cytokines to activate and recruit phagocytic cells, and (b) lysis of bacterially infected cells by either the Fas-ligand or the granzyme pathway. The Fas-dependent pathway controls lymphocyte activity and the granzyme pathway involves introduction of cytotoxic enzymes to the infected cells and the initiation of the caspase cascade inducing apoptosis (27,30).

The B-cell response has two main functions, that of (a) maturation, and (b) antibody production. As indicated above, T lymphocytes direct the maturation of plasma cells into fully functional and antibody-producing B lymphocytes. In the case of biofilms, direct phagocytosis and release of cytotoxic enzymes is not effective at eliminating the pathogen. Biofilms do, however, secrete polysaccharides that are immunogenic. This immunogenicity stimulates production of IgM and IgG antibody subtypes that can further activate complement. Complement serves as the opsonizing mechanism that coats the offending organism for phagocytic cell recognition and removal. Fc receptors found on phagocytic cells are bound by IgG antibody subtypes to also assist in the opsonization process (27).

Although this multi-step, complex process is geared to recognize offending organisms and remove them, the host's immune system can, additionally, use its own cells and soluble factors to injure itself. Cytokines' recruitment of inflammatory cells with varied functionality can result in the progression of disease. With continued inflammatory processes, the granulocytes dump proteolytic enzymes, the monocytic population forms granulomas and fibroblasts migrate into the area inducing scarring (27). These host "defenses" contribute to the biofilm's persistence. Specific examples for various organ systems and biofilm pathogenecity are highlighted below.

16.5   ORGAN SYSTEM DEFENSE

Organ systems within higher organisms have developed functional defenses against planktonic bacterial invaders. Within the systemic circulation, circulating leucocytes, complement, enzymes and other proteins are vigilant sentries that safeguard well-perfused tissues against bacterial colonization. The urinary tract's physical traits, including urinary flow rate, pH and poor nutrient availability prevent many errant invaders from establishing an infection. This physical "cleansing" occuring in the urinary tract is also seen in other organ systems such as the gastrointestinal tract. Beginning with the oral cavity, the physical disruption through mastication and exposure to proteolytic enzymes is the first line of defense. The physical flow of digesta flow and, instead of ammonia and uric acid as in the urinary tract preventing colonization, hydrochloric acid, lysozyme, trypsin, chymotrypsin, pepsin and other host proteolytic enzymes prevent pathogens from adherence and colonization. Proteolytic enzymes also play a similar role in the eye. For example, the cornea is consistently washed with tears containing lysozyme and immunoglobulins. With all these protective measure in place, how then, is it possible for planktonic bacteria to adhere, colonize, and proliferate into the multi-cellular, multi-functional biofilm?

16.6   HOST CONTRIBUTION TO PATHOGENESIS

Despite the defenses that the host has developed for planktonic bacteria, biofilms flourish in the host. Ironically, host "defenses" significantly contribute to the disease progression. The following are examples of host interactions in the various tissues.

16.6.1   RESPIRATORY TRACT

Adherence to the respiratory tract mucosa is the first step for biofilm development. In the case of *P. aeruginosa*, the pili and flagella bind to terminal n-acetyl-galactosamine residues found in the gangliotetrosylceramide (asialo-GM_1) receptor on host cells. The asialo-GM_1 receptor number is reportedly increased in cystic fibrosis patients. However, more recent evidence has shown that clinical biofilm isolates may have altered binding sites to that of laboratory strains (31). Therefore, multiple or differential host receptors may occur depending upon the disease state and clinical isolate. Identification of the specific adhesin is critical to inhibition of that initial step in biofilm formation, adhesion. Simply targeting the pili, flagella or other bacterial cell wall components is not completely effective in preventing biofilm-associated disease.

Ironically, the starvation of invading microorganisms as a host defense, is an inducement for biofilm formation. Colonization initiates production of the glycocalyx, the outer polysaccharide slime layer that protects the emerging biofilm. This glycocalyx secretion serves two purposes; first as a thick protective layer against host phagocytes and second, as a strong adherent to host mucous membranes or implanted devices (23). The glycocalyx is a mucoid alginate, specific to the bacterial strains. The host immune system recognizes this alginate layer as a foreign antigen. Experimental infection with inhaled *P. aeruginosa* (biofilm-producing strain) promotes lymphocyte proliferation in the peripheral airways after one day (32). Following the lymphocyte infiltration, antibodies to alginate are detected in the serum of these animals. Regretably, airway cell infiltrates and consequent antibody formation contribute to chronic respiratory disease. As the antibody titers increase, they bind to alginate and lipopolysaccharide found in the biofilm or released planktonic cells and form a circulating immune complex (33–35). Clinical disease severity is closely linked with the concentration of circulating immune complexes, i.e., increased levels of immune complexes are associated with more severe symptoms. With time, the circulating immune complexes deposit in the lung tissues, causing deformation of the parenchymal architecture and attracting inflammatory cells. As inflammatory cell infiltrates accumulate, iterative tissue destruction occurs.

Monocytic cell infiltrates commonly induce granulomas and the airways are no exception. The local lymph nodes become hyperplastic, further deforming the airway and compromising air exchange. The influx of monocytes attracts other leucocytes such as the neutrophils, which, in response to the offending organism, degranulate releasing chemotactic factors, attracting more cells and stimulating disease progression. With recurrence of cellular infiltration, fibrosis occurs, making the process irreversible and decreasing the overall tissue viability, elasticity and function.

Alginate formation also affects the immune function in other ways. Virulence factors are decreased once the bacteria form a biofilm because adherence and colonization have are complete (35–37). Alginate-forming strains are associated not only with reduced neutrophil chemotaxis but also reduced functionality. The neutrophil respiratory burst that is associated with microbiocidal activity against planktonic cells is impaired when exposed to alginate-containing biofilms. Neutrophils exposed to alginate show decreased superoxide production (38). Additionally, the hydrophilic, anionic alginate is an effective scavenger of microbiocidal free oxygen radicals, thus rendering them ineffective (39,40). Alginate further decreases the function of neutrophils by suppressing the expression of C3 receptor and ultimately inhibiting pathogen binding. Decreased C3 receptors have an impact on the ability of the phagocytes to recognize and bind the foreign pathogen. Although opsonic antibodies can decrease bacterial numbers in vitro, antibodies produced in chronic lung infections are ineffective (41,42). Chronically infected patients, including cystic fibrosis patients, are capable of eliciting an immune reaction, however, the functional series of reactions necessary for microbe recognition, binding, and killing are impaired.

Cystic fibrosis patients suffering with chronic colonization may develop yet another syndrome. Bactericidal permeability-increasing protein (BPI) is associated with phagocytic activity of neutrophils and, indeed, increased levels of this neutrophilic protein are recognized in patients with Gram-negative pneumonia. In cystic fibrosis patients, an additional antibody is recognized circulating in these infected patients. Anti-neutrophil cytoplasmic auto-antibody (ANCA) against BPI is found in the plasma of these patients and those with chronic airway and intestinal diseases. This antibody has been demonstrated to inactivate BPI activity thus neutralizing the neutrophil function (32,43,44). The production of this antibody contributes significantly to the already debilitating disease.

16.6.2   ORAL CAVITY

The oral cavity is replete with proteolytic enzymes, pH alterations, shear forces and high flow rates. Despite this relatively hostile environment, oral bacteria have developed survival mechanisms by utilizing the host milieu to its advantage. The consequent survival product is dental plaque, a biofilm consisting of host components, polysaccharides and bacteria (45). Examples of oral bacteria capable of forming dental plaque include *Streptococcus mutans*, *Porphyromonas gingivalis*, *Actinobacillus actinomycetemcomitans* and *Bacteroides forsythus* (46). *S. mutans* synthesizes glucans from scavenged sucrose by extracellular glucosyltransferase (47). These polysaccharides form a fuzzy coat and allow the bacteria to be extremely adherent to the surface of the tooth. Further, the *S. mutans* is able to withstand acidic conditions while producing additional acidic products, thus enhancing virulence and dental caries potential. Selectin deficiency induces increased susceptibility to bacterial colonization and disease progression (48). As the buildup of bacteria occurs in the crevices between teeth and below the gum line, host neutrophils are recruited to the area in an effort to control and remove the offending organisms. Despite the active recruitment (49), neutrophils are unproductive in biofilm bacteria elimination (50). *S. mutans* grown in the presence of sucrose is better able to survive exposure to neutrophils (51). Sucrose-preincubated bacteria only induced a 1.5-fold increase in neutrophilic oxygen products in contrast to untreated bacteria (51). Sucrose preincubation with *Streptococcus* is associated with increased lysosomal release and neutrophil phagocytosis (52).

The sucrose-derived outer polysaccharide layer of the *Streptococcus* biofilm diminishes the neutrophil's microbiocidal activity. The polysaccharide/bacterial aggregate may induce a "non-phagocytosable stimulus" (53) thus inducing a "frustrated phagocytosis" (51). Polysaccharides may not be the only component of the oral cavity that induce biofilm formation. Urea, minerals, carbon dioxide along with proteins and glycoprotein, serve as energy and nutrient supplies for dental caries-causing bacteria (54,55). The impact of these additional factors on host phagocytosis requires exploration.

Other host responses to bacterial colonization include antibody production and cytokine release. The cytokines involved in the response to dental caries include transforming growth factor-β (TGF-β), Interleukin-4 (IL-4), IL-10 and IL-12 (56). These cytokines recruit still more inflammatory cells that release pro-inflammatory cytokines, proteases and prostanoids that promote periodontal disease, i.e., gingival destruction and bone resorption (56,57). One approach to treat or prevent disease progression may be the modulation of the cytokine release profile. IL-4 reduces the number of infiltrating inflammatory cells and up regulates IL-1 receptor antagonists, thus decreasing the extent of destructive inflammation (58).

Neutrophils and macrophages secrete zinc- and calcium-dependent endopeptidases or matrix metalloproteinases (MMPs). MMPs are enzymes that degrade collagen, gelatin, laminin, fibronectin and proteoglycans. These enzymes found in dental caries increase connective tissue remodeling and bone resorption; again, contributing significantly to the severity of periodontal disease (59). Interestingly, bacterial proteases do not significantly contribute to the disease progression, but rather, the bacterial biofilm induces excessive host responses damaging its own tissues.

Once the host tissue is damaged, phospholipids from the cell membrane are converted to arachadonic acid through phospholipase A_2 and metabolized via the cyclooxygenase or lipoxygenase pathway. In the case of periodontal disease, cyclooxygenase 2 (COX-2) is an inducible enzyme, upreglulated by cytokines. Activation of these pathways produces prostaglandins, prostacyclin, thromboxane, and leukotrienes. Indeed, elevated levels of these products have been recognized in gingivitis, periodontitis, and infection of implants (60).

Biofilm-induced pathogenesis and subsequent periodontal disease progression are not limited to the oral cavity but can progress to systemic sequelae. Progressive bacterial infection can cause low-level sepsis. Additionally, chronic periodontal disease associated with inflammation, may induce the release of pro-inflammatory cytokines, IL-1β and TNF-α, into the systemic circulation (61,62). With increased levels of these cytokines, alterations in lipid metabolism resulting in increased circulating low density lipoprotein (LDL) and triglyceride (TG) levels occur (63). The change in lipid metabolism is caused by (a) production of various cytokines, (b) hemodynamic changes and amino acid utilization by various tissues involved in lipid metabolism, and (c) modification of the hypothalamic/pituitary/adrenal axis escalating the hormones involved in lipid metabolism, namely adrenocorticotropic hormone, cortisol, adrenaline, noradrenaline, and glucagon (64). Therefore, when IL-1β and TNF-α are increased, potentially due to chronic periodontal disease, altered lipid metabolism may result. It doesn't end there, however. Type 1 diabetes can be induced through increased levels of IL-1β prompting protein kinase C to initiate apopototic mechanistic destruction of pancreatic beta islet cells. Besides destroying beta cells through PKC activation, IL-1β is capable of increasing nitric oxide levels and depleting cellular energy stores in these cells (65).

Similarly, TNF-α induces insulin resistance and type 2 diabetes through alteration of the tyrosine kinase activity mediated by the insulin receptor. Further mechanisms include reduction of the insulin-responsive glucose transporter synthesis and causing macrophage-dependent cytotoxicity of pancreatic islets (66).

The host response plays a role locally in exacerbating the disease in the oral cavity and contributing to the chronicity and progression. Although significant, the potential sequelae of this process are more severe. For those patients already plagued by diabetes or those predisposed to developing the illness, a local disease can severely affect a the patient's quality of life.

16.6.3   URINARY TRACT

Not unlike the oral cavity, the urinary tract has intrinsic physical attributes that should prevent clinical disease. There is a high flow rate, a paucity of nutrients and pH alterations. Despite these characteristics, biofilm-forming bacteria have evolved mechanisms that not only work within these conditions but also adapt to a changing environment to utilize the host cells and responses enhancing survival and proliferation. Isolates of *E. coli* from urinary tract infections or uropathogenic *E. coli* have filamentous structures referred to as Type 1 pili (67). The pili are helical, polymerized FimA subunits that are joined to the bacteria via three FimF, FimH, and adhesin (68). Adhesin and host mannose-containing glycoprotein receptors found on the lumenal side of the bladder epithelium form a very adherent association. Once adherent, the uropathogenic *E. coli* is able to colonize and initiate clinical disease (69). The bacteria go one step further, and invade into the host epithelium. These bacteria, mediated by FimH, utilize the host tyrosine phosphorylation pathway to invade the cell. Commandeered by the invading bacteria, the host cell actin cytoskeleton through the phosphoinositide 3-kinase and FAK pathways become enveloped within host cell organelles (70). In this relatively hostile environment, adhesion is critical for disease initiation. Additionally, the bacteria must adapt quickly to changing conditions in order to persist in this environment. FimH has been shown to rapidly accommodate to environmental hydrodynamic flow. Utilizing an *in vitro* system to alter hydrodynamic flow, FimH expression and production of variants could vary depending upon the conditions. The amino acid substitutions of these variants were similar to some aggressive uropathogenic strains (71). Another adaptation that FimH exhibits is its binding affinity to host mannose receptors (72–73). This allows the bacteria to persist in response to the host cellular adaptations that may occur during infection.

Once adherent, *E. coli* expresses proteins responsible for initial colonization. Even though the host environment is nutrient poor and oxygen-limited, the biofilm thrives. Oxygen limitation induces production of antigen 43, curli and fimbriae, proteins that mediate auto aggregation (74–76). The host may attempt to limit nutrients further through influx of inflammatory cells into the area, but biofilm bacteria respond to this stress by upregulating nutrient transport and antibiotic resistance genes, including *ylcB* (copper resistance protein), b2146 and b2147 (oxidoreductases), *malE* (maltose-binding protein) and *oppA* (oligopeptide uptake protein) (1).

As the biofilm structure accumulates, there comes a time when the structure outgrows its nutrient network. Quorum sensing detects this environment and promotes dispersal of biofilm components in response to yet another set of inducible genes. This process is a continuum driven by a complex interaction of biofilm status and host environmental conditions. One of the regulatory genes in *E. coli*, *csrA* encodes an RNA binding protein that activates glycolysis, acetate metabolism and motility (77–79). Alternatively, CsrB activates biofilm formation and is a repressor of CsrA activity. CsrA's major role in biofilm formation and consequent dispersal of the colony involves glycogen synthase activity (80). Theoretically, glycogen levels may provide the switch for *csrA* expression and biofilm dispersal.

16.6.4   EYE

The host applies innate immunity for ophthalmic protection. Besides complement fixation and recruitment of inflammatory cells, defensins secreted by macrophages are protective. The eye is one of the few immune privileged sites, due to its anatomical features that include avascularity of the vitreous and barriers to the systemic circulation (81). Pathogenic organisms capable of forming biofilms include *Staphylococcus* spp. and *Bacillus* spp. among others. Although these organisms are capable of producing toxins that cause damage, a variety of inflammatory factors are also secreted that significantly contribute to the disease potential (82). Once introduced into the eye, the organisms are subject to innate immune responses of the eye. If the organisms survive, become adherent, and form a biofilm, the immune reaction that is initially slow suddenly becomes substantial. Using staphylococci as an example, once colonized, the *agr* regulatory sequence directs the expression of toxins and capsule. The host recognizes the bacterial peptidoglycan and white blood cells migrate to the area. These macrophages and neutrophils release cytokines and peptides such as C-reactive protein, IL-1, IL-6, and TNF-α (83). This protein release attracts more inflammatory cells to the area and induces vasodilation. Both staphylococci and streptococci induce massive responses within days such that the eye can be irreversibly destroyed. In fact, staphylococcal-mediated recruitment of neutrophils can substantially fill the posterior vitreous chamber and actually impair vision (84).

16.6.5   OTHER ORGAN SYSTEMS

Listeria monocytogenes is a food-borne pathogen inducing septicemia, encephalitis, and abortions in animals and humans. *L. monocytogenes* in its planktonic form is quite adept at evading phagocytic killing (85). Through cell surface proteins, the *Listeria* attach to host cells and gain entry. Once inside the host cell the bacteria are enclosed within a membrane-bound vesicle acidified by the host as a defensive mechanism. In response to acidification the pathogen expresses listeriolysin O to survive the acidic conditions (86). Biofilm formation by these organisms only enhances its already well-developed persistence within the host. As with other organisms, the signal for biofilm formation is nutrient deprivation. *L. monocytogenes* expresses two genes, *hpt* and *relA*, in response to nitrogen starvation. Mutants of these two genes do not adhere or proliferate within animal models of infection (87). Although the exact mechanism is unknown, the gene products of *hpt* and *relA* appear to be crucial for evasion of the host response.

Another adaptation is seen in the bile duct of fish infected with *Vibrio anguillarum*. The ompU gene is regulated ~5-fold with exposure to bile salts (88). OmpU is a porin that allows nutrients to cross the cell membrane. This inducible gene allows survival and adherence, the next step in biofilm formation within the bile duct. Although first thought to play a role in adhesion, studies have shown that adherence is mediated by the products of another set of inducible genes.

One etiologic agent for persistent otitis media and chronic obstructive pulmonary disease is *Haemophilus influenzae*. The organisms express pili to facilitate biofilm formation. Differential production of outer membrane lipooligosaccharides in biofilm structures as compared to its planktonic counterpart is also observed (89). These adaptations allow the bacteria to survive and evade the host immune system.

16.7   FOREIGN IMPLANTS AND DEVICES

Invasive therapy has become commonplace as a method for treatment of various clinical conditions. Materials that comprise implants or indwelling devices do not by themselves induce an immune response; however, these foreign materials can serve as a scaffold for biofilm formation. As seen with implanted silastic (a commonly used medical plastic) materials, a conditioning film comprised of erythrocytes and neutrophils is temporally formed, followed by deposition of fibrin, monocytes, and vascular connective tissue (90). Titanium is also used for orthopedic implants. Lysozyme, an anti-microbial protein found in the serum and contained within the cellular lysosomal compartment, shows a strong adherence to titanium (91). Stainless steel is similarly coated when exposed to host proteins (92). Lysozyme also readily binds to the anionic surface of hydrogels, from which soft contact lenses are fashioned, as does lactoferrin, immunogluobulins and glycoproteins (93). These endogenous proteins, found as host defense against invading pathogenic organisms, paradoxically may serve as the scaffold to which bacteria initially adhere and subsequently form the biofilm.

Polysaccharide formation by the biofilm-forming bacteria is critical for adherence and colonization. *Bacteroides fragilis* induces intra-abdominal abscesses in a rat model following intraperitoneal administration through its capsular polysaccharide complex (CPC) (94). CPC consists of two high molecular weight polysaccharides, A and B. Polysaccharide A has a tetrasaccaride repeating unit with a balanced ionic charge i.e., an amino and carboxyl group. Polysaccharide B has a hexasaccharide repeating unit which has a 2-aminoethylphosphonate subgroup containing both a free amino and negatively charged phosphate groups. Polysaccharides A and B are tightly adherent because of the ionic interactions (95). Separation of the two polysaccharides and testing for abscess formation demonstrated that polysaccharide A is the more likely of the two to induce abscess formation, that is a 30-fold or greater inducer than either the native CPC or polysaccharide B. Comparisons of other bacterial polysaccharides that have opposite charges within the molecule show a similar trend for abscess induction (94). Although the mechanism is unknown, this chemical interaction may be a potent trigger of the host response.

With biofilm formation, virulence factor production often decreases as compared to the planktonic counterpart. This is a logical progression when one considers that the bacterial cells are already adherent, protected and beginning to colonize. The purpose of many virulence factors, expressed as secreted enzymes and proteases by bacteria such as *Pseudomonas,* is to facilitate host tissue invasion and to provide an environment for proliferation. Once biofilms have formed, proliferation ensues and virulence factors are no longer required.

Bacteria within the biofilm produce specific signal molecules dependent upon the environmental milieu. Some of these environmental cues regulate metabolism and nutritional requirements, and responses to oxidative stress or exogenous factors including antibiotic treatment and host immune response. Gram-negative bacteria utilize homoserine lactone signaling (HSL) molecules to do this in a cell-density dependent manner. Quorum sensing, as the name implies, depends upon cell number for specific gene expression (96). At high cell density, the HSL molecules bind to promoters that direct virulence factor synthesis regulated by the LasR-autoinducer complex. A second signaling complex also is directed by the LasR auto-inducer complex which, in turn, directs the RHII auto-inducer complex regulating genes including the stationary phase sigma factor and biosurfactants (96).

Gram-positive bacteria utilize secreted peptides to detect the cell density. The peptides are secreted under the direction of the ATP-binding cassette (ABC) transporter. As the peptide secretion increases, a phosphorylation cascade is initiated to specifically bind DNA sequences and alter transcription of quorum-sensing genes (96,97).

The major processes involved in virulence and protection against host defenses can be down-regulated on a single-cell level because the overall process is now assumed by the larger multi-cellular structure. Examples of these down-regulated enzymes include catalase and manganese co-factored superoxide dismutase (98). Iron and oxygen play a critical role in the regulation of these virulence factors and enzymes. Iron is sequestered by the host through various transport proteins, and may be sequestered within the outer polysaccharide layer of the biofilm. Further, oxygen tension is reduced in bacterial biofilms (99,100). However, a sharp decrease in iron may not specifically decrease the catalase activity of *Pseudomonas*. Iron supplementation increases the catalase activity markedly in planktonic and biofilm cells (101). This observation was further confirmed by measuring a downstream iron-regulated enzyme in these bacteria, manganese induced-superoxide dismutase (Mn-SOD). In planktonic cells, Mn-SOD was active because iron was not in limited supply. In biofilm cells, however, where the flow of nutrients to various parts of the biofilm structure may be limited or controlled, Mn-SOD was not expressed (101). Oxygen, also in limited supply within the biofilm structure, does not affect catalase activities or Mn-SOD, as *Pseudomonas* is a facultative anaerobe utilizing nitrate as the electron acceptor or arginine for substrate-level phosphorylation (102). However, *sodA*, the gene encoding Mn-SOD is inducible via anaerobic exposure. Anaerobic arginine metabolism produces ornithine through arginine deaminase. Ornithine is a component of many microbial iron chelators and may be a mechanism of sequestering or providing this critical element to individual cells (103–105). Arginine plays a central role in *Pseudomonas, Mycobacterium, Staphylococcus,* and *Helicobacter* strains through a twin-arginine translocase (TAT) secretion pathway. Virulence factor production and biofilm formation depend upon the arginine–arginine signal sequence to direct the secretion and translocation of these proteins to their proper periplasmic, outer membrane or extracellular locale. The proteins utilizing this secretory and translocation system include those involved in phospholipases, iron uptake, anaerobic respiration and choline degradation. Mutants in this translocation system are unable to produce virulence factors and biofilms (106).

Physical barriers formed by the polysaccharide surrounding the biofilm are only a part of the biofilm's defense against the host's response. The biofilm has evolved both a physical and chemical defense to host phagocytic activity (98). Although the alginate outer layer prevents penetration of these unwanted agents, there likely are components of the biofilm that specifically assist in battling host defenses. Hydrogen peroxide produced by host phagocytes is very effective in killing planktonic bacteria, but is unable to penetrate or kill *Pseudomonas* biofilms even when continuously exposed to high concentrations (107). This effect is mediated in part by bacterial catalase in the biofilm as only wild-type biofilm-forming *katF* positive strains were fully resistant to hydrogen peroxide (107).

The highly evolved functional structure of the biofilm also applies to antibiotic resistance. Originally, it was believed that biofilms were resistant to antibiotics in part because of the drug's inability to penetrate the glycocalyx. Efflux pumps were also believed to play a role in the resistance of biofilms; however, expression of these pumps decreases within the biofilm bacteria dependent on time and location of the cells within the community, as compared to the planktonic bacterial form. Furthermore, utilizing mutants of these pumps, no differences in antibiotic suscepti- bility were noted from the wild-type strain, suggesting a minimal role for the pumps in this process (108).

16.8   STAPHYLOCOCCAL BIOFILMS

In the case of indwelling medical devices as the substrates for the biofilms, one bacterial species tends to predominate, that being *S. epidermidis* (109). Adherence of the bacteria initiates the production of a slime layer. The slime layer consists of galactose-rich capsular polysaccharide adhesin (Ps/A) and teichoic acid (110,111). Two polysaccharide fractions are isolated from the slime layer, Polysaccharide I and Polysaccharide II or polysaccharide intercellular adhesin (PIA). PIA has multiple functions; it can induce hemagglutinnation, intercellular adhesion and, perhaps, phagocytosis inhibition (109). The role of PIA in colonization is confirmed through the use of mutants defective in the *ica* (intercellular adhesion) operon. These *ica* mutants cannot form biofilms, do not produce PIA or do not induce intercellular aggregation (112,113). The use of *ica* mutants suggest that PIA may be the initial mediator for adherence.

Bacterial proteins that contribute to biofilm formation include Atl, teichoic acid, accumulation-associated protein (AAP), biofilm associated protein, MSCRAMM (microbial surface components recognizing adhesive matrix molecules), and Sdr proteins. These proteins, secreted or anchored in the peptidoglycan, are critical for adherence. Teichoic acid is a highly charged cell wall component that has a very strong adherence capability (114). Virulence factors associated with biofilm production are under the regulation of the accessory gene regulator (Agr) and staphylococcal accessory regulator (Sar). Recently, it has been determined that Bap interferes with the activity of the MSCRAMMM proteins, resulting in decreased adherence to immobilized fibrinogen/fibronectin and decreased internalization by epithelial cells (115). This interaction results in decreased adherence and colonization of *Staphylococcus* to host cells. Other evidence has demonstrated that Bap inhibits the number of phagocytic cells in a chronic infection (116). Further, Bap and the homologous proteins from other Gram-positive bacteria, namely, Esp, Pls, and Aap are necessary for biofilm formation (115). The resolution of this apparent dichotomy remains to be seen.

The polysaccharides secreted by staphylococcal biofilm strains are immunogenic. The host immune system mounts an antibody response with involvement of T-lymphocytes. Similar to other capsular polysaccharides, the staphylococcal polysaccharides induce plasma cells to elicit an IgM response that lasts for five weeks, then the antibodies become predominantly IgG (117). Similar IgG responses were elicited in rabbits immunized with the polysaccharides plus adjuvants. The IgG antibodies were capable of opsonic killing, a feature of protective antibodies against the bacterial antigen (118).

Slime-protected organisms are resistant to host cellular defenses by neutrophils. Planktonic staphylococci induce 1- to 3-fold more superoxide anion production by the neutrophils that those bacteria that are adherent (119). Interestingly, the slime layer does not itself inhibit or decrease the efficiency of opsonization by complement (120). Flow cytometry studies, utilizing isolated polymorphonuclear leucocytes subjected to both a slime-producing and nonproducing co-variant, demonstrated that the slime-producing strain was much more adherent and induced a greater oxidative burst response from the neutrophils (121). Both opsonic and nonopsonic mechanisms may be involved. Nonopsonic mechanisms such as lectinophagocytosis may play a larger role because of the increase in polysaccharides associated with the biofilm (122). Others have also shown that isolated slime added to neutrophils increased the n-formyl-methionyl-leucyl-phenylalanine-induced superoxide production as assessed by superoxide dismutase-reduction of cytochrome c (123). Neutrophils show increased degranulation of lactoferrin and myeloperoxidase (121). Eliciting an immune response to slime-covered biofilms is not the issue, but rather effective phagocytosis and removal of the pathogen.

16.9   FUTURE THERAPEUTIC DIRECTIONS

Damage to host tissues and production of prostaglandins and leukotrienes have led to the suggestion that non-steroidal anti-inflammatory drugs (NSAIDS) may be useful in controlling the host's contribution to the biofilm-induced pathogenesis. A number of NSAIDS have been tested in the ligature-induced canine periodontitis model and naturally occurring periodontal disease models with positive outcomes (124,125). Triclosan (2,4,4-trichloro-2-hydroxy-diphenyl ether) is an antibacterial and anti-inflammatory agent that purportedly inhibits both cyclooxygenase and lipooxygenase pathways (126). Combination therapies utilizing triclosan may be an effective treatment for progressive gingivitis.

Anti-inflammatory activity is not limited to exogenously administered compounds. Endogenously produced anti-inflammatory activity is seen with lipoxins, which are oxygenated arachadonic acid derivatives (127). Lipoxins are produced by neutrophils from patients that have severe periodontal disease (128). Further, in a mouse model of periodontitis, stable lipoxin analogues were effective at inhibiting neutrophil infiltration and reducing the PGF_2 levels in these animals (128). It is unknown, however, whether lipoxins play a dual role in pathogenesis and as being anti-inflammatory. If lipoxins are to be used as an endogenous anti-inflammatory, the balance between potential pathogenesis and antagonism must be an important consideration for host-modulation.

The balance of differential cytokine expression may be one method in which to prevent chronic inflammation and boost the beneficial effects of the host immune response. An example that has been described is the administration of soluble factors involved in the host response. Exogenous IL-4 administration in experimental arthritis reduces inflammation and suggests that this may be a useful treatment option for gingivitis and periodontitis (129). Advanced periodontal disease progression entails tissue destruction and, ultimately, bone resorption. As with osteoporosis, inhibition of osteoclast activity with bis-phosphonates such as alendronate may be a viable treatment option for advanced periodontitis (130).

Potential treatment for chronic periodontal disease targeting endogenous metalloproteases (MMP) inhibition including tissue inhibitors of MMP or TIMP and α_2-microglobulin may prove useful because tissue destruction plays a major role in many biofilm diseases. Tetracycline antibiotics inhibit the activity of MMPs via chelation and pro-MMP molecules antagonism (131). Suboptimal (antibiotic activity) levels of the tetracycline class administered for 9 to 18 months inhibit MMP activity but do not result in an increase in resistant bacteria or a change in the oral microflora (132,133).

The systemic effects subsequent to a local biofilm infection may have serious consequences for the host, as was described for diabetic patients (65). Host immune system modulation may be an effective means for controlling chronic biofilm-induced pathogenesis. Therefore, potential interventions include reduction of serum cytokine levels or lipid levels. Cytokine levels could be reduced through monoclonal antibodies specifically directed against these cytokines or receptor antagonists of IL-1β or TNF-α. Lipid reduction utilizing pharmaceutical intervention or dietary alterations is well-documented.

Inadequate antibiotic penetration into biofilms and differential growth rates of single or multiple bacterial species contained within the glycocalyx have made biofilms notorious for their antibiotic resistance. Recent strategies have been developed for combination therapy that enhances the host response. Combination of toxithromycin and imipenem therapies in a *Staphylococcus* biofilm model decreased the number of viable bacteria and induced a higher neutrophil invasion into the biofilm structure (134). Continued exploration of alternative therapy combinations and penetration studies may reveal unique antibacterial therapies for these resolute structures.

Once the bacteria are adherent and colonized, conventional therapies are inadequate, therefore a viable alternative is prevention. Prevention through vaccines is one method. Bacterial polysaccharides are highly immunogenic and are a potential target antigen. For example, vaccination with the major polysaccharide component of *Streptococcus* or *Staphylococcus* may be an effective preventive for adhesion and colonization with these organisms. Vaccination with adhesion factors from enteropathogenic bacteria effectively prevents urinary tract infections (67).

Medical implants are often associated with biofilm formation because the foreign material serves as a foundation for bacterial adherence and the material is immuno-logically segregated. Many pretreatments for various medical polymers have been suggested. However, to date, there is not a synthetic polymer or biocompatible metal onto which *Staphylococcus* spp. do not adhere (135). *In vivo* resistance to coagulase-negative *Staphylococcus* biofilm infection has been benefited by gelatine-impregnated grafts (136).

Quorum sensing facilitates protection of the cells contained with the biofilm. Through inhibition of this process, the biofilm survival and subsequent dispersal to distant sites might be prevented. One potential protein is RNAII inhibiting peptide (RIP) that inhibits cell–cell communication by interfering with quorum sensing. Prevention of biofilm formation by RIP may decrease the dispersion rate after chronic biofilm infection is recognized. *In vitro*, RIP inhibited adherence of *S. aureus* to host cells and catheter polymers (137,138). Exploration into interspecies commu-nication may reveal additional inhibitors of the quorum sensing and auto-induction cascade supplementing the host's response to biofilm infection.

Our current understanding of biofilm formation and persistence is incomplete. Further knowledge of these complex structures, and the regulation thereof, will contribute to disease prevention and intervention. As described above, the host response plays a major role in pathogenesis. Therefore, prospective therapies must focus not only on the microorganism and the associated structure but also on the host response.

REFERENCES

1. Schembri, M.A., Kjaergaard, K., and Klemm, P., Global gene expression in *Escherichia coli* biofilms. *Molecular Microbiology* 48, 253–267, 2003.
2. Donlan, R.M., and Costerton, J.W., Biofilms: survival mechanisms of clinically relevant organisms. *Clin. Microbiol. Rev.* 15, 2137–2142, 2002.
3. Medzhitov, R., and Janeway, C.A. Jr., Innate immunity: impact on the adaptive immune response. *Curr. Opin. Immunol.* 9, 4–9, 1997.
4. Aderem, A., and Underhill, D.M., Mechanisms of phagocytosis in macrophages. *Annu. Rev. Immunol.* 17, 593–623, 1999.
5. Gordon, S., ed., *Phagocytosis: The Host,* vol 5. JAI Press Inc., Stamford, Conn, 1999.
6. Ofek, I., Goldhar, J., Keisari, Y., and Sharon, N., Nonopsonic phagocytosis of microorganisms. *Annu. Rev. Microbiol.* 49, 239–276, 1995.
7. Gresham, B.H., Dale, B.M., Potter, J.W., Chang, P.W., Vines, C.M., Lowell, C.A., Lagenaur, C.F., and Willman, C.L., Negative regulation of phagocytosis in murine macrophages by the src kinase family member, *Fgr. J. Exp. Med.* 191, 515–528, 2000.
8. Swanson, J.A., and Baer, S.C., Phagocytosis by zippers and triggers *Trends Cell Biol.* 5, 89–93, 1995.

9. Turner, M.E., Schweighoffer, E., Colucci, F., DiSanto, J.P., and Tybulewicz, V.L., Tyrosine kinase SYK: essential functions for immunoreceptor signaling. *Immunol. Today* 21, 148–154, 2000.

10. Cossart, P., Boquet, P., Normark, S., and Rappuoli, R., ed., *Cellular Microbiology.* ASM Press, Washington D.C., 2000.

11. Sunder-Plassmann, G.S., Patruta, S.I., and Horl, W.H., Pathobiology of the role of iron in infection. *Am. J. Kidney Dis.* 34, S25–S29, 1999.

12. Blackwell, J.M., and Searle, S., Genetic regulation of macrophage activation; understanding the function of Nramp1. *Immunol. Lett.* 65, 73–80, 1999.

13. Canonne-Hergaux, F., Gruenheid, S., Govoni, G., and Gros, P., The Nramp1 protein and its role in resistance to infection and macrophage function. *Proc. Assoc. Am. Physicians* 111, 283–289, 1999.

14. Gruenheid, S., Pinner, E., Desjardins, J., and Gros, P., Natural resistance to infection with intracellular pathogens: the Nramp1 protein is recruited to the membrane of the phagosome. *J. Exp. Med.* 185, 717–730, 1997.

15. Makui, H.E., Roig, E., Cole, S.T., Jelmann, J.D., Gros, P., and Cellier, M., Identification of *Escherichia coli* K-12 Nramp (mntH) as a selective divalent metal ion transporter. *Mol. Microbiol.* 35, 1065–1078, 2000.

16. Clark, R.A., Activation of the neutrophil respiratory burst oxidase. *J. Infect. Dis.* 179, S309-S317, 1999.

17. Hampton, M.B.A., Kettle, J., and Winterbourn, C.C., Inside the neutrophil phagosome: oxidants, myeloperoxidase and bacterial killing. *Blood* 92, 3007–3017, 1998.

18. Jensen, E.T., Kharazmi, A., Garred, P., Kronborg, G., Fomsgaar, A., Mollnes, T.E., and Hoiby, N., Complement activation by *Pseudomonas aeruginosa* biofilms. *Microbiol. Pathol.* 15, 377–381, 1993.

19. Jensen, E.T., Kharazmi, A., Lam, K., Costerton, J.W., and Hoiby, N., Human polymorphonuclear leukocyte response to *Pseudomonas aeruginosa* grown in biofilms. *Infect. Immun.* 58, 2383–2388, 1990.

20. Kharazmi, A., Mechanisms involved in the evasion of the host defense by *Pseudomonas aeruginosa*. *Immunology Lett.* 30, 201–206, 1991.

21. Jensen, E.T., Kharazmi, A., Noiby, N., and Costerton, J.W., Some bacterial parameters influencing the neutrophil oxidative burst response to *Pseudomonas aeruginosa* biofilms. *APMIS* 100, 727–733, 1992.

22. Eftekhar, F., and Speert, D.P., Alginase treatment of mucoid *Pseudomonas aeruginosa* enhances phagocytosis by human monocytes-derived macrophages. *Infect. Immun.* 56, 2788–2793, 1988.

23. Hoyle, B.D., Jass, J., and Costerton, J.W., The biofilm glycocalyx as a resistance factor. *J. Antimicrob. Chemother.* 26, 1–6, 1990.

24. Singh, P.K., Parsek, M.R., Greenberg, E.P., and Welsh, M.J., A component of innate immunity prevents bacterial biofilm development. *Nature* 417, 552–555, 2002.

25. Darby, C., and Hultgren, S.J., Innate defense evicts bacterial squatters. *Nat. Immunol.* 3, 602–604, 2002.

26. U.S. Food and Drug Administration. Lactoferrin considered safe to fight *E. coli*. *FDA News* 2003; P03-62.

27. Stefan, H., Kaufmann, E., Sher, A., and Ahmed, R., *Immunology of Infectious Diseases.* ASM Press, Washington, D.C., 2002.

28. Pieters, J., Processing and presentation of phagocytosed antigens to the immune system. *Adv. Cell. Mol. Biol. Membr. Organs* 5, 379–406, 1999.

29. Morel, P.A., and Oriss, T.B., Cross regulation between Th1 and Th2 cells. *Crit. Rev. Immunol.* 18, 275–303, 1998.

30. Stenger, S., Rosat, J.P., Bloom, B.R., Krensky, A.M., and Modlin, R.L., Granulysin: a lethal weapon of cytolytic T cells. *Immunol. Today* 20, 390–394, 1999.

31. Schroeder, T.H., Zaidi, R., and Pier, G.B., Lack of adherence of clinical isolates of *Pseudomonas aeruginosa* to Asialo-GM1 on epithelial cells. *Infect. Immun.* 69, 719–729, 2001.

32. Kobayashi, H., Airway biofilm disease. *Int. J. Antimicrob. Agents* 17, 351–356, 2001.

33. Hoiby, N., Doring, G., and Shiotz, P.O., The role of immune complex in the pathogenesis of bacterial infections. *Ann. Rev. Microbiol.* 40, 29–53, 1986.

34. Kronborg, G., Shand, G.H., and Fomsgaard, A., Lipopolysaccharide is present in immune complex isolated from sputum in patients with cystic fibrosis and chronic *Pseudomonas aeruginosa* lung infection. *APMIS* 100, 175–180, 1992.

35. Luzar, M.A., and Montie, T.C., Avirulence and altered physiological properties of cystic fibrosis strains of *Pseudomonas aeruginosa*. *Infect. Immun.* 50, 572–576, 1985.

36. Woods, D.E., Sokal, D.A., and Bryan, L.F., *In vitro* regulation of virulence in *Pseudomonas aeruginosa* associated with genetic rearrangement. *J. Infect. Dis.* 163, 143–149, 1991.

37. Fagan, M., Francis, P., and Hayward, A.C., Phenotypic conversion of *Pseudomonas aeruginosa* in cystic fibrosis. *J. Clin. Microbiol.* 28, 1143–1146, 1990.

38. Ohgaki, N., Bacterial biofilm in chronic airway infection. *J. Am. Infect. Dis.* 68, 138–151, 1994.

39. Simpson, J.A., Smith, S.E., and Dean, R.T., Alginate inhibition of the uptake of *Pseudomonas aeruginosa* by macrophages. *J. Gen. Microbiol.* 134, 29–36, 1988.

40. Learn, D.B., Brestel, E.P., and Seetharama, S., Hypochlorite scavenging by *Pseudomonas aeruginosa* alginate. *Infect. Immun.* 55, 1813–1818, 1987.

41. Meluleni, G.J., Grout, M., Evans, D.J., and Pier, G.B., Mucoid *Pseudomonas aeruginosa* growing in a biofilm *in vitro* are killed by opsonic antibodies to the mucoid exopolysaccharide capsule but not by antibodies produced during chronic lung infection in cystic fibrosis patients. *J. Immunol.* 155, 2029–2038, 1995.

42. Baltimore, R.S., and Mitchell, M., Immunologic investigations of mucoid strains of *Pseudomonas aeruginosa*: comparison of susceptibility to opsonic antibody in mucoid and nonmucoid strains. *J. Infect. Diseases* 141, 238–247, 1980.

43. Zhao, M.H., Jayne, D.R.W., Ardiles, L.F., and Lockwood, C.M., Autoantibodies against bacterial permeability increasing protein in patients with cystic fibrosis. *QJ. Med.* 89, 259–265, 1996.

44. Kobayashi, O., Clinical role of autoantibody against bacterial permeability increasing protein in chronic airway infection. *J. Infect. Chemother.* 4, 83–93, 1998.

45. Liljemark, W.R., and Bloomquist, C., Human oral microbial ecology and dental caries and periodontal diseases. *Crit. Rev. Oral Biol. Med.* 7, 180–198, 1996.

46. Consensus report on periodontal diseases. Pathogenesis and microbial factors. *Ann. Periodontol.* 1, 926–932, 1996.

47. Hamada, S., and Slade, H.D., Biology, immunology and cariogenicity of *Streptococcus mutans*. *Microbiol. Rev.* 44, 331–384, 1980.

48. Niederman, R., Westernoff, T., Lee, C., Mark, L.L., Kawashima, N., Ullman-Culler, M., Dewhirst, F.E., Paster, B.J., Wagner, D.D., Mayadas, T., Hynes, R.O., and Stashenko, P., Infection-mediated early-onset periodontal disease in P/E-selectin-deficient mice. *J. Clin. Periodontol.* 28, 569–575, 2001.

49. Thompson, H.L., and Wilton, J.N.A., Effects of anaerobiosis and aerobiosis on interactions of human polymorphonuclear leucocytes with the dental plaque bacteria *Streptococcus mutans, Capnocytophagaochracea,* and *Bacteroides gingivalis. Infect and Immun.* 59, 932–940, 1991.

50. Schroeder, H.E., and Attstrom, R., Pocket formation: an hypothesis. Lehner, T., Gimasoni, G., eds, *The Borderline between Caries and Periodontal Diseases.* Academic Press, London, 1980, 99–123.

51. Steinberg, D., Poran, S., and Shapira, L., The effect of extracellular polysaccharides from *Streptococcus mutans* on the bactericidal activity of human neutrophils. *Arch. Oral Biol.* 44, 437–444, 1999.

52. Bachni, P., Listgarten, M.A., Taichman, N.S., and McArthur, W.P., Electron microscopic study of the interaction of oral microorganisms with polymorphonuclear leukocytes. *Arch. Oral Biol.* 22, 685–692, 1977.

53. Taichman, N.S., Tsai, C.C., Baehni, P.C., Stoller, N., and McArthur, W.P., Interaction of inflammatory cell-sand oral microorganisms, IV. *In vitro* release of lysosomal constituents from polymorphonuclear leukocytes exposed to supragingival and subgingival bacterial plaque. *Infect. Immun.* 16, 1013–1023, 1977.

54. Wong, L., and Sissions, C.H., A comparison of human dental plaque microcosm biofilms grown in an undefined medium and a chemically defined artificial saliva. *Arch. Oral Biol.* 46, 477–486, 2001.

55. Bradshaw, D.J., Homer, K.A., Marsh, P.D., and Beighton, D., Metabolic cooperation in oral microbial communities during growth on mucin. *Microbiol.* 140, 3407–3412, 1994.

56. Page, R.C., The pathobiology of periodontal diseases may affect systemic diseases: Inversion of a paradigm. *Ann. Periodontol.* 3, 108–120, 1998.

57. Oringer, R.J., Research, Science, and Therapy Committee of the American Academy of Periodontology. Modulation of the host response in periodontal therapy. *J. Periodontol.* 73, 460–470, 2002.

58. Wong, H.L., Cost, G.L., Lotze, M.T., and Wahl, S.M., Interleukin-4 differentially regulates monocytes IL-1 family gene expression and synthesis *in vitro* and *in vivo*. *J. Exp. Med.* 177, 775–781, 1993.

59. Golub, L., Sorsa, T., Lee, H.M., Ciancio, S., Sorbi, D., and Ramamurthy, N., Doxycycline inhibits neutrophil (PMN)-type matrix metalloproteinases in human adult periodontitis gingival. *J. Clin. Periodontol.* 21, 1–9, 1996.

60. Paquette, D.W., and Williams, R.C., Modulation of host inflammatory mediators as a treatment strategy for periodontal diseases. *Periodontol.* 24, 239–252, 2000.

61. Iacopino, A.M., Periodontitis and diabetes interrelationships: role of inflammation. *Ann. Periodontol.* 6, 125–137, 2001.

62. Prabhu, A., Michalowicz, B.S., and Matheur, A., Detection of local and systemic cytokines in adult periodontitis. *J. Periodontol.* 67, 515–522, 1996.

63. Samra, J.S., Summers, L.K.M., and Frayn, K.N., Sepsis and fat metabolism. *Br. J. Surg.* 83, 1186–1196, 1996.

64. Imura, H., Fukata, J., and Mori, T., Cytokines and endocrine function: An interaction between the immune and neuroendocrine systems. *Clin. Endocrinol.* 35, 107–115, 1991.

65. Sjoholm, A., Aspects of the involvement of interleukin-I and nitric oxide in the pathogenesis of insulin-dependent diabetes mellitus. *Cell Death Diff.* 5, 461–468, 1998.

66. Moller, D.E., Potential role of TNF-alpha in the pathogenesis of insulin resistance and type 2 diabetes. *Trends Endocinol. Metab.* 11, 212–217, 2000.

67. Langermann, S., Palaszynski, S., Barnhart, M., Auguste, G., Pinkner, J.S., Burlein, J., Barren, P., Koenig, S., Leath, S., Jones, C.H., and Hultgren, S.J., Prevention of mucosal *Escherichia coli* infection by FimH-adhesin-based systemic vaccination. *Science* 276, 607–611, 1997.

68. Jones, C.H., Pinkner, J.S., Roth, R., Heuser, J., Nicholes, A.V., Abraham, S.N., and Hultgren, S.J., FimH adhesin of type I pili is assembled into a fibrillar tip structure in the Enterobactericaceae. *Proc. Natl. Acad. Sci. USA*, 92, 1995.

69. Mulvey, M.A., Lopez-Boado, U.S., Wilson, C.L., Roth, R., Parks, W.C., Heuser, J., and Hultgren, S.J., Induction and evasion of host defenses by type-1-piliated uropathogenic *Escherichia Coli. Science* 282, 1494–1497, 1998.

70. Martinez, J.J., Mulvey, M.A., Schilling, J.D., Pinkner, J.S., and Hultgren, S.J., Type 1 pilus-mediated bacterial invasion of bladder epithelial cells. *EMBO J.* 12, 2803–2812, 2000.

71. Schembri, M.A., and Klemm, P., Biofilm formation in a hydrodynamic environment by novel FimH variants and ramifications for virulence. *Infect. Immun.* 69, 1322–1328, 2001.

72. Sokurenko, E.V., Courtney, H.S., Maslow, J., Sitonen, A., and Hasty, D.L., Quantitation differences in adhesiveness of type 1 fimbriated *Escherichia coli* due to structural differences in fimH genes. *J. Bacteriol.* 177, 3680–3686, 1995.

73. Sokurenko, E.V., Chesnokova, V., Dykhuizen, D.E., Ofek, I., Wu, X,-R., Krogfelt, K.A., Struve, C., Schembri, M.A., and Hasty, D.L., Pathogenic adaptation of *Escherichia coli* by natural variation of the FimH adhesin. *Proc. Natl. Acad. Sci. USA* 95, 8922–8926, 1998.

74. Hasman, H., Schembri, M.A., and Klemm, P., Antigen 43 and type 1 fimbriae determine colony morphology of *Escherichia coli* K-12. *J. Bacteriol.* 183, 1089–1095, 2000.

75. Prigent-Combaret, C., Vidal, O., Dorel, C., and Lejeune, P., Abiotic surface sensing and biofilm-dependent regulation of gene expression in *Escherichia coli. J. Bacteriol.* 181, 5993–6002, 1999.

76. Schembri, M.A., Christiansen, G., and Klemm, P., FimH-mediated autoaggregation of *Escherichia coli. Mol. Microbiol.* 41, 1419–1430, 2001.

77. Sabnis, N., Yang, H., and Romeo, T., Pleiotropic regulation of central carbohydrate metabolism in *Escherichia coli* via the gene csrA. *J. Biol. Chem.* 270, 29096–29104, 1995.

78. Wei, B., Shin, S., LaPorte, K., Wolfe, A.J., and Romeo, T., Global regulatory mutations in csrA and rpoS cause severe central carbon stress in *Escherichia coli* in the presence of acetate. *J. Bacteriol.* 182, 1632–1640, 2000.

79. Wei, B.L., Brun-Zinkernagel, A.M., Simecka, J.W., Pruss, B.M., Babitzke, P., and Romeo, T., Positive regulation of motility and *flhDC* expression by the RNA-binding protein CsrA of *Escherichia coli. Mol. Microbiol.* 40, 245–256, 2001.

80. Jackson, D.W., Suzuki, K., Oakford, L., Simecka, J.W., Hart, M.E., and Romeo, T., Biofilm formation and dispersal under the influence of the global regulator CsrA of *Escherichia coli. J. Bacteriol.* 184, 290–301, 2002.

81. Streilein, J.W., Immunologic privilege of the eye. *Springer Semin. Immunopathol.* 21, 95–111, 1999.

82. Johnson, H.M., Russell, J.K., and Pontzer, C.H., *Staphylococcal* enterotoxin, superantigens. *Proc. Soc. Exp. Biol. Med.* 198, 765–771, 1991.

83. Leid, J.G., Costerton, J.W., Shirtliff, M.E., Gilmore, M.S., and Engelbert, M., Immunology of staphylococcal biofilm infections in the eye: new tools to study biofilm endophthalmitis. *DNA and Cell Biology* 21, 405–413, 2002.

84. Pleyer, U., Mondino, B.J., Adamu, S.A., Pitchekian-Halabi, H., Engstrom, R.E., and Glasgow, B.J., Immune response to *Staphylococcus epidermidis*-induced endophthalmitis in a rabbit model. *Invest. Ophthalmol. Vis. Sci.* 33, 2650–2663, 1992.

85. Roberts, A.J., and Wiedmann, M., Pathogen, host and environmental factors contributing to the pathogenesis of listeriosis. *Cell Mol. Life Sci.* 60, 904–918, 2003.

86. Glomski, I.J., Gedde, M.M., Tsang, A.W., Swanson, J.A., and Portnoy, D.A., The *Listeria monocytogenes* hemolysin has an acidic pH optimum to compartmentalize activity and prevent damage to infected host cells. *J. Cell Biol.* 156, 1029–1038, 2002.

87. Taylor, C.M., Beresford, M., Epton, H.A.S., Sigee, D.C., Shama, G., Andrew, P.W., and Roberts, I.S., *Listeria monocytogenes relA* and *hpt* mutants are impaired in surface-attached growth and virulence. *J. Bacteriol.* 184, 621–628, 2002.

88. Wang S-Y., Lauritz, J., Jass, J., and Milton, D.L., Role for the major outer-membrane protein from *Vibrio anguillarum* in bile resistance and biofilm formation. *Microbiology* 149, 1061–1071, 2003.

89. Murphy, T.F., and Kirkham, C., Biofilm formation by nontypeable *Haemophilus influenzae*: strain variability, outer membrane antigen expression and role of pili. *BMC Microbiol.* 2, 1–8, 2002.

90. Buret, A., Ward, K.H., Olson, M.E., and Costerton, J.W., An *in vivo* model to study the pathobiology of infectious biofilms on biomaterial surfaces. *J. Biomed. Material Res.* 25, 865–874, 1991.

91. Mondon, M., Berger, S., and Ziegler, C., Scanning-force techniques to monitor time-dependent changes in topography and adhesion force of proteins on surfaces. *Anal. Bioanal. Chem.* 375, 849–855, 2003.

92. Parkar, S.G., Flint, S.H., Palmer, J.S., and Brooks, J.D., Factors influencing attachment of thermophilic bacilli to stainless steel. *J. Appl. Microbiol.* 90, 901–908, 2001.

93. Sack, R.A., Jones, B., Antignani, A., Libow, R., and Harvey, H., Specificity and biological activity of the protein deposited on the hydrogel surface. *Invest. Ophthalmol. Vis. Sci.* 28, 842–849, 1987.

94. Tzianabos, A.O., Onderdonk, A.B., Rosner, B., Cisneros, R.L., and Kasper, D., Structural features of polysaccharides that induce intra-abdominal abscesses. *Science* 262, 416–419, 1993.

95. Tzianabos, A.O., Pantosti, A., Brisson, J.R., Jennings, H.J., and Kasper, D.L., The capsular polysaccharide of *Bacteroides fragilis* comprises two ionically linked polysaccharides. *J. Biol. Chem.* 267, 18230–18235, 1992.

96. Miller, M.B., and Bassler, B.L., Quorum Sensing in Bacteria. *Annu. Rev. Microbiol.* 55, 165–199, 2001.

97. Ji, G., Beavis, R.C., and Novick, R.P., Bacterial interference caused by autoinducing peptide variants. *Science* 276, 2027–2030, 1997.

98. Elkins, J.G., Hassett, D.J., Stewart, P.S., Schweizer, H.P., and McDermott, T.R., Protective role of catalase in *Pseudomonas aeruginosa* biofilm resistance to hydrogen peroxide. *Appl. Environl. Microbiol.* 65, 4594–4600, 1999.

99. Costerton, J.W., Lewandowski, Z., Caldwell, D.E., Korber, D.R., and Lappin-Scott, H.M., Microbial Biofilms. *Annu. Rev. Microbiol.* 49, 711–45, 1995.

100. Xu, K., Sterwart, P.S., Xia, F., Huang, C.T., and McFeters, G.A., Physiological heterogeneity in *Pseudomonas aeruginosa* biofilm is determined by oxygen availability. *Appl. Environ. Microbiol.* 64, 4035–4039, 1998.

101. Frederick, J.R., Elkins, J.G., Bollinger, N., Hassett, D.J., and McDermott, T.R., Factors affecting catalase expression in *Pseudomonas aeruginosa* biofilms and planktonic cells. *Appl. Environ. Microbiol.* 67, 1375–1379, 2001.

102. Hassett, D.J., Anaerobic production of alginate by *Pseudomonas aeruginosa*: alginate restricts the diffusion of oxygen. *J. Bacteriol.* 178, 7322–7325, 1996.

103. Cunin, R., Glandsdorff, N., Pierard, A., and Stalon, V., Biosynthesis and metabolism of arginine in bacteria. *Microbiol. Rev.* 50, 314–352, 1986.

104. Matzanke, B.F., Structures, coordination chemistry and functions of microbial iron chelates. *CRC Handbook of Microbial Iron Chelates.* CRC Press, Inc., Boca Raton, FLA, 1991, 15–64.

105. Mercenier, A., Simon J-P., VanderWauven, C., Haas, D., and Stalon, V., Regulation of enzyme synthesis in the arginine deaminase pathway of *Pseudomonas aeruginosa*. *J. Bacteriol.* 144, 159–163, 1980.

106. Ochsner, U.A., Snyder, A., Vasil, A.I., and Vasil, M.L., Effects of the twin-arginine translocase on secretion of virulence factors, stress response and pathogenesis. *Proc. Natl. Acad. Sci. USA* 99, 8312–8317, 2002.

107. Stewart, P.S., Roe, F., Rayner, J., Elkins, J.G., Lewandowski, Z., Ochsner, U.A., and Hassett, D.J., Effect of catalase on hydrogen peroxide penetration into *Pseudomonas aeruginosa* biofilms. *Appl. Environ. Microbiol.* 66, 836–838, 2000.

108. DeKievit, T.R., Parkins, M.D., Gillis, R.J., Srikumar, R., Ceri, H., Poole, K., Iglewski, B.H., and Storey, D.G., Multidrug efflux pumps: Expression patterns and contribution to antibiotic resistance in *Pseudomonas aeruginosa* biofilms. *Antimicrob. Agents Chemother.* 45, 1761–1770, 2001.

109. Gotz, F., *Staphylococcus* and biofilms. *Mol. Microbiol.* 43, 1367–1378, 2002.

110. Hussain, M., Wilcox, M.H., and White, P.J., The slime of coagulase-negative staphylococci: biochemistry and relation to adherence. *FEMS Microbiol. Rev.* 10, 191–207, 1993.

111. Kojima, J., Ohta, T., Uchiyama, I., Baba, T., Yuzawa, H., and Kobayashi, I., Whole genome sequencing of methicillin-resistant *Staphylococcus aureus*. *Lancet* 357, 1225–1240, 2001.

112. Heilmann, C., Schweitzer, O., Gerke, C., Vanittanakom, N., Mack, D., and Gotz, F., Molecular basis of intercellular adhesion in the biofilm-forming *Staphylococcus epidermidis*. *Mol. Microbiol.* 20, 1083–1091, 1996.

113. Rupp, M.E., Ulphani, J.S., Fey, P.D., Bartscht, K., and Mack, D., Characterization of the importance of polysaccharide intercellular adhesin/hemagglutinin of *Staphylococcus epidermidis* in the pathogenesis of biomaterial-based infection in a mouse foreign body infection model. *Infect. Immun.* 67, 2627–2632, 1999.

114. Gross, M., Cramton, S.E., Gotz, F., and Peschel, A., Key role of teichoic acid net charge in *Staphylococcus aureus* colonization of artificial surfaces. *Infect. Immun.* 69, 3423–3426, 2001.

115. Cucarella, C., Tormo, M.A., Knecht, E., Amorena, B., Lasa, I., Foster, T.J., and Penades, J.R., Expression of the biofilm-associated protein interferes with host protein receptors of *Staphylococcus aureus* and alters the infective process. *Infect. Immun.* 70, 3180–3186, 2002.

116. Miyamoto, Y.J., Wann, E.R., Fowler, T., Duffield, E., Hook, M., and Mcintyre, B.W., Fibronectin binding protein A of *Staphylococcus aureus* can mediate human T lymphocyte adhesion and coactivation. *J. Immunol.* 166, 5129–5138, 2001.

117. Maira-Litran, T., Kropec, A., Abeygunawardana, C., Joyce, J., Mark, G. III, Goldmann, D.A., and Pier, G.B., Immunochemical properties of the staphylococcal poly-N-acetylglucosamine surface polysaccharide. *Infect. Immun.* 70, 4433–4440, 2002.

118. McKenney, D., Hubner, H., Muller, E., Wang, Y., Goldmann, D.A., and Pier, G.B., The *ica* locus of *Staphylococcus epidermidis* encodes production of the capsular polysaccharide/adhesin. *Infect. Immun.* 66, 4711–4720, 1998.

119. Riber, U., Espersen, F., Skinhoj, P., and Kharazmi, A., Induction of oxidative burst response in human neutrophils by adherent staphylococci. Comparison between *Staphylococcus epidermidis* and *Staphylococcus aureus*. *APMIS* 101, 55–60, 1993.

120. Kristinsson, K.G., Hastings, J.G., and Spencer, R.C., The role of extracellular slime in opsonophagocytosis of *Staphylococcus epidermidis*. *J. Med. Microbiol.* 27, 207–213, 1988.

121. Heinzelmann, M., Herzig, D.O., Swain, B., Mercer-Jones, M.A., Bergamini, T.M., and Polk, H.C. Jr., Phagocytosis and oxidative-burst response of planktonic *Staphylococcus epidermidis* RP62A and its non-slime-producing variant in human neutrophils. *Clin. Diagn. Lab. Immunol.* 4, 705–710, 1997.

122. Ofek, L., and Doyle, R.J., *Bacterial Adhesion to Cells and Tissues,* 1 ed., 1–40. Chapman & Hall, New York, NY, 1994.

123. Johnson, G.M., Regelmann, W.E., Gray, E.D., Peters, G., and Quie, P.G., Staphylococcal slime and host defenses. Effect on polymorphonuclear granulocytes. *Zentralbl. Bakteriol. Suppl.* 16, 33–44, 1987.

124. Paquette, D.W., Fiorellini, J.P., and Martuscelli, G., Enantiospecific inhibition of ligature-induced periodontitis in beagles with topical (S)-ketoprofen. *J. Clin. Periodontol.* 24, 521–528, 1997.

125. Heasman, P.A., Offebacher, S., Collins, J.G., Edwards, G., and Seymour, R.A., Flurbiprofen in the prevention and treatment of experimental gingivitis. *J. Clin. Periodontol.* 20, 732–738, 1993.

126. Gaffar, A., Scherl, D., Affitto, J., and Coleman, E.J., The effect of triclosan on mediators of gingival inflammation. *J. Clin. Periodontol.* 22, 480–484, 1995.

127. Serhan, C.N., Lipoxins and novel spirin-triggered 15-epilipoin (ATL): A jungle of cell-cell interactions or a therapeutic opportunity? *Prostaglandins* 53, 107–137, 1997.

128. Pouliot, M., Clish, C.B., Petasis, N.A., VanDyke, T.E., and Serhan, C.N., Lipoxin A4 analogues inhibit leukocyte recruitment to *Porphyromonas gingivalis*: a role for cyclooxygenase-2 and lipoxins in periodontal disease. *Biochemistry* 39, 4761–4768, 2000.

129. Allen, J.B., Wong, H.L., Cost, G.L., Bienkowski, M., and Wahl, S.M., Suppression of monocytes function and differential regulation of IL-1 and IL-1ra by IL-4 contribute to resolution of experimental arthritis. *J. Immunol.* 151, 4344–4351, 1993.

130. Jeffcoat, M.K., and Reddy, M.S., Alveolar bone loss and osteoporosis: evidence for a common mode of therapy using the bisphosphonate alendronate. Davidovitch, Z., Norton, L., eds., *The Biologic Mechanism of Tooth Resorption and Replacement by Implants.* Harvard Society for the Advancement of Orthodontics, Boston, 1996, 365–373.

131. Golub, L.M., Lee, H.M., and Lehrer, G., Minocycline reduces gingival collagenolytic activity during diabetes: preliminary observations and a proposed new mechanism of action. *J. Periodont. Res.* 18, 516–526, 1983.

132. Thomas, J., Walker, C., and Bradshaw, M., Long-term use of subantimicrobial dose doxycycline doe not lead to changes in antimicrobial susceptibility. *J. Periodontol.* 71, 1472–1483, 2000.

133. Walker, C., Thomas, J., Nango, S., Lenon, J., Wetzel, J., and Powala, C., Long-term treatment with subantimicrobial dose doxycycline exerts no antibacterial effect on the subgingival microflora associated with adult periodontitis. *J. Periodontol.* 71, 1465–1471, 2000.

134. Yamasaki, O., Akiyama, H., Toi, Y., and Arata, J., A combination of roxithromycin and imipenem as an antimicrobial strategy against biofilms formed by *Staphylococcus aureus. J. Antimicrob. Chemother.* 48, 573–577, 2001.

135. Gotz, F., and Peters, G., Colonization of medical devices by coagulase-negative staphylococci. In: Walfvogel, F.A., and Visno, A.L., eds., *Infections Associated with Indwelling Medical Devices.* American Society of Microbiology Press, Washington, D.C., 2000, 55–88.

136. Farooq, M., Freischlag, J., Kelly, H., Seabrook, G., Cambria, R., and Towne, J., Gelatin-sealed polyester resists *Staphylococcus epidermidis* biofilm infection. *J. Surg. Res.* 87, 57–61, 1999.

137. Balaban, N., Gov, Y., Bilter, A., and Boelaert, J.R., Prevention of *Staphylococcus aureus* biofilm on dialysis catheters and adherence to human cells. *Kidney Int.* 63, 340–345, 2003.

Section IV

*Overview of Antiinfective
Agents and Clinical Therapy*

17 Pharmacodynamics and the Treatment of IMD-Related Infections

Roger Finch and Sarah Gander

CONTENTS

17.1 INTRODUCTION

The efficacy of antimicrobial therapy is dependent upon a complex set of interactions between the drug, microorganism, and host. To be clinically effective it is essential that therapeutic drug concentrations are achieved at the site of infection in addition to inhibiting microbial growth. Likewise, the host plays a critical role, not only with regard to innate defences to counter infections, but in the manner in which the drug is handled and eliminated by the body. These pharmacokinetic characteristics have importance in predicting the efficiency of an agent. The classical pharmacokinetic parameters include the characteristics of absorption, distribution, metabolism, and elimination of a particular drug. Combined with other factors such as the dosing regimen these parameters are useful in determining the time course of the antibiotic in serum and other body fluids.

The pharmacokinetics of an antibiotic, however, do not adequately define the activity of the drug at the site of the infection. Another set of parameters, the pharmacodynamics, describe the relationship between the inhibitory effect of the antibiotic on the infecting pathogen in relation to the concentration of drug in the serum or at the site of infection. In order to appreciate the emerging literature on the antimicrobial effects of drugs against biofilm associated organisms, it is necessary to place this in the context of current knowledge of the pharmacokinetic and pharmacodynamic principles that govern mode of action, dosage regimens, and the relationship these have or might have in determining therapeutic outcome as well as encouraging or preventing the emergence of resistant organisms, matters which are fundamental to the therapeutic control of infectious disease.

17.2   PHARMACODYNAMIC PARAMETERS

The major pharmacodynamic parameters relevant to the interaction between a target microorganism and drug are the minimum inhibitory concentration (MIC) and the minimum bactericidal concentration (MBC), which provide a standardized assessment of *in vitro* activity of an antibiotic against an infecting bacterium. However, they do not describe the time course of antibiotic activity (1). Consequently, other pharmacodynamic parameters have been developed which apply conventional pharmacokinetic information and relate this to the known *in vitro* susceptibility of the microorganism. These parameters include time (T) above the MIC ($T > $ MIC), area under the concentration curve at 24 hours (AUC_{24}) to MIC ratio (AUC_{24}/MIC) and maximum serum concentration (C_{max}) to MIC ratio (C_{max}/MIC). These extra parameters have been increasingly employed to describe the efficiency of different antibiotics at the site of infection (Figure 17.1). Additional pharmacodynamic parameters that describe the effect of an agent on a pathogen following a period of exposure include the post-antibiotic effect (PAE), post-antibiotic sub-MIC effect (PAE-SME) and post-antibiotic leukocyte enhancement (PALE) (2). However, these

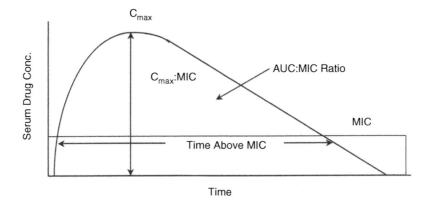

FIGURE 17.1   Pharmacodynamic/pharmacokinetic relationship of antibiotics.

latter two parameters have had less clinical utility in predicting the performances of an agent but have been more recently linked to the pharmacodynamic characteristics described above.

With regard to pharmacodynamic characteristics, antibiotics are divided into two main groups dependent upon whether their bactericidal activity (2) are described as being concentration or time dependent. With concentration dependent antibiotics the rate and extent of bactericidal activity increases with increasing drug concentration. Here the pharmacodynamic parameters which best predict their bactericidal effect are AUC_{24}/MIC and C_{max}/MIC. Agents in this group include the fluoroquinolones and the aminoglycosides. With time dependent antibiotics increasing drug concentration does not enhance the bactericidal effect and the extent of killing is largely dependent on the time of exposure above the MIC (2). The time-dependent antibiotics have been divided into two further groups depending on the persistent effects, such as PAE and PAE-SME, exhibited by the drugs (1). For those antibiotics which exhibit only mild or moderate persistence effects, for example the β-lactams and the oxazolidinone linezolid, the pharmacodynamic parameter which best predict their effect is $T > MIC$. For antibiotics which exhibit prolonged persistent effects, which include the streptogramins and tetracyclines, the pharmacodynamic parameter best linked to bacteriological eradication has been shown to be AUC_{24}/MIC (2).

As the information available on the pharmacodynamics of antibiotics increases the link between drug exposure and clinical outcome has become clearer. There has been a move away from a reliance on conventional pharmacokinetic characteristics to inform dosage regimen selection to one characterized by evidence based dosing strategies (3). Knowing whether an antibiotic is time or concentration dependent has important influences in supporting the effectiveness of dosing regimens. With concentration dependent antibiotics, for example the aminoglycosides and the fluoroquinolones, the aim is to increase the concentration of the drug at the site of infection to as high as possible, without causing toxicity to the patient. With time dependent antibiotics such as the β-lactams and the oxazolidinone linezolid, bacterial killing is not enhanced by increasing the concentration of the drug at the site of infection. Here the aim is to keep the concentration of the drug above the MIC for as long as possible. Studies have shown that maximal killing is achieved with β-lactams when the drug concentration is above the MIC ($T > MIC$) for 60 to 70% of the dosing interval (4,5).

Although there are a number of factors that can contribute to the development of resistance to antibiotics, there is a clear association between exposure to antimicrobial agents and the emergence of resistance (6). However, pharmacodynamics are beginning to be used to establish dosing regimens which minimize the risk of the emergence of resistance. Studies have been performed to demonstrate the relationship between antimicrobial dosing and the risk of the emergence of resistance which include clinical trials as well as animal models of infection (6). Unfortunately, relatively few of these studies have examined the relationship between antibiotic dosing and resistance in patients and most of the data comes from animal or *in vitro* models (7). One study using a neutropenic rat model looked at fluoroquinolone pharmacodynamics in nosocomial pneumonia caused by *Pseudomonas aeruginosa* (8). It was found that AUC_{24}/MIC ratios greater than 100 and C_{max}/MIC ratios of >8 were important in preventing the emergence of resistance.

The pharmacodynamic data on the major classes of antibiotics will now be reviewed before considering their application to models of biofilm infections.

17.2.1   FLUOROQUINOLONES

Fluoroquinolones exhibit concentration dependent bactericidal activity and a prolonged PAE against both Gram-negative and Gram-positive bacteria (9–11). They were the first class of antibiotic subject to extensive pharmacodynamic studies which were incorporated into the drug development programme (12) and influenced licensed dosage regimens. As with most antibiotics the majority of this data is generated from *in vitro* studies or animal models. In 1987 it was suggested that the C_{max}/MIC ratio was an important parameter in predicting the bactericidal activity of the fluoroquinolone enoxacin against both Gram-negative and Gram-positive bacteria (13). Using an *in vitro* pharmacodynamic model they found that unless C_{max}/MIC ratios were 8 or above, regrowth of the test organisms occurred and furthermore the regrowing bacteria had MICs four to eight times higher than those examined prior to exposure. This led Blaser et al. to conclude that a C_{max}/MIC ratio of above eight was also important in the prevention of emergence of resistant isolates (13). Further *in vitro* work on the exposure of *Pseudomonas aeruginosa* to ciprofloxacin by Dudley et al. confirmed that the selection of resistant isolates occurred with C_{max}/MIC ratios >8 (14). More recently work has been done to determine whether AUC_{24}/MIC or C_{max}/MIC was more important in the antibiotic performance of ciprofloxacin and ofloxacin against *P. aeruginosa* (15). This was achieved by altering the C_{max} concentrations and half-lives of the drugs, but keeping the AUC_{24} values constant. Madaras-Kelly et al. (15) found that both drugs demonstrated similar antibacterial activity at AUC_{24}/MIC ratios of ≥100. The data suggest that AUC_{24}/MIC ratio is the best parameter for predicting the antibacterial activity of the fluoroquinolones against *P. aeruginosa*. Many of the newer fluoroquinolones (e.g., moxifloxacin, levofloxacin, gatifloxacin) show activity against Gram-positive as well as Gram-negative bacteria. In contrast to their effects against Gram-negative pathogens, the fluoroquinolones exert a concentration independent effect against Gram-positive organisms (16–19). One of the more recent fluoroquinolones, moxifloxacin, has, however, been shown to exhibit concentration dependent killing against Gram-positive pathogens (16,20,21). Lacy et al. (22) investigated the pharmacodynamics of levofloxacin against penicillin-resistant *S. pneumoniae* (PRSP) and found that effective AUC_{24}/MIC ratios ranged from 30 to 55. These values are significantly lower than those suggested for *P. aeruginosa,* of 100 to 125 (15) and indicates that the pharmacodynamics of the fluoroquinolones are organism specific.

A small number of animal infection models have been employed in the study of the pharmacodynamics of the fluoroquinolones. Much of the data have been extrapolated from that for the aminoglycosides as they exhibit similar concentration-dependent activity (23). A substantial amount of the animal infection model work has involved Gram-positive organisms. Using normal and neutropenic mice, Vesga and Craig (24) evaluated levofloxacin against PRSP and found that AUC_{24}/MIC ratios of 22 to 59 were required to produce a static effect. In a similar study Vesga et al. (25) looked at the effects of sparfloxacin on pathogens including *S. pneumoniae* and found that to produce a static effect a mean AUC_{24}/MIC of 29 was necessary. In a neutropenic murine thigh infection model AUC_{24}/MIC ratios required to produce a bacteriostatic effect for Enterobacteriaceae, *S. pneumoniae* and *S. aureus* were observed to be 41, 52 and 36 respectively (26).

There have been relatively few human studies undertaken to confirm the data generated by the *in vitro* and animal model investigations. One of the earliest studies looked at the effect of ciprofloxacin against nosocomial pneumonia caused by Gram-negative bacteria (27). It was found that elevated C_{max}/MIC and AUC_{24}/MIC ratios and also extended $T > $ MIC were associated with successful eradication of pathogens from the lower respiratory tract. Forrest et al. (28) also examined the pharmacodynamics of ciprofloxacin in a retrospective study of patients, chiefly with lower respiratory tract infections, but also soft tissue wounds, bacteraemias and complicated urinary tract infections caused by *P. aeruginosa,* other aerobic Gram-negative bacilli and *S. aureus*. They demonstrated that an AUC_{24}/MIC of 125 had a significantly higher probability of treatment success than did lower values, with 250 to 500 being optimal for bacterial eradication regardless of the species of bacteria (28). A subsequent study assessing the bactericidal effect of ciprofloxacin in serum ultrafiltrates from healthy volunteers also found higher AUC_{24}/MIC values to be more effective in bacterial eradication (29); the maximal killing effect for ciprofloxacin was seen at 15 to 40 × MIC for the Gram-positive organisms, *S. pneumoniae* and *S. aureus*, and 20 to 50 × MIC for *P. aeruginosa*. These data support the inverse relation of MIC and bacterial killing rates, when at equal levels of exposure to antimicrobial agents (29). A further study of one of the more recently developed respiratory quinolones, grepafloxacin, administered orally to patients with acute bacterial exacerbations of chronic bronchitis (30), found the response to be strongly related to AUC_{24}/MIC. At values <75 the probability of clinical cure was 71%, at 75 to 175 80% and >175 98%. In 1998 Preston et al. (31) suggested that successful clinical outcome and microbial eradication by levofloxacin was best predicted by the pharmacodynamic parameter C_{max}/MIC. For patients with respiratory tract, skin or urinary tract infections a C_{max}/MIC value of >12.2 correlated with a successful outcome.

17.2.2 AMINOGLYCOSIDES

The aminoglycoside antibiotics have been shown to exhibit concentration dependent activity against Gram-negative bacteria in a number of studies (13,32–38). In contrast the aminoglycosides may show concentration independent killing against Gram-positive organisms, although there have been too few studies (12). Early *in vitro* studies demonstrated that a netilmicin C_{max}/MIC value of >8 was required to prevent regrowth of resistant sub-populations of bacteria (13). Using a mouse thigh infection model to assess the effects of amikacin against Gram-negative bacilli it was demonstrated that amikacin exhibits concentration-dependent killing and prolonged PAEs similar to other aminoglycosides (39). Mice with renal impairment were used as they exhibit amikacin serum half-lives much closer to those observed in humans than those with normal renal function and it was found that AUC_{24}/MIC correlated best with efficacy against the four strains of bacteria tested (39). Leggett et al. (40) found the parameter AUC_{24}/MIC to be the best predictor of antibacterial activity in a mouse pneumonitis model. A C_{max}/MIC value of >8 has been associated with improved clinical outcomes in early clinical trials of aminoglycosides (41,42). A retrospective study also found increasing C_{max}/MIC ratios were strongly associated with a favorable clinical outcome (33). A prospective study designed to test dosing regimens where C_{max}/MIC value was ≥10 produced very favorable results with regard to both clinical outcome and reduction in toxicity (43). However, interpatient variability in aminoglycoside pharmacokinetics and toxicity concerns have made the application of this technique controversial (44,45).

17.2.3 β-Lactams

β-lactams exhibit time dependent, concentration independent bactericidal activity, with the pharmacodynamic parameter T>MIC being the best predictor of activity (1,2,46,47). The goal when dosing these drugs is therefore to optimize the duration of the antibiotic concentration at or above the MIC (11). The results from both *in vitro* models and *in vivo* studies have confirmed the importance of T>MIC in optimizing β-lactam bactericidal activity (4,40,46,48–56). Nishida et al. (49) evaluated the effects of the three cephalosporins, cefazolin, cephaloridine and cephalothin, against *Escherichia coli* in an *in vitro* model simulating intramuscular administration. Maximal killing occurred at 1 to 4 times the MIC and concentrations above this did not increase the effect. Using an *in vitro* capillary model, Zinner et al. (57) investigated the effects of different dosage regimens of cefoperazone against *P. aeruginosa, S. aureus, Klebsiella pneumoniae,* and *E. coli.* The drug was administered at various dosing intervals. No difference in the rate of killing was noted for *E. coli* and *K. pneumoniae* with the different regimens. With the less susceptible *S. aureus and P. aeruginosa* it was observed that smaller doses administered more frequently increased T>MIC which resulted in greater bactericidal activity. Another study (54), utilizing a kinetic dilution model, examined the effect of ampicillin against *E. coli.* The varying drug concentrations used were designed to simulate varying dosing intervals and elimination half-lives in humans. The important pharmacodynamic parameter for preventing the emergence of resistant sub-populations was found to be AUC. The involvement of T>MIC was not detected in this study. It was concluded that simply maintaining the concentration above the MIC was not important in ascertaining the effect of the drug on the bacteria. The parameters AUC_{24} and C_{max} are thought to be as important as T>MIC in predicting the activity of β-lactams in certain situations. Lavoie and Bergeron (58), using an *in vivo* fibrin clot model, demonstrated that higher doses of ampicillin produced greater bactericidal activity against *Haemophilus influenzae* compared to continuous infusion. It appears that higher doses are needed to increase the concentration of the drug in the dense bacterial growth found with this model.

These findings demonstrate that the relationship between the optimal pharmacodynamic parameters and site of infection is not straight forward. Most *in vivo* studies however confirm the importance of T>MIC in optimising β-lactam activity (40,46,53,55,56,59,60). Many *in vivo* animal studies have utilized the neutropenic mouse thigh model. Initial studies in this model demonstrated that the β-lactam ticarcillin was more effective in reducing bacterial counts when administered hourly rather than the total daily dose being given every 3 hours (61). Using this model, $T > \text{MIC}$ has been shown to be the best predictor of efficacy for penicillin against *S. pneumoniae*, cefazolin against *S. aureus* and *E. coli* and ticarcillin against *P. aeruginosa* (53). This study also demonstrated that for β-lactam/bacteria combinations where there is no or a minimal PAE, the drug must be maintained at $T > \text{MIC}$ for 90 to 100% of the dosing interval to achieve maximum killing. Where the β-lactam/ bacteria combination does exhibit a PAE, maximum killing is achieved when the $T > \text{MIC}$ is only 50 to 60% of the dosing interval (53). Other investigations utilizing animal models have also demonstrated $T > \text{MIC}$ to be the best predictor of β-lactam activity. Leggett et al. (40) looked at the effect of ceftazidime against *K. pneumoniae* in a murine peritonitis and thigh infection models, and Fantin et al. (60) investigated the effects of ceftazidime against *E. coli* in a thigh infection. Continuous infusion or frequent administration of penicillin was found to result in better efficacy against *S. pneumoniae* in a rat pneumonia model, especially when immunodeficient animals were used (62,63). Similar results were reported using the same model with *K. pneumoniae* and ceftazidime (52,59). Evidence that $T > \text{MIC}$ is an important pharmacodynamic parameter in predicting the activity of β-lactams has been documented in studies of animal models of endocarditis. A rabbit model for *S. aureus* endocarditis was used to compare cure rates of four dosing regimens or methicillin (64). Four or eight hourly dosing was found to be more effective than 12 hourly dosing or continuous infusion. The failure of the continuous infusion may be due to problems with the drug penetrating into the vegetations. Joly et al. (65) and Pangon et al. (66) studied three cephalosporins in a rabbit model of *E. coli* endocarditis and found that significant killing of the bacteria in the vegetations only occurred when the drug concentrations in the vegetations were >200 times higher than the MBC. Few clinical studies have investigated the effects of different β-lactam dosage regimens on clinical or bacteriological outcome (47), consequently most of the pharmacodynamic data is derived from *in vitro* or animal models.

The available clinical data, however, does tend to agree with the *in vitro* findings. Studies have found that continuous infusion is as effective as the more conventional intermittent dosing regimens (67–69), although there is insufficient data to make firm recommendations for continuous infusion. One study examined the influence of various pharmacodynamic parameters of cefmenoxime on the time to bacterial eradication in patients with nosocomial pneumonia (50). A significant relationship was observed between $T > \text{MIC}$ and bacterial eradication. The lack of data has led to debate over the magnitude and time the drugs should be maintained above the MIC. Experimental data suggests that the MIC should be exceeded by 1 to 5 times for between 40 and 100% of the dosage interval (2,46,47,56,70–73). Clearly much more clinical data are required to validate the importance of $T > \text{MIC}$ for β-lactams across the spectrum of infectious diseases (12).

17.2.4 MACROLIDES

The macrolide antibiotics include erythromycin, clarithromycin and azithromycin among others. They are generally believed to perform as concentration-independent, or time-dependent drugs (74,75). However, due to the differing pharmacokinetic profiles of the individual macrolides, the pharmacodynamic outcome parameters are more difficult to characterize than for other antimicrobial agents (12). Macrolides concentrate in both the extracellular space and in macrophages and consequently their activity relies on the intracellular drug levels as well as serum concentrations (74,75). There is a limited amount of data from *in vitro* and animal models on the macrolide antibiotics, partly due to the difficulties in developing suitable models that are representative of the activity of these drugs *in vivo* (12). One *in vitro* study, using a mouse thigh model, found that $T > MIC$ was the most significant parameter in determining the efficacy of erythromycin against *S. pneumoniae* (53,76), investigated the effects of clarithromycin against *S. pneumoniae* and compared the relationship between the number of bacteria remaining in the mouse thigh after 24 hours and the three pharmacodynamic parameters; AUC_{24}/MIC, C_{max}/MIC and $T > MIC$. The correlation between the bacterial counts and $T > MIC$ was found to be highly significant, but for the other two parameters, AUC_{24}/MIC and C_{max}/MIC, the correlation was poor. A further study using the same strain of *S. pneumoniae* and the macrolide azithromycin noted a different relationship between 24 hour bacterial counts and the three pharmacodynamic parameters (77) in which the best correlation was seen with AUC_{24}/MIC and C_{max}/MIC. This discrepancy in the results of the two studies was attributed to the prolonged persistent effects exhibited by azithromycin (77). In summary, it appears that the optimal pharmacodynamic parameter for erthromycin and clarithromycin is $T > MIC$, but not for azithromycin where C_{max}/MIC or AUC_{24}/MIC appears most predictive.

17.2.5 GLYCOPEPTIDES

The data available on the glycopeptide antibiotics indicate that they exhibit time-dependent bactericidal activity. As they tend to show sub-MIC and post-antibiotic effects the serum concentrations need not exceed the MIC for all of the dosing interval and the important pharmacodynamic parameter in predicting outcome may not be $T > MIC$, but AUC_{24}/MIC (7,78–80). The concentration effect of vancomycin and teicoplanin was examined against *S. aureus in vitro* by Peetermans et al. (79). It was found that for both drugs concentration dependence was only observed when the concentrations were at or below the MIC and above this value there was no increase in the rate of killing. A further *in vitro* study looking at the pharmacodynamics of the novel glycopeptide oritavancin (LY333328) found that it exhibited concentration-dependent killing against a strain of vancomycin resistant *Enterococcus faecium* (80). It is thought that the difference in pharmacodynamic outcome parameter may be due to high protein binding. Another recent *in vitro* study also found that teicoplanin exhibited a concentration-dependent bactericidal effect against *S. epidermidis* (81). The data derived from animal models also suggests that there is a link between $T > MIC$ and outcome. Chambers and Kennedy (82) using an endocarditis model evaluated the effects of teicoplanin administered intravenously or intramuscularly. The intramuscular injection, the most effective of the two regimens, was found to result in minimum serum concentrations (C_{min}) that remained above the MIC longer than for the intravenous regimen. A mouse peritonitis model of a *S. pneumoniae* infection was used to investigate a wide spectrum of treatment regimens with vancomycin and teicoplanin (83). They found that the pharmacodynamic parameters AUC_{24}/MIC and C_{max}/MIC best explained the effects of vancomycin and the parameters $T > MIC$ and C_{max}/MIC teicoplanin.

There is a small amount of clinical data on the pharmacodynamics of the glycopeptides. In a study of 20 patients with staphylococcal endocarditis (84) the patients that had a successful outcome tended to have a higher C_{max} of teicoplanin than those who failed. Identification of the most important pharmacodynamic parameter was difficult as no measurements of other parameters were made and as C_{max} increases so will AUC_{24}/MIC and $T > MIC$. Hyatt et al. (85) investigated the treatment of Gram-positive infections with vancomycin and found that successful clinical and microbiological outcome was linked to both AUC_{24} and MIC. Patients with an $AUC_{24} > 125$ had a higher probability of success as did those with pathogens with a MIC < 1 µg/ml. The results indicate that both concentration and $T > MIC$ are important pharmacodynamic parameters for predicting vancomycin efficacy.

In conclusion, from the available data, it appears that a single pharmacodynamic parameter cannot be used to predict the outcome of treatment with the glycopeptide antibiotics; indeed the pharmacodynamic outcome parameters are often specific to different drug/bacterial strain combinations.

17.3   PHARMACOKINETIC AND PHARMACODYNAMIC MODELS OF INFECTION

The best method for evaluating specific drug dosing regimens is the controlled trial in humans. However, apart from being expensive, clinical trials are time consuming, tend to involve relatively small numbers of patients and are restricted in the dosing variations used for practical as well as ethical considerations (86). As a result the use of *in vitro* and *in vivo* animal models have been widely adopted. *In vitro* models have the advantages of being relatively inexpensive, more easily controlled than *in vivo* studies and are used to evaluate a wider range of dosing regimens of many antimicrobial agents against a variety of bacterial strains. There are limitations however when using *in vitro* models. The absence of an immune system does not allow any evaluation of the contribution of the cellular defence mechanism. Furthermore, the influence of protein binding by antibiotics is often not taken into consideration (87). Additionally, bacteria *in vitro* tend to have a faster rate of growth and exhibit decreased virulence (88). *In vivo* animal models are very useful for detailing the relationships between tissue and serum drug concentrations (89) but again suffer from inter-species differences that may affect pharmacodynamic behavior and the immune response to infection.

17.3.1   *IN VITRO* MODELS

The simplest *in vitro* model is the static model. This model is used to evaluate the effects of fixed concentrations of drugs on growing cultures of bacteria. The MIC and MBC are the pharmacodynamic parameters determined using this method. More complex models, often referred to as dynamic *in vitro* models, are commonly used to mimic fluctuating antibacterial levels as seen *in vivo* throughout a drug's dosing regimen.

Single or one compartment models are the most basic of the dynamic models. They consist of a single fixed volume chamber containing a liquid culture of bacteria. The growth medium is pumped through the chamber at a fixed rate, usually simulating the half-life of the test drug. Antimicrobial agents are added to the vessel either as a single dose at the beginning of the experiment or as several doses during the course of the experiment. These models attempt to mimic the pharmacokinetic profiles of drugs in humans, representing dosing regimens which are monitored by tracking the eradication of the bacteria over time (90–94). These models however suffer from a dilutional loss of bacteria. Although correctional methods can be used, two compartment dynamic models prevent the loss of bacteria through dilution and also mimic human pharmacokinetics more closely than the single compartment model (22,81,95–98).

There are a variety of designs of two compartment models. A relatively standard one consists of a chamber, the central compartment, with intake and outflow ports, to allow the medium to flow through. Inserted into this central compartment is the peripheral compartment separated by a filter or dialysis membrane which allows the diffusion of small solutes such as the antimicrobial agent and the broth, but not the bacteria (Figure 17.2). The central compartment corresponds to the serum and the peripheral compartment to the infected tissue. The two compartment model functions by placing a broth culture of growing bacteria into the peripheral compartment and uninoculated broth into the central compartment. Generally, antibiotic is added at the beginning of the experiment and is present at a high concentration in both compartments. Over time as fresh drug free medium is pumped into the central compartment the drug concentration decreases in both the central and peripheral compartments (86). By controlling the flow rate of the medium it is possible to mimic the human elimination half-life of the test antibiotic. As well as evaluating monotherapy these models have been adapted to allow the study of drug combinations (99–102).

17.3.2 BIOFILM MODELS

A very small number of models have been developed to evaluate the pharmacodynamics of antimicrobial agents against adherent biofilm bacteria. Blaser et al. (103) compared the bactericidal activity of 28 treatment regimens in an established *in vivo* model of device-related infection and a pharmacodynamic *in vitro* model. In the *in vitro* model bacterial biofilms were grown on sinter glass beads and in the animal model perforated Teflon tubes were implanted into guinea pigs. The results for the two models were found to correlate well; however, the pharmacodynamic ratios C_{max}/MIC and AUC$_{24}$/MIC and also T>MIC were not predictive for therapeutic outcome in either model. A study carried out by ourselves investigated the effects of exposure of resistant and susceptible strains of staphylococci to five antibiotics (telavancin, vancomycin, teicoplanin, linezolid, and moxifloxacin) using an *in vitro* pharmacokinetic biofilm model (Figure 17.3). The bacterial biofilms were exposed to the antibiotics using both exponentially decreasing and constant drug concentrations. The experiments performed with exponentially decreasing concentrations of the drugs were designed to simulate the parenteral administration of the antibiotics to humans; the rate of decrease was calculated to reflect the half-lives of the various

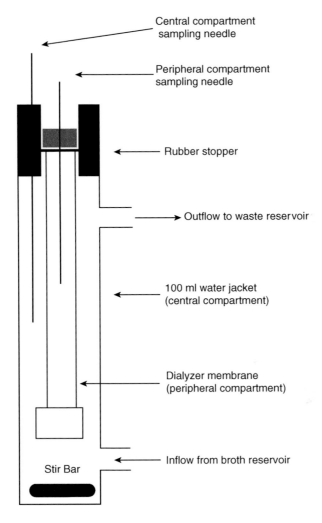

FIGURE 17.2   An example of a two-component *in vitro* pharmacodynamic model.

drugs tested. In contrast, exposure to constant antibiotic concentrations simulated the administration of drug by continuous infusion over 2 hours. The results obtained indicate that the predictive pharmacodynamic parameters determined using planktonic broth cultured bacteria do apply to the organisms when grown as biofilms (104).

However, many studies especially of endocarditis have inadvertently looked at bacteria growing as biofilms. The fibrin clots, which generally contain human cryoprecipitate and bovine thrombin (92,105,106), have bacteria added during the preparation of the clot, these organism then grow on and within the clot and this constitutes a bacterial biofilm. Palmer and Rybak (105) compared the pharmacodynamic activities of levofloxacin in an *in vitro* model with infected platelet-fibrin clots simulating vegetations. Against two strains of *S. aureus*, one methicillin sensitive

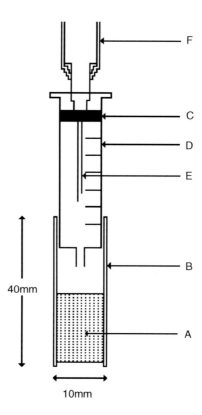

FIGURE 17.3 Schematic representation of a Sorbarod model. The Sorbarod filter (A) contained within a length of PVC tubing (B) has cells loaded onto it from a syringe. The plunger is withdrawn from a 2 ml syringe, leaving only the rubber seal within the syringe lumen (C). The syringe (D) is introduced into the PVC tubing containing the Sorbarod and a sterile needle (E) inserted through the rubber seal. The filter unit is then clamped upright and media inlet tubing (F) attached to the needle through which fresh medium is delivered.

and one resistant, it was found that killing activity for levofloxacin appeared to correlate better with the C_{max}/MIC ratio than with the AUC_{24}/MIC ratio. These results contrast with the more recent data which suggest that the newer fluoroquinolones exert a concentration-independent effect against Gram-positive organisms where the important pharmacodynamic parameter appears to be AUC_{24}/MIC (16–19). The bactericidal activities of various fluoroquinolones were studied using fibrin clots infected with penicillin-resistant *S. pneumoniae* (92). Although there was no clear association, they concluded that AUC_{24}/MIC and T>MIC were better predictors of activity than C_{max}/MIC, which agrees with the findings of the more recent studies mentioned above. In 1997 Rybak et al. (106) evaluated the bactericidal activity of quinupristin/dalfopristin against fibrin-platelet clots infected with strains of *S. aureus*. One strain was constitutively erythromycin and methicillin resistant and the other susceptible. Killing was not achieved against the resistant isolate, however a 99.9%

kill was demonstrated against the susceptible strain and the AUC_{24} was found to correlate significantly with bactericidal activity. The results from this study agree with another evaluation of quinupristin/dalfopristin (107) which indicated concentration-dependent killing. An *in vivo* rabbit endocarditis model was used to study the efficacy of linezolid against *S. aureus* (108). Of the three different dosing regimens used (25, 50 and 75 mg/kg) only the higher doses produced trough (minimum serum concentrations) that were above the MIC and it was only these doses that produced significant antibacterial effects. The results indicate that the important pharmacodynamic parameter for predicting bactericidal activity is $T > MIC$ which is supported by subsequent studies (3,109).

Determination of the pharmacodynamic parameter which best relates to bactericidal activity in the endocarditis models mentioned above are generally in agreement with the results of investigations carried out against planktonic cells. This suggests that the same pharmacodynamic parameters are important in predicting outcome whether the bacteria are growing planktonically or as biofilms.

17.4 THE FUTURE

The subject of pharmacodynamics is expanding rapidly. To date most of the information has been generated using *in vitro* models and, to a lesser extent, animal models of infection. What is often lacking are clinical trials which validate this data. The pharmacodynamic data produced so far *in vitro*, in animals and in humans, however, does appear overall to be in general agreement with one another (12). Pharmacodynamics are, at the moment, chiefly used as an aid to define and optimize antimicrobial therapy dosing regimens in order to maximize bacteriological eradication.

More recently the advent and availability of sophisticated mathematical and statistical techniques has revolutionized the ability to delineate exposure-response and exposure-toxicity relationships for antibiotics (110), most notably through use of Monte Carlo simulations. The aim of antimicrobial treatment is to achieve the highest probability of a positive outcome (clinical cure, bacterial eradication etc.). The ability of a fixed antibiotic dose to achieve a specific target such as an AUC_{24}/MIC ratio, is influenced by a number of factors. The variability of the drugs pharmacokinetics in the target population, the range of MIC values, the protein binding of the drug, and the desired exposure target. With Monte Carlo simulations firstly an endpoint must be determined for example total bacterial eradication or bacterial stasis. Generally a large number of subjects are used, allowing for a larger range of drug exposures to be tested, and measurements of various pharmacokinetic and pharmacodynamic parameters taken. The data generated from the clinical information can then be used to determine the target value that produces the desired effect, for example achieving a free drug AUC_{24}/MIC value of 35 to 40 for therapy of pneumococcal pneumonia with fluoroquinolones, which has a high predictive probability for achieving pathogen eradication (111). This approach can be used as a basis for rational decisions regarding the choice of dosage regimens. Using data produced by Preston et al. (31) from a population model examining the pharmacokinetics of levofloxacin in 272 patients with community-acquired infections, Drusano (110) constructed a 10,000 subject Monte Carlo simulation. AUC_{24}/MIC exposure targets of 27.5 for bacterial stasis, and 34.5 for a 1 $\log_{10}$ CFU/ml drop in the number of bacteria recovered, were defined using a mouse thigh infection model. Using these figures it was found that the overall probability of attaining the stasis target was 0.978 and the probability of attaining the 1 $\log_{10}$ CFU/ml drop target was 0.947.

Pharmacodynamics are increasingly used as a cost effective aid that supports the development of new antibiotics. It is also increasingly adopted to develop regimens that prevent the development of resistance during therapy. It has already been noted that the mutant prevention concentration (MPC), defined as the lowest antibiotic concentration at which resistant mutants do not arise on antibiotic containing agar plates, when linked to pharmacodynamic parameters of an antibiotic, could provide a very useful mechanism in support of the prevention of the emergence of resistant organisms.

As mentioned above, very little work has been carried out on the pharmacodynamic activity of antibiotics against bacteria growing as adherent biofilms. From the little data that does exist it appears that the mode of growth of the bacteria, whether planktonic or as an attached biofilm, does not affect the pharmacodynamic characteristics of established antibiotics in predicting bactericidal activity. They could also have an increasing role in the assessment of the performance of new agents in relation to biofilm associated pathogens. Further studies are needed to determine whether pharmacodynamic parameters used to predict activity of antibiotics against planktonically grown bacteria apply to the same bacteria cultured as adherent biofilms. To carry out *in vitro* studies it would be useful to have a standard model that would be capable of taking into consideration factors such as the age of the biofilm, the pharmacokinetics of the drugs and the range of organisms involved. An additional consideration with bacterial biofilms is antimicrobial resistance. Biofilm cells have been shown to exhibit reduced susceptibility to antibiotics and the role of pharmacodynamics in preventing the emergence of resistant organisms may be very pertinent when dealing with biofilms. The potential value of such studies is considerable. Medical device associated infections are an increasing healthcare burden. Infection is among the major complications, often with organisms of low virulence but characterized by multi-drug resistance and biofilm formation. Models that can assess the performance of established agents and predict that of future agents could have an important place in experimental therapeutics. If sufficiently predictive they could better define the need and likelihood of success of clinical studies which are difficult to perform and generally costly.

REFERENCES

1. Zhanel, G.G., Influence of pharmacokinetic and pharmacodynamic principles on antibiotic selection. *Curr. Infect. Dis. Rep.* 3, 29–34, 2001.
2. Craig, W.A., Pharmacokinetic/pharmacodynamic parameters: rationale for antibacterial dosing of mice and men. *Clin. Infect. Dis.* 26, 1–10, 1998.
3. Goldberg, J., and Owens, Jr., R.C., Optimizing antimicrobial dosing in the critically ill patient. *Curr. Opin. Crit. Care* 8, 435–440, 2002.
4. Craig, W.A., and Andes, D., Pharmacokinetics and pharmacodynamics of antibiotics in otitis media. *Pediat. Infect. Dis. J.* 15, 255–259, 1996.

5. Andes, D., and Craig, W.A., *In vivo* activities of amoxicillin and amoxicillin-clavulanate against *Streptococcus pneumoniae:* application to breakpoint determinations. *Antimicrob. Agents Chemother.* 42, 2375–2379, 1998.

6. Andes, D., Pharmacokinetic and pharmacodynamic properties of antimicrobials in the therapy of respiratory tract infections. *Curr. Opin. Infect. Dis.* 14, 165–172, 2001.

7. Craig, W.A., Does the dose matter? *Clin. Infect. Dis.* 33(Suppl 3), S233–237, 2001

8. Drusano, G.L., Johnson, D.E., Rosen, M., and Standiford, H.C., Pharmacodynamics of a fluoroquinolone antimicrobial agent in a neutropenic rat model of *Pseudomonas sepsis. Antimicrob. Agents Chemother.* 37, 483–490, 1993.

9. Turnidge, J., Pharmacokinetics and pharmacodynamics of fluoroquinolones. *Drugs* 58(Suppl 2), 29–36, 1999.

10. Woodnutt, G., Pharmacodynamics to combat resistance. *J. Antimicrob. Chemother.* 46(Suppl T1), 25–31, 2000.

11. Rodvold, K.A., Pharmacodynamics of antiinfective therapy: taking what we know to the patient's bedside. *Pharmacotherapy* 21, 319S–330S, 2001.

12. Gunderson, B.W., Ross, G.H., Ibrahim, K.H., and Rotschafer, J.C., What do we really know about antibiotic pharmacodynamics? *Pharmacotherapy* 21, 302S–318S, 2001.

13. Blaser, J., Stone, B.B., Groner, M.C., and Zinner, S.H., Comparative study with enoxacin and netilmicin in a pharmacodynamic model to determine importance of ratio of antibiotic peak concentration to MIC for bactericidal activity and emergence of resistance. *Antimicrob. Agents Chemother.* 31, 1054–1060, 1987.

14. Dudley, M.N., Blaser, J., Gilbert, D., Mayer, K.H., and Zinner, S.H., Combination therapy with ciprofloxacin plus azlocillin against *Pseudomonas aeruginosa:* effect of simultaneous versus staggered administration in an *in vitro* model of infection. *J. Infect. Dis.* 164, 499–506, 1991.

15. Madaras-Kelly, K.J., Ostergaard, B.E., Hovde, L.B., and Rotschafer, J.C., Twenty-four-hour area under the concentration-time curve/MIC ratio as a generic predictor of fluoroquinolone antimicrobial effect by using three strains of *Pseudomonas aeruginosa* and an *in vitro* pharmacodynamic model. *Antimicrob. Agents Chemother.* 40, 627–632, 1996.

16. Dalhoff, A., Pharmacodynamics of fluoroquinolones. *J. Antimicrob. Chemother.* 43(Suppl B), 51–59, 1999.

17. Ross, G.H., Wright, D.H., Hovde, L.B., Peterson, M.L., and Rotschafer, J.C., Fluoroquinolone resistance in anaerobic bacteria following exposure to levofloxacin, trovafloxacin, and sparfloxacin in an *in vitro* pharmacodynamic model. *Antimicrob. Agents Chemother.* 45, 2136–2140, 2001.

18. Peterson, M.L., Hovde, L.B., Wright, D.H., Brown, G.H., Hoang, A.D., and Rotschafer, J.C., Pharmacodynamics of trovafloxacin and levofloxacin against Bacteroides fragilis in an *in vitro* pharmacodynamic model. *Antimicrob. Agents Chemother.* 46, 203–210, 2002.

19. Ibrahim, K.H., Hovde, L.B., Ross, G., Gunderson, B., Wright, D.H., and Rotschafer, J.C., Microbiologic effectiveness of time- or concentration-based dosing strategies in Streptococcus pneumoniae. *Diagnost. Microbiol. Infect. Dis.* 44, 265–271, 2002.

20. Klugman, K.P., and Capper, T., Concentration-dependent killing of antibiotic-resistant pneumococci by the methoxyquinolone moxifloxacin. *J. Antimicrob. Chemother.* 40, 797–802, 1997.

21. Boswell, F.J., Andrews, J.M., and Wise, R., Pharmacodynamic properties of BAY 12-8039 on gram-positive and gram-negative organisms as demonstrated by studies of time-kill kinetics and postantibiotic effect. *Antimicrob. Agents Chemother.* 41, 1377–1379, 1997.

22. Lacy, M.K., Lu, W., Xu, X., Tessier, P.R., Nicolau, D.P., Quintiliani, R., and Nightingale, C.H., Pharmacodynamic comparisons of levofloxacin, ciprofloxacin, and ampicillin against *Streptococcus pneumoniae* in an *in vitro* model of infection. *Antimicrob. Agents Chemother.* 43, 672–677, 1999.

23. Wright, D.H., Brown, G.H., Peterson, M., and Rotschafer, J.C., Application of fluoroquinolone pharmacodynamics. *J. Antimicrob. Chemother.* 46, 669–683, 2000.

24. Vesga, O., and Craig, W.A., Activity of levofloxacin against penicillin-resistant Streptococcus pneumoniae in normal and neutropenic mice. *36th Interscience Conference on Antimicrobial Agents and Chemotherapy.* New Orleans, 1996.

25. Vesga, O., Conklin, R., Stamstad, T., and Craig, W.A., *In vivo* pharmacodynamic activity of sparfloxacin against multiple bacterial pathogens. *36th Interscience Conference on Antimicrobial Agents and Chemotherapy.* New Orleans, 1996.

26. Andes, D., and Craig, W.A., Pharmacodynamics of the new fluoroquinolone gatifloxacin in murine thigh and lung infection models. *Antimicrob. Agents Chemother.* 46, 1665–1670, 2002.

27. Peloquin, C.A., Cumbo, T.J., Nix, D.E., Sands, M.F., and Schentag, J.J., Evaluation of intravenous ciprofloxacin in patients with nosocomial lower respiratory tract infections. Impact of plasma concentrations, organism, minimum inhibitory concentration, and clinical condition on bacterial eradication. *Arch. Intern. Med.* 149, 2269–2273, 1989.

28. Forrest, A., Nix, D.E., Ballow, C.H., Goss, T.F., Birmingham, M.C., and Schentag, J.J., Pharmacodynamics of intravenous ciprofloxacin in seriously ill patients. *Antimicrob. Agents Chemother.* 37, 1073–1081, 1993.

29. Hyatt, J.M., Nix, D.E., and Schentag, J.J., Pharmacokinetic and pharmacodynamic activities of ciprofloxacin against strains of *Streptococcus pneumoniae, Staphylococcus aureus,* and *Pseudomonas aeruginosa* for which MICs are similar. *Antimicrob. Agents Chemother.* 38, 2730–2737, 1994.

30. Forrest, A., Chodosh, S., Amantea, M.A., Collins, D.A., and Schentag, J.J., Pharmacokinetics and pharmacodynamics of oral grepafloxacin in patients with acute bacterial exacerbations of chronic bronchitis. *J. Antimicrob. Chemother.* 40(Suppl A), 45–57, 1997.

31. Preston, S.L., Drusano, G.L., Berman, A.L., Fowler, C.L., Chow, A.T., Dornseif, B., Reichl, V., Natarajan, J., Wong, F.A., and Corrado, M., Levofloxacin population pharmacokinetics and creation of a demographic model for prediction of individual drug clearance in patients with serious community-acquired infection. *Antimicrob. Agents Chemother.* 42, 1098–1104, 1998.

32. MacArthur, R.D., Lolans, V., Zar, F.A., and Jackson, G.G., Biphasic, concentration-dependent and rate-limited, concentration-independent bacterial killing by an aminoglycoside antibiotic. *J. Infect. Dis.* 150, 778–779, 1984.

33. Moore, R.D., Lietman, P.S., and Smith, C.R., Clinical response to aminoglycoside therapy: importance of the ratio of peak concentration to minimal inhibitory concentration. *J. Infect. Dis.* 155, 93–99, 1987.

34. Kapusnik, J.E., Hackbarth, C.J., Chambers, H.F., Carpenter, T., and Sande, M.A., Single, large, daily dosing versus intermittent dosing of tobramycin for treating experimental pseudomonas pneumonia. *J. Infect. Dis.* 158, 7–12, 1988.

35. Daikos, G.L., Jackson, G.G., Lolans, V.T., and Livermore, D.M., Adaptive resistance to aminoglycoside antibiotics from first-exposure down-regulation. *J. Infect. Dis.* 162, 414–420, 1990.

36. Jackson, G.G., Lolans, V.T., and Daikos, G.L., The inductive role of ionic binding in the bactericidal and postexposure effects of aminoglycoside antibiotics with implications for dosing. *J. Infect. Dis.* 162, 408–413, 1990.

37. Begg, E.J., Peddie, B.A., Chambers, S.T., and Boswell, D.R., Comparison of gentamicin dosing regimens using an in-vitro model. *J. Antimicrob. Chemother.* 29, 427–433, 1992.

38. Staneva, M., Markova, B., Atanasova, I., and Terziivanov, D., Pharmacokinetic and pharmacodynamic approach for comparing two therapeutic regimens using amikacin. *Antimicrob. Agents Chemother.* 38, 981–985, 1994.

39. Craig, W.A., Redington, J., and Ebert, S.C., Pharmacodynamics of amikacin *in vitro* and in mouse thigh and lung infections. *J. Antimicrob. Chemother.* 27(Suppl C), 29–40, 1991.

40. Leggett, J.E., Fantin, B., Ebert, S., Totsuka, K., Vogelman, B., Calame, W., Mattie, H., and Craig, W.A., Comparative antibiotic dose-effect relations at several dosing intervals in murine pneumonitis and thigh-infection models. *J. Infect. Dis.* 159, 281–292, 1989.

41. Moore, R.D., Smith, C.R., and Lietman, P.S., The association of aminoglycoside plasma levels with mortality in patients with Gram-negative bacteremia. *J. Infect. Dis.* 149, 443–448, 1984.

42. Moore, R.D., Smith, C.R., and Lietman, P.S., Association of aminoglycoside plasma levels with therapeutic outcome in Gram-negative pneumonia. *Am. J. Med.* 77, 657–662, 1984.

43. Nicolau, D.P., Freeman, C.D., Belliveau, P.P., Nightingale, C.H., Ross, J.W., and Quintiliani, R., Experience with a once-daily aminoglycoside program administered to 2, 184 adult patients. *Antimicrob. Agents Chemother.* 39, 650–655, 1995.

44. Zaske, D.E., Cipolle, R.J., Rotschafer, J.C., Solem, L.D., Mosier, N.R., and Strate, R.G., Gentamicin pharmacokinetics in 1,640 patients: method for control of serum concentrations. *Antimicrob. Agents Chemother.* 21, 407–411, 1982.

45. Rotschafer, J.C., and Rybak, M.J., Single daily dosing of aminoglycosides: a commentary. *Ann. Pharmacother.* 28, 797–801, 1994.

46. Craig, W.A., Interrelationship between pharmacokinetics and pharmacodynamics in determining dosage regimens for broad-spectrum cephalosporins. *Diagn. Microbiol. Infect. Dis.* 22, 89–96, 1995.

47. Turnidge, J.D., The pharmacodynamics of β-lactams. *Clin. Infect. Dis.* 27, 10–22, 1998.

48. Grasso, S., Meinardi, G., de Carneri, I., and Tamassia, V., New *in vitro* model to study the effect of antibiotic concentration and rate of elimination on antibacterial activity. *Antimicrob. Agents Chemother.* 13, 570–576, 1978.

49. Nishida, M., Murakawa, T., Kamimura, T., and Okada, N., Bactericidal activity of cephalosporins in an *in vitro* model simulating serum levels. *Antimicrob. Agents Chemother.* 14, 6–12, 1978.

50. Schentag, J.J., Smith, I.L., Swanson, D.J., DeAngelis, C., Fracasso, J.E., Vari, A., and Vance, J.W., Role for dual individualization with cefmenoxime. *Am. J. Med.* 77, 43–50, 1984.

51. Frimodt-Moller, N., Bentzon, M.W., and Thomsen, V.F., Experimental infection with *Streptococcus pneumoniae* in mice: correlation of *in vitro* activity and pharmacokinetic parameters with *in vivo* effect for 14 cephalosporins. *J. Infect. Dis.* 154, 511–517, 1986.

52. Roosendaal, R., Bakker-Woudenberg, I.A., van den Berghe-van Raffe, M., and Michel, M.F., Continuous versus intermittent administration of ceftazidime in experimental *Klebsiella pneumoniae* pneumonia in normal and leukopenic rats. *Antimicrob. Agents Chemother.* 30, 403–408, 1986.

53. Vogelman, B., Gudmundsson, S., Leggett, J., Turnidge, J., Ebert, S., and Craig, W.A., Correlation of antimicrobial pharmacokinetic parameters with therapeutic efficacy in an animal model. *J. Infect. Dis.* 158, 831–847, 1988.

54. White, C.A., Toothaker, R.D., Smith, A.L., and Slattery, J.T., *In vitro* evaluation of the determinants of bactericidal activity of ampicillin dosing regimens against *Escherichia coli. Antimicrob. Agents Chemother.* 33, 1046–1051, 1989.

55. Onyeji, C.O., Nicolau, D.P., Nightingale, C.H., and Quintiliani, R., Optimal times above MICs of ceftibuten and cefaclor in experimental intra-abdominal infections. *Antimicrob. Agents Chemother.* 38, 1112–1117, 1994.

56. Lutsar, I., Ahmed, A., Friedland, I.R., Trujillo, M., Wubbel, L., Olsen, K., and McCracken, G.H. Jr., Pharmacodynamics and bactericidal activity of ceftriaxone therapy in experimental cephalosporin-resistant pneumococcal meningitis. *Antimicrob. Agents Chemother.* 41, 2414–2417, 1997.

57. Zinner, S.H., Dudley, M.N., Gilbert, D., and Bassignani, M., Effect of dose and schedule on cefoperazone pharmacodynamics in an *in vitro* model of infection in a neutropenic host. *Am. J. Med.* 85, 56–58, 1988.

58. Lavoie, G.Y., and Bergeron, M.G., Influence of four modes of administration on penetration of aztreonam, cefuroxime, and ampicillin into interstitial fluid and fibrin clots and on *in vivo* efficacy against *Haemophilus influenzae. Antimicrob. Agents Chemother.* 28, 404–412, 1985.

59. Roosendaal, R., Bakker-Woudenberg, I.A., van den Berghe-van Raffe, M., and Michel, M.F., Continuous versus intermittent administration of ceftazidime in experimental *Klebsiella pneumoniae* pneumonia in normal and leukopenic rats. *Antimicrob. Agents Chemother.* 30, 403–408, 1986.

60. Fantin, B., Leggett, J., and Ebert, S., Craig, W.A., Correlation between *in vitro* and *in vivo* activity of antimicrobial agents against gram-negative bacilli in a murine infection model. *Antimicrob. Agents Chemother.* 35, 1413–1422, 1991.

61. Gerber, A.U., Craig, W.A., Brugger, H.P., Feller, C., Vastola, A.P., and Brandel, J., Impact of dosing intervals on activity of gentamicin and ticarcillin against *Pseudomonas aeruginosa* in granulocytopenic mice. *J. Infect. Dis.* 147, 910–917, 1983.

62. Bakker-Woudenberg, I.A., van Gerwen, A.L., and Michel, M.F., Efficacy of antimicrobial therapy in experimental rat pneumonia: antibiotic treatment schedules in rats with impaired phagocytosis. *Infect. Immunity* 25, 376–387, 1979.

63. Bakker-Woudenberg, I.A., van den Berg, J.C., Fontijne, P., and Michel, M.F., Efficacy of continuous versus intermittent administration of penicillin G in *Streptococcus pneumoniae* pneumonia in normal and immunodeficient rats. *Eur. J. Clin. Microbiol.* 3, 131–135, 1984.

64. Gengo, F.M., Mannion, T.W., Nightingale, C.H., and Schentag, J.J., Integration of pharmacokinetics and pharmacodynamics of methicillin in curative treatment of experimental endocarditis. *J. Antimicrob. Chemother.* 14, 619–631, 1984.

65. Joly, V., Pangon, B., Vallois, J.M., Abel, L., Brion, N., Bure, A., Chau, N.P., Contrepois, A., and Carbon, C., Value of antibiotic levels in serum and cardiac vegetations for predicting antibacterial effect of ceftriaxone in experimental *Escherichia coli* endocarditis. *Antimicrob. Agents Chemother.* 31, 1632–1639, 1987.

66. Pangon, B., Joly, V., Vallois, J.M., Abel, L., Bure, A., Brion, N., Contrepois, A., and Carbon, C., Comparative efficacy of cefotiam, cefmenoxime, and ceftriaxone in experimental endocarditis and correlation with pharmacokinetics and *in vitro* efficacy. *Antimicrob. Agents Chemother.* 31, 518–522, 1987.

67. Brewin, A., Arango, L., Hadley, W.K., and Murray, J.F., High-dose penicillin therapy and pneumococcal pneumonia. *JAMA* 230, 409–413, 1974.

68. Bodey, G.P., Ketchel, S.J., and Rodriguez, V., A randomized study of carbenicillin plus cefamandole or tobramycin in the treatment of febrile episodes in cancer patients. *Am. J. Med.* 67, 608–616, 1979.

69. Lagast, H., Meunier-Carpentier, F., and Klastersky, J., Treatment of gram-negative bacillary septicemia with cefoperazone. *Eur. J. Clin. Microbiol.* 2, 554–558, 1983.

70. Cars, O., Efficacy of β-lactam antibiotics: integration of pharmacokinetics and pharmacodynamics. *Diagnost. Microbiol. Infect. Dis.* 27, 29–33, 1997.

71. Lamp, K.C., and Vickers, M.K., Pharmacodynamics of ampicillin-sulbactam in an *in vitro* infection model against *Escherichia coli* strains with various levels of resistance. *Antimicrob. Agents Chemother.* 42, 231–235, 1998.

72. MacGowan, A.P., and Bowker, K.E., Continuous infusion of β-lactam antibiotics. *Clin. Pharmacokinetics* 35, 391–402, 1998.

73. Andes, D.R., and Craig, W.A., Pharmacokinetics and pharmacodynamics of antibiotics in meningitis. *Infect. Dis. Clin. North America* 13, 595–618, 1999.

74. Rapp, R.P., Pharmacokinetics and pharmacodynamics of intravenous and oral azithromycin: enhanced tissue activity and minimal drug interactions. *Ann. Pharmacother.* 32, 785–793, 1998.

75. McConnell, S.A., and Amsden, G.W., Review and comparison of advanced-generation macrolides clarithromycin and dirithromycin. *Pharmacotherapy* 19, 404–415, 1999.

76. Ebert, S., Rikardsdottir, S., and Craig, W.A., Pharmacodynamic comparison of clarithromycin vs erythromycin. *31st Interscience Conference on Antimicrobial Agents and Chemotherapy.* Washington DC, 1991.

77. Craig, W., Rikardsdottir, S., and Watanabe, Y., *In vivo* and *in vitro* postantibiotic effects (PAEs) of azithromycin. *32nd Interscience Conference on Antimicrobial Agents and Chemotherapy.* Washington DC, 1992

78. Greenberg, R.N., and Benes, C.A., Time-kill studies with oxacillin, vancomycin, and teicoplanin versus *Staphylococcus aureus. J. Infect. Dis.* 161, 1036–1037, 1990.

79. Peetermans, W.E., Hoogeterp, J.J., Hazekamp-van Dokkum, A.M., van den Broek, P., and Mattie, H., Antistaphylococcal activities of teicoplanin and vancomycin *in vitro* and in an experimental infection. *Antimicrob. Agents Chemother.* 34, 1869–1874, 1990.

80. MacGowan, A.P., Pharmacodynamics, pharmacokinetics, and therapeutic drug monitoring of glycopeptides. *Therapeutic Drug Monitoring* 20, 473–477, 1998.

81. Odenholt, I., Lowdin, E., and Cars, O., *In vitro* studies of the pharmacodynamics of teicoplanin against *Staphylococcus aureus, Staphylococcus epidermidis* and *Enterococcus faecium. Clin. Microbiol. Infect.* 9, 930–937, 2003.

82. Chambers, H.F., and Kennedy, S., Effects of dosage, peak and trough concentrations in serum, protein binding, and bactericidal rate on efficacy of teicoplanin in a rabbit model of endocarditis. *Antimicrob. Agents Chemother.* 34, 510–514, 1990.

83. Knudsen, J.D., Fuursted, K., Raber, S., Espersen, F., and Frimodt-Moller, N., Pharmacodynamics of glycopeptides in the mouse peritonitis model of *Streptococcus pneumoniae* or *Staphylococcus aureus* infection. *Antimicrob. Agents Chemother.* 44, 1247–1254, 2000.

84. Leport, C., Perronne, C., Massip, P., Canton, P., Leclercq, P., Bernard, E., Lutun, P., Garaud, J.J., and Vilde, J., Evaluation of teicoplanin for treatment of endocarditis caused by gram-positive cocci in 20 patients. *Antimicrob. Agents Chemother.* 33, 871–876, 1989.

85. Hyatt, J.M., McKinnon, P.S., Zimmer, G.S., and Schentag, J.J., The importance of pharmacokinetic/pharmacodynamic surrogate markers to outcome. Focus on antibacterial agents. *Clin. Pharmacokinetics* 28, 143–160, 1995.

86. White, R.L., What *in vitro* models of infection can and cannot do. *Pharmacotherapy* 21, 292S–301S, 2001.

87. Li, R.C., and Zhu, Z.Y., *In vitro* models for prediction of antimicrobial activity: a pharmacokinetic and pharmacodynamic perspective. *J. Chemother.* 9(Suppl 1), 55–63, 1997.

88. Li, R.C., New pharmacodynamic parameters for antimicrobial agents. *Int. J. Antimicrob. Agents* 13, 229–235, 2000.

89. Andes, D., and Craig, W.A., Animal model pharmacokinetics and pharmacodynamics: a critical review. *Int. J. Antimicrob. Agents* 19, 261–268, 2002.

90. Lewis, R.E., Klepser, M.E., Ernst, E.J., Lund, B.C., Biedenbach, D.J., and Jones, R.N., Evaluation of low-dose, extended-interval clindamycin regimens against *Staphylococcus aureus* and *Streptococcus pneumoniae* using a dynamic *in vitro* model of infection. *Antimicrob. Agents Chemother.* 43, 2005–2009, 1999.

91. Aeschlimann, J.R., Allen, G.P., Hershberger, E., and Rybak, M.J., Activities of LY333328 and vancomycin administered alone or in combination with gentamicin against three strains of vancomycin-intermediate *Staphylococcus aureus* in an *in vitro* pharmacodynamic infection model. *Antimicrob. Agents Chemother.* 44, 2991–2998, 2000.

92. Hershberger, E., and Rybak, M.J., Activities of trovafloxacin, gatifloxacin, clinafloxacin, sparfloxacin, levofloxacin, and ciprofloxacin against penicillin-resistant Streptococcus pneumoniae in an *in vitro* infection model. *Antimicrob. Agents Chemother.* 44, 598–601, 2000.

93. Coyle, E.A., and Rybak, M.J., Activity of oritavancin (LY333328), an investigational glycopeptide, compared to that of vancomycin against multidrug-resistant *Streptococcus pneumoniae* in an *in vitro* pharmacodynamic model. *Antimicrob. Agents Chemother.* 45, 706–709, 2001.

94. Thorburn, C.E., and Edwards, D.I., The effect of pharmacokinetics on the bactericidal activity of ciprofloxacin and sparfloxacin against Streptococcus pneumoniae and the emergence of resistance. *J. Antimicrob. Chemother.* 48, 15–22, 2001.

95. Garrison, M.W., Vance-Bryan, K., Larson, T.A., Toscano, J.P., and Rotschafer, J.C., Assessment of effects of protein binding on daptomycin and vancomycin killing of *Staphylococcus aureus* by using an *in vitro* pharmacodynamic model. *Antimicrob. Agents Chemother.* 34, 1925–1931, 1990.

96. Lowdin, E., Odenholt, I., Bengtsson, S., and Cars, O., Pharmacodynamic effects of sub-MICs of benzylpenicillin against *Streptococcus pyogenes* in a newly developed *in vitro* kinetic model. *Antimicrob. Agents Chemother.* 40, 2478–2482, 1996.

97. Firsov, A.A., Vostrov, S.N., Shevchenko, A.A., and Cornaglia, G., Parameters of bacterial killing and regrowth kinetics and antimicrobial effect examined in terms of area under the concentration-time curve relationships: action of ciprofloxacin against *Escherichia coli* in an *in vitro* dynamic model. *Antimicrob. Agents Chemother.* 41, 1281–1287, 1997.

98. Vostrov, S.N., Kononenko, O.V., Lubenko, I.Y., Zinner, S.H., and Firsov, A.A., Comparative pharmacodynamics of gatifloxacin and ciprofloxacin in an *in vitro* dynamic model: prediction of equiefficient doses and the breakpoints of the area under the curve/MIC ratio. *Antimicrob. Agents Chemother.* 44, 879–884, 2000.

99. Blaser, J., In-vitro model for simultaneous simulation of the serum kinetics of two drugs with different half-lives. *J. Antimicrob. Chemother.* 15(Suppl A), 125–130, 1985.

100. Shah, P.M., Simultaneous simulation of two different concentration time curves *in vitro*. *J. Antimicrob. Chemother.* 15(Suppl A), 261–264, 1985.

101. Zinner, S.H., Blaser, J., Stone, B.B., and Groner, M.C., Use of an in-vitro kinetic model to study antibiotic combinations. *J. Antimicrob. Chemother.* 15(Suppl A), 221–226, 1985.

102. Lister, P.D., Prevan, A.M., and Sanders, C.C., Importance of β-lactamase inhibitor pharmacokinetics in the pharmacodynamics of inhibitor-drug combinations: studies with piperacillin-tazobactam and piperacillin-sulbactam. *Antimicrob. Agents Chemother.* 41, 721–727, 1997.

103. Blaser, J., Vergeres, P., Widmer, A.F., and Zimmerli, W., *In vivo* verification of *in vitro* model of antibiotic treatment of device-related infection. *Antimicrob. Agents Chemother.* 39, 1134–1139, 1995.

104. Gander, S., Hayward, K., and Finch, R., An investigation of the antimicrobial effects of linezolid on bacterial biofilms utilizing an *in vitro* pharmacokinetic model. *J. Antimicrob. Chemother.* 49, 301–308, 2002.

105. Palmer, S.M., and Rybak, M.J., Pharmacodynamics of once- or twice-daily levofloxacin versus vancomycin, with or without rifampin, against *Staphylococcus aureus* in an *in vitro* model with infected platelet-fibrin clots. *Antimicrob. Agents Chemother.* 40, 701–705, 1996.

106. Rybak, M.J., Houlihan, H.H., Mercier, R.C., and Kaatz, G.W., Pharmacodynamics of RP 59500 (quinupristin-dalfopristin) administered by intermittent versus continuous infusion against *Staphylococcus aureus*-infected fibrin-platelet clots in an *in vitro* infection model. *Antimicrob. Agents Chemother.* 41, 1359–1363, 1997.

107. Aeschlimann, J.R., and Rybak, M.J., Pharmacodynamic analysis of the activity of quinupristin-dalfopristin against vancomycin-resistant *Enterococcus faecium* with differing MBCs via time-kill-curve and postantibiotic effect methods. *Antimicrob. Agents Chemother.* 42, 2188–2192, 1998.

108. Oramas-Shirey, M.P., Buchanan, L.V., Dileto-Fang, C.L., Dailey, C.F., Ford, C.W., Batts, D.H., and Gibson, J.K., Efficacy of linezolid in a staphylococcal endocarditis rabbit model. *J. Antimicrob. Chemother.* 47, 349–352, 2001.

109. MacGowan, A.P., Pharmacokinetic and pharmacodynamic profile of linezolid in healthy volunteers and patients with Gram-positive infections. *J. Antimicrob. Chemother.* 51(Suppl 2), ii17–ii25, 2003.

110. Drusano, G.L., Pharmacodynamics of anti-infectives: target delineation and target attainment. In: Finch, R.G., Greenwood, D., Norrby, S.R., and Whitley, R.J., eds., *Antibiotic and Chemotherapy.* Churchill Livingstone: London, 2003:48–58.

111. Ambrose, P.G., Grasela, D.M., Grasela, T.H., Passarell, J., Mayer, H.B., and Pierce, P.F., Pharmacodynamics of fluoroquinolones against Streptococcus pneumoniae in patients with community-acquired respiratory tract infections. *Antimicrob. Agents Chemother.* 45, 2793–2797, 2001.

18 Protein Synthesis Inhibitors, Fluoroquinolones, and Rifampin for Biofilm Infections

Steven L. Barriere

CONTENTS

18.1   INTRODUCTION

Bacterial biofilms are a common, yet complex, biological phenomenon. They are associated with infections of foreign bodies (vascular and urinary catheters, endotracheal tubes, prosthetic devices), chronic respiratory infection (cystic fibrosis), otitis media, bone and joint infections, endocarditis, and such common clinical entities as periodontal disease and dental caries (1).

A biofilm is, on the surface, a relatively simple structure of aggregates of organisms and their extracellular products. However, recent research has revealed the complexity of the formation and structure of biofilms, which relates directly to the difficulty in preventing or eradicating biofilms with antimicrobial therapy (1,2).

Numerous papers have been published over the past 10 years on the activity of antibiotics in biofilms. Extensive investigations have demonstrated the activity of various antibacterial agents, especially aminoglycosides and fluoroquinolones vs. *Pseudomonas aeruginosa* in biofilms. Although less extensively studied, the effects of various antimicrobials against staphylococci (especially *Staphylococcus epidermidis*) have been evaluated.

Three factors appear to be most important in determining antimicrobial action in biofilms:

(a) Bacteria within the biofilms have slower growth rates than organisms external to the biofilms. Bacteria in slow growth or dormant phase are unaffected by many antibiotics.

(b) Antimicrobial penetration into the biofilm is reduced, partly due to the structure of the biofilm matrix and partly due to the physicochemical characteristics of the drugs.

(c) The extracellular polymeric substances that are formed by the bacteria forming the biofilms act as a barrier to penetration by oxygen. Hence, the activity of antibiotics is compromised due to the oxygen-poor environment (3).

In addition to the aforementioned factors, it is evident that the colonies of organisms within biofilms are heterogeneous, with a small but significant portion of the population highly resistant to antibiotics (4). These resistant organisms persist despite exposure to bactericidal antibacterials. These persisters explain the commonly observed phenomenon observed in biofilms models of a lack of additional bactericidal effect beyond a degree of killing following exposure to bactericidal drugs; and diminished or absent responses to subsequent drug exposure (4).

Adding further to the complexity of the antibiotic–biofilm interaction is the finding that the chemical composition of biomaterials can affect the activity of certain antibiotics against the bacteria within the biofilm (5).

Therefore, the activity of an antimicrobial agent in biofilm infections depends upon its physicochemical properties in addition to inherent antibacterial activity. For example, different members of the same class of compounds (fluoroquinolones) may have disparate activity against the organisms within a biofilm infection. An additional critical factor in the pathogenesis of biofilm formation is the ability of the organisms to adhere to a surface (5). This propensity is associated with the ability of organisms to produce the exopolysaccharide "slime" as well as a variety of proteins that contribute to form the biofilm matrix. For example, some of these proteins identified in staphylococci include clumping Factor A, staphylococcal surface protein and biofilm-associated protein (6).

A variety of different organisms are associated with biofilm infections, but since a key factor in pathogenesis is the ability to adhere to surfaces (production of adhesion factors), staphylococci (particularly *S. epidermidis*) are common pathogens. Staphylococci are the most common organisms found in bloodstream infections (7). This high prevalence coupled with staphylococcal adherence capabilities result in their being responsible for most infections associated with intravascular catheters and prosthetic materials due to seeding from the bloodstream (5).

Other important biofilm-associated infections are cystic fibrosis and related diseases such as diffuse panbronchiolitis (1). These conditions are characterized by colonization and relapsing infection of the lung with *P. aeruginosa*. While this organism is capable of slime formation, it is believed that physiologic defects found in cystic fibrosis patients (elevated salt content in airway surface fluids that inhibits endogenous antimicrobial peptides) are responsible for the chronic colonization with these bacteria (1).

Understanding of the mechanisms of resistance in biofilms to the effects of antibiotics supports the observations that some antibiotics appear to be more effective than others despite apparently adequate *in vitro* activity. Additionally, the combination of resistance factors operative in biofilm infections, as outlined above, strongly suggests that combinations of antimicrobial agents are needed for effective treatment.

This chapter will survey the effects of protein synthesis inhibitors such as aminoglycosides, macrolides, quinupristin/dalforpristin (Synercid®), and linezolid (Zyvox®). Additionally it will cover the use of fluoroquinolones (topoisomerase II inhibitiors), as well as inhibitors of DNA-dependent RNA polymerase (rifampin and analogues).

18.2   AMINOGLYCOSIDES

This class of drugs is primarily used for the treatment of aerobic Gram-negative infections. Aminoglycosides are rapidly bactericidal in contrast to other inhibitors of protein synthesis, which are generally bacteriostatic. It is unclear why aminoglycosides, which bind to the 30S and 50S subunits of the ribosome and cause misreading of mRNA, are bactericidal (8). In combination with cell wall-active agents (e.g., β-lactams, glycopeptides), they produce synergistic bactericidal activity. This is used clinically in the treatment of enterococcal infections and for infections due to *P. aeruginosa*.

In biofilm infections, aminoglycosides are used for the management of infections due to Gram-negative bacteria, especially *P. aeruginosa*, in cystic fibrosis (CF) and other chronic lung disorders (1). They are used systemically, generally in combination with an antipseudomonal β-lactam, to treat acute infectious episodes. However, a primary role in CF is suppressive with inhalation therapy as a mainstay. An inhaled preparation of tobramycin (TOBI®, Chiron Corporation) has been shown to be effective in diminishing acute exacerbations of CF disease (9). It is believed that delivery of high concentrations of the drug locally via nebulization into the airway of afflicted individuals is the key factor in inhibiting biofilm formation as well as providing sufficient concentrations of antibiotic within the biofilm to exert bactericidal effects.

Two randomized, placebo-controlled trials were conducted with nebulized tobramycin in CF patients ≥ 6 years of age (9). The treatment groups were well matched with regard to baseline FEV_1 (between 25% and 75%), as well as in the receipt of other standard modalities of therapy for CF such as systemic antimicrobial therapy, β_2 agonists, cromolyn, inhaled steroids, and airway clearance techniques. Approximately 75% of the patients also received dornase alfa (Pulmozyme®, Genentech).

Two hundred and fifty-eight patients were randomized and treated. The tobramycin-treated patients experienced a 7 to 11% increase in $FEV_1\%$ from baseline during the 24 weeks of study compared to no change for placebo (8). Additionally, tobramycin therapy resulted in a significant reduction in the sputum bacterial burden during the on drug periods. Sputum bacterial counts returned to baseline levels during off drug periods, and smaller reductions in bacterial density were achieved with subsequent drug exposures. Patients who received tobramycin were hospitalized for a mean of ~5 days during the 24 week treatment period vs. ~8 days for placebo, and required four fewer days of systemic antipseudomonal therapy during that period. However, reductions in susceptibility to tobramycin were observed during the 24 weeks, possibly explaining the diminished response to repeated study drug exposures. This is reflective of the aforementioned mechanism of drug resistance in biofilms.

An analogous clinical situation exists in patients with endotracheal tubes, who are being mechanically ventilated. Bacteria colonize the plastic tubing and establish a biofilm. Organisms shed from this biofilm and are responsible for the development of ventilator-associated pneumonia.

The results of an exploratory clinical study (10) suggest potential benefits of locally delivered high concentrations of aminoglycosides in this setting. Thirty-six intubated patients received one of three different antibiotics administered via nebulizer. The drugs used were cefuroxime, cefotaxime, or gentamicin. Following extubation, the endotracheal tubes were examined for biofilm formation. Biofilms were found in 8/12, 7/12 and 5/12 of the endotracheal tubes from the patients who had received cefuroxime, cefotaxime, and gentamicin, respectively. However, in contrast to the cephalosporin-exposed tubes, none of the gentamicin-exposed tubes contained bacteria considered to be pathogenic in ventilator-associated pneumonia (10).

Gram-negative bacteria such as *P. aeruginosa* are also found in biofilms associated with urinary catheters. Recurrent urinary tract infection is an enormous problem in chronically catheterized patients, leading to substantial morbidity and mortality. An experimental *in vitro* model of catheter-associated biofilm was used to evaluate the effects of various antimicrobials on a *P. aeruginosa* biofilm (11). The biofilm was exposed to drug concentrations up to 128-fold the MBC for each respective agent. The drugs used were piperacillin, ceftazidime, paripenem [a carbapenem], amikacin, ciprofloxacin, and levofloxacin. The β-lactams were found to be effective only at high concentrations whereas amikacin and the fluoroquinolones produced more rapid killing at lower concentrations (11). This difference may be reflective of the differing pharmacodynamic effects of β-lactams compared to aminoglycosides: time-dependency vs. concentration-dependency. Sterilization of the biofilm was only achieved with amikacin and the fluoroquinolones.

In summary, the use of aminoglycosides in biofilm disease has been primarily in the use of high concentrations delivered locally in patients with cystic fibrosis. While the achieved clinical results appear to be modest, reductions in hospitalization and systemic antimicrobial use result in substantial cost saving and reductions in morbidity.

18.3 MACROLIDES

Like aminoglycosides, macrolides inhibit protein synthesis in bacteria by reversible binding to the 50S ribosome subunit, and subsequent inhibition of translocation of peptidyl tRNA (8). However, unlike aminoglycosides, macrolides are generally bacteriostatic *in vitro* and *in vivo*, and have useful activity primarily vs. Gram-positive bacteria.

Despite this limitation in *in vitro* activity, there is substantial evidence that macrolides such as erythromycin, clarithromycin, and azithromycin have salutary effects in chronic respiratory disease associated with *P. aeruginosa* (12). It has been suggested that the 14- or 15-membered ring macrolides exert an anti-inflammatory effect, resulting in long term reductions in symptoms of patients with a disease similar to CF, diffuse panbronchiolitis (DPB). Kobayashi observed a good correlation between serum levels of immune complexes and disease activity in this condition (13). These investigators suggest that inhibition of the formation of antigen (alginate – a component of *P. aeruginosa* biofilm) antibody complexes by macrolides results in this salutary effect (14). However, it would appear that a reduction in immune complexes is a secondary effect due to macrolide-induced reduction in the production of alginate by *P. aeruginosa*.

This was demonstrated *in vitro* (15) with macrolides, revealing a dose-dependent reduction in alginate production by *P. aeruginosa*, as well as *in vivo* in a murine model (16). Kobayashi suggests that the macrolides, despite minimal antibacterial activity vs. *P. aeruginosa*, appear to be able to inhibit the activity of guanosine diphosphomannose dehydrogenase, resulting in decreased alginate formation (13). Yosuda confirmed this theory and also demonstrated increased penetration of various antimicrobials into biofilms exposed to clarithromycin (14). Electron micrographs revealed dramatic disruptions in the structure of biofilms following clarithromycin exposure (14). The biofilm penetration of fluoroquinolones and aminoglycosides was enhanced with increasing concentrations of clarithromycin. These findings have been reproduced *in vitro* and *in vivo* by other investigators (17).

These early observations have led to clinical investigations in both DPB and CF, which share many of the same clinical characteristics (12). Long term, low dosage erythromycin has been shown to improve symptoms of DPB and increase 10 year survival from ~12% to ≥90% (13). An excellent review of the use of macrolides in CF is available (12). Macrolide use in CF has some clinical utility, but improvements in disease severity and progression have not been as dramatic.

As noted previously, biofilms form on urinary catheters, primarily with Gram-negative bacteria and are a major problem associated with significant morbidity and mortality. Macrolides have been evaluated in this setting as well, despite their minimal activity vs. Gram-negative bacteria, because of their salutary effects in airway biofilm disease. Vranes performed an *in vitro* experiment examining the effects of azithromycin on the adherence of *P. aeruginosa* to polystyrene (a common component of urinary catheters) (18). Sub-MIC concentrations of azithromycin (0.06–0.5 × MIC) reduced the adherence by 50 to 75%.

Similarly, patients requiring hemodialysis for renal failure have either native or synthetic arteriovenous fistulas created for vascular access. These fistulas frequently become infected and biofilms are formed. Gascon et al. (19) reported a case of dialysis-catheter infection due to *P. aeruginosa*, with bacteremia, that was refractory to repeated courses of systemic antibacterial therapy. Clinical cure was achieved when oral clarithromycin was added to the regimen.

In summary, the use of macrolides has shown substantial promise in the treatment of biofilm-associated infections, primarily in chronic airway disease.

Chronic administration of low dosages of these agents has substantially improved survival in DPB and limited data reveal improvements in respiratory function in patients with CF.

18.4   QUINUPRISTIN/DALFOPRISTIN (Q/D)

This combination product (Synercid, King Pharma) was introduced to clinical use in 1999. The combination is bactericidal vs. most Gram-positive bacteria, including strains of MRSA and GISA, but is limited by substantial toxicity. It is approved for use primarily for the treatment of vancomycin-resistant enterococcal (Enterococcus faecium only) infections. Q/D is comprised of two streptogramins, which are a group of natural cyclic peptides (8). The mechanism of action of Q/D is similar to that of macrolides, explaining the cross-resistance found in isolates of staphylococci that carry the *ermB* gene, conferring constitutive resistance to macrolides.

An *in vitro* biofilm model employing various strains of staphylococci with differing oxacillin and vancomycin susceptibilities, was used to compare the effects of antimicrobial agents, including Q/D (20). Two methods of exposure were assessed: 2-hour exposure to constant drug concentrations or exponentially decreasing concentrations of drug mimicking human pharmacokinetic disposition. The total amount of drug exposure was the same for both methods. Q/D produced a substantial decrease in biofilm-associated bacteria in both experiments, but was the most effective agent in the constant exposure method.

In a different *in vitro* model, the activity of Q/D was compared with ciprofloxacin against several strains of *S. epidermidis*, with varying MIC to glycopeptides, macrolides and ciprofloxacin (21). All strains were susceptible to Q/D (MIC ≤0.25 µg/mL). Extents of killing by Q/D were similar to those observed for the fluoroquinolone, but the killing rates were somewhat slower.

18.5   LINEZOLID

Linezolid (Zyvox, Pfizer) is a newly introduced protein synthesis inhibitor, with a mechanism of action very similar to that of chloramphenicol, resulting in inhibition of peptide bond formation (8). Linezolid is bacteriostatic and, like chloramphenicol, produces reversible bone marrow toxicity (primarily thrombocytopenia) in a substantial number of patients receiving prolonged (≥10–14 days) therapy.

The activity of linezolid has been studied in biofilms. Linezolid was studied in the experiment described above (20) wherein various staphylococci were exposed to either constant or exponentially decreasing concentrations of antibacterials. Like Q/D, linezolid produced substantial reductions in organisms associated with the biofilm, regardless of their susceptibility to glycopeptides or oxacillin.

An *in vitro* model of biofilm infection simulating a central venous catheter infection with *Staphylococcus epidermidis* was developed to assess the activity of various antimicrobials (22). In this experiment, biofilm formation was confirmed by electron microscopy and quantified by bacterial counts on polyurethane coupons. The antibacterials tested included vancomycin, gentamicin, and linezolid. Drug exposures were carried out for up to 10 days. Linezolid achieved eradication of

S. epidermidis biofilms more rapidly than the other drugs, after only 3 days of exposure compared to 10 days for vancomycin (22). Viable bacteria were recovered from gentamicin-exposed biofilms throughout the experiment, with minimal reduction in counts.

In summary, linezolid show some promise for the treatment of biofilm infections due to staphylococci. Clinical studies are needed to validate the *in vitro* findings.

18.6 FLUOROQUINOLONES

Numerous fluoroquinolones are available for clinical use. These drugs exert their antibacterial effects by inhibition of topoisomerases (Types II and IV), which are responsible for double-stranded breaks and ATP-relaxation of bacterial (but not human) DNA, respectively (8). Like aminoglycosides, fluoroquinolones exert concentration-dependent bactericidal effects, and are widely used for the treatment of various Gram-negative bacterial infections, as well as respiratory infections caused by Gram-positive and Gram-negative bacteria and atypical pathogens. As noted previously, fluoroquinolones have substantial activity against biofilm-associated bacteria. Their potential for use has been established *in vitro* and *in vivo*.

Chronic suppressive therapy with oral ciprofloxacin has become a mainstay of treatment in the management of patients with CF (23). This use is part of consensus guidelines for the treatment of *P. aeruginosa* infection in CF (23). The mechanism of fluoroquinolone action in biofilms is unclear, but may be related to the ability of these drugs to penetrate biofilms and subsequent reduction of bacterial burden, similar to aminoglycosides.

Ciprofloxacin was studied in the experimental biofilm model described above (20) wherein various staphylococci were exposed to either constant or exponentially decreasing concentrations of antibacterials. Like the protein synthesis inhibitors, ciprofloxacin produced the greatest reductions in organisms associated with the biofilm, but only against non-drug resistant strains.

Fluoroquinolones are excreted in urine to a significant degree and have been shown to concentrate in renal tissue (8). The beneficial effects of a fluoroquinolone on urinary catheter-associated biofilms were demonstrated in a double-blind randomized study in quadriplegic patients (24). Ofloxacin was compared to trimethoprim-sulfamethoxazole (TMP-SMX), a long-standing drug of choice for the treatment of urinary tract infection. Clinical cure rates at the end of therapy were 90% for ofloxacin vs. 57% for TMP-SMX ($p=0.015$). Ofloxacin therapy also led to significantly greater biofilm eradication (67% vs. 35%, $p=0.014$).

In summary, fluoroquinolones have an established role in chronic airway disease associated with biofilms, and show some potential for the treatment of biofilm disease caused by staphylococci, but with poor activity vs. drug resistant strains (i.e., MRSA, GISA).

18.7 RIFAMPIN

Rifampin and its analogues are highly active and bactericidal vs. staphylococci. Rifampin exerts antibacterial effects by inhibition of DNA-dependent RNA polymerase (8).

However, only a single step mutation renders bacteria resistant to the drug. Hence, clinical use of rifampin necessitates it be used in combinations of two or more active compounds, to minimize this mutational resistance.

Extensive literature exists demonstrating the *in vitro* and *in vivo* effects of combining rifampin with other anti-staphylococcal agents for the treatment of a variety of infections (25). This includes impressive data in the management of biofilm infections associated with *P. aeruginosa* (despite no inherent *in vitro* activity) as well as staphylococci.

A biofilm model with mucoid *P. aeruginosa* strains was used to evaluate the effects of various agents (26). In this model, the addition of rifampin to ceftazidime ± gentamicin resulted in synergistic bactericidal activity. Rifampin + ciprofloxacin has been shown to be very effective in the treatment of prosthetic implant infections (27), and in tricuspid valve infective endocarditis due to *S. aureus* (28).

Blaser et al. evaluated the effects of several antibiotics in an *in vivo* model of biofilm infection (29). The drugs tested included glycopeptides and fluoro-quinolones with or without rifampin. Organisms used were isolates of MSSA and MRSE. Rifampin combinations were uniformly more bactericidal in the biofilm than any of the drugs administered alone or in combination.

In summary, rifampin is a useful adjunct to other antibacterial regimens in the management of biofilm infections. Rapid emergence of resistance precludes its use as monotherapy, but synergistic bactericidal activity appears to be associated with important clinical benefit when combined with other agents.

18.8   CONCLUSIONS

Protein synthesis inhibitors and fluoroquinolones have been examined extensively *in vitro* and *in vivo* for the management of biofilm infections. Aminoglycosides are routinely used in the treatment of patients with cystic fibrosis and similar conditions, both systemically and locally applied (inhalation). The beneficial effects of amino-glycosides in biofilm infections appear to be related to their ability to reduce organism burden via their bactericidal effects. However, these drugs are not as dramatically effective as other protein synthesis inhibitors in this regard.

Macrolides appear to have dual effects in biofilm infections, but these effects are distinct from their antibacterial activity. The drugs inhibit the production of components of the slime by Gram-negative bacteria and may produce an anti-inflammatory effect.

Linezolid and quinupristin/dalfopristin are the newest protein synthesis inhibitors to become clinically available, and appear promising for the treatment of biofilm infections based upon *in vitro* observations. Clinical investigations are required to validate these findings.

Fluoroquinolones, like aminoglycosides, are generally used in biofilm infection to reduce bacterial burden, and may be more effective in this regard owing to their superior penetration into the biofilm matrix. Chronic suppressive therapy with oral ciprofloxacin has become part of routine therapy for the management of cystic fibrosis.

Finally, rifampin exerts synergistic bactericidal effects in combination with various agents against Gram-positive and Gram-negative bacteria. This activity, coupled with excellent intracellular penetration makes this a potentially valuable agent for the management of biofilm infections.

REFERENCES

1. Costerton, J.W., Stewart, P.S., and Greenberg, E.P., Bacterial biofilms: a common cause of persistent infections. *Science* 284, 1318–1322, 1999.
2. Habash, M., and Reid, G., Microbial biofilms: their development and significance for medical device-related infections. *J. Clin. Pharmacol.* 39, 887–898, 1999.
3. Walters, M.C. 3rd., Roe, F., Bugnicourt, A., Franklin, M.J, and Stewart, P.S., Contributions of antibiotic penetration, oxygen limitation, and low metabolic activity to tolerance of *Pseudomonas aeruginosa* biofilms to ciprofloxacin and tobramycin. *Antimicrob. Agents Chemother.* 47, 317–323, 2003.
4. Lewis, K., Riddle of biofilm resistance. *Antimicrob. Agents Chemother.* 45, 999–1007, 2001.
5. O'Gara, J.P., and Humpreys, H., *Staphylococcus epidermidis* biofilms: importance and implications. *J. Med. Microbiol.* 50, 582–587, 2001.
6. Gotz, F. *Staphylococcus* and biofilms. *Mol. Microbiol.* 43, 1367–1378, 2002.
7. National Nosocomial Infection Surveillance System. Data Summary for January 1992–June 2002. *Am. J. Infect. Control.* 30, 458–475, 2002.
8. Scholar, E.M., and Pratt, W.B., *The Antimicrobial Drugs*, Second Edition. Oxford University Press, 2000.
9. TOBI® Product Information. Chiron Corp. 2001
10. Adair, C.G., Gorman, S.P., Byers, L.M., et al., Eradication of endotracheal tube biofilm by nebulized gentamicin. *Intensive Care Med.* 28, 426–431, 2002.
11. Goto, T., Nakame, Y., Nishida, M., and Ohi, Y., *In vitro* bactericidal activities of beta-lactams, amikacin, and fluoroquinolones against *Pseudomonas aeruginosa* biofilm in artificial urine. *Urology* 53, 1058–1062, 1999.
12. Gaylor, A.S., and Reilly, J.C., Therapy with macrolides in patients with cystic fibrosis. *Pharmacother.* 22, 227–239, 2002.
13. Kobayashi, H. Biofilm diseases: It's clinical manifestation and therapeutic possibilities of macrolides. *Am. J. Med.* 99(6A), 26S–30S, 1995.
14. Yasuda, H., Ajiki, Y., Koga, T., Kawada, H., and Yokota, T., Interaction between biofilms formed by *Pseudomonas aeruginosa* and clarithromycin. *Antimicrob. Agents Chemother.* 37, 1749–1755, 1993.
15. Ichimiya, T., Takeoka, K., Hiramatsu, K., et al., The influence of azithromycin on the biofilm formation of *Pseudomonas aeruginosa in vitro*. *Chemotherapy* 42, 186–191, 1996.
16. Yanagihara, K., Tomono, K., Imamura, Y., et al. Effect of clarithromycin on chronic respiratory infection caused by *Pseudomonas aeruginosa* with biofilm formation in an experimental murine model. *J. Antimicrob. Chemother.* 49, 867–870, 2002.
17. Bui, K.Q., Banevicius, M.A., Nightingale, C.H., Quintiliani, R., and Nicolau, D.P., *In vitro* and *in vivo* influence of adjunct clarithromycin on the treatment of mucoid *Pseudomonas aeruginosa*. *J. Antimicrob. Chemother.* 45, 57–62, 2000.
18. Vranes, J., Effect of subminimal inhibitory concentrations of azithromycin on adherence of *Pseudomonas aeruginosa* to polystyrene. *J. Chemother.* 12, 280–285, 2000.
19. Gascon, A., Iglesias, E., Zabala, S., and Belvis, J.J., Catheter salvage in a patient on hemodialysis with a catheter-related bacteremia by *Pseudomonas aeruginosa*. *Am. J. Nephrol.* 20, 496–497, 2000.
20. Gander, S., Hayward, K., and Finch, R., An investigation of the antimicrobial effects of linezolid on bacterial biofilms utilizing and *in vitro* pharmacokinetic model. *J. Antimicrob. Chemother.* 49, 301–308, 2002.
21. Hamilton-Miller, J.M.T., and Shah, S., Activity of quinupristin-dalfopristin against *Staphylococcus epidermidis* in biofilms, a comparison with ciprofloxacin. *J. Antimicrob. Chemother.* 39(A), 103–108, 1997.

22. Curtin, J., Cormican, M., Fleming, G., Keelehan, J., and Colleran, E., Linezolid compared with eperezolid, vancomycin and gentamicin in an *in vitro* model of antimicrobial lock therapy for *Staphylococcus epidermidis* central venous catheter-related biofilm infection. *Antimicrob. Agents Chemother.* 47, 3145–3148, 2003.

23. Doring, G., Conway, S.P., Heijerman, H.G., et al., Antibiotic therapy against *Pseudomonas aeruginosa* in cystic fibrosis: a European consensus. *Eur. Resp. J.* 16, 749–767, 2000.

24. Reid, G., Poter, P., Delaney, G., et al., Ofloxacin for the treatment of urinary tract infections and biofilms in spinal cord injury. *Int. J. Antimicrob. Agents* 13, 305–307, 2000.

25. Vesely, J.J., Pien, F.D., and Pien, B.C., Rifampin, a useful drug for non-mycobacterial infections. *Pharmacotherapy*, Mar–Apr 18(2), 345–357, 1998.

26. Ghani, M., and Soothill, J.S., Ceftazidime, gentamicin and rifapicin in combination, kill biofilms of mucoid *Pseudomonas aeruginosa. Can. J. Microbiol.* 43, 999–1004, 1997.

27. Konig, D.P., Schierholz, J.M., Munnich, U., and Rutt, J., Treatment of staphylococcal implant infection with rifampicin-ciprofloxacin in stable implants. *Arch. Orthop. Trauma. Surg.* 121, 297–299, 2001.

28. Heldman, A.W., Hartert, T.V., Ray, S.C., et al., Oral antibiotic treatment of right-sided staphylococcal endocarditis in injection drug users: prospective randomized comparison with parenteral therapy. *Am. J. Med.* 101, 68–76, 1996.

29. Blaser, J., Vergeres, P., Widmer, A.F., and Zimmerli, W., *In vivo* verification of an *in vitro* model of antibiotic treatment of device-related infection. *Antimicrob. Agents Chemother.* 39, 1134–1139, 1995.

19 β-Lactams for the Treatment of Biofilm-Associated Infections

Ingrid L. Dodge, Karen Joy Shaw, and Karen Bush

CONTENTS

19.1   INTRODUCTION

Biofilms are increasingly recognized as a medically relevant manifestation of persistent microbiological infection. As other chapters in this volume demonstrate, the biofilm field has made striking advances in the understanding of biofilm structure, etiology, ecology, and antimicrobial resistance, at least for a few model organisms.

However, more work is needed to fully appreciate the clinical impact of biofilms. For some infections, such as urinary tract infections and implantable medical devices, there is clear evidence from both clinical results and animal models that biofilms play an important role in persistence and recurrence of the infection. For other infections, such as those found in otitis media and cystic fibrosis, biofilms have been implicated in playing a role in the infectious process, but the importance of that role remains unclear. While the role of biofilms in many infections is yet to be determined, the growing problem of antimicrobial resistance makes the examination of all resistance mechanisms an imperative.

This chapter focuses on the use of β-lactam antibiotics in the treatment of biofilm-associated infections. Therefore, we ignore for the purposes of this chapter biofilms caused by nonbacterial pathogens such as *Candida albicans* that have well-documented medical importance. The first section reviews the structure and mechanism of action of β-lactam antibiotics and describes how bacteria gain resistance to β-lactams. The following section briefly covers the concepts of biofilm structure and antimicrobial resistance, highlighting areas that affect biofilm resistance to β-lactams. Specific disease processes are then examined, with discussions of the benefits of, liabilities of, and opportunities for employing β-lactam antibiotics in these cases. It is important to stress that the amount of evidence supporting a role for biofilms in disease etiology varies greatly in these sections, from a clear role with a strong animal model (urinary tract infections) to a less well-documented role (otitis media). Finally, the last section addresses the potential uses of β-lactams in preventing biofilm-associated infections. This may prove to be the greatest avenue of opportunity for employing β-lactam antibiotics, which target rapidly-dividing organisms.

19.2   β-LACTAM ANTIBIOTICS, MECHANISM OF ACTION AND RESISTANCE

Penicillin, the first β-lactam antibiotic, dramatically improved the survival of injured soldiers after its introduction during World War II, which set the stage for its widespread use in the civilian population in the postwar years. Its early safety and efficacy established penicillin as a major breakthrough in the treatment of bacterial disease. Derivatives of penicillin, and the closely related cephalosporins, are still considered to be first line therapy for many bacterial infections, and may hold promise in the treatment of biofilm-associated infections.

β-Lactam antibiotics are characterized by their chemically activated four-membered β-lactam ring, which can be readily hydrolyzed enzymatically. Opening of the β-lactam ring occurs during the covalent attachment of the β-lactam to its bacterial enzymatic target. Hydrolysis of the acyl enzyme results in a microbiologically inactive drug (1). β-lactams target the penicillin binding proteins (PBP) involved in the terminal steps of cell wall biosynthesis, thereby preventing the formation of peptide bonds necessary to crosslink peptidoglycans of the bacterial cell wall (2,3). Each bacterial species has numerous PBPs, ranging in number from three to eight, with varying molecular masses and activities (4). In general, targeting of the higher molecular weight PBPs leads to cell death (1,5). Depending upon the class of β-lactam, inactivation of multiple PBPs may be essential for bactericidal activity (1,6). β-lactam-containing agents are most frequently reported to kill only during the logarithmic growth phase of bacteria, although there are occasional references to killing during stationary phase (1,7).

It is not only the prevention of peptidoglycan crosslinking that can lead to the β-lactam bactericidal effect, however. In some bacteria, it has been proposed that the crosslinking action of transpeptidases may be balanced by the action of autolysins, which can hydrolyze transpeptide bonds (8,9). However, this proposal is not currently regarded to be a major contributing factor to the mechanism of β-lactam action.

As a result of prolonged and widespread β-lactam usage for a variety of infections, bacteria have developed resistance to this safe and efficacious family of antibacterial agents. Resistance can result from any of the following mechanisms (Table 19.1), either singly or in combination: mutation of the target protein(s), acquisition of new PBPs, expression of β-lactamases, porin changes, increased efflux, or, possibly, downregulation of autolysins (4,8,10–13).

Gram-negative bacteria most frequently develop resistance due to increased β-lactamase activity, combined with porin changes or increased efflux (12,14). Through these combinations of events, decreased amounts of β-lactam-containing agents are able to reach the target PBPs, thus allowing continued bacterial growth (15). In Gram-positive bacteria, the major β-lactam resistance mechanisms involve mutant PBPs with or without β-lactamase expression (especially true of staphylococci) (4,16). Autolysin regulation has also been implicated in β-lactam resistance, although to

TABLE 19.1
Major Resistance Mechanisms for β-Lactam-Containing Agents in Gram-Positive and Gram-Negative Bacteria, Listed in Order of Importance

Bacteria	Resistance Mechanism	Examples
Gram-negative	β-Lactamase (26)	Group 1/Class C AmpC cephalosporinases in Enterobacteriaceae and *Pseudomonas* spp.
		Group 2/Class A penicillinases and extended spectrum β-lactamases in Enterobacteriaceae and *Pseudomonas* spp.
		Group 3/Class B metallo-β-lactamases in *Bacteroides, K. pneumoniae* and *Pseudomonas* spp.
	Porin alterations (103,104)	Loss of OmpA in *Klebsiella* spp.
		Loss of OprD in *P. aeruginosa*
	Efflux (105,106)	AcrA in Enterobacteriaceae
		MexCDOprJ in *P. aeruginosa*
	PBPs with reduced affinity for β-lactams (31,32)	PBP1 in *H. pylori*
		PBP3 in *H. influenzae*
Gram-positive	Mosaic PBPs with reduced affinity for β-lactams (34,36,37)	PBP2a in *S. aureus*
		PBP2x in *S. pneumoniae*
	β-Lactamase (26)	Group 2/Class A penicillinases in *S. aureus*

a lesser degree than the other mechanisms. When bacteria downregulate autolysin expression, the degradation of the cell wall and ensuing lysis may not occur, and β-lactams could be converted from a bacteriolytic drug to a bacteriostatic drug. In this case the bacteria cannot productively divide, but are not killed by the action of the β-lactam, a phenomenon referred to as tolerance (17). Although tolerance may be a mechanism of β-lactam resistance, genetic studies have so far been unable to clearly demonstrate a role for autolysins in β-lactam resistance (18).

Of these resistance mechanisms, bacterial β-lactamase expression is the most worrisome phenomenon. Since 1940, over 425 structurally or functionally unique β-lactamases have been identified, many of which are plasmid encoded (Bush, ICAAC 2003). Like the structurally-related PBPs (2,15), β-lactamases inactivate the β-lactam-containing drug by hydrolyzing the β-lactam ring, but at a much faster rate than PBPs (15,19). PBP-acyl enzyme complexes can typically exist for 10 min to 9 h before a single hydrolyzed β-lactam molecule is released from the PBP (19,20), whereas β-lactamases can hydrolyze substrates like penicillin with turnover numbers (k_{cat}) as high as 1,000 molecules of β-lactam per sec per molecule of enzyme (21,22). β-Lactamases can be grouped molecularly by their sequence similarity (molecular classes A–D) or, functionally (functional groups 1–3) by their β-lactam substrate(s) of penicillins, carbapenems, or cephalosporins and their sensitivity to clavulanic acid or tazobactam, two β-lactamase inhibitors with similar inhibitory profiles (23–27). There is much commonality between the two classifications, as shown below.

The most common β-lactamases include the functional group 1 β-lactamases (structural class C enzymes, Table 19.1), which are cephalosporinases not inhibited by clavulanate, and the functional group 2 β-lactamases that include clavulanate-inhibitable penicillinases and extended-spectrum β-lactamases (28). Enzymes from functional group 2 mostly fall into molecular class A, except for the cloxacillinases, which have been grouped into molecular class D (26). Functional group 3 is unique, as it is comprised of the metallo-β-lactamases (molecular class B), which require zinc for enzymatic activity. Carbapenem hydrolysis and insensitivity to inhibition by commercial β-lactamase inhibitors are characteristics of these enzymes. Interestingly, these group 3 enzymes most frequently appear in organisms that express at least one other β-lactamase from a different functional group, thereby providing the producing organism with the ability to hydrolyze virtually any β-lactam to which it is exposed (29). All non-group 3 β-lactamases are serine β-lactamases that form acyl enzymes through the active site serine before hydrolysis. Many β-lactamases from Gram-negative bacteria are now reported to be plasmid-encoded, allowing them to be rapidly disseminated among bacterial species (28,30).

Finally, bacteria become resistant to β-lactam action through alterations of the PBPs, by decreasing the affinity of β-lactam binding to the target PBP, and subsequent covalent modification. For example, in Gram-negative bacteria, mutations in PBP1 lead to amoxicillin resistance in *Helicobacter pylori*, while mutations in the PBP3-encoding gene *ftsI* lead to ampicillin and cefuroxime resistance in *Haemophilus influenzae* (31–33). Although mutations in a PBP can render bacteria resistant to that agent, these mutants may be susceptible to other β-lactams, leaving opportunity for alternate therapy. There is no cross-resistance to other classes of antimicrobial agents in the event of PBP alterations.

More important, however, is the β-lactam resistance in Gram-positive bacteria caused by acquisition of new genetic material to encode an entirely new PBP or to construct hybrid PBPs. Methicillin resistance in *Staphylococcus aureus* (MRSA) is caused by the importation of the *mec*A gene that encodes PBP2a (or, PBP2′) with a low affinity for common β-lactams (34). This results in resistance to all β-lactams currently in clinical practice, although some new cephalosporins such as BAL9141 and RWJ-54428 have increased binding to PBP2a and are being investigated for the possible treatment of staphylococcal infections caused by MRSA (35). Penicillin-resistant *Streptococcus pneumoniae* is also caused by low affinity mosaic PBPs, resulting from the presence of hybrid PBPs containing nucleotide sequences acquired from other streptococci (36,37).

19.3 BIOFILMS: DEVELOPMENT AND ANTIBIOTIC RESISTANCE

Biofilms are sessile multicellular structures formed by a wide variety of micro-organisms, including bacteria and fungi. A biofilm can be comprised of either a single bacterial or fungal species, or can be generated by a complex mix of organisms, as is seen in microbial mats found near hot springs and within municipal water pipes. Individual microorganisms within the biofilm are encased in an extracellular matrix, which is pierced by water channels, allowing both nutrient exchange and waste removal. Although the water channels of biofilms provide access of antimicrobial agents to the individual cells comprising the biofilm, biofilms are notoriously resistant to antimicrobial action. Indeed, resistance of the biofilm can be 1000-fold higher than that seen in planktonic cells (38).

Biofilm formation is a highly choreographed process that can be broken down into four stages: initial interaction, adhesion, maturation of biofilm, and dispersion. Stoodley et al. (39) adds a fifth stage between stage two and three, discriminating between early and late biofilm maturation. Although these stages were initially described in reference to *Pseudomonas aeruginosa*, they appear to be broadly applicable to biofilms formed by various bacterial species, as well as fungi. During the first stage of biofilm formation, the microorganism must interact with the surface. This is thought to be a somewhat non-specific process, and may be due to electrostatic and/or hydrophobic interactions between the microorganism surface and the attachment surface. At this stage the bacteria are free-living and metabolically active, and should thus be susceptible to killing by β-lactams in an environment supporting multiplication.

Once the organisms have become associated with the colonization surface, the second phase of biofilm formation ensues. This irreversible adhesion step involves altered expression of genes and/or their products, such as those encoding type IV pili (40). For example, elaboration of an extracellular matrix, usually polysaccharide-based, is critical during this stage of biofilm development. The extracellular matrix serves a dual function for the biofilm. First, it provides structural support during the three-dimensional growth of the biofilm, and second, it can act as an impediment to the intrusion of both antibiotics and immune cells.

Conflicting data exist concerning penetration of antibiotics into biofilms. Zahller and Stewart (41) reported that ampicillin was unable to penetrate wild-type *Klebsiella pneumoniae* biofilms, while ciprofloxacin penetrated the entire biofilm within 20 minutes. However, the putative ampicillin impermeability was most likely due to ampicillin hydrolysis by the chromosomal *K. pneumoniae* β-lactamase in the biofilm, as a β-lactamase-deficient *K. pneumoniae* strain allowed full penetration of the drug. Notably, although ampicillin penetrated the biofilm, it still failed to kill the β-lactamase-negative cells within the interior of the biofilm, perhaps due to the fact that cells were in stationary phase, and thus not sensitive to the action of β-lactams (41,42). In other studies, Hoyle et al. (43) demonstrated reduced diffusion of piperacillin through *P. aeruginosa* biofilms. This difference in diffusion result may be due to binding of the piperacillin to the *P. aeruginosa* biofilm extracellular matrix, differences in the two drugs, or may be due to differences in the biofilm structures themselves. *P. aeruginosa* forms thick biofilms, as described below, which may provide a greater impediment to antibiotic diffusion.

Once the organisms are firmly attached, the biofilm can grow and mature in one of three ways: by division of attached cells, by lateral aggregation of attached cells, or by recruitment of planktonic cells from the fluid phase. This maturation of the biofilm, as cells are added, layer on layer, is accompanied by a number of architectural and gene expression changes in the constituent cells. Morphologically, mature biofilms can range from pod-like structures typically seen with *P. aeruginosa*, to undulating structures seen with *Escherichia coli* and *S. aureus*, to isolated bead-like structures seen with *Salmonella enterica*. Mature biofilm structure is impacted by the organism(s) forming the biofilm, as well as external components, such as characteristics of the colonization surface and shear stress imposed by flowing fluids. Although all biofilms have water channels that allow nutrient and waste exchange, interior layers of the biofilm are both oxygen and nutrient deprived, and appear to be metabolically quiescent. Some researchers have proposed that it is this environment, rather than any biofilm-specific adaptation or gene expression, which accounts for antimicrobial resistance in biofilms. This notion is supported by two observations. First, dispersed biofilms rapidly reacquire susceptibility to a wide range of antibiotics. Second, when antimicrobial agents are applied to biofilms *in vitro*, a zone of killing near the biofilm-liquid interface is observed, corresponding with the aerobic/nutrient-rich zone (44). This hypothesis does not explain, however, the ability of biofilms to persist in the face of long courses of antimicrobial agents, in which one would expect the entire biofilm to be killed in waves as each layer is exposed to nutrients and oxygen. It also does not explain the biofilm-specific expression of various genes, many with as yet undefined action.

Another phenomenon that may account for biofilm resistance is development of the so-called "persister" phenotype. Persister cells arise during mid-exponential growth phase, and are resistant to antimicrobial killing (45). Indeed, Spoering and Lewis have asserted that the resistance of *P. aeruginosa* biofilms to antibiotics may depend on the presence of persister cells within the biofilm, as they detected no difference in resistance between *P. aeruginosa* biofilms and stationary phase cultures (46).

Intriguingly, Keren et al. (45) found that the persister phenotype is a physiological adaptation, not accumulation of mutated cells, as persistence can be reverted by reinoculation into dilute culture. Antimicrobial resistance of biofilms is likely contributed by synergy of a number of factors, including quiescent cell state, persister phenotype, adsorption by the extracellular matrix, and biofilm-specific adaptations.

The final stage of the biofilm cycle is dispersal. Much interest has recently focused on this phase of biofilm development, with the hope that if dispersal can be inhibited, metastatic spread of the biofilm can be blocked. Controlled induction of dispersion may also be a method to render biofilming bacteria susceptible to antimicrobial agents, such as β-lactams. This phase of biofilm development has again been most intensively studied in *P. aeruginosa*. At later stages in the pseudomonal biofilm, or when the biofilm is stressed, the centers of the pod-like structures become hollowed-out, and free-swimming bacteria can be visualized. Over time, the shell of cells on the outside of the pod ruptures, releasing the free-swimming *P. aeruginosa* into the environment. It is now believed that this rupture event may be due to a programmed cell death process occurring in the lining cells (47). As mentioned above, rupture of the biofilm pods, or mere sloughing of segments of the biofilm under shear or environmental stress, allow the biofilm to propagate itself to distant sites, forming new niduses of infection (48,49).

19.4   β-LACTAM USE IN BIOFILM-ASSOCIATED INFECTIONS

19.4.1   INTRODUCTION

β-Lactam antibiotics, which include the penicillins, carbapenems, cephamycins, monobactams, and cephalosporins, are first-line therapy for many infections, including acute otitis media, respiratory infections, and streptococcal pharyngitis (48,49). It is now appreciated within the medical and microbiological communities that many of these "simple" infections may involve a biofilm component. Although penicillin was one of the first antibiotics discovered, relatively little research has directly examined the efficacy of β-lactams against biofilm-mediated infections.

Biofilms and their role in the infectious process exhibit relationships that differ according to the specific disease state. In infections such as those associated with implantable medical devices, the role of biofilms in the infectious process is clearly documented. For other infections, such as otitis media, biofilms have been suggested to play a role in persistence and recurrence of infection, but more evidence is needed before the clinical relevance of biofilms in these infections is fully understood. An important concept to keep in mind during reading of this section is the difference between controlling an infection and eradicating an infection. For many diseases, such as cystic fibrosis, patients are continually colonized with organisms in their lungs. On a routine basis, this colonization does not overwhelmingly impact the quality of life of cystic fibrosis patients. During an exacerbation, however, rampant bacterial growth leads to overt clinical symptoms of infection. For cystic fibrosis patients, then, what may be critical is not eradicating all the bacteria in their lungs (which is likely impossible), but controlling bacterial growth such that clinical symptoms are minimized.

Another important element to consider in the following sections is the fact that many therapies used in the clinic have been selected empirically, due in part to the lack of reliably predictive animal models of biofilm infection. Although a therapy may be effective alone or in combination, the biological mechanism by which the therapy works may not be clear. In the following sections infections that have well-documented evidence for biofilm involvement are discussed first, followed by those infections where biofilms are only suggested to play a role in the infectious process. In this manner, we hope to provide a broad overview of potentially biofilm-associated infections, and the role of β-lactam antibiotics in the treatment of these infections.

19.4.2   FOREIGN BODY/IMPLANTED MEDICAL DEVICE INFECTIONS

Biofilms are uncontroversially associated with infections at the site of foreign bodies or implanted medical devices. Over 200 million intravascular devices are implanted annually in the United States. As well as intravascular devices, a large number of foreign bodies (e.g., stents, pacemakers, joint replacements, and plastic surgery augmentation devices) are routinely introduced into the human body. All of these foreign bodies are susceptible to infection by biofilming bacteria. Current therapy for foreign-body biofilm infections is to remove the infected device and treat the patient with antibiotics. This methodology, while effective, results in morbidity and hospital costs for the patient, due to repeat surgeries or catheter placement, as well as infectious sequelae. The most common agent causing foreign body infections is *Staphylococcus epidermidis*, but the most serious is *S. aureus*. Patients infected with *S. aureus* require longer hospital stays, stronger antimicrobial treatment, and still experience higher mortality rates than patients infected by other organisms.

Given the severity of the complications associated with foreign-body infections, much research has centered around understanding the nature of staphylococcal biofilms, as well as investigating antimicrobial therapies to eradicate them. Some of these therapies are unusual, such as the use of laser-induced shock waves or low electric current to enhance penetration of antimicrobial agents into biofilms, but these techniques have not been attempted with β-lactams (50,51).

In general, β-lactams have been found to be ineffective when used alone against foreign-body related infections. For example, in an *in vitro* model of biofilm eradication, Gander and Finch (52) demonstrated that flucloxacillin was unable to eradicate *S. aureus* biofilms, while quinupristin/dalfopristin demonstrated some efficacy. Likewise, Monzon et al. (53) showed that cephalothin exhibited poor efficacy against *S. epidermidis* clinical isolate biofilms, while rifampicin, tetracycline, and erythromycin possessed a greater killing effect.

Although β-lactams may be ineffective alone in eradicating device-related biofilms, there are some suggestions that β-lactams may be useful in combination therapy. For example, in experimental infective endocarditis, using *S. aureus* as the infectious agent, nafcillin in combination with gentamicin was found to enhance outcomes *in vivo* (54). Likewise, a number of clinical trials (reviewed in 55) have demonstrated efficacy of double β-lactam or β-lactam combination therapy in the treatment of infective endocarditis. These results, combined with a report of synergy of cefalothin with rifampicin against young biofilms of *S. epidermidis*, suggests that β-lactam combination therapy should be investigated further for treatment of device-related infections (56).

19.4.3 Urinary Tract Infections

Urinary tract infections are a common complaint of women, with 11% incidence per year in the United States. Over half of all women will experience a urinary tract infection during their lifetime. After the initial infection, many women (25 to 50%) will present with another infection within a year, and 2 to 5% of women will have chronic recurring urinary tract infections (57). In addition, patients undergoing urinary tract catheterization due to spinal cord injury, hospitalization, or incontinence, are also at risk. An article by Siroky (58) in the *American Journal of Medicine* cited a rate of 2.5 urinary infections per year in spinal cord injury patients. The most common agent causing urinary tract infections is *E. coli*, accounting for 75 to 90% of cases (57).

Recent work by Hultgren and colleagues (59) demonstrated that uropathogenic *E. coli* can form biofilms, albeit unusual ones. These authors showed that, while infecting the bladder umbrella cells, *E. coli* can invade into the cells and form biofilms intracellularly, consisting of small, coccoid-like cells. Protected by the mammalian cellular membrane, *E. coli* are sequestered from the phagocytic action of neutrophils. Intracellular *E. coli* may also be protected from antibiotic therapy. Eventually, however, the infected cell will apoptose and slough into the lumen of the bladder. During this process, the uropathogenic *E. coli* adds another twist – it changes morphology from small, almost coccoid cells to long, filamentous cells. These long cells are unable to be efficiently phagocytosed by neutrophils, and allow the uropathogenic *E. coli* to invade new umbrella cells, starting the cycle anew. This finding of intracellular biofilms, as well as being a description of a novel biofilm mechanism, also explains the persistence and recurrence of urinary tract infections.

Surprisingly, in his review of treatment of bacteriuria in spinal cord injured patients, Sirosky (58) makes no mention or recommendation for β-lactam usage in urinary tract infections in this population. Likewise, the Infectious Diseases Society of America advocates the use of trimethoprim-sulfamethoxazole in the case where resistance against these agents is less than 20%, and otherwise advocates a 3-day course of fluoroquinolones. However, panipenem/betamipron has shown good efficacy against urinary tract infections, although this combination is unavailable in the United States (60). Also, Goto et al. (61,62) demonstrated that a *P. aeruginosa* biofilm grown in artificial urine could be completely destroyed within 48 hours by treatment with panipenem at 64 times the MBC (minimum bactericidal concentration). These authors still advocated the use of fluoroquinolones, however, as both ciprofloxacin and levofloxacin eradicated the biofilm within 24 hours at 32 times the MBC. Preferential use of fluoroquinolones for urinary tract infections in catheterized patients may also be related to findings that fluoroquinolones can be adsorbed to ureteral stents, enhancing antimicrobial activity (63). Also, fluoroquinolones are excreted primarily through the kidney, delivering them effectively to the site of infection.

The discovery that uropathogenic *E. coli* form unusual biofilms in the urinary tract, coupled with the use of new β-lactams and the efficacy of β-lactams in treating acute urinary tract infections, may lead to the greater use of β-lactams for urinary tract infections in the future.

19.4.4 OTITIS MEDIA

Otitis media is a common ailment of children, resulting in more than 30 million clinic visits per year in the United States (64). There are three forms of otitis media: acute otitis media, otitis media with effusion, and chronic suppurative otitis media. Acute otitis media is usually preceded by an upper respiratory tract infection, and results from blockage of the eustacian tube, allowing accumulation of middle ear fluid. This middle ear fluid provides a rich medium for bacterial growth. The incidence of otitis media has risen due to increasing use of day care centers, where young children are continually exposed to each other. Other risk factors include male gender, exposure to cigarette smoke, fall or winter season, age younger than two, and prior history of acute otitis media. Otitis media carries significant morbidity, with many children ultimately requiring shunt implantation and/or typanostomy tubes to treat chronic otitis media. Even so, some children suffer hearing loss, with attendant speech, language, and socalization deficits.

The high incidence of recurrence of otitis media, along with "sterile" cultures of middle ear fluid, has caused some researchers to hypothesize that recurrent otitis media is often caused by biofilms (65,66). The most common organisms causing acute otitis media are *S. pneumoniae*, *H. influenzae* and *Moraxella catarrhalis*. During the first incidence of acute otitis media, amoxicillin (50 mg/kg/day) is the therapy of choice (67). During subsequent incidences and/or in areas where drug-resistant *S. pneumoniae* is prevalent, high-dose (70 to 100 mg/kg/day) amoxicillin is the treatment of choice. Indeed, a recent study by Piglansky et al. (68) found that high-dose amoxicillin is still an effective empiric therapy for acute otitis media, with 80% of patients achieving sustained clinical cure with this regimen. The authors of this study recommended amoxicillin-clavulanate for cases where amoxicillin was ineffective, a recommendation validated by another study (69). The fact that success is achieved with β-lactam antibiotics in acute otitis media is interesting, given the growing evidence that otitis media may be caused by biofilms, and biofilms are not thought to be well-targeted by β-lactams. This could imply that a low level of biofilm in the middle ear is not pathogenic and clinical disease is only seen when bacterial growth passes a certain threshold. It could also be that otitis media results from bacterial growth in both biofilm and planktonic modes. Conversely, the clinical efficacy of β-lactams against acute otitis media could imply that β-lactams are more efficacious against biofilms than previously thought, or may simply reflect the fact that many cases of otitis media will resolve without treatment. In fact, current CDC guidelines advocate "watchful waiting" in the treatment of otitis media, where antibiotic therapy is withheld for 24 to 72 hours while the child is monitored (49). More research into biofilm clearance after β-lactam treatment, such as confirming eradication by PCR or LPS testing of middle ear fluid, is needed to determine the efficacy of β-lactams in treating biofilm-mediated otitis media.

19.4.5 CYSTIC FIBROSIS

According to some experts, the lung pathology seen in cystic fibrosis patients is related to biofilm formation (38,70,71). Cystic fibrosis (CF) is an autosomal recessive disorder with a prevalence of 1:3,000 in Caucasian populations (72). The disease results from mutations in the cystic fibrosis transmembrane regulator (CFTR). Mutations in CFTR, which possesses multiple membrane-spanning domains, results in intracellular aggregation of the protein. Without a functional CFTR, a cloride ion transporter, CF patients are afflicted with unusually viscous and salty mucous. These thick, salty secretions impair the action of the mucociliary clearance system, resulting in mucous plugs in the lungs and exocrine organs, such as the pancreas.

The mucous-plugged lung environment is a rich medium for the growth of pathogenic bacteria, and CF patients are almost continuously infected. *P. aeruginosa* is a well-studied opportunistic pathogen that forms dense biofilms in CF lungs. Over time, inflammation directed ineffectively against the colonizing bacteria leads to profound lung damage, including bronchiectasis and consolidation, ultimately resulting in poor pulmonary function, catastrophic pulmonary bleeds, and death. While the life expectancy of CF patients has improved dramatically with the introduction of pancreatic enzyme replacement therapy and aggressive antibiotic treatment, it is still well below that of unaffected individuals (31.6 years, 2002 CF Foundation Patient Registry Annual Report).

Two bacterial indices are associated with poor outcome in CF. First is the conversion in colonization from nonalginate-producing to alginate-producing *P. aeruginosa*. Counterintuitively, alginate-producing *P. aeruginosa* are somewhat more susceptible to antibiotics and exhibit less β-lactamase activity during *in vitro* culture, but this is counterbalanced by greater biofilming efficiency and reduced immune clearance (71,73,74). Once in the biofilm, however, all *P. aeruginosa* clinical isolates are remarkably resistant to antibiotics, and even multiple combinations of antibiotics (70,75,76). Two factors work against the efficacy of β-lactam antibiotics in CF *P. aeruginosa* infections. First, *P. aeruginosa* is capable of producing a number of β-lactamases, and expression of these enzymes in the surface layers of the biofilm can prevent effective penetration of β-lactam antibiotics into the deeper layers of the biofilm (77). Second, interior sections of the biofilm are both nutrient and oxygen-deprived, leading to a slow/no growth phenotype in these sections of the biofilm. As β-lactam antibiotics are most effective against rapidly-growing bacteria, the interior of the biofilm is not likely to be susceptible to β-lactam treatment, and may explain the ineffectiveness of this class of antibiotics in eradicating *P. aeruginosa* biofilms (46,70).

Despite these drawbacks, however, some investigators have noted efficacy of β-lactams in combination with other antimicrobial agents in the treatment of *P. aeruginosa* biofilms. For example, Bui et al. (78) found that a combination of ceftazidime and clarithromycin was effective in increasing survival in a murine *P. aeruginosa* pneumonia model, while clarithromycin was ineffective alone. In another study, Aaron et al. (75) found that combinations of tobramycin and a β-lactam, or tobramycin plus a β-lactam and a third agent (azithromycin, ciprofloxacin, or a second β-lactam) were effective against biofilm-grown mucoid *P. aeruginosa*, while combinations lacking a β-lactam (e.g., tobramycin and ciprofloxacin) were ineffective. This result was surprising, as other investigators have demonstrated that biofilming *P. aeruginosa* produce periplasmic glucans that interact with, and possibly sequester, tobramycin (38). Furthermore, Walters et al. (79) demonstrated that while both ciprofloxacin and tobramycin are capable of penetrating *P. aeruginosa* biofilms, they are incapable of killing cells within the biofilm.

In a recent review, Gibson et al. (80) recommend β-lactam treatment for all CF pulmonary exacerbations resulting from *S. aureus*, *P. aeruginosa*, *Stenotrophomonas maltophilia*, *Burkholderia cepacia* and *Achromobacter xylosoxidans*, with the exception of methicillin-resistant *S. aureus* (MRSA). This is in spite of the fact that ceftazidime is completely ineffective in killing *S. maltophilia* biofilms *in vitro* (81). This may reflect the fact that CF pulmonary exacerbations are mostly caused by rapidly-growing bacteria, and these are more sensitive to β-lactam antibiotics. This group also advocates the use of β-lactams for the outpatient management of CF, which again may serve to keep rapidly-growing populations of bacteria in check.

The second microbial indication of poor outcome in CF is coinfection with *B. cepacia* complex. *B. cepacia* complex infection is associated with rapid lung function deterioration and increased mortality (82,83). The carbapenem meropenem is currently one of the few antibiotics with activity against *B. cepacia,* and is thus of great value to CF clinicians (80,82–86). It is a useful drug in combination therapy, although the prevalence of resistance is increasing.

Overall, even though β-lactams have shown marginal effectiveness during *in vitro* biofilm testing with *P. aeruginosa*, they do appear to have efficacy both against pulmonary exacerbations in CF patients and in multi-drug combinations *in vitro*. The emergence of new β-lactamase inhibitors gives new hope for the use of β-lactam antibiotics in the treatment of *P. aeruginosa* lung infections.

19.4.6 STREPTOCOCCAL PHARYNGITIS

Pharyngitis, or sore throat, accounts for more than 40 million adult clinic visits in the US annually (87). In addition, pharyngitis accounts for a large number of pediatric clinic visits. In the majority of cases, pharyngitis is caused by viral infection, often subsequent to an upper respiratory tract infection, and antibiotics are not indicated. The most common bacterial agent of pharyngitis, and the only case where antibiotics are indicated, is group A streptococci (GAS, *Streptococcus pyogenes*). Indeed GAS are responsible for 20 to 40% of all cases of pediatric exudative pharyngitis (72). Antibiotic therapy for pharyngitis due to GAS is necessary to prevent complications such as rheumatic fever and rash, as antibiotics do little to shorten the course of the infection or ameliorate the pain associated with pharyngitis (88).

Although not clinically described as a biofilm, GAS pharyngitis exhibits the characteristics of a biofilm-mediated disease, as the infecting organism grows in dense, attached, macroscopic colonies on the pharyngeal surface. Also, GAS pharyngitis has a high incidence of recurrence, and of treatment failure (up to one third of cases), again consistent with a possible biofilm origin. In support of this idea, a recent paper by Conley et al. (89) reported that clinical GAS isolates were able to form biofilms *in vitro*, but *in vitro* biofilming capability was not correlated with penicillin resistance.

Penicillin is a common therapy for GAS pharyngitis, although two recent reviews (90,91) have advocated the use of amoxicillin or other β-lactams, due to shorter courses and higher cure rates.

The use of penicillin, and possibly other β-lactams, may also have another unexpected benefit in the treatment of GAS. *S. pyogenes* (GAS) possesses a surface antigen that is recognized by antibodies to CD15s. CD15s, also known as sialyl-Lewis^x, binds to the selectin family of mammalian adhesion molecules. Expression of this molecule may allow *S. pyogenes* to adhere to mammalian cells, and/or may act as an immunological decoy, preventing eradication by the immune system. Hirota et al. (92) found that treatment of *S. pyogenes* with benzylpenicillin at sub-MIC levels significantly reduced the expression of antigens recognized by anti-CD15s antibodies. The authors also found that treatment with fosfomycin, another cell wall synthesis inhibitor, altered *S. pyogenes* biofilm structure, although the impact of benzylpenicillin on *S. pyogenes* biofilm architecture was not investigated. Results such as these point to the fact that killing is not the only impact an antimicrobial agent can have on a biofilm, and multiple indices should be considered when evaluating antimicrobial action on biofilms. Indeed, agents targeting cell wall synthesis, such as β-lactams, may have an impact on bacterial adhesion, bacteria–bacteria interactions, and overall biofilm architecture.

19.4.7 SKIN INFECTIONS

Skin and soft tissue bacterial infections are one of the most common causes of emergency room visits. These infections can be the result of skin trauma, or can represent cellullitis, folliculitis, or furunculosis. It is recommended that these infections be treated empirically with oral cephalosporins, β-lactamase stable penicillins, or macrolides, which is borne out by a recent epidemiological study of soft tissue infectious agents in the US and Europe (93). Skin and soft tissue infections are likely caused at least partially by biofilms, as the bacteria are attached to the skin surface and grow in dense colonies. Experimental infections in both neutropenic and immune competent mice also confirm this possibility (94,95). In a study performed using immunocompetent animals, where a mixture of *S. aureus* and latex beads was injected intradermally, a combination of roxithromycin and the carbapenem imipenem was effective against *S. aureus* biofilms. Given the rise in prevalence of MRSA, the future utility of current β-lactams for the treatment of *S. aureus* skin infections is in doubt, especially among patients in institutional settings, including prisons (MMWR Oct. 17, 2003). However, there are new cephalosporins and carbapenems under development that have anti-MRSA activities and may prove useful in the future (96,97).

19.5 β-LACTAM USE IN PREVENTING BIOFILM INFECTIONS

Given the difficulty of eradicating biofilms once established, and the morbidity associated with replacing infected medical devices, much research has focused on the prevention of biofilm infections. To this end, various antimicrobial lock solutions instilled into the lumen of a catheter to prevent antimicrobial infection, coatings, and biomaterials have been tried, with some success. The broad-spectrum efficacy of β-lactam antibiotics against rapidly growing bacteria make them attractive agents for inclusion in such anti-biofilming measures. Indeed Schierholz et al. (98) found that flucloxacillin, among other antibiotics, could be readily incorporated into polyurethane and provided sustained delivery from the polyurethane for more than 14 days. Rupp and Hamer also showed that cefazolin prevented adherence and subsequent biofilm formation by *S. epidermidis* on intravascular catheters *in vitro* (99). This may be due, in part, to the failure of *S. epidermidis* to upregulate PIA, a molecule necessary for biofilm matrix production, during treatment with β-lactams. While treatment of *S. epidermidis* with quinupristin-dalfopristin induced a 9 to 11 fold increase in PIA expression, treatment with penicillin and oxacillin had no effect, along with chloramphenicol, clindamycin, gentamicin, ofloxacin, teicoplanin, and vancomycin (100). Given the ability of β-lactams to alter cell wall targets, and thus possibly adhesion and/or biofilm architecture, they may prove to be important agents in preventing biofilm formation.

Krishnasami et al. (101) have also reported some efficacy with cefazolin used as an antimicrobial lock solution in hemodialysis patients, although these authors failed to disclose what proportion of the patients in their study received this particular treatment and what the success rate was. In any case, this study suggests that β-lactams might be efficacious when used as antimicrobial lock solutions. β-Lactams could also be combined with newly discovered agents, such as the putisolvins recently isolated from *Pseudomonas putida* (102). Putisolvins have been shown to break down the biofilm matrix, which could facilitate entry of antimicrobial agents. A combined β-lactam could also be useful in killing bacteria elaborated from biofilms through any means, such as through shock waves or electrical currents, as it has been abundantly shown that organisms released from biofilms rapidly regain antibiotic susceptibility.

19.6   CONCLUSIONS

While limited research has directly examined the impact of β-lactam antibiotics on biofilm formation and biofilm killing, some reports describe efficacy of β-lactam antibiotics against biofilm-mediated infections. As a deeper understanding is gained of biofilm biology and the diseases caused by biofilms, greater opportunities for intervention with β-lactam antibiotics may be found, especially with the development of new agents or novel formulations specifically aimed at treating established biofilms. More direct research is needed on the impact of β-lactams on biofilm structure and function, as well as on the efficacy of β-lactams and β-lactam combination therapies against biofilm-mediated infections. Combinations of different classes of agents may also provide an effective approach to this difficult-to-treat microbiological phenomenon. The emerging levels of high resistance to multiple antibiotic classes in bacterial species that form biofilms, as well as the intrinsic antimicrobial resistance conferred by the biofilm state, stresses the need for new antimicrobial agents and new antiinfective strategies in the fight against biofilms.

ACKNOWLEDGMENTS

We would like to thank the infectious diseases groups at Johnson & Johnson PRD in La Jolla, CA and Raritan, NJ for their helpful discussions on this topic. In particular, we would like to thank Raul Goldschmidt and Brian Morrow for critical reading and insightful comments on this manuscript.

REFERENCES

1. Spratt, B.G., and Cromie, K.D., Penicillin-binding proteins of Gram-negative bacteria. *Rev. Infect. Dis.* 10, 699–711, 1988.
2. Massova, I., and Mobashery, S., Kinship and diversification of bacterial penicillin-binding proteins and β-lactamases. *Antimicrob. Agents Chemother.* 42, 1–17, 1998.
3. Yocum, R.R., Waxman, D.J., Rasmussen, J.R., and Strominger, J.L., Mechanism of penicillin action: penicillin and substrate bind covalently to the same active site serine in two bacterial D-alanine carboxypepdidases. *Proc. Natl. Acad. Sci. USA* 76, 2730–2734, 1979.
4. Georgopapadakou, N.H., Penicillin-binding proteins and bacterial resistance to β-lactams. *Antimicrob. Agents Chemother.* 37, 2045–2053, 1993.
5. Ghuysen, J.M., Serine β-lactamases and penicillin-binding proteins. In: Ornston, L.N., Ballows, A., Greenberg, E.P., eds., *Annual Reviews of Microbiology*, Vol. 45. Palo Alto, CA: Annual Reviews Inc., 1991:37–67.
6. Pinho, M.G., Lencastre, H.D., and Tomasz, A., An acquired and a natvie penicillin-binding protein cooperate in building the cell wall of drug-resistant staphylococci. *Proc. Natl. Acad. Sci. USA* 98, 10886–10891, 2001.
7. Tuomanen, E., and Schwartz, J., Penicillin-binding protein 7 and its relationship to lysis of nongrowing *Escherichia coli*. *J. Bacteriol.* 169, 4912–4915, 1987.
8. Kitano, K., and Tomasz, A., Triggering of autolytic cell wall degradation in *Escherichia coli* by beta-lactam antibiotics. *Antimicrob. Agents Chemother.* 16, 838–848, 1979.
9. Kitano, K., Tuomanen, E., and Tomasz, A., Transglycosylase and endopeptidase participate in the degradation of murein during autolysis of *Escherichia coli*. *J. Bacteriol.* 167, 759–765, 1986.
10. Lakaye, B., Dubus, A., Lepage, S., Groslambert, S., and Frere, J-M., When drug inactivation renders the target irrelevant to antibiotic resistance: a case story with beta-lactams. *Mol. Microbiol.* 31, 89–101, 1999.
11. Lakaye, B., Dubus, A., Joris, B., and Frere, J-M., Method for estimation of low outer membrane permeability to beta-lactam antibiotics. *Antimicrob. Agents Chemother.* 46, 2901–2907, 2002.
12. Bradford, P.A., Urban, C., Mariano, N., Projan, S.J., Rahal, J.J., and Bush, K., Imipenem resistance in *Klebsiella pneumoniae* is associated with the combination of ACT-1, a plasmid-mediated AmpC β-lactamase, and the loss of an outer membrane protein. *Antimicrob. Agents Chemother.* 41, 563–569, 1997.
13. Li, X-Z., Ma, D., Livermore, D.M., and Nikaido, H., Role of efflux pump(s) in intrinsic resistance of *Pseudomonas aeruginosa*: Active efflux as a contributing factor to β-lactam resistance. *Antimicrob. Agents Chemother.* 38, 1742–1752, 1994.
14. Mazzariol, A., Cornaglia, G., and Nikaido, H., Contributions of the AmpC beta-lactamase and the AcrAB multidrug efflux system in intrinsic resistance of *Escherichia coli* K-12 to beta-lactams. *Antimicrob. Agents Chemother.* 44, 1387–1390, 2001.
15. Matagne, A., Dubus, A., Galleni, M., and Frere, J-M., The beta-lactamase cycle: a tale of selective pressure and bacterial ingenuity. *Nat. Prod. Rep.* 16, 1–19, 1999.

16. Franciolli, M., Bille, J., Glauser, M.P., and Moreillon, P., Beta-lactam resistance mechansisms of methicillin-resistant *Staphylococcus aureus. J. Infect. Dis.* 163, 514–523, 1990.

17. Goodell, E.W., Lopez, R., and Tomasz, A., Suppression of lytic effect of beta lactams on *Escherichia coli* and other bacteria. *Proc. Natl. Acad. Sci. USA* 73, 3293–3297, 1976.

18. Mitchell, L.S., and Tuomanen, E.I., Molecular analysis of antibiotic tolerance in pneumococci. *Int. J. Med. Microbiol.* 292, 75–9, 2002.

19. Waxman, D.J., and Strominger, J.L., Pencillin-binding proteins and the mechanism of action of beta-lactam antibiotics. *Ann. Rev. Biochem.* 52, 825–869, 1983.

20. Frère, J-M., and Joris, B., Penicillin-sensitive enzymes in peptidoglycan biosynthesis. *CRC Crit. Rev. Microbiol.* 11, 299–396, 1985.

21. Matagne, A., Misselyn-Baudin, A-M., Joris, B., Erpicum, T., Graniwer, B., and Frere, J-M., The diversity of the catalytic properties of class A β-lactamases. *Biochem. J.* 265, 131–146, 1990.

22. Bush, K., and Sykes, R.B., Methodology for the study of β-lactamases. *Antimicrob. Agents Chemother.* 30, 6–10, 1986.

23. Ambler, R.P., Coulson, A.F.W., Frère, J-M., et al., A standard numbering scheme for the Class A β-lactamases. *Biochem. J.* 276, 269–272, 1991.

24. Huovinen, P., Huovinen, S., and Jacoby, G.A., Sequence of PSE-2 beta-lactamase. *Antimicrob. Agents Chemother.* 32, 134–136, 1988.

25. Jaurin, B., and Grundstrom, T., *amp C* cephalosporinase of *Escherichia coli* K-12 has a different evolutionary origin from that of β-lactamases of the penicillinase type. *Proc. Natl. Acad. Sci. USA* 78, 4897–4901, 1981.

26. Bush, K., Jacoby, G.A., and Medeiros, A.A., A functional classification scheme for β-lactamases and its correlation with molecular structure. *Antimicrob. Agents Chemother.* 39, 1211–1233, 1995.

27. Payne, D.J., Cramp, R., Winstanley, D.J., and Knowles, D.J.C., Comparative activities of clavulanic acid, sulbactam, and tazobactam against clinically important β-lactamases. *Antimicrob. Agents Chemother.* 38, 767–772, 1994.

28. Bush, K., New beta-lactamases in Gram-negative bacteria: diversity and impact on the selection of antimicrobial therapy. *Clin. Infect. Dis.* 32, 1085–1089, 2001.

29. Rasmussen, B.A., and Bush, K., Carbapenem-hydrolyzing β-lactamases. *Antimicrob. Agents Chemother.* 41, 223–232, 1997.

30. Bradford, P.A., Extended-spectrum beta-lactamases in the 21st century: characterization, epidemiology, and detection of this important resistance threat. *Clin. Microbiol. Rev.* 14, 933–951, 2001.

31. Kwon, D.H., Dore, M.P., Kim, J.J., et al., High-level beta-lactam resistance associated with acquired multidrug resistance in *Helicobacter pylori. Antimicrob. Agents Chemother.* 47, 2169–2178, 2003.

32. Ubukata, K., Shibasaki, Y., Yamamoto, K., et al., Association of amino acid substitutions in penicillin-binding protein 3 with beta-lactam resistance in beta-lactamase-negative ampicillin-resistant *Haemophilus influenzae. Antimicrob. Agents Chemother.* 45, 1693–1699, 2001.

33. Straker, K., Wootton, M., Simm, A.M., Bennett, P.M., MacGowan, A.P., and Walsh, T.R., Cefuroxime resistance in non-beta-lactamase *Haemophilus influenzae* is linked to mutations in *fts*I. *J. Antimicrob. Chemother.* 51, 523–530, 2003.

34. Berger-Bachi, B., Rohrer, S., Factors influencing methicillin resistance in staphylococci. *Arch. Microbiol.* 178, 165–171, 2002.

35. Abbanat, D., Macielag, M., and Bush, K., Novel antibacterial agents for the treatment of serious Gram-positive infections. *Exp. Opin. Investigational Drugs* 379–399, 2003.

36. du, Plessis, M., Bingen, E., and Klugman, K.P., Analysis of penicillin-binding protein genes of clinical isolates of *Streptococcus pneumoniae* with reduced susceptibility to amoxicillin. *Antimicrob. Agents Chemother.* 46, 2349–2357, 2002.

37. Reichmann, P., Konig, A., Marton, A., and Hakenbeck, R., Penicillin-binding proteins as resistance determinants in clinical isolates of *Streptococcus pneumoniae*. *Microg. Drug Resistance* 2, 177–181, 1996.

38. Mah, T.F., Pitts, B., Pellock, B., Walker, G.C., Stewart, P.S., and O'Toole, G.A., A genetic basis for *Pseudomonas aeruginosa* biofilm antibiotic resistance. *Nature* 426, 306–310, 2003.

39. Stoodley, P., Sauer, K., Davies, D.G., and Costerton, J.W., Biofilms as complex differentiated communities. *Annu. Rev. Microbiol.* 56, 187–209, 2002.

40. Ichikawa, J.K., Norris, A., Bangera, M.G., et al., Interaction of *Pseudomonas aeruginosa* with epithelial cells: identification of differentially regulated genes by expression microarray analysis of human cDNAs. *Proceedings of the National Academy of Sciences of the United States of America* 97:9659–9664, 2000.

41. Zahller, J., and Stewart, P.S., Transmission electron microscopic study of antibiotic action on *Klebsiella pneumoniae* biofilm. *Antimicrob. Agents Chemother.* 46, 2679–2683, 2002.

42. Anderl, J.N., Franklin, M.J., and Stewart, P.S., Role of antibiotic penetration limitation in *Klebsiella pneumoniae* biofilm resistance to ampicillin and ciprofloxacin. *Antimicrob Agents Chemother.* 44, 1818–1824, 2000.

43. Hoyle, B.D., Alcantara, J., and Costerton, J.W., *Pseudomonas aeruginosa* biofilm as a diffusion barrier to piperacillin. *Antimicrob. Agents Chemother.* 36, 2054–2056, 1992.

44. Stewart, P.S., *Slow Growth* in *P. aeruginosa Biofilms*, Biofilms 2003, Victoria, BC, Canada, November 1–6. American Society for Microbiology, 2003.

45. Keren, I., Kaldalu, N., Spoering, A., Wang, Y., and Lewis, K., Persister cells and tolerance to antimicrobials. *FEMS Microbiol. Lett.* 230, 13–18, 2004.

46. Spoering, A.L., and Lewis, K., Biofilms and planktonic cells of *Pseudomonas aeruginosa* have similar resistance to killing by antimicrobials. *J. Bacteriol.* 183, 6746–6751, 2001.

47. Webb, J.S., Thompson, L.S., James, S., et al., Cell death in *Pseudomonas aeruginosa* biofilm development. *J. Bacteriol.* 185, 4585–4592, 2003.

48. Bisno, A.L., Gerber, M.A., Gwaltney, J.M., Jr., Kaplan, E.L., and Schwartz, R.H., Practice guidelines for the diagnosis and management of group A streptococcal pharyngitis. Infectious Diseases Society of America. *Clin. Infect. Dis.* 35, 113–125, 2002.

49. Semchenko, A., *Management of Acute Sinusitis and Otitis Media*. American Academy of Family Physicians Monograph 2001.

50. Nigri, G.R., Tsai, S., Kossodo, S., et al., Laser-induced shock waves enhance sterilization of infected vascular prosthetic grafts. *Lasers Surg. Med.* 29, 448–454, 2001.

51. Arnold, J.W., Boothe, D.H., and Mitchell, B.W., *Electrostatic Space Charge System, An Alternative Intervention Strategy for Biofilm Reduction,* Biofilms 2003, Victoria, BC, Canada, November 1–6. American Society for Microbiology, 2003.

52. Gander, S., and Finch, R., The effects of exposure at constant (1 h) or exponentially decreasing concentrations of quinupristin/dalfopristin on biofilms of Gram-positive bacteria. *J. Antimicrob. Chemother.* 46, 61–67, 2000.

53. Monzon, M., Oteiza, C., Leiva, J., Lamata, M., and Amorena, B., Biofilm testing of *Staphylococcus epidermidis* clinical isolates: low performance of vancomycin in relation to other antibiotics. *Diagn. Microbiol. Infect. Dis.* 44, 319–324, 2002.

54. Sande, M.A., and Courtney, K.B., Nafcillin-gentamicin synergism in experimental *staphylococcal endocarditis. J. Lab. Clin. Med.* 88, 118–124, 1976.

55. Le, T., and Bayer, A.S., Combination antibiotic therapy for infective endocarditis. *Clin. Infect. Dis.* 36, 615–621, 2003.

56. Monzon, M., Oteiza, C., Leiva, J., and Amorena, B., Synergy of different antibiotic combinations in biofilms of *Staphylococcus epidermidis*. *J. Antimicrob. Chemother.* 48, 793–801, 2001.

57. Fihn, S.D., Clinical practice. Acute uncomplicated urinary tract infection in women. *N. Engl. J. Med.* 349, 259–266, 2003.

58. Siroky, M.B., Pathogenesis of bacteriuria and infection in the spinal cord injured patient. *Am. J. Med.* 113 Suppl 1A, 67S–79S, 2002.

59. Anderson, G.G., Palermo, J.J., Schilling, J.D., Roth, R., Heuser, J., and Hultgren, S.J., Intracellular bacterial biofilm-like pods in urinary tract infections. *Science* 301, 105–107, 2003.

60. Goa, K.L., and Noble, S., Panipenem/betamipron. *Drugs* 63, 913–925, discussion 926, 2003.

61. Goto, T., Nakame, Y., Nishida, M., and Ohi, Y., *In vitro* bactericidal activities of beta-lactamases, amikacin, and fluoroquinolones against *Pseudomonas aeruginosa* biofilm in artificial urine. *Urology* 53, 1058–1062, 1999.

62. Goto, T., Nakame, Y., Nishida, M., and Ohi, Y., Bacterial biofilms and catheters in experimental urinary tract infection. *Int. J. Antimicrob. Agents* 11, 227–231, discussion 237–239, 1999.

63. Reid, G., Habash, M., Vachon, D., Denstedt, J., Riddell, J., and Beheshti, M., Oral fluoroquinolone therapy results in drug adsorption on ureteral stents and prevention of biofilm formation. *Int. J. Antimicrob. Agents* 17, 317–319, discussion 319–320, 2001.

64. Rothman, R., Owens, T., and Simel, D.L., Does this child have acute otitis media? *JAMA* 290, 1633–1640, 2003.

65. Rayner, M.G., Zhang, Y., Gorry, M.C., Chen, Y., Post, J.C., and Ehrlich, G.D., Evidence of bacterial metabolic activity in culture-negative otitis media with effusion. *JAMA* 279, 296–299, 1998.

66. Post, J.C., Direct evidence of bacterial biofilms in otitis media. *Laryngoscope.* 111, 2083–2094, 2001.

67. Hoberman, A., Marchant, C.D., Kaplan, S.L., and Feldman, S., Treatment of acute otitis media consensus recommendations. *Clin. Pediatr. (Phila)* 41, 373–390, 2002.

68. Piglansky, L., Leibovitz, E., Raiz, S., et al., Bacteriologic and clinical efficacy of high dose amoxicillin for therapy of acute otitis media in children. *Pediatr. Infect. Dis. J.* 22, 405–413, 2003.

69. Dagan, R., Hoberman, A., Johnson, C., et al., Bacteriologic and clinical efficacy of high dose amoxicillin/clavulanate in children with acute otitis media. *Pediatr. Infect. Dis. J.* 20, 829–837, 2001.

70. Drenkard, E., and Ausubel, F.M., Pseudomonas biofilm formation and antibiotic resistance are linked to phenotypic variation. *Nature* 416, 740–743, 2002.

71. Ciofu, O., Fussing, V., Bagge, N., Koch, C., and Hoiby, N., Characterization of paired mucoid/non-mucoid *Pseudomonas aeruginosa* isolates from Danish cystic fibrosis patients: antibiotic resistance, beta-lactamase activity and RiboPrinting. *J. Antimicrob. Chemother.* 48, 391–396, 2001.

72. Fauci, A.S., Braunwald, E., Isselbacher, K.J., Wilson, J.D., Martin, J.B., Kasper, D.L., Hauser, S.L., and Longo, D.L., eds., *Harrison's Principles of Internal Medicine*, 14th Ed. San Francisco, CA: McGraw-Hill, 1998:2569.

73. Song, Z., Wu, H., Ciofu, O., et al., *Pseudomonas aeruginosa* alginate is refractory to Th1 immune response and impedes host immune clearance in a mouse model of acute lung infection. *J. Med. Microbiol.* 52, 731–740, 2003.

74. Head, N.E., and Yu, H., Cross-sectional analysis of clinical and environmental isolates of *Pseudomonas aeruginosa*: biofilm formation, virulence, and genome diversity. *Infect. Immun.* 72, 133–144, 2004.

75. Aaron, S.D., Ferris, W., Ramotar, K., Vandemheen, K., Chan, F., and Saginur, R., Single and combination antibiotic susceptibilities of planktonic, adherent, and biofilm-grown *Pseudomonas aeruginosa* isolates cultured from sputa of adults with cystic fibrosis. *J. Clin. Microbiol.* 40, 4172–4179, 2002.

76. Coquet, L., Junter, G.A., and Jouenne, T., Resistance of artificial biofilms of *Pseudomonas aeruginosa* to imipenem and tobramycin. *J. Antimicrob. Chemother.* 42, 755–760, 1998.

77. Ciofu, O., Beveridge, T.J., Kadurugamuwa, J., Walther-Rasmussen, J., and Hoiby, N., Chromosomal beta-lactamase is packaged into membrane vesicles and secreted from *Pseudomonas aeruginosa. J. Antimicrob. Chemother.* 45, 9–13, 2000.

78. Bui, K.Q., Banevicius, M.A., Nightingale, C.H., Quintiliani, R., and Nicolau, D.P., *In vitro* and *in vivo* influence of adjunct clarithromycin on the treatment of mucoid *Pseudomonas aeruginosa. J. Antimicrob. Chemother.* 45, 57–62, 2000.

79. Walters, M.C., 3rd, Roe, F., Bugnicourt, A., Franklin, M.J., and Stewart, P.S., Contributions of antibiotic penetration, oxygen limitation, and low metabolic activity to tolerance of *Pseudomonas aeruginosa* biofilms to ciprofloxacin and tobramycin. *Antimicrob. Agents Chemother.* 47, 317–323, 2003.

80. Gibson, R.L., Burns, J.L., and Ramsey, B.W., Pathophysiology and management of pulmonary infections in cystic fibrosis. *Am. J. Respir. Crit. Care Med.* 168, 918–951, 2003.

81. Di Bonaventura, G., Spedicato, I., D'Antonio, D., Robuffo, I., and Piccolomini, R., Biofilm formation by *Stenotrophomonas maltophilia*: modulation by quinolones, trimethoprim-sulfamethoxazole, and ceftazidime. *Antimicrob. Agents Chemother.* 48, 151–160, 2004.

82. Ledson, M.J., Gallagher, M.J., Robinson, M., et al., A randomized double-blinded placebo-controlled crossover trial of nebulized taurolidine in adult cystic fibrosis patients infected with *Burkholderia cepacia. J. Aerosol. Med.* 15, 51–57, 2002.

83. Soni, R., Marks, G., Henry, D.A., et al., Effect of *Burkholderia cepacia* infection in the clinical course of patients with cystic fibrosis: a pilot study in a Sydney clinic. *Respirology* 7, 241–245, 2002.

84. Aaron, S.D., Ferris, W., Henry, D.A., Speert, D.P., and Macdonald, N.E., Multiple combination bactericidal antibiotic testing for patients with cystic fibrosis infected with *Burkholderia cepacia. Am. J. Respir. Crit. Care. Med.* 161, 1206–1212, 2000.

85. Conway, S.P., Brownlee, K.G., Denton, M., and Peckham, D.G., Antibiotic treatment of multidrug-resistant organisms in cystic fibrosis. *Am. J. Respir. Med.* 2, 321–332, 2003.

86. Wilson, D.L., Owens, R.C., Jr., and Zuckerman, J.B., Successful meropenem desensitization in a patient with cystic fibrosis. *Ann. Pharmacother.* 37, 1424–1428, 2003.

87. Bourbeau, P.P., Role of the microbiology laboratory in diagnosis and management of pharyngitis. *J. Clin. Microbiol.* 41, 3467–3472, 2003.

88. Bisno, A.L., Acute pharyngitis. *N. Engl. J. Med.* 344, 205–211, 2001.

89. Conley, J., Olson, M.E., Cook, L.S., Ceri, H., Phan, V., and Davies, H.D., Biofilm formation by group a streptococci: is there a relationship with treatment failure? *J. Clin. Microbiol.* 41, 4043–4048, 2003.

90. Curtin-Wirt, C., Casey, J.R., Murray, P.C., et al., Efficacy of penicillin vs. amoxicillin in children with group A beta hemolytic streptococcal tonsillopharyngitis. *Clin. Pediatr. (Phila)* 42, 219–225, 2003.

91. Brook, I., Antibacterial therapy for acute group a streptococcal pharyngotonsillitis: short-course versus traditional 10-day oral regimens. *Paediatr. Drugs* 4, 747–754, 2002.

92. Hirota, K., Murakami, K., Nemoto, K., et al., Fosfomycin reduces CD15s-related antigen expression of *Streptococcus pyogenes. Antimicrob. Agents Chemother.* 42, 1083–1087, 1998.

93. Jones, M.E., Karlowsky, J.A., Draghi, D.C., Thornsberry, C., Sahm, D.F., and Nathwani, D., Epidemiology and antibiotic susceptibility of bacteria causing skin and soft tissue infections in the USA and Europe: a guide to appropriate antimicrobial therapy. *Int. J. Antimicrob. Agents* 22, 406–419, 2003.

94. Akiyama, H., Kanzaki, H., Tada, J., and Arata, J., *Staphylococcus aureus* infection on cut wounds in the mouse skin: experimental staphylococcal botryomycosis. *J. Dermatol. Sci.* 11, 234–238, 1996.

95. Yamasaki, O., Akiyama, H., Toi, Y., and Arata, J., A combination of roxithromycin and imipenem as an antimicrobial strategy against biofilms formed by *Staphylococcus aureus. J. Antimicrob. Chemother.* 48, 573–577, 2001.

96. Yanagihara, K., Okada, M., Tashiro, M., Ohno, H., Miyazaki, Y., Hirakata, Y., Tashiro, T., and Kohno, S., Efficacy of S-3578 against MRSA in hematogenous pulmonary infection model. 43rd Annual Interscience Conference on Antimicrobial Agents and Chemotherapy, Chicago, IL, September 14–17. American Society for Microbiology, 2003.

97. Bush, K., Abbanat, D., Davies, T., Dudley, M., Blais, J., Hilliard, J., Wira, E., and Foleno, B., *In vitro* and *in vivo* antibacterial activity of RWJ-442831, a prodrug of the anti-MRSA cephalosporin RWJ-54428. 43rd Annual Interscience Conference on Antimicrobial Agents and Chemotherapy, Chicago, IL, September 14–17. American Society for Microbiology, 2003.

98. Schierholz, J.M., Steinhauser, H., Rump, A.F., Berkels, R., and Pulverer, G., Controlled release of antibiotics from biomedical polyurethanes: morphological and structural features. *Biomaterials* 18, 839–844, 1997.

99. Rupp, M.E., and Hamer, K.E., Effect of subinhibitory concentrations of vancomycin, cefazolin, ofloxacin, L-ofloxacin and D-ofloxacin on adherence to intravascular catheters and biofilm formation by *Staphylococcus epidermidis. J. Antimicrob. Chemother.* 41, 155–161, 1998.

100. Rachid, S., Ohlsen, K., Witte, W., Hacker, J., and Ziebuhr, W., Effect of subinhibitory antibiotic concentrations on polysaccharide intercellular adhesin expression in biofilm-forming *Staphylococcus epidermidis. Antimicrob. Agents Chemother.* 44, 3357–3363, 2000.

101. Krishnasami, Z., Carlton, D., Bimbo, L., et al., Management of hemodialysis catheter-related bacteremia with an adjunctive antibiotic lock solution. *Kidney Int.* 61, 1136–1142, 2002.

102. Kuiper, I., Lagendijk, E.L., Pickford, R., et al., Characterization of two *Pseudomonas putida* lipopeptide biosurfactants, putisolvin I and II, which inhibit biofilm formation and break down existing biofilms. *Mol. Microbiol.* 51, 97–113, 2004.

103. Chevalier, J., Pages, J.M., Eyraud, A., and Mallea, M., Membrane permeability modifications are involved in antibiotic resistance in *Klebsiella pneumoniae. Biochem. Biophys. Res. Commun.* 274, 496–499, 2000.

104. Quinn, J.P., Darzins, A., Miyashiro, D., Ripp, S., and Miller, R.V., Imipenem resistance in *Pseudomonas aeruginosa* PAO: mapping of the OprD2 gene. *Antimicrob. Agents Chemother.* 35, 753–755, 1991.

105. Bornet, C., Chollet, R., Mallea, M., et al., Imipenem and expression of multidrug efflux pump in *Enterobacter aerogenes. Biochem. Biophys. Res. Commun.* 301, 985–990, 2003.
106. Mao, W., Warren, M.S., Black, D.S., et al., On the mechanism of substrate specificity by resistance nodulation division (RND)-type multidrug resistance pumps: the large periplasmic loops of MexD from *Pseudomonas aeruginosa* are involved in substrate recognition. *Mol. Microbiol.* 46, 889–901, 2002.

20 Glycopeptide Antibacterials and the Treatment of Biofilm-Related Infections

John L. Pace, Roasaire Verna, and Jan Verhoef

CONTENTS

20.1 INTRODUCTION

Vancomycin and teicoplanin are glycopeptide antibacterials that act through inhibition of cell-wall (i.e., peptidoglycan) synthesis in Gram-positive bacteria (1). The history of vancomycin is remarkable because it has received more use in the past 20 years than in the first 20 years after its approval in the late-1950s. In fact, current annual sales of this antibiotic are approximately US$370M. The principal reason for this phenomenon is that vancomycin has come to be relied upon, and in some cases considered the drug of last recourse, for the treatment of multidrug-resistant staphylococci and some other Gram-positive pathogens (2,3).

In 1986 resistance to vancomycin was first observed among the enterococci (4–6). The delay in development of resistance is also unique in that 30 years of clinical use had passed and belies the complexity of the nature of vancomycin resistance, which acquires multi-enzymatic synthesis of a novel low-affinity substrate (7). Due to the substantial role that vancomycin has come to play, vancomycin-resistance was seen as a matter of great concern because it was believed that this type of high-level resistance in vancomycin-resistant enterococci (VRE) would be readily transferred into more pathogenic bacteria like *Staphylococcus aureus,* leading to a public health calamity.

Since 1997 vancomycin-intermediate susceptible *S. aureus* (VISA) have been isolated, as well as bacteria tolerant to the bactericidal effects of glycopeptide antibiotics (8–11). However, the low-level resistance among these bacteria is distinct from that mediated by acquisition of the *van* operon as for enterococci, and while VISA are more difficult to treat with vancomycin, and have lead to clinical failures, generally the problem largely could be controlled.

Finally, *S. aureus* clinical isolates exhibiting frank vancomycin resistance (VRSA) have been detected (12–15). If VRSA do become widely disseminated, this indeed could pose a much more substantial problem. However, the frequency of VRSA isolated to date still appears limited (16,17). Nevertheless, the question whether the newer clinically approved agents like linezolid or quinupristin/dalfopristin might be successfully utilized to treat serious highly invasive *S. aureus*-based diseases, or whether any significant susceptibility to other older antibacterial classes may be expected need to be resolved as soon as possible (18–20). At the same time, the frequency and virulence of the VRSA isolated to date appear limited.

One of the challenges in treating staphylococcal infections is the biofilm-associated infection where staphylococci firmly adhere to foreign material. This challenge is now even greater than the current state of resistance to glycopeptide antibiotics (21–24). As with reduced effects of other drug classes, the cause of diminished glycopeptide efficacy against biofilm-ensconced bacteria is largely derived from phenotypic tolerance, although a multifactorial mechanism is likely (25–31). This challenge is reinforced in that bacterial variants both tolerant to vancomycin, and readily able to form biofilms have arisen in the clinic. Notably, these isolates more effectively interact with the surfaces of indwelling-medical devices (IMD) and increasingly are the cause of associated infections (32–35). Obviously, a variety of problems surround the continued effective use of vancomycin and teicoplanin. In this chapter we will discuss the central issues, and address approaches to the looming challenges facing the use of glycopeptide antibacterials.

20.2   VANCOMYCIN

Vancomycin is the only glycopeptide antibiotic approved for clinical use throughout much of the World. This is striking considering the analoging approach long favored for pharmaceutical drug discovery, and the large numbers of antibacterial agents from other structural classes. Chemical characteristics of this Group I glycopeptide antibiotic have been reviewed previously (36).

Vancomycin minimal inhibitory concentration (MIC) breakpoints are: susceptible, <4 μg/ml; intermediate, 8 to 16 μg/ml; and resistant, >32 μg/ml (37,38). MICs of susceptible bacteria typically range from 0.25 to 2 μg/ml for *Staphylococcus* spp., 0.13 to 2 μg/ml for *Streptococcus* spp., 0.25 to 4 μg/ml for *Enterococcus* spp., and 0.13 to 4 μg/ml for miscellaneous other Gram-positive pathogens. *Lactobacillus* spp. *Pediococcus* spp., *Leuconostoc* spp., and some anaerobes appear to be intrinsically less susceptible (1,39). While vancomycin is bactericidal for staphylococci and many other species, this agent exerts only bacteriostatic activity against even susceptible enterococci (1). Further, β-lactam antibiotics have received preferred use status due to vancomycin's weaker concentration-independent bactericidal activity. Thus nafcillin is sometimes co-administered with vancomycin for therapy of suspected staphylococcal infection prior to receipt of the initial susceptibility report, and penicillins (i.e., ampicillin) may be employed even against beta-lactam resistant enterococci albeit at very high dosage prior to introduction of vancomycin therapy. Obviously this strategy is not used in cases of β-lactam allergy.

Vancomycin is often utilized in combination with other agents with the hope of achieving antibacterial synergy, particularly in the case of serious difficult to treat infections (40). Aminoglycocides, most often gentamicin, are frequently utilized for this purpose. These synergistic interactions are often abrogated in cases where the bacterial pathogen is resistant to either antibiotic (40).

Vancomycin mediates its antibacterial activity via hydrogen-bond formation with the D-alanyl-D-alanine terminating pentapeptide moiety of the lipid II substrate, primarily preventing transglycosylation and secondarily transpeptidation steps of cell-wall synthesis through steric hindrance (4,7,41). Vancomycin resistance as originally characterized from resistant enterococci is of five classes termed VanA, VanB, VanC, VanD, and VanE (4–6). The VanB and VanA phenotypes characterized by inducible resistance to vancomycin or to both vancomycin and teicoplanin, respectively, are clinically significant (4–6). The VanA phenotype is conferred by a transposon bearing a group of seven genes (4–6). Resistance is affected by replacement of the D-alaynyl-D-alanine substrate by a D-alanyl-D-lactate substrate for which glycopeptide antibiotics exhibit low affinity (4,5,7). In contrast to VRE, VISA appear to synthesize a thickened cell wall with a lessened degree of cross-linking. This results in higher levels of distal free D-alanyl-D-alanine groups that bind the antibiotics, and reduce free concentrations of the drugs near the surface of the cell membrane where transglycosylation and transpeptidation occur (7,42).

20.3 TEICOPLANIN

Teicoplanin is the second glycopeptide antibiotic approved in European countries including France, Germany, and Italy (43). This agent is a class 4 glycopeptide with a ristocetin-like core, and a fatty acid moiety appended to the amino sugar (24). Indications for the use of teicoplanin are similar to those of vancomycin in countries where it is used (43). Primary differences from vancomycin include lower MICs against some bacterial species, but higher frequency of reduced susceptibility for teicoplanin particularly among the coagulase-negative staphylococci (CoNS) (43). Teicoplanin is active against VanB enterococci because it is not an inducer (4). It likely mediates its activity through both substrate binding and membrane anchoring, but not antibiotic dimerization in contrast to vancomycin (44,45). There is some evidence that teicoplanin penetrates less readily the biofilm glycocalyx, fibrin clots, and cardiac vegetations (46). It exhibits greater serum protein binding, and has a longer half-life. Bolus administration may be beneficial, and can be utilized due to a reduced tendency for histamine release and red-man syndrome sometimes observed with vancomycin (1).

20.4 NEWER GLYCOPEPTIDES IN DEVELOPMENT

Three new semi-synthetic glycopeptides are in development for the treatment of Gram-positive bacterial infections. These include dalbavancin (BI-397), oritavancin (LY 333328), and telavancin (TD-6424) (47–62). Dalbavancin is a glycoppetide similar in spectrum to teicoplanin, but with an exceptionally long half-life (48,49,55). Dalbavancin pharmacokinetics have been described previously (63). Phase II trials of this agent for complicated skin and skin structure infections (CSSSI) with once-weekly dosing in comparison to vancomcyin have been completed, and other studies are ongoing (64).

Oritavancin and telavancin are distinct in exhibiting *in vitro* activity against all VRE, and against both the Michigan and Pennsylvania VRSA isolates (18–20). These two agents exert uniquely exquisite antibacterial activity against streptococci, including *Streptococcus pneumoniae* where MICs are as low as 0.003 µg/ml (50–62). Oritavancin and telavancin also deliver superior bactericidal activity *in vitro* as compared to vancomycin. They are bactericidal for enterococci, and are more active than vancomycin or linezolid in animal infection models (65,66). In fact, telavancin reduced MRSA counts by 3 logs within 4 hours in contrast to 8 hours for vancomycin when the inoculum level was 10^6 CFU/ml, and within 8 and 24 hours respectively when the inoclum was increased to 10^7 CFU/ml (66). Telavancin exhibits post-antibiotic effects 2–4 fold greater than for vancomycin (53,54). Oritavancin, like dalbavancin, has an exceptionally long half-life, and has been evaluated in numerous human trials (67). Achievable serum concentration and an 8–10 hour half-life allow for once-daily dosing of telavancin (58,59,68). While elevated serum protein binding is observed with telavancin, as well as for oritavancin and dalbavancin, impressive serum static and bactericidal titers have been observed from human volunteers (68). Evaluation of telavancin in a human Phase II CSSSI study has been completed, and a *S. aureus* bacteremia study is ongoing (59).

20.5 CHARACTERIZATION OF GLYCOPEPTIDE ANTIBIOTICS AGAINST BIOFILM BACTERIA

The numbers of infections caused by biofilm-forming bacteria have increased dramatically as indwelling-medical devices (IMD) have come to play an increasingly important role in modern medicine (23). More than 30 million urinary-tract catheters, five million central venous catheters, two million fracture fixation devices, and numerous other miscellaneous implants are utilized annually in the United States alone (23). Associated infection rates range from 1 to 50%.

Staphylococcus epidermidis is the leading cause of many of the infections (23,69). Mucoid slime, which has been confirmed as the polysaccharide intercellular adhesin in some *S. epidermidis*, is an integral part of biofilms (70). The antibacterial activities of vancomycin and teicoplanin are inhibited by slime in contrast to that of rifampin (71,72). Synergistic effect of vancomycin and gentamicin is also antagonized by slime (71). This inhibitory effect in part explains why it is difficult to eradicate *S. epidermidis* infections associated with IMD (71,72).

Darouiche et al. (73) found that the reduced antibacterial effect was likely not due to lack of antibiotic penetration into the biofilm. Vancomycin penetrates to levels exceeding the MBC, and in fact higher levels of vancomycin than linezolid were quantified in luminal biofilms from hemodialysis catheters (74,75). Vancomycin may be bound by the bacterial slime, particularly in CoNS biofilms, resulting in higher levels but antibacterial effect is reduced (75).

While IMD niche infections play a substantial role in the rise of the CoNS from the status of commensal or contaminant to pathogen, it should be kept in mind that all staphylococci readily form biofilms on native tissue as well as devices (76,77). *S. aureus* is another major cause of these infections (78). This pathogen interacts strongly with host ligands such as host extracellular matrix components and con- ditioning films laid down on devices (78,79). *S. aureus* infections are associated with a poorer prognosis than CoNS infections. Their greater virulence means higher morbidity and mortality, and more aggressive therapy is required.

Many other Gram-positive pathogens including the enterococci also produce device- associated infections (23,78,80). Other common pathogens include *Pseudomonas aeruginosa*, *Burkholderia cepacia*, *Acinetobacter* spp., *Escherichia coli*, *Providencia stuartii*, and *Candida* spp., but these microorganisms are not susceptible to glycopeptide antibiotics and are beyond the scope of this chapter (23,80).

20.5.1 FINDINGS WITH *IN-VITRO* MODELS

S. epidermidis biofilms formed on sinter-glass beads have been tested for suscepti- bility to antibiotics (21). Rifampin, in contrast to teicoplanin, actively reduced bacterial counts from the biofilms, and levofloxacin was also active (21). In related animal models a correlation was shown for efficacy if the *in vitro* MBC was lower than achiev- able serum levels. In a 96-well polystyrene tissue-culture plate model fosfomycin, cefuroxime, rifampicin, and cefazolin were more active than vancomycin against *in vitro* methicillin-susceptible *S. aureus* biofilms (3). Vancomycin did reduce biofilm counts, but was substantially less active at four-fold minimal bactericidal concentration (MBC) than against aged biofilms. Telavancin was more active than vancomycin or teicoplanin in the *in vitro* Sorbarod biofilm model with *S. aureus* and *S. epidermidis* (81). Telavancin reduced bacterial counts by 1 to 1.5 logs in contrast to 0 to 0.5 logs for the other glycopeptides. While findings with oritavancin were similar, total reduction in cell counts was greater with telavancin (81). These findings are consistent with the bactericidal properties of telavancin (53–55,58,59,66).

Studies have been performed with materials found in IMD. Combinations of vancomycin and amikacin or rifampin were able to sterilize *S. epidermidis* on the surface of Vialon and polyvinylchloride catheters (82). However, vancomycin treatment of *S. epidermidis* biofilm formed on a silicone elastomer (peritoneal catheter) only decreased counts by 0.5 log CFU/cm^2, consistent with findings of phenotypic tolerance in biofilm bacteria (83). *S. aureus* susceptibility was similarly decreaed for vancomycin and benzylpenicillin when the bacteria were adherent to silicone catheter surfaces (84). Ten-fold greater concentration of vancomycin was required to kill adherent than suspended *S. aureus* (84). Studies to differentiate the effect of surface adherence from biofilm formation and antibiotic action suggested that the two events are distinct, and that antibiotics including vancomycin, beta-lactams, and fluroquinolones, while not able to inhibit adherence, do reduce biofilm formation even at sub-inhibitory concentrations (85).

Vancomycin reduced bacterial counts by two logs in *S. aureus* infected fibrin-platelet clots simulating vegetations associated with endocaridtis (86). No substantial differences were noted between different vancomycin dosing regimes consistent with the antibiotic's concentration-independent bactericidal activity. Combination use of vancomycin and gentamicin increased bactericidal activity (86).

When vancomycin-susceptible enterococcal biofilms were perfused with therapeutic trough levels of either vancomycin or teicoplanin, an initial 1 to 2 log reduction in viable bacterial cells released from the biofilm was noted (87). However, within 24 hours bacterial numbers increased consistent with antibiotic tolerance often observed for biofilm bacteria. Quinupristin/dalfopristin was more active than vancomycin or teicoplanin against vancomycin-susceptible enterococcal biofilms (88).

20.5.2 FINDINGS WITH *IN-VIVO* MODELS

A variety of animal models have been described for the evaluation of antibiotics against biofilm-associated bacterial infections. A combination of rifampin and vancomycin was found to be superior to either agent alone in treated mice following implantation of *S. epidermidis* biofilms preformed on catheter segments (89). Vancomycin but not teicoplanin was able to reduce MRSA counts in a chronic rat tissue cage infection model even though levels of teicoplanin exceeded the MBC (90). This finding of *in vivo* tolerance for teicoplanin but not vancomycin was curious because antibiotic concentrations achieved should have negated the effect of higher serum protein binding for teicoplanin, and similar activities for the two antibiotics were observed in a guinea pig prophylaxis model (90).

Cefuroxime was more effective than vancomycin, tobramycin, or ciprofloxacin following 21 days of therapy in reducing *S. aureus* counts in rat osteomyelitis models, using either direct infection at surgery or pre-colonized implants (91,92). Findings correlated with results from *in vitro* biofilm studies (91,92). In a rabbit model of orthopedic device infection, vancomycin plus rifampin (90%) provided the highest cure rate in contrast to minocycline plus rifampin (70%), or vancomycin alone (20%) (93).

The rabbit model of endocarditis is often considered to be one of the most challenging tests of glycopeptide efficacy. Vancomycin is more effective than linezolid against MRSA in this model (94). Vancomycin reduces valvular vegetations to a greater degree, and sterilizes more aortic vegetations in contrast to linezolid. Oritavancin has been characterized both in the endocarditis model, and against vancomycin-resistant *Enterococcus faecium* in a rat central venous catheter (CVC) associated infection model (95–97). Oritavancin (25 mg/kg Q24, 4d) was as effective as vancomycin (25 mg/kg Q8, 4d) in reducing MRSA in vegetation and cardiac tissue in the rabbit left-sided endocarditis model (95). Moreover, oritavancin in contrast to vancomycin was effective against all enterococcal strains tested, including VanB and VanA positive isolates, in reducing aortic bacterial counts (96). In the CVC model, oritavancin reduced peripheral bacteremia from 75% for control animals to 0% of treated animals, and similarly CVC- associated bacteria were reduced from 87.5 to 12.5% (97). Telavancin reduced mean VISA aortic-valve vegetations in the rabbit by 5.5 logs and sterilized four of six animals as compared to no reduction or sterilizations by vancomycin when animals received eight doses of either agent at 30 mg/kg IV BID (98). In the rat staphylococcal endocarditis model, dalbavancin was as active as vancomycin (48).

20.6 CLINICAL USE OF GLYCOPEPTIDE ANTIBIOTICS

Indications for the use of vancomycin typically include treatment of serious Gram-positive bacterial infections due to susceptible organisms, including oxacillin (methicillin)-susceptible and –resistant *S. aureus*, and *Streptococcus* spp. and *Enterococcus* spp. in patients allergic to β-lactam antibiotics (37,39). Dosing is 1 g or 15 mg/kg IV, Q12, and may be administered 7 days to 8 weeks dependent on disease site, and clinical status (36,37). Trough serum levels may be monitored and should be maintained at a level of 10 to 12 µg/ml. Adjustment of dosage may be required in patients with renal insufficiency, i.e., nomograms based on reduced creatinine clearance are available, or where rapid clearance is expected (36,37). Renal- and oto-toxicity are infrequently observed, but monitoring of renal function should be considered particularly during combination use of vancomycin with antibiotics of known nephrotoxic potential (36,37). Vancomycin also can be administered orally for therapy of *S. aureus* enterocolitis, and *Clostridium difficile*-associated pseudomembranous colitis (36,37). This should however be discouraged because of emergence of VRE, and use of metronidazole would be preferable in these clinical situations.

20.6.1 INTRAVASCULAR CATHETER-ASSOCIATED INFECTIONS

Approximately 30% of hospital acquired bacteremias are associated with intra-vascular catheters. Therapy is largely driven by clinical signs, pathogen virulence, and level of risk for catheter-related blood stream infection (CRBSI) (see chapter 22) (99–102). Vancomycin is the drug of choice for empirical therapy of CRBSI due to the preponderance of infections caused by methicillin-resistant staphylococci, but therapy should be adjusted accordingly based on culture identification and susceptibility (101–105). Typical antibiotic therapy continues for 7 to 10 days, but prolonged courses of systemic therapy (4 to 6 weeks) are required in patients suffering from continued bacteremia or those at risk of endocarditis. In the case of low-risk infections lock therapy in addition to systemic antibiotics can be attempted as salvage for treatment of central-venous catheter-related sepsis (106). A variety of antibiotics have been tried in combination with heparin, but minocycline-EDTA has exhibited superior activity as compared with glycopeptides (106,107). Enterococcal CRBSI often necessitates removal of the catheter but the combination of vancomycin or ampicillin, plus an aminoglycoside, has proven sufficiently effective that removal of the device may not always be required (108).

20.6.2   ORTHOPEDIC DEVICE-RELATED INFECTIONS

Nearly six million people in the United States have an internal fixation device or artificial joint (109). While there is a low frequency of infection, they are still numerous and often require removal of the device. Antibiotic therapy plays an essential role in the treatment of infections associated with orthopedic devices and in particular prosthetic joints (see Chapter 24). Bacterial cultures should be obtained and susceptibility determined prior to initiating therapy unless overwhelming sepsis is obvious and empiric treatment is mandatory. Staphylococci cause the preponderance of orthopedic-device related infections and account for approximately 50% of cases (109–111). Vancomycin should be reserved for first-line use in the case of oxacillin-resistant staphylococci, or penicillin-resistant enterococci (110–114). Otherwise β-lactams, including nafcillin, cefazolin, or penicillin are preferred, depending on the Gram-positive bacterial pathogen present. Vancomycin should also be used as first line therapy when a susceptible Gram-positive pathogen is present and the patient exhibits beta-lactam sensitivity. Therapy duration is typically 4 to 6 weeks to several months. Alternatives to the use of vancomycin include linezolid, and the combination of levofloxacin plus rifampin.

20.6.3   INFECTIOUS ENDOCARDITIS

The highest rates of morbidity and mortality are associated with bacterial endocarditis and infections of cardiac assist devices from biofilm-forming bacteria (23,115,116). Staphylococci again are a major cause of these infections. It is important to confirm susceptibility of the infecting pathogen. Vancomycin is the drug of choice when treating native-valve infections due to oxacillin-resistant bacteria (see Chapter 23). However, this glycopeptide antibiotic is not optimal due to slow bactericidal activity and lower penetration as compared with beta-lactams (117–119). For susceptible bacteria, penicillinase-resistant penicillins such as nafcillin are preferred. Rare treatment failures with first-generation cephalosporins have been observed, and are likely due to an inoculum effect from high-density vegetations of penicillinase-producing oxacillin-susceptible *S. aureus* (120,121). Treatment is for 4 to 6 weeks. Aminoglycosides like gentamicin are often used in combination with the primary antibiotic during the first few days to a week of therapy to reduce the likelihood of nephrotoxicity.

Six to eight weeks of therapy with vancomycin in combination with rifampin is used to treat prosthetic-valve endocarditis due to oxacillin-resistant *S. aureus*.

Again, gentamicin is included in the regimen during the first two weeks of therapy. This regimen is similarly useful for infections due to oxacillin-resistant CoNS (122,123). Vancomycin combined with gentamicin is also recommended in cases due to streptococcal infection where the patient is allergic to beta-lactams, and in infections with vancomycin-susceptible high-level penicillin-resistant enterococci (124–126).

20.7   CONCLUSIONS

The use of IMD likely will continue to rise over the coming decade due to the aging population, and their prevalent role in the treatment of a variety of diseases. Related infections due to biofilm-forming bacteria even if maintained at the same frequency will necessarily increase. Experience suggests that the newly isolated VRSA unfortunately will increase in prevalence. This is of great concern, particularly in light of their multi-drug resistance, association with IMD-related disease, and reduced industrial efforts toward creation of new antibiotics to address the problems. The important role for glycopeptide antibacterials, and particularly vancomycin, is likely to continue for the foreseeable future.

REFERENCES

1. Nagarajan, R., ed., *Glycopeptide Antibiotics*. New York: Marcel Dekker, 1994.
2. Chambers, H.F., Methicillin resistance in staphylococci: molecular and biochemical basis and clinical implications. *Clin. Microbiol. Rev.* 10, 781–791, 1997.
3. Amorena, B., Gracia, E., Monzon, M., Leiva, J., Oteiza, C., Perez, M., Alabart, J.L., and Hernandez-Yago, J., Antibiotic susceptibility assay for *Staphylococcus aureus* in biofilms developed *in vitro*. *J. Antimicrob. Chemother.* 44, 43–55, 1999.
4. Arthur, M., Reynolds, P., and Courvalin, P., Glycopeptide resistance in enterococci. *Trends Microbiol.* 4, 401–407, 1996.
5. Cetinkaya, Y., Falk, P., and Mayhall, C.G., Vancomycin-resistant enterococci. *Clin. Microbiol. Rev.* 13, 686–707, 2000.
6. DeLisle, S., and Perl, T.M., Vancomycin-resistant enterococci: A road map on how to prevent the emergence and transmission of antimicrobial resistance. *Chest* 123, 504S–518S, 2003.
7. Bugg, T.D.H., Wright, G.D., Dutka-Malen, S., Arthur, M., Courvalin, P., and Walsh, C.T., Molecular basis for vancomycin resistance in *Enterococcus faecium* BM4147: biosynthesis of a depsipeptide peptidoglycan precursor by vancomycin resistance proteins VanH and VanA. *Biochemistry* 30, 10408–10415, 1991.
8. Hiramatsu, K., Hanaki, H., Ino, T., Yabuta, K., Oguri, T., and Tenover, F.C., Methicillin-resistant *Staphylococcus aureus* clinical strain with reduced vancomycin susceptibility. *J. Antimicrob. Chemother.* 40, 135–136, 1997.
9. May, J., Shannon, K., King, A., and French, G., Glycopeptide tolerance in *Staphylococcus aureus*. *J. Antimicrob. Agents Chemother.* 42, 189–197, 1998.
10. Tenover, F.C., Lancaster, M.V., Hill, B.C., Steward, C.D., Stocker, S.A., Hancock, G.A., O'Hara, C.M., Clark, N.C., and Hiramatsu, K., Characterization of staphylococci with reduced susceptibilities to vancomycin and other glycopeptides. *J. Clin. Microbiol.* 36, 1010–1027, 1998.

11. Liu, C., and Chambers, H.F., *Staphylococcus aureus* with heterogeneous resistance to vancomycin: epidemiology, clinical significance, and critical assessment of diagnostic methods. *Antimicrob. Agents Chemother.* 47, 3040–3045, 2003.

12. Sievert, D.M., Boulton, M.L., Stoltman, G., Johnson, D., Stobierski, M.G., Downes, F.P., Sonsel, P.A., Rudrik, J.T., Brown, W., Hafeez, W., Lundstrom, T., Flanagan, E., Johnson, R., Mitchell, J., and Chang, S., *Staphylococcus aureus* resistant to vancomycin – United States, 2002. *MMWR* 51, 565–567, 2002.

13. Miller, D., Urdaneta, V., Weltman, A., and Park, S., Vancomycin-resistant *Staphylococcus aureus* – Pennsylvania, 2002. *MMWR* 51, 902, 2002.

14. Chang, S., Sievert, D.M., Hageman, J.C., Boulton, M.L., Tenover, F.C., Downes, F.P., Shah, S., Rudrik, J.T., Pupp, G.R., Brown, W.J., Cardo, D., and Fridkin, S.K., Vancomycin-resistant *Staphylococcus aureus* investigative team: infection with vancomycin-resistant *Staphylococcus aureus* containing the *vanA* resistance gene. *N. Engl. J. Med.* 348, 1342–1347, 2003.

15. Kacica, M., and McDonald, L.C., Brief report: Vancomycin-resistant *Staphylococcus aureus* – New York, 2004. *MMWR* 53, 322–323, 2004.

16. Bush, K., Vancomycin-resistant *Staphylococcus aureus* in the clinic: not quite armageddon. *Clin. Infect. Dis.* 38, 1056–1057, 2004.

17. Gemmell, C.G., Glycopeptide resistance in *Staphylococcus aureus*: is it a real threat? *J. Infect. Chemother.* 10, 69–75, 2004.

18. Bozdogan, B., Esel, D., Whitener, C., Browne, F.A., and Appelbaum, P.C., Antibacterial susceptibility of a vancomycin-resistant *Staphylococcus aureus* strain isolated at the Hershey Medical Center. *J. Antimicrob. Chemother.* 52, 864–868, 2003.

19. Weigel, L.M., McDougal, L.K., Clark, N., Killgore, G., Tenover, F.C., Appelbaum, P.C., and Bozdogan, B., Molecular characterization of a vancomycin-resistant clinical isolate of *Staphylococcus aureus* from Pennsylvania. 43rd ICAAC, Chicago, IL, Sept 14–17, 2003.

20. Tenover, F.C., Weigel, L.M., Appelbaum, P.C., McDougal, L.K., Chaitram, J., McAllister, S., Clark, N., Killgore, G., O'Hara, C.M., Jevitt, L., Patel, J.B., and Bozdogan, B., Vancomycin-resistant *Staphylococcus aureus* isolate from a patient in Pennsylvania. *Antimicrob. Agents Chemother.* 48, 275–280, 2004.

21. Schwank, S., Rajacic, Z., Zimmerli, W., and Blaser, J., Impact of bacterial biofilm formation on *in vitro* and *in vivo* activities of antibiotics. *Antimicrob. Agents Chemother.* 42, 895–898, 1998.

22. Schierholz, J.M., Beuth, J., and Pulverer, G., Adherent bacteria and activity of antibiotics. *J. Antimicrob. Chemother.* 43, 158–160, 1999.

23. Darouiche, R.O., Device-associated infections: A macroproblem that starts with microadherence. *Clin. Infect. Dis.* 33, 1567–1572, 2001.

24. Donlan, R.M., Biofilm formation: A clinically relevant microbiological process. *Clin. Infect. Dis.* 33, 1387–1392, 2001.

25. Costerton, J.W., Chang, K.J., Geesey, G.G., Ladd, T.I., Nickel, J.C., Dasgupta, M., and Marrie, T.J., Bacterial biofilms in nature and disease. *Ann. Rev. Microbiol.* 41, 435–464, 1987.

26. Consterton, J.W., Lewandowski, Z., Caldweel, D.E., Korber, E.R., and Lappin-Scott, H.M., Microbial biofilms. *Ann. Rev. Microbiol.* 49, 711–745, 1995.

27. Shapiro, J.A., Thinking about bacterial populations as multicellular organisms. *Annu. Rev. Microbiol.* 52, 81–104, 1998.

28. Davey, M.E., and O'Toole, G.A., Microbial biofilms: from ecology to molecular genetics. *Microbiol. Mol. Biol. Rev.* 64, 847–867, 2000.

29. Mah, T.F., and O'Toole, G.A., Mechanisms of biofilm resistance to antimicrobial agents. *Trends Microbiol.* 9, 34–39, 2001.

30. Donlan, R.M., and Costerton, J.W., Biofilms: Survival mechanisms of clinically relevant microorganisms. *Clin. Microbiol. Rev.* 15, 167–193, 2002.
31. Dunne, W.M. Jr., Bacterial Adhesion: Seen any good biofilms lately? *Clin. Microbiol. Rev.* 15, 155–166, 2002.
32. Vuong, C., Saenz, H.L., Gotz, F., and Otto, M., Impact of the *agr* quorum-sensing system on adherence to polystyrene in *Staphylococcus aureus. J. Infect. Dis.* 182, 1688–1693, 2000.
33. Sakoulas, G., Eliopoulos, G.M., Moellering, R.C. Jr., Wennersten, C., Venkataraman, L., Novick, R.P., and Gold, H.S., Accessory gene regulator (*agr*) locus in geographically diverse *Staphylococcus aureus* isolates with reduced susceptibility to vancomycin. *Antimicrob. Agents Chemother.* 46, 1492–1502, 2002.
34. Verdier, I., Reverdy, M.E., Etienne, J., Lina, G., Bes, M., and Vandensech, F., *Staphylococcus aureus* isolates with reduced susceptibility to glycopeptides belong to accessory gene regulator group I or II. *Antimicrob. Agents Chemother.* 48, 1024–1027, 2004.
35. Sakoulas, G., Moise-Broder, P.A., Schentag, J., Forrest, A., Moellering, R.C. Jr., and Eliopoulos, G.M., Relationship of MIC and bactericidal activity to efficacy of vancomycin for treatment of methicillin-resistant *Staphylococcus aureus* bacteremia. *J. Clin. Microbiol.* 42, 2398–2402, 2004.
36. Yao, R.C., and Crandall, L.W., Glycopeptides: Classification, occurrence, and discovery. In: Nagarajan, R., ed., *Glycopeptide Antibiotics.* New York: Marcel Dekker, 1994:1–28.
37. http://www.cai.mcgill.ca/meded/drugdb/vancomycin/vancomycin_db.htm
38. NCCLS. Performance standards for antimicrobial susceptibility testing, Fourth Information Supplement. NCCLS Document M100–S4. Villanova: NCCLS, 1992.
39. Zeckel, M.L., and Woodworth, J.R., Vancomycin: A clinical overview. In: Nagarajan, R., ed., *Glycopeptide Antibiotics.* New York: Marcel Dekker, 1994:309–409.
40. Monzon, M., Oteiza, C., Leiva, J., and Amorena, B., Synergy of different antibiotic combinations in biofilms of *Staphylococcus epidermidis. J. Antimicrob. Chemother.* 48, 792–801, 2001.
41. Allen, N.E., LeTourneau, D.L., Hobbs, J.N., and Thompson, R.C., Hexapeptide derivatives of glycopeptide antibiotics: tools for mechanism of action studies. *Antimicrob. Agents Chemother.* 46, 2344–2348, 2002.
42. Hanaki, H., Labischinski, H., Sasaki, K., Kuwahara-Arai, K., Inaba, Y., and Hiramatsu, K., Mechanism of vancomycin resistance in MRSA strain Mu50. *Jpn. J. Antibiot.* 51, 237–247, 1998.
43. Goldstein, B.P., Rosina, R., and Parenti, F., Teicoplanin. In: Nagarajan, R., ed., *Glycopeptide Antibiotics.* New York: Marcel Dekker, 1994:273–307.
44. Beauregard, D.A., Williams, D.H., Gwynn, M.N., and Knowles, D.J.C., Dimerization and membrane anchors in extracellular targeting of vancomycin group antibiotics. *Antimicrob. Agents Chemother.* 39, 781–785, 1995.
45. Sharman, G.J., and Williams, D.H., Common factors in the mode of action of vancomycin group antibiotics active against resistant bacteria. *Chem. Commun.* 7, 723–724, 1997.
46. Carbon, C., Experimental endocarditis: a review of its relevance to human endocarditis. *J. Antimicrob. Chemother.* 31(Suppl D), 71–85, 1993.
47. Biavasco, F., Vignaroll, C., Lupidi, R., Manso, E., Facinell, B., and Veraldo, P.E., *In vitro* antibacterial activity of LY 333328, a new semisynthetic glycopeptide. *Antimicrob. Agents Chemother.* 41, 2165–2172, 1997.
48. Candiani, G., Abbondi, M., Borogonovi, M., Romano, G., and Parenti, F., *In-vitro* and *in vivo* antibacterial activity of BI 397, a new semi-synthetic glycopeptide antibiotic. *J. Antimicrob. Chemother.* 44, 179–192, 1999.

49. Malabarba, A., Donadio, S. BI 397. *Drugs of the Future* 24, 839–846, 1999.
50. Barrett, J.F., Oritavancin Eli Lilly & Co. *Curr. Opin. Investigat. Drugs* 2, 1039–1044, 2001.
51. Woodford, N., Novel agents for the treatment of resistant Gram-positive infections. *Expert. Opin. Investigat. Drugs* 12, 117–137, 2003.
52. Allen, N.E., and Nicas, T.I., Mechanism of action of oritavancin and related glycopeptide antibiotics. *FEMS Microbiol. Rev.* 26, 511–532, 2002.
53. Pace, J.L., TD-6424: A novel multifunctional antibiotic. 42nd ICAAC, San Diego, CA, Sept 27–30, 2002.
54. Pace, J., Judice, K., Hegde, S., Leadbetter, M., Linsell, M., Kaniga, K., Reyes, N., Farrington, L., Debabov, D., Nodwell, M., and Christensen, B., Activity of TD-6424 against methicillin-resistant and –susceptible Gram-positive bacteria. 10th Internat. Symp. *Staphylococci Staphylococcal Infect.*, Tsukuba, Japan, Oct 18–20, 2002.
55. Steirt, M., Schmitz, F., Dalbavancin Biosearch Italia/Versicor. *Curr. Opin. Investigat. Drugs* 3, 229–233, 2002.
56. Debabov, D., Pace, J., Nodwell, M., Trapp, S., Campbell, B., Karr, D., Wu, T., Krause, K., Johnston, D., Lane, C., Schmidt, D., Higgins, D., Christensen, B., Judice, K., and Kaniga, K., TD-6424, a novel rapidly bactericidal concentration-dependent antibiotic, acts through a unique dual mode of action. 43rd ICAAC, Chicago, IL, Sept 14–17, 2002.
57. Pace, J., Krause, K., Johnston, D., Debabov, D., Wu, T., Farrington, L., Lane, C., Higgins, D., Christensen, B., Judice, K., and Kaniga, K., *In vitro* activity of TD-6424 against *Staphylococcus aureus. Antimicrob. Agents Chemother.* 47, 3602–3604, 2003.
58. Judice, J.K., and Pace, J.L., Semi-synthetic glycopeptide antibacterials. *Bioorg. Med. Chem. Lett.* 13, 4165–4168, 2003.
59. Pace, J.L., and Judice, J.K., Telavancin (Theravance) and other semi-synthetic glycopeptide clinical candidates. *Curr. Opin. Investigat. Drugs.* 6, 216–225, 2005.
60. Goldstein, E.J.C., Citron, D.M., Merriam, C.V., Warren, Y.A., Tyrell, K.L., and Fernandez, H.T., *In vitro* activities of the new semisynthetic glycopeptide telavancin (TD-6424), vancomycin, daptomycin, linezolid, and four comparator agents against anaerobic Gram-positive species and *Corynebacterium* spp. *Antimicrob. Agents Chemother.* 48, 2149–2152, 2004.
61. King, A., Phillips, I., Farrington, L., Pace, J., and Kaniga, K., Comparative *in vitro* activity of TD-6424, a raidly bactericidal, concentration-dependent antibiotic with multiple mechanisms of action against Gram-positive bacteria. 13th ECCMID, Glasgow, UK, May 10–13, 2003.
62. Leadbetter, M.R., Linsell, M.S., Fatheree, P.R., Trapp, S.G., Lam, B.M.T., Nodwell, M.B., Pace, J.L., Bazzini, B., Krause, K.M., Quast, K., Soriano, E., Wu, T.X., Shaw, J.P., Adams, S.M., Karr, D.E., Villena, J.D., and Judice, J.K., Hydrophobic vancomycin derivatives with improved ADME properties: The discovery of TD-6424. *J. Antibiotics.* 57, 326–336, 2004.
63. Leighton, A., Gottlieb, A.B., Dorr, M.B., Jabes, D., Mosconi, G., VanSaders, C., Mroszczak, E.J., Campbell, K.C.M., and Kelly, E., Tolerability, pharmacokinetics, and serum bactericidal activity of intravenous dalbavancin in healthy volunteers. *Antimicrob. Agents Chemother.* 48, 940–945, 2004.
64. Seltzer, E., Dorr, M.B., Goldstein, B.P., Perry, M., Dowell, J.A., and Henkel, T., the Dalbavancin Skin and Soft-Tissue Infection Study Group. Once-weekly dalbavancin versus standard-of-care antimicrobial regimens for treatment of skin and soft-tissue infections. *Clin. Infect. Dis.* 37, 1298–1303, 2003.
65. Hedge, S.S., Reyes, N., Wiens, T., Vanasse, N., Skinner, R., McCullough, Kaniga, K., Pace, J., Thomas, R., Shaw, J.P., Obedencio, G., and Judice, J.K., Pharmacodynamics

of Telavancin (TD-6424) a novel bactericidal agent, against Gram-positive bacteria. *Antimicrob Agents Chemother.* 48, 3043–3050, 2004.

66. Kaniga, K., Krause, K., Johnston, D., Lane, C., White, L., Wu, T., Debabov, D.V., Pace, J.L., and Higgins, D.L., Effect of pH, media and inoculum size on telavancin (TD-6424) *in vitro* activity. 14th ECCMID, Prague, Czech Republic, May 13–16, 2004.

67. de la Pena, A., Chien, J., Geiser, J., Brown, T., Farlow, D., Weerakkody, G., and Wasilewski, M., Microbiological outcomes and pharmacokinetics of oritavancin in patients with Gram-positive bacteremia. 12th ECCMID, Milan, Italy, May 9–12, 2002.

68. Barriere, S., Shaw, J., Seroogy, J., Kaniga, K., Pace, J., Judice, K., and Mant, T., Pharmacokintic disposition and serum bactericidal activity following IV infusion of a single and multiple ascending doses of TD-6424 in healthy male subjects. 13th ECCMID, Glasgow, UK, May 10–13, 2003.

69. Bailey, E.M., Constance, T.D., Albrecht, L.M., and Rybak, M.J., Coagulase-negative staphylococci: incidence, pathogenicity, and treatment in the 1990s. *Ann. Pharmacother.* 24, 714–720, 1990.

70. Ammendolia, M.G., Di Rosa, D., Montanaro, L., Arciola, C.R., and Baldassarri, L., Slime production and expression of the slime-associated antigen by staphylococcal clinical isolates. *J. Clin. Microbiol.* 37, 3235–3238, 1999.

71. Farber, B.F., Kaplan, M.H., and Clogston, A.G., *Staphylococcus epidermidis* extracted slime inhibits the antimicrobial action of glycopeptide antibiotics. *J. Infect. Dis.* 161, 37–40, 1990.

72. Souli, M., and Giammerellou, H., Effects of slime produced by clinical isolates of coagulase-negative staphylococci on activities of various antimicrobial agents. *Antimicrob. Agents Chemother.* 42, 939–941, 1998.

73. Darouiche, R.O., Dhir, A., Miller, A.J., Landon, G.C., Raad, I.I., and Musher, D.M., Vancomycin penetration into biofilm covering infected prostheses and effect on bacteria. *J. Infect. Dis.* 170, 720–723, 1994.

74. Dunne, W.M. Jr., Mason, E.O. Jr., and Kaplan, S.L., Diffusion of rifampin and vancomycin through a *Staphylococcus* biofilm. *Antimicrob. Agents Chemother.* 37, 2522–2526, 1993.

75. Wilcox, M.H., Kite, P., Mills, K., and Sugden, S., In situ measurement of linezolid and vancomycin concentrations in intravascular catheter-associated biofilm. *J. Antimicrob. Chemother.* 47, 171–175, 2001.

76. Gotz, F., Staphylococcus and biofilms. *Mol. Microbiol.* 43, 1367–1378, 2002.

77. Mack, D., Fischer, W., Krokotsch, A., Leopold, K., Hartmann, R., Egge, H., and Laufs, R., The intercellular adhesion involved in biofilm accumulation of *Staphylococcus epidermidis* is a linera beta-1,6-linked glucosaminoglycan: purification and structural analysis. *J. Bacteriol.* 178, 175–183, 1996.

78. Steinberg, J.P., Clark, C.C., and Hackman, B.O., Nosocomial and community-acquired *Staphylococcus aureus* bacteremias from 1980 to 1993: impact of intravascular devices and methicillin resistance. *Clin. Infect. Dis.* 23, 255–259, 1996.

79. Fitzpatrick, F., Humphreys, H., Smyth, E., Kennedy, C.A., and O'Gara, J.P., Environmental regulation of biofilm formation in intensive care unit isolates of *Staphylococcus epidermidis. J. Hosp. Infect.* 42, 212–218, 2002.

80. Jarvis, W.R., and Martone, W.J., Predominant pathogens in hospital infections. *J. Antimicrob. Chemother.* 29 (Suppl A), 19–24, 1992.

81. Gander, S., Kinnaird, A., Baklavadaki, L., and Finch, R., The effect of the novel glycopeptide, TD-6424, on biofilms of susceptible and resistant staphylococci. 43rd ICAAC, Chicago, IL, Sept 14–17, 2003.

82. Pascual, A., Ramirez de Arellano, R., and Perea, E.J., Activity of glycopeptides in combination with amikacin or rifampin against *Staphylococcus epidermidis* biofilms on plastic catheters. *Eur. J. Clin. Microbiol. Infect. Dis.* 13, 515–517, 1994.

83. Evans, R.C., and Holmes, C.J., Effect of vancomycin hydrochloride on *Staphylococcus epidermidis* biofilm associated with silicone elastomer. *Antimicrob. Agents Chemother.* 31, 889–894, 1987.

84. Williams, I., Venables, W.A., Lloyd, D., Paul, F., and Critchley, I., The effects of adherence to simicone surfaces on antibiotic susceptibility in *Staphylococcus aureus*. *Microbiology* 143, 2407–2413, 1997.

85. Rupp, M.E., and Hamer, K.E., Effect of subinhibitory concentrations of vancomycin, cefazolin, ofloxacin, L-ofloxacin, and D-ofloxacin on adherence to intravascular catheters and biofilm formation by *Staphylococcus epidermidis*. *J. Antimicrob. Chemother.* 41, 155–161, 1998.

86. Houlihan, H.H., Mercier, R.C., and Rybak, M.J., Pharmacodynamics of vancomycin alone and in combination with gentamicin at various dosing intervals against methicillin-resistant *Staphylococcus aureus*-infected fibrin-platelet clots in an *in vitro* infection model. *Antimicrob. Agents Chemother.* 41, 2497–2501, 1997.

87. Foley, I., and Gilbert, P., In-vitro studies of the activity of glycopeptide combinations against *Enterococcus faecalis biofilms*. *J. Antimicrob. Chemother.* 40, 667–672, 1997.

88. Gander, S., and Finch, R., The effects of exposure at constant (1h) or expotentially decreasing concentrations of quinupristin/dalfopristin on biofilms of Gram-positive bacteria. *J. Antimicrob. Chemother.* 46, 63–67, 2000.

89. Gagnon, R.F., Richards, G.K., and Subang, R., Experimental *Staphylococcus epidermidis* implant infection in the mouse. Kinetics of rifampin and vancomycin action. *ASAIO J.* 38, M596–M599, 1992.

90. Schaad, H.J., Chuard, C., Vaudaux, P., Waldvogel, F.A., and Lew, D.P., Teicoplanin alone or combined with rifampin compared with vancomycin for prophylaxis and treatment of experimental foreign body infection by methicillin-resistant *Staphylococcus aureus*. *Antimicrob. Agents Chemother.* 38, 1703–1710, 1994.

91. Gracia, E., Lacleriga, A., Monzon, M., Leiva, J., Oteiza, C., and Amorena, B., Application of a rat osteomyelitis model to compare *in vivo* and *in vitro* the antibiotic efficacy against bacteria with high capacity to form biofilms. *J. Surg. Res.* 79, 146–153, 1998.

92. Monzon, M., Gracia-Alvarez, F., Lacleriga, A., Gracia, E., Leiva, J., Oteiza, C., and Amorena, B., A simple infection model using pre-colonized implants to reproduce rat chronic *Staphylococcus aureus* osteomyelitis and study antibiotic treatment. *J. Orthop. Res.* 19, 820–826, 2001.

93. Isiklar, Z.U., Darouiche, R.O., Landon, G.C., and Beck, T., Efficacy of antibiotic alone for orthopedic device related infections. *Clin. Orthop.* 332, 184–189, 1996.

94. Chaing, F.Y., and Climo, M., Efficacy of linezolid alone or in combination with vancomycin for treatment of experimental endocarditis due to methicillin-resistant *Staphylococcus aureus*. *Antimicrob. Agents Chemother.* 47, 3002–3004, 2003.

95. Kaatz, G.W., Seo, S.M., Aeschlimann, J.R., Houlihan, H.H., Mercier, R.C., and Rybak, M.J., Efficacy of LY333328 against experimental methicillin-resistant *Staphylococcus aureus* endocarditis. *Antimicrob. Agents Chemother.* 42(4), 981–983, 1998.

96. Saleh-Mghir, A., Lefort, A., Petegnief, Y., Dautrey, S., Vallois, J.M., Le Guludec, D., Carbon, C., and Fantin, B., Activity and diffusion of LY333328 in experimental endocarditis due to vancomycin-resistant *Enterococcus faecalis*. *Antimicrob. Agents Chemother.* 43, 115–120, 1999.

97. Rupp, M.E., Fey, P.D., and Longo, G.M., Effect of LY333328 against vancomycin-resistant *Enterococcus faecium* in a rat central venous catheter-associated infection model. *J. Antimicrob. Chemother.* 47, 705–707, 2001.

98. Bausino, L., Madrigal, A., and Chambers, H., Evaluation of TD-6424 in a rabbit model of aortic valve endocarditis (AVE) due to methicillin-resistant *Staphylococcus aureus* (MRSA) or vancomycin-intermediate *Staphylococcus aureus* (VISA). 43rd ICAAC, Chicago, IL, Sept 14–17, 2003.

99. Jarvis, W.R., Edwards, J.R., Culver, D.H., et al., Nosocomial infection rates in adult and pediatric intensive care units in the United States. National Nosocomial Infections Surveillance System. *Am. J. Med.* 91, 185S–191S, 1991.

100. Banerjee, S.N., Emori, T.G., Culver, D.H., et al., Secular trends in nosocomial primary bloodstream infections in the United States, 1980–1989. National Nosocomial Infections Surveillance System. *Am. J. Med.* 91, 86S–89S, 1991.

101. Raad, I.I., and Hanna, H.A., Intravascular catheter-related infections: new horizons and recent advances. *Arch. Intern. Med.* 162, 871–878, 2002.

102. Mermel, L.A., Farr, B.M., Sheretz, R.J., Raad, I.I., O'Grady, N., Harris, J.S., and Craven, D.E., Guidelines for the management of intravascular catheter-related infections. *Clin. Infect. Dis.* 32, 1249–1272, 2001.

103. Raad, I., Davis, S., Khan, A., Tarrand, J., Elting, L., and Bodey, G.P., Impact of central venous catheter removal on the recurrence of catheter-related coagulase-negative staphylococcal bacteremia. *Infect. Control. Hosp. Epidemiol.* 13, 215–221, 1992.

104. Raad, I., Narro, J., Khan, A., Tarrand, J., Vartivarian, S., and Bodey, G.P., Serious complications of vascular catheter-related *Staphylococcus aureus* bacteremia in cancer patients. *Eur. J. Clin. Microbiol. Infect. Dis.* 11, 657–682, 1992.

105. Chang, F.Y., Peacock, J.E. Jr., Musher, D.M., et al., *Staphylococcus aureus* bacteremia: recurrence and the impact of antibiotic treatment in a prospective multicenter study. *Medicine* 82, 333–339, 2003.

106. Droste, J.C., Jeraj, H.A., MacDonald, A., and Farrington, K., Stability and *in vitro* efficacy of antibiotic-heparin lock solutions potentially useful for treatment of central venous catheter-related sepsis. *J. Antimicrob. Chemother.* 51, 849–855, 2003.

107. Raad, I., Chatzinikolaou, I., Chaiban, G., Hanna, H., Hachem, R., Dvorak, T., Cook, G., and Costerton, W., *In vitro* and ex vivo activities of minocycline and EDTA against microorganisms embedded in biofilm on catheter surfaces. *Antimicrob. Agents Chemother.* 47, 3580–3585, 2003.

108. Sandoe, J.A.T., Witherden, I.R., Au-Yeung, H.K.C., Kite, P., Kerr, G., and Wilcox, M.H., Enterococcal intravascular catheter-related bloodstream infection: management and outcome of 61 consecutive cases. *J. Antimicrob. Chemother.* 50, 577–582, 2002.

109. Widmer, A.F., New developments in diagnosis and treatment of infection in orthopedic implants. *Clin. Infect. Dis.* 33(Suppl 2), S94–S106, 2001.

110. Darley, E.S.R., and MacGowan, A.P., Antibiotic treatment of Gram-positive bone and joint infections. *J. Antimicrob. Chemother.* 53, 928–935, 2004.

111. Allen, D.M., Orthopedic implant infections: current management strategies. *Ann. Acad. Med. Singapore* 26, 687–690, 1997.

112. Jansen, B., and Peters, G., Foreign body associated infection. *J. Antimicrob. Chemother.* 32 (Suppl A), 69–75, 1993.

113. Steckelberg, J., and Osmon, D.R., *Prosthetic joint infections*. In: Bisno, A.L., and Waldvogel, F.A., eds., *Infections Associated with Indwelling Medical Devices*. Washington: American Society for Microbiology, 1994:259–290.

114. Zimmerli, W., and Ochsner, P.E., Management of infection associated with prosthetic joints. *Infection* 31, 99–108, 2003.

115. Cabell, C.H., Jollis, J.G., Peterson, G.E., et al., Changing patient characteristics and the effect on mortality in endocarditis. *Arch. Intern. Med.* 162, 90–94, 2002.

116. Cabell, C.H., Heidenreich, P.A., Chu, V.H., et al., Increasing rates of cardiac device infections among Medicare beneficiaries: 1990–1999. *Am. Heart J.* 147, 582–586, 2004.

117. Levine, D.P., Fromm, B., and Reddy, R., Slow response to vancomycin plus rifampin in methicillin-resistant *Staphylococcus aureus* endocarditis. *Ann. Intern. Med.* 115, 674–680, 1991.

118. Small, P., and Chambers, H., Vancomycin for *Staphylococcus aureus* endocarditis in intravenous drug users. *Antimicrob. Agents Chemother.* 34, 1227–1231, 1990.

119. Fortun, J., Navas, E., Martinez-Beltran, J., et al., Short-course therapy for right-side endocarditis due to *Staphylococcus aureus* in drug abusers: cloxacillin versus glycopeptides in combination with gentamycin. *Clin. Infect. Dis.* 33, 120–125, 2001.

120. Stechelberg, J.M., Rouse, M.S., Tallan, B.M., Henry, N.K., and Wilson, W.R., Relative efficacies of broad-spectrum cephalosporins for treatment for treatment of methicillin-susceptible *Staphylococcus aureus* experimental infective endocarditis. *Antimicrob. Agents Chemother.* 37, 554–558, 1993.

121. Nannini, E.C., Singh, K.V., and Murray, B.E., Relapse of type A beta-lactamase-producing *Staphylococcus aureus* native valve endocarditis during cefazolin therapy: revisiting the issue. *Clin. Infect. Dis.* 37, 1194–1198, 2003.

122. Caputo, G.M., Archer, G.L., Calderwood, S.B., DiNubile, M.J., and Karschmer, A.W., Native valve endocarditis due to coagulase-negative staphylococci. Clinical and microbiological features. *Am. J. Med.* 83, 619–625, 1987.

123. Drinkovic, D., Morris, A.J., Pottumarthy, S., MacCulloch, D., and West, T., Bacteriological outcome of combination versus single-agent treatment for staphylococcal endocarditis. *J. Antimicrob. Chemother.* 52, 820–825, 2003.

124. Watanakunakorn, C., and Bakie, C., Synergism of vancomycin-gentamycin and vancomycin-streptomycin against enterococci. *Antimicrob. Agents Chemother.* 4, 120–124, 1973.

125. Mergan, D.W., Enterococcal endocarditis. *Clin. Infect. Dis.* 15, 63–71, 1992.

126. Gruneberg, R.N., Antunes, F., Chambers, H.F., Garau, J., Graninger, W., Menichetti, F., Peetermans, W.E., Pittet, D., and Shah, P.M., Vogelaers. The role of glycopeptide antibiotics in the treatment of infective endocarditis. *Internat. J. Antimicrob. Agents* 12, 191–198, 1999.

21 Antibiotic Resistance in Biofilms

Nafsika H. Georgopapadakou

CONTENTS

21.1 INTRODUCTION

Biofilms are microbial communities encased in polysaccharide-rich extracellular matrices and living in association with surfaces (1). Biofilm formation is an important process for survival of microbial pathogens in the environment (e.g., *Vibrio cholerae*) or in the mammalian host (e.g., *Pseudomonas aeruginosa*). Free-swimming (planktonic) microbial cells attach, first transiently and then permanently as a single layer, to an inert surface or to a tissue. This monolayer gives rise to larger cell clusters that eventually develop into a highly structured biofilm, consisting of mushroom-shaped bacterial microcolonies separated by fluid-filled channels. The channels allow nutrients to reach all levels of the biofilm and toxic waste products to diffuse out (2–8). Biofilms are formed particularly in high-shear environments: the respiratory and urinary tracts, the oropharynx, and native heart valves. Thus, biofilm formation is an important aspect of many chronic human infections: dental caries, middle ear infections, medical device-related infections (ocular/cochlear implants, orthopedic devices, indwelling catheters, IUDs), native/prosthetic valve endocarditis, osteomyelitis, prostatitis, and chronic lung infections in cystic fibrosis patients (9–18). For bacterial biofilm formation, cell-to-cell communication (quorum sensing) may be required (19).

Signal molecules produced by bacteria accumulate locally, triggering elaboration of virulence factors and biofilms.

Biofilms have a characteristic architecture and biofilm organisms have phenotypic and biochemical properties distinct from their free-swimming, planktonic counterparts (7). One such biofilm-specific property is antibiotic resistance which can be as high as 1,000-fold over planktonic cells. The biofilm matrix may be a diffusion barrier to some antibiotics; other factors are the altered microbial physiology and the biofilm environment (20–24).

21.2 COMMON BIOFILM-FORMING MICROBIAL PATHOGENS

Biofilms may involve a single bacterial or fungal species, or a mixture of species such as *P. aeruginosa, Escherichia coli, Klebsiella pneumoniae, Staphylococcus* spp., *Streptococcus mutans, Candida* spp. and others (25). Of the common biofilm-forming microbial pathogens listed in Table 21.1, all but one form biofilms in the mammalian host. *V. cholerae*, the causative agent of cholera, forms biofilms in the aquatic environment which contribute to its persistence. The organism first attaches to a surface through its pili forming a monolayer (26). Subsequently, in response to environmental signals, it switches phenotype, overproduces exopolysaccharide and increases intercellular adhesion, forming a biofilm. Formation of a biofilm correlates

---

TABLE 21.1

Common Human Bacterial and Fungal Infections Caused by Biofilm-Forming Microorganisms

Infection/Site	Causative organism
Dental caries	*Streptococcus mutans*
Esophagus (AIDS)	*Candida* spp.
Otitis media	*Haemophilus influenzae*
Respiratory tract (CAP)	*Streptococcus pneumoniae*
Respiratory tract (ICU)	Gram-negative rods
Respiratory tract (cystic fibrosis)	*Pseudomonas aeruginosa, Burkholderia cepacia*
Native valve endocarditis	viridans group streptococci
Prosthetic valve endocarditis	*S. aureus, S. epidermidis, Candida* spp., *Aspergillus* spp.
Peritonitis (peritoneal dialysis)	Miscellaneous bacteria, fungi
Urinary tract (catheter)	*Escherichia coli, Proteus mirabilis,* other Gram-negative rods
Osteomyelitis	*Staphylococcus aureus,* other bacteria
Intraocular lens, contact lens	*P. aeruginosa, Serratia marcescens,* Gram-positive cocci
Orthopedic devices	*S. aureus* and *S. epidermidis*
Prostatitis	*Escherichia coli,* other Enterobacteriaceae, *Chlamydia trachomatis, Mycoplasma*
Central venous catheters	*S. epidermidis,* others
Vagina	*Candida* spp.
IUDs	*S. epidermidis,* enterococci, lactobacilli, β-Hemolytic streptococci

---

with environmental survival: increased resistance to osmotic and oxidative stresses as well as to killing by chlorine (27).

P. aeruginosa is probably the most prominent, and best studied, biofilm-forming microorganism as it is associated with cystic fibrosis. *S. mutans,* associated with dental caries and *Candida* spp, associated with oropharyngeal candidiasis in AIDS patients and vaginal candidiasis in healthy women, are other common biofilm-forming organisms.

Recent advances in medicine, particularly the increased use of indwelling medical devices, have caused implant-related infections, which typically involve biofilms, to become common disease entities. Host factors also play a role. For example, valvular heart disease is the necessary underlying condition for fungal endocarditis, which commonly involves *Candida* (predisposing factors, intravenous devices and antibiotic use) and *Aspergillus* (predisposing factor, immunosuppression) species (28,29).

21.3 ANTIBIOTIC RESISTANCE IN BIOFILMS

Antibiotic activity against biofilm microorganisms cannot be accurately determined using standard NCCLS broth microdilution methods for susceptibility testing, since these techniques are based on exposing *planktonic* organisms to the antimicrobial agent. Instead, the biofilm is exposed to the antimicrobial agent, removed from the attached substratum, homogenized and quantitated as viable cell counts (7). In the development of a model biofilm system, substratum and hydrodynamics are factors to be considered in addition to culture medium and inoculum (7).

Antibiotic resistance in biofilms is due to multiple mechanisms (Table 21.2): intrinsic resistance of the microorganisms involved (multidrug-resistant *Pseudomonas*, methicillin-resistant Staphylococci, azole-resistant *Candida* sp.); decreased antibiotic diffusion through, or inactivation within, the extracellular matrix; decreased growth of the organism due to nutrient limitation and activation of stress response.

21.3.1 Intrinsic Resistance of Microorganisms

Intrinsic resistance of the microorganism is a factor with *Pseudomonas*, methicillin-resistant staphylococci, *Candida,* and *Aspergillus*, and may be exacerbated in the biofilm by environmental factors. For example, efflux pumps in *Pseudomonas* and *Candida* may be upregulated in response to cell density. The reported high imipenem

---

TABLE 21.2
Antibiotic Resistance in Some Common Biofilm-Forming Pathogens

Organism	Antibiotic class	Examples
Staphylococcus	β-Lactam antibiotics (MRS)	Ampicillin
(aureus, epidermidis)	Quinolones	Ciprofloxacin
	Glycopeptides	Vancomycin
Streptococcus mutans		
Pseudomonas	β-Lactam antibiotics	Imipenem, ceftazidime
Candida spp.	Azoles	Fluconazole
Aspergillus	Azoles	Fluconazole, itraconazole

resistance of *Pseudomonas* in biofilms (7) could be due to increased efflux in addition to slow growth.

21.3.2 DECREASED DIFFUSION OF THE ANTIMICROBIAL AGENT THROUGH THE BIOFILM MATRIX

Diffusion of ciprofloxacin into a *P. aeruginosa* biofilm is substantially decreased relative to dispersed cells (30) and *P. aeruginosa* in biofilms is 15 times less susceptible to tobramycin than as dispersed cells (31). Significantly, a 2% alginate suspension isolated from *P. aeruginosa* inhibited diffusion of tobramycin (and gentamicin), an effect reversed by alginate lyase (32). Similarly, tobramycin was less active against *Staphylococcus epidermidis* in biofilms (33). In a related study (34) the diffusion rate of several β-lactams (ceftazidime, cefsulodin, piperacillin) and aminoglycosides (gentamicin, tobramycin) through alginate gels was found to be higher for β-lactams. The results are bolstered by additional data with another combination of organism and antimicrobial agents (35).

21.3.3 DECREASED GROWTH OF BIOFILM ORGANISM: NUTRIENT LIMITATION, STRESS RESPONSE

S. epidermidis biofilm growth rates strongly influence susceptibility; the faster the rate of cell growth, the more rapid the rate of growth inhibition by ciprofloxacin (36). Similarly, older (10-day-old) chemostat-grown *P. aeruginosa* biofilms were significantly more resistant to tobramycin and piperacillin than were younger (2-day-old) biofilms (37). Exposure to 500 μg of piperacillin plus 5 μg of tobramycin per ml completely inhibited planktonic and young biofilm cells but only by 20% older biofilm cells. The results are bolstered by additional data with other combinations of organisms and antimicrobial agents (38–40).

In *E. coli*, sigma factors under the control of the *rpoS* regulon regulate the transcription of genes whose products mitigate the effects of stress. It was found (1) that the *rpoS*+ *E. coli* biofilms had higher cell densities and a higher number of viable cells than *rpoS*− *E. coli*. Since *rpoS* is activated during slow growth of this organism, conditions that induce slow growth, such as nutrient and oxygen limitation or build-up of toxic metabolites, favor the formation of biofilms. Such conditions might be particularly acute within the depths of established biofilms. For example, agar-entrapped *E. coli* cells were more resistant to an aminoglycoside as oxygen tension was decreased which was attributed to lowered uptake (41). Thus, even single-species biofilms are heterogeneous in terms of genes expressed (i.e., phenotype) by cells near the surface or in the center of the biofilm (42,43).

21.4 CONCLUSIONS AND FUTURE DIRECTIONS

The distinct phenotype of microbial biofilms makes them resistant to antibiotics, and their matrix makes them resistant to the antimicrobial molecules and cells mobilized by the host (44–46). The increasing use of indwelling medical devices has increased the incidence of persistent, implant-associated infections which invariably involve biofilms and often microorganisms (*Staphylococcus, Pseudomonas, Candida* species) intrinsically resistant to multiple antibiotics. The chronic nature of biofilm infections increases their potential to act as reservoirs for acute exacerbations and to promote immune complex sequelae.

Due to their indolent nature, biofilm-related infections are usually diagnosed well after they have been established. This makes targeting biofilm formation less attractive than removing the biofilm already in place. This is currently accomplished by replacing the medical implant; enzymatically removing the biofilm might be a more appealing prospect. In this respect, the age of the biofilm might be important (47–49).

Another approach likely to be used increasingly in the future is making medical devices that discourage bacterial adhesion and colonization. Controlling the composition of biomaterials or incorporating antibiotics are obvious areas to explore (50–53).

REFERENCES

1. Watnick, P., and Kolter, R., Biofilm, city of microbes. *J. Bacteriol.* 182, 2675–2679, 2000.
2. Davey, M.E., and O'Toole, G.A., Microbial biofilms: from ecology to molecular genetics. *Microbiol. Mol. Biol. Rev.* 64, 847–867, 2000.
3. Kolenbrander, P.E., and London, J., Adhere today, here tomorrow: oral bacterial adherence. *J. Bacteriol.* 175, 3247–3252, 1993.
4. Douglas, L.J., *Candida* biofilms and their role in infection. *Trends Microbiol.* 11, 30–36, 2003.
5. DeBeer, D., Stoodley, P., and Lewandowski, Z., Liquid flow in heterogeneous biofilms. *Biotech. Bioeng.* 44, 636, 1994.
6. Costerton, J.W., Stewart, P.S., and Greenberg, E.P., Bacterial biofilms: a common cause of persistent infections. *Science* 284, 318–322, 1999.
7. Donlan, R.M., and Costerton, J.W., Biofilms: survival mechanisms of clinically relevant microorganisms. *Clin. Microbiol. Rev.* 15, 167–193, 2002.
8. Jefferson, K.K., What drives bacteria to produce a biofilm? *FEMS Microbiol. Lett.* 236, 163–173, 2004.
9. Richards, M.J., Edwards, J.R., Culver, D.H, and Gaynes, R.P., Nosocomial infections in medical intensive care units in the United States. National Nosocomial Infections Surveillance System. *Crit. Care Med.* 27, 887–892, 1999.
10. Kodjikian, L., Burillon, C., Chanloy, C., Bostvironnois, V., Pellon, G., Mari, E., Freney, J., and Roger, T., *In vivo* study of bacterial adhesion to five types of intraocular lenses. *Invest. Ophthalmol. Vis. Sci.* 43, 3717–3721, 2002.
11. Darouiche, R.O., Device-associated infections: a macroproblem that starts with microadherence. *Clin. Infect. Dis.* 33, 1567–1572, 2001.
12. Corona, M.L., Peters, S.G., Narr, B.J., and Thompson, R.L., Subspecialty clinics: critical care medicine. Infections related to central venous catheters. *Mayo. Clin. Proc.* 65, 979–986, 1990.
13. Raad, I., Intravascular-catheter-related infections. *Lancet* 351, 893–898, 1998.
14. Domingue, G.J., and Hellstrom, W.J.G., Prostatitis. *Clin. Microbiol. Rev.* 11, 604–613, 1998.
15. Stickler, D.J., Morris, N.S., McLean, R.J.C., and Fuqua, C., Biofilms on indwelling urethral catheters produce quorum-sensing signal molecules in situ and *in vitro*. *Appl. Environ. Microbiol.* 64, 3486–3490, 1998.
16. Wolf, A.S., and Kreiger, D., Bacterial colonization of intrauterine devices (IUDs). *Arch. Gynecol.* 239, 31–37, 1986.

17. Koch, C., and Hoiby, N., Pathogenesis of cystic fibrosis. *Lancet* 341, 1065–1069, 1993.
18. Govan, J.R.W., and Deretic, V., Microbial pathogenesis in cystic fibrosis: mucoid *Pseudomonas aeruginosa* and *Burkholderia cepacia. Microbiol. Rev.* 60, 539–574, 1996.
19. Davies, D.G., Parsek, M.R., Pearson, J.P., Iglewski, B.H., Costerton, J.W., and Greenberg, E.P., The involvement of cell-to-cell signals in the development of a bacterial biofilm. *Science* 280, 295–298, 1998.
20. Hoyle, B.D., and Costerton, W.J., Bacterial resistance to antibiotics: the role of biofilms. *Prog. Drug Res.* 37, 91–105, 1991.
21. Lewis, K., Riddle of biofilm resistance. *Antimicrob. Agents Chemother.* 45, 999–1007, 2001.
22. Prince, A.S., Biofilms, antimicrobial resistance and airway infection. *N. Engl. J. Med.* 347, 1110–1111, 2002.
23. Mah, T.F., and O'Toole, G.A., Mechanisms of biofilm resistance to antimicrobial agents. *Trends Microbiol.* 9, 34–39, 2001.
24. Stewart, P.S., Mechanisms of antibiotic resistance in bacterial biofilms. *Int. J. Med. Microbiol.* 292, 107–113, 2002.
25. Wimpenny, J., Manz, W., and Szewzyk, U., Heterogeneity in biofilms. *FEMS Microbiol. Rev.* 24, 661–671, 2000.
26. Moorthy, S., and Watnick, P.I., Genetic evidence that the *Vibrio cholerae* monolayer is a distinct stage in biofilm development. *Mol. Microbiol.* 52, 573–587, 2004.
27. Yildiz, F.H., and Schoolnik, G.K., *Vibrio cholerae* O1 El Tor: identification of a gene cluster required for the rugose colony type, exopolysaccharide production, chlorine resistance, and biofilm formation. *Proc. Natl. Acad. Sci. USA* 96, 4028–4033, 1999.
28. Kojic, E.M., and Darouiche, R.O., *Candida* infections of medical devices. *Clin. Microbiol. Rev.* 17, 255–267, 2004.
29. Paterson, D.L., New clinical presentations of invasive aspergillosis in non-conventional hosts. *Clin. Microbiol. Infect;* 10(Suppl 1), 24–30, 2004.
30. Suci, P.A., Mittelman, M.W., Yu, F.P., and Geesey, G.G., Investigation of ciprofloxacin penetration into *Pseudomonas aeruginosa* biofilms. *Antimicrob. Agents Chemother.* 38, 2125–2133, 1994.
31. Hoyle, B.D., Wong, C.K.W., and Costerton, J.W., Disparate efficacy of tobramycin on $Ca^{2+}-$, $Mg^{2+}-$, and HEPES-treated *Pseudomonas aeruginosa* biofilms. *Can. J. Microbiol.* 38, 1214–1218, 1992.
32. Hatch, R.A., and Schiller, N.L., Alginate lyase promotes diffusion of aminoglycosides through the extracellular polysaccharide of mucoid *Pseudomonas aeruginosa. Antimicrob. Agents Chemother.* 42, 974–977, 1998.
33. DuGuid, I.G., Evans, E., Brown, M.R.W., and Gilbert, P., Effect of biofilm culture on the susceptibility of *Staphylococcus epidermidis* to tobramycin. *J. Antimicrob. Chemother.* 30, 803–810, 1992.
34. Gordon, C.A., Hodges, N.A., and Marriott, C., Antibiotic interaction and diffusion through alginate and exopolysaccharide of cystic fibrosis-derived *Pseudomonas aeruginosa. J. Antimicrob. Chemother.* 22, 667–674, 1988.
35. Anderl, J.N., Franklin, M.J., and Stewart, P.S., Role of antibiotic penetration limitation in *Klebsiella pneumoniae* biofilm resistance to ampicillin and ciprofloxacin. *Antimicrob. Agents Chemother.* 44, 1818–1824, 2000.
36. DuGuid, I.G., Evans, E., Brown, M.R.W., and Gilbert, P., Growth-rate-dependent killing by ciprofloxacin of biofilm-derived *Staphylococcus epidermidis;* evidence for cell-cycle dependency. *J. Antimicrob. Chemother.* 30, 791–802, 1990.

37. Anwar, H., Strap, J.L., Chen, K., and Costerton, J.W., Dynamic interactions of biofilms of mucoid *Pseudomonas aeruginosa* with tobramycin and piperacillin. *Antimicrob. Agents. Chemother.* 36, 1208–1214, 1992.

38. Amorena, B., Gracia, E., Monzon, M., Leiva, J., Oteiza, C., Perez, M., Alabart, J.-L., and Hernandez-Yago, J., Antibiotic susceptibility assay for *Staphylococcus aureus* in biofilms developed *in vitro. J. Antimicrob. Chemother.* 44, 43–55, 1999.

39. Chuard, C., Vaudaux, P., Waldovogel, F. A., and Lew, D.P., Susceptibility of *Staphylococcus aureus* growing on fibronectin-coated surfaces to bactericidal antibiotics. *Antimicrob. Agents Chemother.* 37, 625–632, 1993.

40. Desai, M., Buhler, T., Weller, P.H., and Brown, M.R.W., Increasing resistance of planktonic and biofilm cultures of *Burkholderia cepacia* to ciprofloxacin and ceftazidime during exponential growth. *J. Antimicrob. Chemother.* 42, 153–160, 1998.

41. Tresse, O., Jouenne, T., and Junter, G.A., The role of oxygen limitation in the resistance of agar-entrapped, sessile-like *Escherichia coli* to aminoglycoside and β-lactam antibiotics. *J. Antimicrob. Chemother.* 36, 521–526, 1995.

42. Walters, M.C., Roe, F., Bugnicourt, A., Franklin, M.J., and Stewart, P.S. Contributions of antibiotic penetration, oxygen limitation, and low metabolic activity to tolerance of *Pseudomonas aeruginosa* biofilms to ciprofloxacin and tobramycin. *Antimicrob. Agents Chemother.* 47, 317–323, 2003.

43. Sauer, K., Camper, A.K., Ehrlich, G.D., Costerton, J.W., and Davies, D.G., *Pseudomonas aeruginosa* displays multiple phenotypes during development as a biofilm. *J. Bacteriol.* 184, 1140–1154, 2002.

44. May, T.B., Shinabarger, D., Maharaj, R., Kato, J., Chu, L., DeVault, J.D., Roychoudhury, S., Zielinski, N.A., Berry, A., Rothmel, R.K., Misra, T.K., and Chakrabarty, A.M., Alginate synthesis by *Pseudomonas aeruginosa*: a key pathogenic factor in chronic pulmonary infections of cystic fibrosis patients. *Clin. Microbiol. Rev.* 4, 191–206, 1991.

45. Davies, G., Chakrabarty, A.M., and Geesey, G.G., Exopolysaccharide production in biofilms: substratum activation of alginate gene expression by *Pseudomonas aeruginosa. Appl. Environ. Microbiol.* 59, 1181–1186, 1993.

46. Kuhn, D.M., and Ghannoum, M.A., *Candida* biofilms: antifungal resistance and emerging therapeutic options. *Curr. Opin. Investig. Drugs* 5, 186–197, 2004.

47. Stewart, P.S., New ways to stop biofilm infections. *Lancet* 361, 97, 2003.

48. Hartman, G.R., Wise, R., Quorum sensing: potential means of treating Gram-negative infections? *Lancet* 351, 848–849, 1998.

49. Hentzer, M., Riedel, K., Rasmussen, T.B., et al., Inhibition of quorum sensing in *Pseudomonas aeruginosa* biofilm bacteria by a halogenated furanone compound. *Microbiology* 148, 87–102, 2002.

50. Hatch, R.A., and Schiller, N.L., Alginate lyase promotes diffusion of aminoglycosides through the extracellular polysaccharide of mucoid *Pseudomonas aeruginosa. Antimicrob. Agents Chemother.* 42, 974–977, 1998.

51. Dall, L., Barnes, W.G., Lane, J.W., and Mills, J., Enzymatic modification of glycocalyx in the treatment of experimental endocarditis due to viridans streptococci. *J. Infect. Dis.* 156, 736–740, 1987.

52. Johansen, C., Falholt, P., and Gram, L., Enzymatic removal and disinfection of bacterial biofilm. *Appl. Environ. Microbiol.* 63, 3724–3728, 1997.

53. Kamal, G.D., Pfaller, M.A., Rempe, L.E., and Jebson, P.J.R., Reduced intravascular catheter infection by antibiotic bonding. A prospective, randomized, controlled trial. *J. Am. Med. Assoc.* 265, 2364–2368, 1991.

22 Treatment Protocols for Infections of Vascular Catheters

Russell E. Lewis and Issam I. Raad

CONTENTS

22.1   INTRODUCTION

Use of indwelling central venous catheters has increased considerably over the last two decades and is considered an essential component in the care of patients undergoing intensive medical or surgical procedures. The benefits and convenience of prolonged indwelling central venous access, however, is offset by an increased risk of bloodstream infection. Catheter-related bloodstream infections (CRBSI) affect 4 to 8% of all patients with indwelling central venous catheters and are associated with significant morbidity, especially in critically-ill patients (1) Difficulties associated with the diagnosis, and treatment of CRBSI as well as surgical management of implanted catheters can significantly increase the cost of hospital care. The cost of managing a single case of CRSI was estimated to be in excess of $28,000 and is associated with an increased length of hospital stay of 5 to 30 days (2,3).

Treatment of CRBSI varies according to the type of catheter, severity of the patient's acute illness, evidence of infection complications, and the most likely pathogens involved. In most cases of CRBSI, prompt removal of the catheter and appropriate systemic antimicrobial therapy will be the most effective treatment strategy. However, in patients with tunneled (e.g., Hickman, Broviac, Groshong, Quinton) or implanted catheters, surgical removal of the catheter may not be feasible, particularly if the patient is not medically stable or is pancytopenic. Alternative approaches, such as antibiotic lock therapy in addition to systemic antimicrobial therapy, are often considered in low to moderate risk patients to salvage the catheter until removal is necessary or possible.

This chapter will review key concepts in the medical management and prevention of catheter-related infections according to evidence-based guidelines developed by The Infectious Diseases Society of America, the American College of Critical Care Medicine, and Society for Healthcare Epidemiology (4,5). Specific attention will be focused on central venous catheters (CVCs) although many of the concepts presented are applicable for infection of dialysis and urinary catheters.

22.2 PATHOGENESIS AND MICROBIAL FEATURES OF VASCULAR CATHETER INFECTIONS

The risk for developing vascular catheter infection depends on the type of intravascular catheter (tunneled vs. non-tunneled, single vs. multiple lumen), clinical setting, location of the site of insertion, and the duration of catheter placement. For short-term, nontunneled catheters without Dacron cuffs, the skin insertion site is the major source for microbial colonization (Figure 22.1). Organisms migrate along the external surface of the catheter lumen in the intercutaneous and subcutaneous segments leading to colonization of the catheter tip and subsequent bloodstream infection (4). For long-term tunneled catheters with a Dacron cuff just inside the catheter exit site (e.g., Hickman or Broviac) or totally implanted catheters, contamination of the

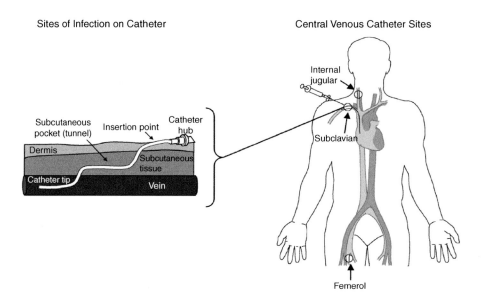

FIGURE 22.1 Central venous catheterization sites and areas of infection in the catheter lumen.

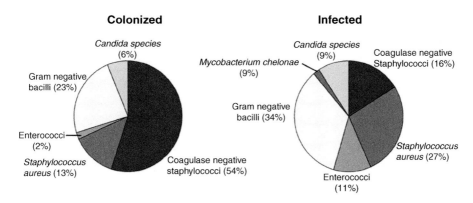

FIGURE 22.2   Frequency of organism recovery from colonized and infected central venous catheters. Adapted from references (1,4,6,7).

catheter hub and intraluminal infection are more common routes of infection. Frequently, these sites are contaminated through the hands of healthcare workers during manipulation of the catheter lumens (4).

Because the skin of the patient or the hands of healthcare workers are the most common sources of contamination of catheters, coagulase-negative staphylococci (e.g., *Staphylococcus epidermidis*) and *Staphylococcus aureus* are the most common colonizers of catheter surfaces (Figure 22.2) (6). *Enterococcus* spp. are frequent colonizers of vascular catheters in patients with a long-term indwelling femoral catheter (4). Gram-positive bacilli, such as *Corynebacterium* (especially *jeikeiem* strains) and *Bacillus* species, can also be introduced from the skin or catheter hub and occasionally cause catheter-related infections (4).

Gram-negative aerobic bacteria associated with catheter infections are generally nonenteric pathogens acquired from the hospital environment or contaminated water or infusates and are frequently resistant to multiple antimicrobials (6). The most common Gram-negative pathogens include *Pseudomonas aeruginosa*, *Stenotrophomonas maltophilia*, and *Acinetobacter* species (3,4). *Candida* species are increasingly common causes of CRBSI infection, particularly in patients who receive total parenteral nutrition or are on broad-spectrum antimicrobial therapy, and can be associated with attributable mortality rates ranging between 30 and 40% (7,8). *Candida albicans* and *Candida parapsilosis* can be found on the hands of healthcare workers, and are common colonizers of catheter tips, urine, respiratory secretions, and wounds in patients in the medical or surgical intensive care units. Since the early 1990s, fluconazole-resistant non-*albicans* species, particularly, *C. glabrata*, have become more prevalent with the empiric or pre-emptive use of fluconazole in neutropenic patients and critically-ill patients (9).

Within hours of catheter insertion, host-derived proteins such as fibrin, fibronectin, thrombospondin and laminin form an adhesive surface for microorganisms on the external and internal surface of the catheter lumen (3,10). Common CRBSI organisms such as coagulase negative staphylococci, *Staphylococcus aureus* and *Candida* species have been shown to avidly bind to these host derived proteins on the catheter surface, thus anchoring themselves to the catheter lumen surface. Studies using electron microscopy have suggested that bacterial colonization of the thrombin sheath occurs in most catheters within hours of insertion, even in the absence of clinical symptoms of infection (11). Clinically-apparent infection and subsequent bacterial seeding, therefore, may be a function of surpassing a "quantitative threshold" of bacterial replication on the catheter surface. This concept is supported by studies of *quantitative* sonicated catheter cultures that have demonstrated a correlation between colonization burden and development of CRBSI (12).

Growth to the quantitative threshold necessary for bloodstream infection and dissemination is facilitated by the formation of an exopolysaccharide matrix, or biofilm, during growth on the catheter lumen. Biofilm production by Staphylococci, *Candida* and some Gram-negative aerobes is well recognized as a key factor involved in chronic infections caused by these pathogens (13). Catheter-related infections are true biofilm-mediated infections in the sense that: (a) direct examination of the catheter reveals bacteria or fungi living in cell clusters or microcolonies encased in an extracellular matrix composed of bacterial, fungal, and host components; (b) the infecting biofilms are adherent to the surface of the catheter lumen; (c) the infection is localized to a specific area and dissemination, when it occurs, is a secondary phenomenon; and (d) the infection is difficult or impossible to eradicate with antibiotics despite the fact that organisms are susceptible to drug killing in the planktonic state. For example, several *in vitro* studies have documented 100 to 10,000-fold higher concentrations of antimicrobials are needed to kill biofilm embedded organsims (14–17). Therefore, disruption of the biofilm mode of growth is an essential factor in the prevention and treatment of CRBSI.

In addition to growth in biofilms, bacteria and fungi can grow to high density inside a thrombus, which commonly form on the catheter surface or in central veins or arteries with prolonged intravascular catheterization. The risk of thrombotic complications with CVCs ranges from 2 to 29% depending on the site of catheterization (18). Femoral catheterization and internal jugular catheterization are associated with the highest rates of thrombotic complications (15 to 30%) compared to subclavian venous catheters (1 to 3%) (4). Although the importance of small thrombi on the catheter lumen remains unknown, all clots have the potential to embolize or become infected with bacteria or fungi. Septic thrombosis is one of the most severe complications encountered with CVCs. Frequently, patients will exhibit high-grade and persistent bacteremia or fungemia due to seeding from infected thrombi that continues following catheter withdrawal (4). Infected thrombi may embolize to the lung, bone, or other distal sites leading to metastatic infection similar to infective endocarditis. Septic thrombosis is also associated with numerous vascular complications including swelling and edema, abscess formation, and pseudoaneurysm (4). Often a combined medical and surgical approach is required to treat large septic thrombosis.

22.3   DIAGNOSIS OF CATHETER ASSOCIATED INFECTION

Diagnosis of catheter infections is difficult in the medically complex patient due to the limited sensitivity or specificity of clinical signs that accompany the infection (3,4).

Fever with or without chills is common in patients with bacteremia, but is not a specific indicator of catheter infection. Inflammation surrounding the catheter exit site is a specific but nonsensitive indicator of infection, especially in patients with suppressed immunity (3,4). Symptoms of evolving systemic inflammatory response syndrome (SIRS) or sepsis including hypotension, oliguria, mental status changes, and multiple organ dysfunction may be seen in patients with acute fulminant *S. aurueus* or Gram-negative bacteremia or fungemia due to *C. albicans.*

Growth of coagulase-negative staphylococcus, *Staphylococcus aureus*, or *Candida* species in blood cultures in the absence of a clearly defined source of infection is frequently the initial indicator of a catheter infection (4). Gram stain of blood sampled through the catheter lumen may provide clues of a localized infection. However, definitive diagnosis often requires removal of the catheter for cultures. Typically, semi-quantitative cultures are performed by rolling a segment of the catheter tip across microbiological agar and colony forming units are counted after overnight incubation (4,19). Because the roll-plate method only cultures the external surface of the catheter lumen, it is less useful for diagnosis of infection in long-term catheters, where the internal surface is the predominant source of infection (3,12). Several quantitative methods for catheter culturing have been proposed as alternatives to the roll-plate method to improve sensitivity and specificity of culture techniques used to diagnose CRBSI. A meta-analysis of culture techniques using receiver operator curve analysis (a comparative analysis of sensitivity and specificity) has suggested that quantitative cultures of sonically-disrupted catheter segments is the most accurate method for catheter segment cultures in short-term and long-term catheters (20).

Both semi-quantitative and quantitative culture methods require removal of the catheter, which often results in unnecessary removal of noninfected catheters to ruleout infection. Newer culture techniques have focused on methods that can differentiate CRBSI prior to catheter removal on the basis of higher organism burden in blood drawn through the catheter vs. peripheral cultures (3). A five-fold higher number of organisms recovered from blood drawn through the CVC lumen versus a peripheral culture is highly suggestive of CRBSI (4,21). Similarly, the time to blood culture positivity is a relative marker of catheter infection (22–24). If simultaneous cultures are drawn from the CVC site and peripheral blood and incubated in a continuously-monitored blood culture systems (e.g., Bac-Tec), growth will be detected in the CVC-drawn blood, on average, 2 hours before the peripheral culture in patients with CRBSI (22–24). This non-invasive technique would probably be the simplest to adapt in most hospitals as continuously monitored blood culture systems are already in widespread use at larger hospitals.

Specialized in situ culture techniques have also been proposed as a method for diagnosis of CRBI without catheter removal. The endoluminal brush technique involves drawing a small brush through the catheter lumen to remove fibrin deposits (25,26). The brush and fibrin deposits can then be cultured to assess the degree of bacterial colonization. The brush method has been reported to have a sensitivity of 95% and a specificity of 84% in diagnosing CRBSI but was associated with induction of transient bacteremia in 6% of patients (26). Other potential complications of endoluminal brushing include release of emboli, arrythmias, and possible catheter rupture.

Thrombotic complications are an important component of CRBSI diagnosis and can be detected in 15 to 33% of patients by ultrasonography with Doppler imaging (18). Localized pain, erythema, edema and less commonly a palpable cord or exudate may be found in patients with septic thrombosis of peripheral veins (4). When the thrombosis involves the great central veins, ipsilateral neck, chest or upper extremity pain may also be present. Similarly, Doppler ultrasonography or flow studies, along with transthoracic or transesophageal echocardiography, are often used to detect infected thrombus and rule out endocarditis in patients with persistent bacteremia.

22.4 MANAGEMENT OF VASCULAR CATHETER INFECTIONS

The key decision in the management of CRBSI is to determine whether the catheter requires immediate removal. Often this decision is made on the basis of the catheter type and whether there is low, moderate or high risk of CRBSI. The risk of CRBSI is determined by the type of organism isolated on culture (high virulence vs. low virulence), clinical signs and symptoms of the patient, and whether the CRBSI is uncomplicated or complicated (Figure 22.3) (3,4).

Low-risk CRBSI are considered infections caused by low-virulence organisms such as coagulase-negative staphylococci that are typically associated with a benign course of infection and rarely develop deep-seated or metastatic complications. Immediate catheter removal is unnecessary in patients with low-risk CRBSI and most patients can be adequately treated with antibiotic therapy alone or in combination with antibiotic lock therapy if intraluminal infection is suspected (e.g., tunneled catheters). Catheter removal may be eventually necessary even in low-risk infections, as recurrent bacteremia occurs in up to 20% of patients within 3 months (27). Catheter removal is recommended however, in patients with prosthetic heart valves (4).

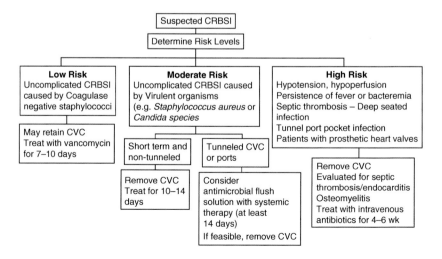

FIGURE 22.3 Algorithm for management of catheter-related bloodstream infection. Adapted from references (3,4)

Moderate risk CRBSI are uncomplicated infections caused by an organism of moderate to high virulence (e.g., *S. aureus* or *Candida* species) that has a tendency to seed metastatic infections. Specifically, removal of vascular catheters infected with *S. aureus* has been associated with more rapid clearance of bacteremia and a higher cure rate of antimicrobial therapy (4). Hence, all nontunneled catheters should be removed immediately if they are found to be the source of bacteremia with *S. aureus*, *Candida*, or other virulent organisms (3,4,28). Tunneled catheters and implanted devices are more difficult to remove and salvage therapy may be initially attempted in uncomplicated infection in a stable patient that responded rapidly (< 3 days) to antimicrobial therapy. Salvage therapy consists of systemic antimicrobial therapy in combination with antibiotic flush therapy for at least 14 days (Figure 22.3). Salvage therapy should not be attempted in any patient with valvular dysfunction or vegetations by transesophageal echocardiogram, or evidence of septic embolization. Evidence of infective endocarditis is treated as complicated CRBSI requiring immediate catheter removal and prolonged (e.g., 4 to 6 weeks) antimicrobial therapy (4).

High-risk CRBSI is a complicated infection that often occurs in critically-ill or immunocompromised patients (3,4). Complicated CRBSI may consist one or more of the following factors: (a) a CRBSI associated with hypotension or organ hypo-perfusion consistent with SIRS; (b) persistence of fever or positive blood cultures for more than 48 hours after the initiation of antimicrobial therapy; (c) a septic thrombosis of the great vein, septic emboli, or deep-seated infections such as endocarditis, and (d) the presence of a tunnel or port pocket infection. Removal of the catheter is necessary in any patient with complicated CRBSI, including those with tunnel tract infections (3,4).

Systemic antimicrobial therapy is indicated for most cases of CRBSI and is often started empirically until culture results are available. The initial choice of antibiotics depends on the severity of the patient's clinical disease, risk factors for infection, and the likely pathogens (Table 22.1) (3,4,29). Because Gram-positive organisms account for the bulk of CRBSIs, vancomycin is considered to be the drug of choice for empirical therapy for CRBSI due to the high frequency of methicillin-resistant coagulase-negative staphylococci and *Staphylococcus aureus* isolates (3,4). Once culture and sensitivity results are available, patients should be switched, if possible, to a nafcillin or oxacillin-based regimen as these agents achieve a more rapid bacte-riologic cure and have been associated with lower relapse rates compared to vancomycin-based therapy (27,28,30).

Few studies have specifically examined the duration of systemic antimicrobial therapy needed to adequately treat CRBSI. However, general recommendations can be devised on the risk level of infection (Figure 22.3). Patients with uncomplicated CRBSI caused by cogaulase-negative staphylococci who respond quickly to therapy, are not immunocompromised, and do not have prosthetic heart valves can be treated with 7 to 10 days of systemic antimicrobial therapy (4,27,28). A more prolonged course of antimicrobial therapy is needed in patients with persistent bactermeia or fungemia, patients with risk factors for endocarditis or evidence of septic thrombosis. In these complicated patients, the risk of complications or relapse is substantially higher and a minimum of 4 to 6 weeks of systemic antimicrobial therapy is recommended (4).

TABLE 22.1
Systemic Antimicrobial Therapy for Catheter Related Bloodstream Infections

Pathogen	Preferred Therapy (Typical Intravenous Dose in Adult)	Alternative Therapy	Comments
Gram-positive cocci			
Coagulase-negative staphylococcus			
Methicillin-sensitive	Nafcillin or Oxacillin (2 g q4h)	Cefazolin	Nafcillin, oxacillin, and cefazolin are superior to vancomycin for methicillin-sensitive coagulase-negative staphylococcus and *S.aureus* infections
Methicilllin-resistant	Vancomycin (1 g q12h)	Linezolid, or Quinuprisin/Dalfopristin, or Daptomycin, or TMP/SMX alone (if susceptible)	
Staphylococcus aureus			
Methicillin-sensitive	Nafcillin or Oxacillin (2 g q4h)	Cefazolin	
Methicillin-resistant	Vancomycin (1 g q12h)	Vancomycin + Rifampin ± Gentamicin[a], or Linezolid, or Quinupristin/Dalfopristin, or Daptomycin, or TMP/SMX alone (if susceptible)	Linezolid or TMP/SMX should be used in caution in cytopenic patients. Myalgias may be seen in patients receiving quinuprisin/ dalfopristin and can be reduced with lower doses (7.5. mg/kg q12h)
Vancomycin-resistant	Linezolid (600 mg q12h) or Quinuprisin/Dalfopristin (7.5 mg/kg q8h), or Daptomycin (6 mg/kg q24h)		

Enterococcus spp.,			
Ampicillin-sensitive	Ampicillin (2 g q4h) + Gentamicin[a] (individualized dosing)	Vancomycin	Vancomycin has some dosing advantages over ampicillin + gentamicin but an increased likelihood for selecting vancomycin-resistant enterococcus
Ampicillin-resistant, vancomycin-susceptible	Vancomycin (1 g q12h)	Linezolid, or Quinupristin/Dalfopristin, or Daptomycin	
Vancomycin-resistant	Linezolid (600 mg q12h) or Quinuprisin/Dalfopristin (7.5 mg/kg q8h), or Daptomycin (6 mg/kg q24h)		
Infrequent pathogens			
Corynebacterium (JK-1)	Vancomycin (1 g q12h)	Penicillin G + Gentamicin[a]	Susceptibilities vary for each isolate
Burkholderia capecia	TMP/SMX (3–5 mg/kg q8h), or Imipenem (500 mg q6h), or Meropenem (1 g q8h)	High-dose (5–7 mg/kg) TMP/SMX Piperacillin/tazobactam ± Minocycline	
Alcaligenes species	Imipenem (500 mg q6h) or Meropenem (1 g q8h)	TMP/SMX, or Imipenem, or Meropenem	
Flavobacterium species	Vancomycin (1 g q12h)	Azithromycin + Moxifloxacin	
Mycobacterium spp.,	Clarithromycin (500 mg PO BID) ± Amikacin (5–7.5 mg/kg q8h)		

[a]Or other suitable aminoglycoside dosage individualized for patient's renal function. TMP/SMX – Trimethoprim/sulfamethoxazole.

Treatment courses of 6 to 8 weeks are typically recommended for patients with suspected osteomyelitis (4,27,28)

Initial antimicrobial therapy should be given intravenously until the patient's condition is stable, culture and susceptibility results are known, and the blood cultures have been sterilized. Oral therapy can then be considered to complete the treatment course of infection in patients without complicated infection (4). Linezolid, minocycline, newer quinolones (levofloxacin, gatifloxacin, moxifloxacin), and trimethoprim/ sulfamethoxazole are the most frequently used oral agents in the treatment of CRBSI because of their spectrum, excellent bioavailability, and penetration into deep tissues (4).

Treatment failure of CRBSI typically manifests as recurrence or persistence of fever, persistently positive blood cultures during antimicrobial therapy, or recurrence of infection after antibiotics are discontinued. Treatment failures are a clear indication for catheter removal and an extensive work-up for septic thrombosis, osteomyelitis, and endocarditis to rule-out other reservoirs of persistent bloodstream infection.

Antibiotic lock or flush therapy is an adjunctive technique used to reduce the growth of organisms in the inside of the catheter lumen. As such, this procedure is more useful for infected long-term catheters where the hub and intraluminal colonization are the leading sources for CRBSI (31). The antibiotic lock technique has several advantages for treating CRBSI, including the ability to achieve high drug concentrations, the relatively low toxicity and cost of the technique, ease of administration, and the possibility to continue therapy at home (31). The two major disadvantages of this treatment approach are the lack of activity at distant infection sites and the possible delay in curing the infection (catheter removal) if the therapy fails (31) . Antibiotic lock therapy has less of a role for the treatment of short-term catheter infection where the nidus of organisms is concentrated on the extralumninal surface of the catheter. Additionally, antibiotic lock therapy would be of no benefit in cases of extraluminal infection (inflammation over the tunnel or exit site or pocket of a totally implanted port).

To date, no randomized prospective double-blind study has examined the efficacy of the antibiotic lock of flush technique for treating lower-risk CRBSI. Several small open-label trials have attempted to define the clinical efficacy of antibiotic lock therapy for CRBSI (Table 22.2) (31–42). The studies are difficult to compare due to differences in antibiotic lock solutions utilized, patient populations, use of concomitant systemic antimicrobial therapy, and variability in the time of drug instillation. However, therapeutic success in the range of 50 to 80% is reported in most case series (31). Most failures reported with antibiotic lock therapy have been reported with *Candida* infections, and relapse rates for *S. aureus* are clearly higher than for infections caused by coagulase-negative staphylococci (31). Therefore, a full course of systemic therapy is always necessary for infections caused by more virulent pathogens.

22.5   PREVENTION OF VASCULAR CATHETER
INFECTIONS

Improvements in the understanding of CRBSI pathogenesis have enhanced preventative strategies for catheter infections. Insertion catheters at sites with a lower density

TABLE 22.2
Clinical Experience with Antibiotic-Lock Therapy for CRBSI

Study	Reference	Patient Population	Lock Solution	% Cure (# episodes)
Messing et al. (1988)	(33)	Parenteral nutrition	Amikacin + Minocycline + Vancomycin	90 (22)
Messing et al. (1990)	(34)	Parenteral nutrition	Amikacin	93 (27)
Johnson et al. (1994)	(37)	Parenteral nutrition	Organism/susceptibility directed	83 (12)
Benoit et al. (1995)	(40)	Parenteral nutrition	Vancomycin, Gentamicin, or Amphtoricin B	78 (9)
Krzywda et al. (1995)	(41)	Parenteral nutrition	Organism/susceptibility directed	64 (22)
Capdevila et al. (1993)	(38)	Dialysis	Vancomycin or ciprofloxacin + 5% heparin	100 (13)
Capdevila et al. (1994)	(39)	AIDS	Vancomycin or ciprofloxacin	100 (12)
Longuet et al. (1995)	(35)	AIDS/Cancer	Vancomycin, teicoplanin, or amikacin	43 (12)
Domingo et al. (1999)	(42)	AIDS	Vancomycin or amikacin	81 (27)
Krishnasami et al. (2002)	(36)	Dialysis	Cefazolin, Vancomycin, gentamicin alone or in combination with heparin	51 (79)
Dannenberg et al. (2003)	(32)	Cancer	Ethanol	67% (18)

of skin flora colonization and less phlebitis (i.e., subclavian vs. jugular or femoral) are at lower risk for contamination and subsequent infection (5). Good hand hygiene (i.e., hand washing with waterless alcohol-based product or antibacterial soap with adequate rinsing) and aseptic technique are essential before insertion or manipulation of peripheral venous catheters. Use of designated personnel who are specifically trained for insertion of maintenance of catheters can significantly reduce the prevalence of catheter malfunction and infection (5). Maximal sterile barrier precautions (e.g., cap, mask, sterile gown, sterile gloves, and large sterile drape) are recommended during the insertion of CVCs and significantly reduce the risk of subsequent CRBSI (5). Skin antisepsis with povidone iodine or chlorhexidine can reduce skin colonization at the site of infection. Securing the catheter with suture-less devices and use of transparent dressings may also reduce the risk of bacterial colonization and enhance visualization of the catheter exit site.

Impregnation of catheters, catheter hubs or cuffs with antimicrobials or antiseptics have proven to be an effective method of reducing the risk of CRBSI (5). Antimicrobial coating has only been studied in noncuffed catheters that have remained in place for <30 days. The two best-studied formulations are chlorhexidine/silver sulfadiazine and minocycline/rifampin (43). Maki and colleagues demonstrated that catheters coated on the external surface with chlorhexidine/silver sulfadiazine reduce the risk of colonization by two-fold and the risk for bloodstream infections by four-fold (44). This protective benefit in short-term (<2 week) catheters was further confirmed by two recent meta-analysis (43,45). Similar protective effects, however, have not been documented with prolonged catheterization (>2 weeks) due to only external coating of the catheter (inside lumen is not coated) and the short durability of the chlorhexidine/silver sulfadiazine coating (45). There is also a small risk of severe hypersensitivity reactions, including anaphylaxis to the chlorhexidine coating.

Catheters impregnated with minocycline and rifampin have been developed that are coated on both the external and internal surface of the lumen. In two large, prospective, multicenter, randomized trials these catheters reduced the risk of CRBSI by more than five-fold compared to uncoated polyurethane catheters and were significantly less likely to be associated with CRBSI compared to chlorhexidine/silver sulfadiazine coated catheters (46,47). In a prospective, randomized, multicenter trial catheters coated with minocycline and rifampin had a twelve-fold lower likelihood of CRBSI compared to catheters coated with chlorhexidine-silver sulfadiazine (47). Although there is a theoretical possibility that use of antimicrobial-impregnated catheters could increase the risk of selecting resistant bacteria (especially for rifampin), two large prospective randomized trials failed to demonstrate selection of resistant isolates (46–48). A recent study has even suggested that use of antibiotic-impregnated catheters can reduce the risk of acquiring vancomcyin-resistant enterococcus through reduction of antibiotic (specifically vancomycin) use for empiric therapy of catheter infections (49).

Because antimicrobial impregnated catheters have a higher acquisition cost compared to traditional central venous catheters, use of these catheters is often reserved for patients with a >3% risk of developing a CRBSI. In these patients, prevention of CRBSI can significantly reduce secondary costs of CRBSI treatment including prolonged hospital or ICU stay (50). Patients at higher risk who could potentially benefit from antimicrobial impregnated catheters include (3,5):

- Patients with femoral or internal jugular vein insertion
- Patients with burns
- Patients with neutropenia or undergoing transplantation
- Patients receiving hemodialysis
- Patients with short-bowel syndrome or receiving total parenteral nutrition
- Patients colonized with methicillin-resistant *S. aureus*
- Patients with an open would near the catheter insertion site
- Catheter insertion or exchange in a patient with known bacteremia or fungemia
- Emergency insertion of a catheter

Ionic metals such as silver and platinum possess a broad range of antimicrobial activity and have been incorporated into catheters and cuffs to prevent CRBSIs. A combination platinum/silver impregnated catheter has been used in Europe and was recently approved for use in the United States, although no published studies have documented the clinical efficacy of these catheters to prevent CRBSI. Ionic silver has been used extensively in subcutaneous collagen cuffs attached to central venous catheters to provide enhanced antimicrobial as well as mechanical impedance to organisms that migrate along the external surface of the catheter lumen (51). However, no study has demonstrated evidence of prolonged protection (>20 days) from the development of CRBSI with this preventative approach.

Use of antibiotic lock solutions for prophylaxis of CRBSI has been studied in neutropenic patients with long-term catheters. Carratala et al. compared a lock solution containing heparin (10 U/mL) plus vancomycin (25 µg/ml) in 60 patients versus heparin alone in 57 patients for prophylaxis of nontunneled, multilumen CVCs (31). Insertion sites and catheters hubs were swabbed twice weekly to detect bacterial colonization and patients were monitored for development of CRBSI. Significant bacterial colonization was seen in 15.8% of patients with heparin lock prophylaxis alone, compared to 0/60 (0%) of patients who received vancomycin plus heparin. CRBSI developed in 4 (7%) of patients who received heparin alone vs. none of the patients who received heparin plus vancomycin prophylaxis.

More recently, Henrickson and colleagues evaluated the ability of an antibiotic flush solution of vancomycin, heparin, and ciprofloxacin (VHC) to prevent CRBSI in 126 pediatric oncology patients (52). Patients randomized to the VHC flush arm were significantly less likely to develop a CRBSI due to a Gram-positive or Gram-negative pathogen, and the time to development of a CRBSI was significantly prolonged. Antibiotics could not be detected in the serum of patients who received VHC flush therapy and no evidence of increased risk of antibiotic-resistant bacteria was observed.

Because use of vancomcyin and other first-line antimicrobial therapies as topical prophylaxis has been discouraged by the Centers for Disease Control (53), novel flush solutions incorporating antiseptics and potent third or fourth line antimicrobials have been developed. A flush solution consisting of minocycline and EDTA, which has both anticoagulent, antimicrobial, and possible anti-biofilm properties, was shown to be effective at preventing recurrence of staphylococcal infections in short- and long-term catheters and demonstrated efficacy in preventing CRBSI in patients undergoing hemodialysis in pediatric cancer patients with long-term, indwelling CVCs (54,55). Future studies in this area will likely utilize novel lock solutions that block multiple factors (organism growth, biofilms formation, thrombosis) contributing to CRBSI pathogenesis.

22.6 SUMMARY AND CONCLUSIONS

Despite many questions concerning the optimal prevention and management of vascular catheter infections, enhanced knowledge of the pathogenesis, complications, and effective treatment strategies have allowed the development of more patient-specific and specialized treatment approaches for CRBSI. Ultimately, catheter removal remains the best defense and management strategy for infections of vascular catheters. Yet the medical necessity of central venous access requires that clinicians develop alternative strategies to catheter removal, especially in the lower-risk patient. Advances in the engineering of catheter design that prevent organism adherence, thrombosis, and biofilm formation hold the greatest promise for reducing risk of vascular catheter infections.

REFERENCES

1. Jarvis, W.R., Edwards, J.R., Culver, D.H., et al., Nosocomial infection rates in adult and pediatric intensive care units in the United States. National Nosocomial Infections Surveillance System. *Am. J. Med.* 91, 185S–191S, 1991.
2. Banerjee, S.N., Emori, T.G., Culver, D.H., et al., Secular trends in nosocomial primary bloodstream infections in the United States, 1980–1989. National Nosocomial Infections Surveillance System. *Am. J. Med.* 91, 86S–89S, 1991.
3. Raad, I.I., and Hanna, H.A., Intravascular catheter-related infections: new horizons and recent advances. *Arch. Intern. Med.* 162, 871–878, 2002.
4. Mermel, L.A., Farr, B.M., Sherertz, R.J., et al., Guidelines for the management of intravascular catheter-related infections. *Clin. Infect. Dis.* 32, 1249–1272, 2001.
5. O'Grady, N.P., Alexander, M., Dellinger, E.P., et al., Guidelines for the prevention of intravascular catheter-related infections. Centers for Disease Control and Prevention. *MMWR Recomm. Rep.* 51, 1–29, 2002.
6. Edgeworth, J.D., Treacher, D.F., and Eykyn, S.J., A 25-year study of nosocomial bacteremia in an adult intensive care unit. *Crit. Care Med.* 27, 1421–1428, 1999.
7. Richards, M.J., Edwards, J.R., Culver, D.H., and Gaynes, R.P., Nosocomial infections in combined medical-surgical intensive care units in the United States. *Infect. Control. Hosp. Epidemiol.* 21, 510–515, 2000.
8. Gudlaugsson, O., Gillespie, S., Lee, K., et al., Attributable mortality of nosocomial candidemia, revisited. *Clin. Infect. Dis.* 37, 1172–1177, 2003.
9. McNeil, M.M., Nash, S.L., Hajjeh, R.A., et al., Trends in mortality due to invasive mycotic diseases in the United States, 1980–1997. *Clin. Infect. Dis.* 33, 641–647, 2001.
10. Herrmann, M., Vaudaux, P.E., Pittet, D., et al., Fibronectin, fibrinogen, and laminin act as mediators of adherence of clinical staphylococcal isolates to foreign material. *J. Infect. Dis.* 158, 693–701, 1988.

11. Raad, I., Costerton, W., Sabharwal, U., Sacilowski, M., Anaissie, E., and Bodey, G.P., Ultrastructural analysis of indwelling vascular catheters: a quantitative relationship between luminal colonization and duration of placement. *J. Infect. Dis.* 168, 400–407, 1993.

12. Sherertz, R.J., Raad, I.I., Belani, A., et al., Three-year experience with sonicated vascular catheter cultures in a clinical microbiology laboratory. *J. Clin. Microbiol.* 28, 76–82, 1990.

13. Parsek, M.R., and Singh, P.K., Bacterial biofilms: an emerging link to disease pathogenesis. *Annu. Rev. Microbiol.* 57, 677–701, 2003.

14. Pascual, A., Ramirez de Arellano, E., and Perea, E.J., Activity of glycopeptides in combination with amikacin or rifampin against Staphylococcus epidermidis biofilms on plastic catheters. *Eur. J. Clin. Microbiol. Infect. Dis.* 13, 515–517, 1994.

15. Pascual, A., Martinez-Martinez, L., Ramirez de Arellano, E., and Perea, E.J., Susceptibility to antimicrobial agents of *Pseudomonas aeruginosa* attached to siliconized latex urinary catheters. *Eur. J. Clin. Microbiol. Infect. Dis.* 12, 761–765, 1993.

16. Pascual, A., Ramirez de Arellano, E., Martinez Martinez, L., and Perea, E.J., Effect of polyurethane catheters and bacterial biofilms on the in-vitro activity of antimicrobials against *Staphylococcus epidermidis*. *J. Hosp. Infect.* 24, 211–218, 1993.

17. Ramage, G., VandeWalle, K., Bachmann, S.P., Wickes, B.L., and Lopez-Ribot, J.L., *In vitro* pharmacodynamic properties of three antifungal agents against preformed *Candida albicans* biofilms determined by time-kill studies. *Antimicrob. Agents Chemother.* 46, 3634–3636, 2002.

18. McGee, D.C., and Gould, M.K., Preventing complications of central venous catheterization. *N. Engl. J. Med.* 348, 1123–1133, 2003.

19. Maki, D.G., Weise, C.E., and Sarafin, H.W., A semiquantitative culture method for identifying intravenous-catheter-related infection. *N. Engl. J. Med.* 296, 1305–1309, 1977.

20. Siegman-Igra, Y., Anglim, A.M., Shapiro, D.E., Adal, K.A., Strain, B.A., and Farr, B.M., Diagnosis of vascular catheter-related bloodstream infection: a meta-analysis. *J. Clin. Microbiol.* 35, 928–936, 1997.

21. Capdevila, J.A., Planes, A.M., Palomar, M., et al., Value of differential quantitative blood cultures in the diagnosis of catheter-related sepsis. *Eur. J. Clin. Microbiol. Infect. Dis.* 11, 403–407, 1992.

22. Blot, F., Nitenberg, G., Chachaty, E., et al., Diagnosis of catheter-related bacteraemia: a prospective comparison of the time to positivity of hub-blood versus peripheral-blood cultures. *Lancet* 354, 1071–1077, 1999.

23. Blot, F., Schmidt, E., Nitenberg, G., et al., Earlier positivity of central-venous- versus peripheral-blood cultures is highly predictive of catheter-related sepsis. *J. Clin. Microbiol.* 36, 105–109, 1998.

24. Raad, I., Hanna, H.A., Alakech, B., Chatzinikolaou, I., Johnson, M.M., and Tarrand, J., Differential time to positivity: a useful method for diagnosing catheter-related bloodstream infections. *Ann. Intern. Med.* 140, 18–25, 2004.

25. Kite, P., Dobbins, B.M., Wilcox, M.H., and McMahon, M.J., Rapid diagnosis of central-venous-catheter-related bloodstream infection without catheter removal. *Lancet* 354, 1504–1507, 1999.

26. Kite, P., Dobbins, B.M., Wilcox, M.H., et al., Evaluation of a novel endoluminal brush method for in situ diagnosis of catheter related sepsis. *J. Clin. Pathol.* 50, 278–282, 1997.

27. Raad, I., Davis, S., Khan, A., Tarrand, J., Elting, L., and Bodey, G.P., Impact of central venous catheter removal on the recurrence of catheter-related coagulase-negative staphylococcal bacteremia. *Infect. Control Hosp. Epidemiol.* 13, 215–221, 1992.

28. Raad, I., Narro, J., Khan, A., Tarrand, J., Vartivarian, S., and Bodey, G.P., Serious complications of vascular catheter-related *Staphylococcus aureus* bacteremia in cancer patients. *Eur. J. Clin. Microbiol. Infect. Dis.* 11, 675–682, 1992.

29. Phillips, M.S., and von Reyn, C.F., Nosocomial infections due to nontuberculous mycobacteria. *Clin. Infect. Dis.* 33, 1363–1374, 2001.

30. Chang, F.Y., Peacock, J.E., Jr., Musher, D.M., et al., *Staphylococcus aureus* bacteremia: recurrence and the impact of antibiotic treatment in a prospective multicenter study. *Medicine (Baltimore)* 333–339, 82, 2003.

31. Carratala, J., The antibiotic-lock technique for therapy of 'highly needed' infected catheters. *Clin. Microbiol. Infect.* 8, 282–289, 2002.

32. Dannenberg, C., Bierbach, U., Rothe, A., Beer, J., and Korholz, D., Ethanol-lock technique in the treatment of bloodstream infections in pediatric oncology patients with broviac catheter. *J. Pediatr. Hematol. Oncol.* 25, 616–621, 2003.

33. Messing, B., Peitra-Cohen, S., Debure, A., Beliah, M., and Bernier, J.J., Antibiotic-lock technique: a new approach to optimal therapy for catheter-related sepsis in home-parenteral nutrition patients. *JPEN J. Parenter. Enteral. Nutr.* 12, 185–189, 1988.

34. Messing, B., Catheter-sepsis during home parenteral nutrition: use of the antibiotic-lock technique. *Nutrition* 14, 466–468, 1998.

35. Longuet, P., Douard, M.C., Maslo, C., Benoit, C., Arlet, G., and Leport, C., Limited efficacy of antibiotic lock technique in catheter related bacteremia of totally implanted ports in HIV infected and oncology patients. In: *Programs and Abstracts of the 35th Interscience Conference on Antimicrobial Agents and Chemotherapy.* San Francisco, CA: ASM Press, 1995.

36. Krishnasami, Z., Carlton, D., Bimbo, L., et al., Management of hemodialysis catheter-related bacteremia with an adjunctive antibiotic lock solution. *Kidney Int.* 61, 1136–1142, 2002.

37. Johnson, D.C., Johnson, F.L., and Goldman, S., Preliminary results treating persistent central venous catheter infections with the antibiotic lock technique in pediatric patients. *Pediatr. Infect. Dis. J.* 13, 930–931, 1994.

38. Capdevila, J.A., Segarra, A., Planes, A.M., et al., Successful treatment of haemodialysis catheter-related sepsis without catheter removal. *Nephrol. Dial. Transplant.* 8, 231–234, 1993.

39. Capdevila, J.A., Barbera, J., and Gavalda, J., Diagnosis and conservative management of infection related to long term venous catheterization in AIDS patients. In: *Program and Abstracts of the 34th Interscience Conference on Antimicrobial Agents and Chemotherapy.* Orlando, FL: ASM Press, 1994.

40. Benoit, J.L., Carandang, G., Sitrin, M., and Arnow, P., Intraluminal antibiotic treatment of central venous catheter infections in patients receiving parenteral nutrition at home. *Clin. Infect. Dis.* 24, 743–744, 1997.

41. Krzywda, E.A., Andris, D.A., Edmiston, C.E. Jr., and Quebbeman, E.J., Treatment of Hickman catheter sepsis using antibiotic lock technique. *Infect. Control. Hosp. Epidemiol.* 16, 596–598, 1995.

42. Domingo, P., Fontanet, A., Sanchez, F., Allende, L., and Vazquez, G., Morbidity associated with long-term use of totally implantable ports in patients with AIDS. *Clin. Infect. Dis.* 29, 346–351, 1999.

43. Veenstra, D.L., Saint, S., Saha, S., Lumley, T., and Sullivan, S.D., Efficacy of antiseptic-impregnated central venous catheters in preventing catheter-related bloodstream infection: a meta-analysis. *JAMA* 281, 261–267, 1999.

44. Maki, D.G., Stolz, S.M., Wheeler, S., and Mermel, L.A., Prevention of central venous catheter-related bloodstream infection by use of an antiseptic-impregnated catheter. A randomized, controlled trial. *Ann. Intern. Med.* 127, 257–266, 1997.

45. Walder, B., Pittet, D., and Tramer, M.R., Prevention of bloodstream infections with central venous catheters treated with anti-infective agents depends on catheter type and insertion time: evidence from a meta-analysis. *Infect. Control Hosp. Epidemiol.* 23, 748–756, 2002.

46. Raad, I., Darouiche, R., Dupuis, J., et al., Central venous catheters coated with minocycline and rifampin for the prevention of catheter-related colonization and bloodstream infections. A randomized, double-blind trial. The Texas Medical Center Catheter Study Group. *Ann. Intern. Med.* 127, 267–274, 1997.

47. Darouiche, R.O., Raad, I.I., Heard, S.O., et al., A comparison of two antimicrobial-impregnated central venous catheters. Catheter Study Group. *N. Engl. J. Med.* 340, 1–8, 1999.

48. Chatzinikolaou, I., Hanna, H., Graviss, L., et al., Clinical experience with minocycline and rifampin-impregnated central venous catheters in bone marrow transplantation recipients: efficacy and low risk of developing staphylococcal resistance. *Infect. Control. Hosp. Epidemiol.* 24, 961–963, 2003.

49. Hanna, H.A., Raad, I.I., Hackett, B., et al., Antibiotic-impregnated catheters associated with significant decrease in nosocomial and multidrug-resistant bacteremias in critically ill patients. *Chest* 124, 1030–1038, 2003.

50. Veenstra, D.L., Saint, S., and Sullivan, S.D., Cost-effectiveness of antiseptic-impregnated central venous catheters for the prevention of catheter-related bloodstream infection. *JAMA* 282, 554–560, 1999.

51. Maki, D.G., Cobb, L., Garman, J.K., Shapiro, J.M., Ringer, M., and Helgerson, R.B., An attachable silver-impregnated cuff for prevention of infection with central venous catheters: a prospective randomized multicenter trial. *Am. J. Med.* 85, 307–314, 1988.

52. Henrickson, K.J., Axtell, R.A., Hoover, S.M., et al., Prevention of central venous catheter-related infections and thrombotic events in immunocompromised children by the use of vancomycin/ciprofloxacin/heparin flush solution: A randomized, multicenter, double-blind trial. *J. Clin. Oncol.* 18, 1269–1278, 2000.

53. Spafford, P.S., Sinkin, R.A., Cox, C., Reubens, L., and Powell, K.R., Recommendations for preventing the spread of vancomycin resistance: Recommendations of the Hospital Infection Control Practices Advisory Committee. *MMWR* 44, 1–13, 1994.

54. Raad, I., Buzaid, A., Rhyne, J., et al., Minocycline and ethylenediaminetetraacetate for the prevention of recurrent vascular catheter infections. *Clin. Infect. Dis.* 25, 149–151, 1997.

55. Chatzinikolaou, I., Zipf, T.F., Hanna, H., et al., Minocycline-ethylenediaminetetraacetate lock solution for the prevention of implantable port infections in children with cancer. *Clin. Infect. Dis.* 36, 116–119, 2003.

23 Treatment Protocols for Bacterial Endocarditis and Infection of Electrophysiologic Cardiac Devices

Martin E. Stryjewski and G. Ralph Corey

CONTENTS

23.1   INTRODUCTION

Infective endocarditis (IE) represents an infection on the endocardial surface of the heart produced by a variety of micro-organisms, including bacteria, fungi, and intracellular pathogens such as *chlamydiae, mycoplasma,* and *rickettsiae.* Infective endocarditis is a life-threatening disease and, despite progress over last decade in its diagnosis and treatment, the disease still continues to have unacceptable mortality and morbidity rates.

Changing patterns in IE have been noticed during the last 10 years (1). *Staphylococcus aureus* (*S. aureus*) is rapidly becoming the leading cause of IE in tertiary centers. This observation can be linked to correspondent changes in medical practice. As an example, there has been a significant increase in the proportion of patients with IE who are on chronic hemodialysis or receiving immunosuppressant therapies. The widespread use of pacemakers and implantable cardiac defibrillators is also accompanied by higher rates of intracardiac infections associated with such devices (2).

Considering this scenario it is remarkable that no randomized double blind controlled trials are available to date to guide the treatment of patients with either IE and/or infected implantable electrophysiologic cardiac devices. Most treatment guidelines come from prospective open label and/or observational studies. Unfortunately, many of these trials have included patients with widely varying infections such as native and prosthetic valve endocarditis or right and left sided valvular infection.

From the therapeutic point of view intracardiac infections should be divided into groups based on the etiologic microorganism, the heart valve involved, and the presence of prosthetic material. This classification has a crucial role when determining the antibiotic as well as the surgical treatment.

Recommendations and dosages are presented in Table 23.1.

23.2   MEDICAL TREATMENT

23.2.1   *Staphylococcus Aureus*

23.2.1.1   Native Valve

23.2.1.1.1   Left-Sided

The great majority of *S. aureus*, regardless of their acquisition route (health care vs. community), produce β-lactamase, and therefore are highly resistant to penicillin G. In this scenario the drugs of choice for methicillin-susceptible *S. aureus* (MSSA) are semi-synthetic, penicillinase-resistant penicillins such as nafcillin or oxacillin sodium. In the unusual case of *S. aureus* susceptible to penicillin, this drug, when used in high doses, is the drug of choice (3).

TABLE 23.1
Antimicrobial Therapy for Common Causes of Infective Endocarditis[a]

Pathogen	Regimen	Recommended Dosing and Intervals	Comments
Methicillin susceptible Staphylococci			
Native valve	Nafcillin or oxacillin for 4 to 6 weeks[b] ± Gentamicin for the first 3 to 5 days of therapy	Nafcillin or oxacillin 2 gr IV q/4 hr; Gentamicin 1 mg/kg IM or IV every 8 hr; Cefazolin 2 g IV q/8 hr	Cefazolin is an alternative option to β-lactams.[a] Peak gentamicin level ~3 μg/ml is preferred; trough gentamicin should be <1 μg/ml. Vancomycin is indicated when confirmed hypersensitivity reaction to beta-lactam antibiotics.
Prosthetic valve	Nafcillin or oxacillin[‡] with rifampin for 6 weeks + Gentamicin for the first 2 weeks	Nafcillin or oxacillin 2 gr IV q/4 hr; Rifampin 300 mg orally q/8 hr; Cefazolin 2 gr IV q/8 hr; Gentamicin 1 mg/kg IM or IV every 8 hr	It is prudent to delay initiation of rifampin for 1 or 2 days, until therapy with two other effective antistaphylococcal drugs has been initiated. Vancomycin is indicated when confirmed hypersensitivity reaction to beta-lactam antibiotics.
Methicillin resistant Staphylococci			
Native valve	Vancomycin × 4 to 6 weeks ± Gentamicin for the first 3 to 5 days	Vancomycin 30 mg/kg in 24 hr in two equally divided doses; Gentamicin 1 mg/kg IM or IV every 8 hr	Peak gentamicin level ~3 μg/ml is preferred; trough gentamicin should be <1 μg/ml. Trough vancomycin level should be 10 to 15 μg/ml.
Prosthetic valve	Vancomycin with rifampin for at least 6 weeks + Gentamicin for the first 2 weeks	Vancomycin and gentamicin as above; Rifampin 300 mg orally or IV q/8 hr	In case of resistance to gentamicin, an alternative third agent should be chosen based on in vitro susceptibility testing (e.g., fluoroquinolone).

TABLE 23.1—Cont'd
Antimicrobial Therapy for Common Causes of Infective Endocarditis[a]

Pathogen	Regimen	Recommended Dosing and Intervals	Comments
Right-sided Staphylococcal Native-valve Endocarditis (in selected patients)	Nafcillin or oxacillin for 2 weeks + Gentamicin for 2 weeks	Nafcillin or oxacillin 2 gr IV q/4 hr; Gentamicin 1 mg/kg IM or IV every 8 hr	This 2-weeks regimen has been studied for infections due to oxacillin- and aminoglycoside-susceptible isolates. Exclusions to short-course therapy.[c]
Enterococci Penicillin Resistant Native and prosthetic valve	Vancomycin for 6 weeks + Gentamicin for 6 weeks	Vancomycin 30 mg/kg in 24 hr in two equally divided doses; Gentamicin 1 mg/kg IM or IV every 8 hr	For beta-lactamase producing strains, ampicillin-sulbactam 3 g IV q4h can be substituted for Vancomycin. Regimens combining vancomycin with aminoglycosides should monitor closely the renal function.
Enterococci Aminoglycoside (Streptomycin and Gentamicin) Resistant Native and prosthetic valve	Ampicillin or penicillin for 8 to 12 weeks or Aqueous crystalline penicillin G sodium for 8 to 12 weeks	Ampicillin 2 to 3 gr IV q/4 hr; Penicillin 20 to 40 million units IV daily either continuously or in six equally divided doses	Alternative: third generation cephalosporin + ampicillin (see text). Surgical therapy is often necessary.
Enterococci Vancomycin Resistant Native and prosthetic valve	Ampicillin sodium or aqueous crystalline penicillin G sodium penicillin for 4 to 6 weeks + Gentamicin for 4 to 6 weeks	Ampicillin 2 to 3g IV q4h; Penicillin G 20 to 40 million units IV qd either continuously or in six equally divided doses; Gentamicin 1 mg/kg IM or IV q8h	Peak gentamicin level ~3 µg/ml is preferred; trough gentamicin should be <1 µg/ml.

Enterococci Pencillin, Aminoglycoside, and Vancomysin resistant Native and prosthetic valve	Linezoid or Quinupristin-dalfopristin or Imipenem/cilastatin ± Ampicillin sodium or Ceftriaxone ± Ampicillin sodium	Linezoid 600 mg PO or IV q12h Quinupristin-dalfopristin 7.5 mg/kg IV q8h Imipenem/cilastatin 500 mg IV, q6h Ampicillin 2 g IV q4h Ceftriaxone 2 g IV q24h	Limited experience with these regimens.
Streptococci Penicillin Susceptible (MIC ≤0.12 μg/ml) (*S. viridans, S. bovis,* and other streptococci) Native valve Prosthetic valve	Penicillin G or ceftriaxone for 4 weeks or Penicillin G (or ceftriaxone) and gentamicin for 2 weeks in selected patients Penicillin G for 6 weeks + Gentamicin for the first 2 weeks	Penicillin G sodium 12 to 18 million U/24 hr IV either continuously or in six equally divided doses Ceftriaxone 2 g once daily IV or IM Gentamicin 1 mg/kg IM or IV q8h Same as for native valve	A 2-week regimen not recommended for patients with myocardial abscess, extracardiac foci of infection, or prosthetic-valve endocarditis. 4 weeks of vancomycin is recommended for patients allergic to β-lactams. Vancomycin is recommended for patients allergic to β-lactams.
Streptococci Relatively Resistant to Penicillin (MIC >0.12 to 0.5 μg/ml) Native valve Prosthetic valve	Aqueous crystalline penicillin G sodium or Ceftriaxone for 4 weeks + Gentamicin during the first 2 weeks Aqueous crystalline penicillin G sodium or Ceftriaxone for 6 weeks + Gentamicin for 6 weeks	Penicillin G 24 million units q24h IV either continuously or in six equally divided doses Ceftriaxone 2g IV or IM q24h Gentamicin 1 mg/kg IM or IV q8h Same as for native-valve	For patients allergic to beta-lactams, vancomycin should be substituted for pencillin G or Ceftriaxone. For patients allergic to beta-lactams, vancomycin should be substituted for pencillin G or Ceftriaxone.

TABLE 23.1—Cont'd
Antimicrobial Therapy for Common Causes of Infective Endocarditis[a]

Pathogen	Regimen	Recommended Dosing and Intervals	Comments
Non-Enterococcal Streptococci Resistant to Penicillin (MIC > 0.5 μg/ml) (Including abiotrophia species) Native valve and prosthetic valve	Aqueous crystalline penicillin G (or ampicillin) or vancomycin for 4 to 6 weeks + Gentamicin for 4 to 6 weeks	Penicillin G sodium 18 to 30 millions U/24 hr IV either continuously or in six equally divided doses Ampicillin 2 to 3g IV q4h Vancomycin 30 mg/kg in 24 hr in two equally divided doses Gentamycin 1 mg/kg IM or IV every 8 hr	Peak of approximately 3 μg/ml is preferred; trough should be <1 μg/ml. Trough vancomycin level should be 10 to 15 μg/ml. Regimens combining vancomycin with aminoglycosides should monitor closely the renal function.
HACEK organisms	Ceftriaxone for 4 weeks (6 weeks for prosthetic valves)		Ampicillin and gentamicin for 4 wk is an alternative regimen, but some isolates may produce β-lactamase, thereby reducing the efficacy of this regimen. Ciprofloxacin is also an alternative.

MIC denotes minimal inhibitory concentration; HACEK organisms, haemophilus species (*Haemophilus parainfluenzae, H. aphrophilus, and H. paraphrophilus*), *Actinobacillus actinomycetemcomitans, Cardiobacterium hominis, Eikenella corrodens,* and *Kingella kingae*; and HIV, human immunodeficiency virus.

[a] Cephalosporins should be avoided in patients with confirmed immediate-type hypersensitivity reactions to beta-lactam antibiotics.

[b] For patients who have infective endocarditis due to methicillin-susceptible staphylococci and who are allergic to penicillins, a first-generation cephalosporin or vancomycin can be substituted for nafcillin or oxacillin. Cephalosporins should be avoided in patients with confirmed immediate-type hypersensitivity reactions to beta-lactam antibiotics.

[c] Exclusions to short-course therapy include any cardiac or extracardiac complications associated with infective endocarditis, persistence of fever for 7 days or more, and infectioΩn with HIV. Patients with vegetations greater than 1 to 2 cm on echocardiogram should probably be excluded from short-course therapy.

The role of aminoglycosides in the treatment of native valve *S. aureus* endo-carditis is controversial. The study done by Korzeniowski et al. (4) showed that the addition of gentamicin for the first two weeks of a 6-week course of nafcillin, failed to improve cure rates or other significant clinical outcomes, and was associated with an increase in renal dysfunction. Many experts now advocate the addition of low dose gentamicin for the first 3 to 5 days in the treatment of MSSA IE because the combination did achieve more rapid clearance of bacteremia. Susceptibility to amino-glycosides should be determined and relative contraindications such as age greater than 65 years, renal insufficiency and eight nerve impairment assessed before such agents are used (3).

When penicillinase-resistant penicillins are not available and nonimmediate-type hypersensitivity to penicillin is documented or suspected, first-generation cephalosporins such as cefazolin are indicated. Treatment failures with cefazolin in MSSA endocarditis may be explained by the production of large amount of type A β-lactamase that inac-tivates cefazolin at a rate high enough to overcome its antibacterial effect. This phenomenon, though rare, may become evident in the presence of a large inoculum of bacteria as can be seen inside an infective vegetation (5). Cefazolin was also shown to be less effective in experimental models of MSSA endocarditis, but the clinical implications of these findings are unknown (6).

In patients with MSSA and allergies to β-lactams as well as in patients with methicillin-resistant *S. aureus* (MRSA) IE vancomycin is the drug of choice (3). However, it must be noted that some evidence suggests than vancomycin is an inferior drug in the treatment of either MSSA or MRSA IE predominantly because of its slowly bactericidal activity as well as its poor tissue penetration (7–9). In patients with MRSA IE and serious adverse reactions associated with vancomycin, new anti-bacterial agents are available. Linezolid, an oxazolidinone, has an excellent pharma-cokinetic profile, but it is bacteriostatic and failures in the treatment of MRSA IE have already been reported despite limited clinical experience (10). Quinupristin/ dalfopristin (QD), a semi-synthetic antibiotic comprised of group B and A strep-togramins respectively, has a bactericidal effect against some but not all staphylococci (11). Thus, the efficacy of this drug in the treatment of patients with MRSA IE may be suboptimal (12). Side effects associated with this agent, including arthalgias and myalgias, seem to be higher than initially reported and have limited its use signifi-cantly (13). Neither linezolid nor QD have been approved in the United States for treatment of *S. aureus* bacteremia or endocarditis. A third antibiotic teicoplanin, a glycopeptide antibiotic active against MRSA, had suboptimal outcomes in patients with staphylococcal endocarditis and is not recommended as first line treatment (14). Daptomycin, a lipopetide antibiotic approved by the FDA in 2003 for treatment of skin and soft tissue infection, has activity against Gram-positive cocci, including MRSA. Daptomycin was proven effective in animal models of MRSA endocarditis (15). However, clinical data for treatment of IE is not yet available.

Rifampin is not recommended in the first line treatment of native valve staphy-lococcal IE, though it may play a role in those patients with inadequate response to conventional therapy. In patients with native valve MRSA IE the addition of rifampin to the vancomycin regimen failed to improve clinical outcomes (7).

The recommended route of therapy is intravenous for 4 to 6 weeks. We treat most patients with native valve left-sided *S. aureus* IE for 6 weeks without adding gentamicin. The use of oral therapy after completion of 4 to 6 weeks of intravenous antibiotics is of little to no benefit. Indications for surgery are addressed in a later section.

It is important to note that the mortality rates in *S. aureus* endocarditis remain unacceptably high, ranging from 25 to 46% (16). This fact reflects at least three phenomena. First, the aggressive nature of infections caused by *S. aureus*. Second, factors related to the medical care such as delays that occur in diagnosis, surgical management, or the use of suboptimal antimicrobial therapy. Finally, many of these patients suffer multiple co-morbidities.

23.2.1.1.2   Right-Sided

Right-sided IE (involving almost exclusively the tricuspid valve) commonly occurs in intravenous drug users (IVDU). This type of IE has a lower mortality and higher cures rates when compared to left-sided IE. The standard treatment consists of 4 weeks of monotherapy. For MSSA nafcillin, oxacillin, or cloxacillin are used and first generation cephalosporins such as cefazolin and cephalotin are suitable alternatives. Shorter therapies have been proposed because of the difficulties in maintaining IVDU in the hospitals for extended periods of time, and given the risks of having intravenous catheters available for self-injections. In selected patients with MSSA infections such therapies usually involve the use of a semi-synthetic penicillin concomitantly with an aminoglycoside in low doses for two weeks. Cure rates of 90 to 95% obtained with this strategy are comparable to those obtained with 4 weeks of monotherapy (17–19). For patients with MSSA infection and complicated right-sided IE, fever longer than 7 days, HIV infection, or large (>1 cm) vegetations a full 4 week course of therapy is indicated (16). In all patients with MRSA IE 4 weeks of intravenous therapy with vancomycin alone is the standard treatment regimen (3,8,9,16).

An inpatient oral therapy regimen consisting of 4 weeks of ciprofloxacin and rifampin was quite efficacious in one study of patients with MSSA IE, though emergence of resistance is a theoretical concern using this approach (3,20).

23.2.1.2   Prosthetic Valve Endocarditis

S. aureus prosthetic valve endocarditis (PVE) is associated with high mortality rates and requires aggressive medical and surgical management (21). One factor contributing to the high mortality rates may be less effective antibiotic therapies in the presence of bacterial biofilms. To illustrate, it has been shown that *S. aureus* in biofilms requires ≥10 times the minimal bactericidal concentration (MBC) of vancomycin to effect a 3-log reduction in bacterial population. Another factor may be the fact that patients with prosthetic valve IE are significantly older than patients with native valve IE and as such have many more comorbid conditions such as diabetes and significant vascular disease (22).

Combined antibiotic therapy for staphylococcal PVE is based primarily on experience with coagulase-negative *Staphylococcus* PVE. Therefore, the treatment for *Staphylococcus* PVE is focused on susceptibility rather than the species specific.

The current recommendation for MSSA PVE favors the use of nafcillin or oxacillin, in combination with rifampin for 6 to 8 weeks and low dose gentamicin during the first 2 weeks (3). For MRSA PVE vancomycin is used in combination with rifampin for 6–8 weeks, and gentamicin is again added during the first two weeks. Of course, susceptibility testing of all organisms to each antibiotic must be performed.

Rifampin has shown in animal models to have a unique ability to kill *Staphylococcus* adherent to prosthetic materials but selection of resistant strains is common when a high burden of bacteria is exposed to the drug. In order to minimize resistance to rifampin this medication should be added only after antibiotics active against *Staphylococcus*, such as β-lactam or vancomycin and aminoglycosides, have been started and the infection burden of bacteria significantly reduced. For strains resistant to gentamicin or other aminoglycosides, a fluroquinolone may be used if the strain is susceptible (23,24).

Surgical valve replacement is often required for the management of the infection because mortality in patients with PVE treated with antibiotics alone ranges from 56 to 76% (25). Indeed, performing valve replacement surgery early in the course of antibiotic therapy may be associated with improved outcomes though even with an aggressive management the mortality rates are still almost 40% (21,26). Many medically managed patients who died were not surgical candidates secondary to comorbidities or complications (e.g., stroke).

Anticoagulation should be temporarily discontinued in patients with *S. aureus* PVE.

23.2.1.3   Coagulase-Negative Staphylococci

Coagulase-negative staphylococci are usually methicillin-resistant and associated with prosthetic valve endocarditis or other intracardiac device infections. Evidence from animal models as well as clinical experience suggests that vancomycin combined with gentamicin and rifampin is the most effective therapy. Vancomycin and rifampin are given for at least 6 weeks and gentamicin added during the first 2 weeks of treatment. If the strain is resistant to gentamicin another aminoglycoside to which the microorganism is susceptible can be used. Given the presence of resistance to all aminoglycosides, a fluoroquinolone may be an appropriate alternative when the strain is susceptible (3). As described above, rifampin should be added when at least two effective anti-staphylococcal drugs have already been initiated.

For coagulase-negative staphylococci susceptible to methicillin experts favor the use of a regimen based on nafcillin or oxacilin in combination with gentamicin and rifampin (3). Recommendation and doses are shown in Table 23.1.

Surgery has a key role in the management of PVE. When infection occurs within 12 months of valve replacement surgery or when it involves the aortic valve, perivalvular extension with abscess formation and valvular dysfunction are common (27). Such complications can only be cured with surgical removal of the device along with debridement and valve replacement in combination with antimicrobial therapy.

Coagulase negative staphylococci endocarditis involving native valves is becoming more frequent. Particular strains of *Staphylococcus epidermis* or other coagulase negative staphylococci species, such as *ludgunesis, warneri,* or *capiiti,* may be more frequently associated with native valve infections, suggesting a more aggressive behavior of these organisms (28,29). The literature suggests higher cure rates for combination therapy when compared with monotherapy though randomized trials are lacking (28,30,31). Therefore we recommend applying the same antibiotic principles as for *S. aureus* native valve infections.

23.2.2 *Streptococcus viridans, Streptococcus bovis,* AND OTHER STREPTOCOCCI (NON-ENTEROCOCCUS)

23.2.2.1 Native Valves

The taxonomic classification of *viridans* Group streptococci includes a variety of streptococcal species including *S. sanguis, S. oralis (mitis), S. mutans,* and others. The treatment for streptococcal endocarditis is based on *in vitro* minimum inhibitory concentration for penicillin (MIC).

23.2.2.1.1 Highly Penicillin-Susceptible Viridans Streptococci, S. bovis and other Non-Enterococcal Streptococci (MIC ≤0.12 μg/ml)

Streptococci with an MIC to penicillin ≤0.12 μg/ml are considered highly susceptible and are usually treated with penicillin G or ceftriaxone for 4 weeks. Ceftriaxone, given in a single daily dose, is particularly useful for outpatient therapy. A very high cure rate (98%) is obtained with these regimens. Comparable cure rates can be achieved for uncomplicated cases with a combination of penicillin or ceftriaxone with low dose aminoglycoside for two weeks. The later regimen is not recommended for patients with extracardiac foci of infection or intracardiac abscesses (3,16,32). The addition of gentamicin to the 4 week regimen of penicillin or ceftriaxone is unnecessary (32). When gentamicin is used the MIC of the organism should be determined and blood levels monitored. Relative contraindications for gentamicin are patients older than 65 years, renal insufficiency and eighth nerve impairment (3).

Cefazolin or other first generation cephalosporins may be substituted for penicillin in patients whose penicillin hypersensitivity is not of the immediate type. Vancomycin is recommended for patients allergic to β-lactams. In obese patients the dose of vancomycin should be based on the ideal weight and the infusion rate should be at least 1 hour to avoid "red man" syndrome. Vancomycin levels should be checked to ensure adequacy. Dose should be adjusted to obtain peak levels of 30 to 40 μg/ml and trough of 10 to 15 μg/ml.

Recommendation and doses are shown in Table 23.1.

23.2.2.1.2 Relatively Penicillin-Resistant Streptococci (MIC >0.12 to 0.5 μg/ml)

When IE is due to streptococcal strains with MIC for penicillin >0.12 μg/ml, but less than 0.5 μg/ml combination therapy with penicillin and gentamicin is indicated (3,16). Gentamicin is given for the first 2 weeks of the 4-week course of penicillin. In patients allergic to β-lactams, a 4 week course of vancomycin is recommended.

Penicillin-resistant *Streptococcus pneumoniae* (PRSP) endocarditis has been reported (33,34). A Spanish series including 63 cases of pneumococcal endocarditis identified 24 cases produced by penicillin-resistant *S. pneumoniae*. Patient characteristics and outcomes were similar when comparing cases produced by penicillin-resistant vs. penicillin-susceptible strains. Penicillin-resistant *S. pneumoniae* was treated with either a third-generation cephalosporin or vancomycin based regimens. Roughly 50% of patients in each group (penicillin susceptible vs. penicillin resistant) required valve replacement. Cure was achieved in approximately 60% of cases with the medical-surgical approach.

Treatment of PRSP consists of a 6 week course with a third-generation cephalosporin (e.g., cefotaxime or ceftriaxone), or vancomycin for cephalosporin resistant strains. New flouroquinolones such as levofloxacin and moxifloxacin may be useful as alternatives but data are lacking. For those cases associated with concomitant meningitis, third-generation cephalosporins are recommended. In all cases antibiotic use should be supported by *in vitro* susceptibilities.Before the penicillin-resistant era, the mortality of pneumococcal endocarditis was above 60% for those patients treated with antibiotic therapy alone compared to 32% for patients treated with antibiotics plus surgery. Based on this historical data and regardless of the strain susceptibility to penicillin, valve replacement surgery will be required in a significant number of patients with pneumococcccal endocarditis.

Recommendation and doses are shown in Table 23.1.

23.2.2.1.3 Streptococcus *spp. with MIC of* *penicillin >0.5 µg/ml, or* Abiotrophia *spp.*

When IE is due to streptococcal strains with MIC for penicillin >0.5 µg/ml or nutritionally variant streptococci (now classified as *Abiotrophia* species), a regimen for penicillin-resistant enterococcal endocarditis is appropriate. These regimens consist of 4 to 6 weeks of vancomycin combined with gentamicin. However, renal function should be closely monitored when vancomycin is used in combination with aminoglycosides. To avoid nephrotoxicity some experts advocate the use of penicillin or ampicillin plus gentamicin (3,16). For patients with symptoms longer than 3 months, a 6-week course of treatment is preferred. The length of therapy for the aminoglycoside is debated. Vancomycin is substituted for β-lactams in patients allergic to those compounds. Cephalosporins are not acceptable options for patients allergic to penicillin because of MICs which are correspondingly elevated.

Penicillin-resistant *S. viridans* is an uncommon cause of IE. Among this group *S. mitis* seems to predominate. The experience with the use of vancomycin with or without aminoglycosides is limited to few cases (35).

23.2.2.2 Prosthetic Valves

Patients with viridans streptococci PVE can be treated with antibiotics alone if no other indications for surgery are present (e.g., unstable prosthesis, heart failure, new or progressive paravalvular leak, perivalvular extension of infection or persistent infection after 7 to 10 days of appropriate antibiotic therapy) (16).

For highly penicillin-susceptible streptococci PVE (MIC ≤0.12 µg/ml) penicillin G for 6 weeks and gentamicin for 2 weeks are usually indicated. When PVE is due to relatively penicillin-resistant streptococci (MIC >0.12 to 0.5 µg/ml) penicillin G is recommended for 6 weeks and gentamicin for 4 weeks. Endocarditis due to nutritionally variant streptococci (now classified as *Abiotrophia* spp.) or viridans streptococci with MIC for penicillin >0.5 µg/ml, should be treated with 4 to 6 weeks of penicillin, ampicillin, or vancomycin combined with gentamicin (16). Unfortunately little data is available to support recommendations concerning length of aminoglycoside use. Vancomycin therapy is indicated for patients with confirmed immediate hypersensitivity to beta-lactams antibiotics.

23.2.3 ENTEROCOCCI

Infective endocarditis caused by enterococci is usually associated with *E. faecalis* and uncommonly with *E. faecium.*

Enterococci are resistant to most classes of antibiotics, which makes treatment of intravascular infections produced by this genus quite difficult. Due to a defective bacterial autolytic enzyme system, cell-wall active agents are bacteriostatic against *enterococci* and should not be given alone to treat endocarditis (36). In combination with gentamicin or streptomycin, penicillin G and ampicillin facilitate the intracellular uptake of the aminoglycoside, which subsequently results in a bactericidal effect against *enterococci* (37). Clinical studies in humans have shown a much better outcome, measured by bacteriologic cure or survival, with a synergistic combination of a cell wall-active agent and aminoglycoside than with single drug therapy. As a result, a large body of clinical experience has been accumulated using the combination of penicillin G and streptomycin (38). Though ampicillin is more active than penicillin G *in vitro* against *enterococci* clinical data supporting the use of this antibiotic is not nearly as extensive as with penicillin. Before embarking on therapy susceptibility of the *Enterococcus* should be determined for penicillin (or ampicillin), vancomycin, and aminoglycosides. For strains with intrinsic high-level resistance to penicillin (MIC >16 µg/ml) vancomycin is indicated. Vancomycin is also synergistic with aminoglycosides, particularly with gentamicin, but the clinical data for treatment of enterococcal IE using with this combination is scanty (38,39).

Gentamicin increasingly has become the aminoglycoside of choice since streptomycin has been associated with higher levels of resistance, significant ototoxicity, and more difficult administration (intramuscular) (38). MICs for streptomycin and gentamicin should be measured in order to guide treatment since resistances to streptomycin and gentamicin are encoded by different genes. When high level resistance to aminoglycosides (HLRA) is detected (MIC >2000 µg/ml for streptomycin and 500 to 2000 µg/ml for gentamicin) their combination with cell-wall active agents is no longer synergistic and therefore not recommended. In these cases a long course of an active cell-wall agent is recommended in the highest doses (e.g., ampicillin 20 g/day for 8 to 12 weeks) (3).

Recent data has shown promising *in vitro* and *in vivo* results with the combination of a third-generation cephalosporin (ceftriaxone or cefotaxime) and ampicillin for HLRA enterococci (40,41). The rationale for these studies is based on achieving synergy through the simultaneous saturation of several penicillin binding proteins (PBPs). It should be mentioned that high doses of ceftriaxone (e.g., 4 g/day) were used in such an approach.

23.2.3.1 Vancomycin-Resistant Enterococci

Endocarditis caused by vancomycin-resistant enterococci (VRE) is difficult to treat. The optimal therapy for such infections is unknown. Most vancomycin-resistant strains of *E. faecalis*, as well as a few of *E. faecium*, are susceptible to achievable concentrations of ampicillin. In such cases the recommended therapy is ampicillin or penicillin combined with gentamicin or streptomycin (unless high- level resistance is present) (42). Even when enterococci are considered resistant to ampicillin (MIC ≥ 16 µg/ml), higher doses, in the range of 18 to 30 g/day, can be used in order to achieve sustained plasma levels of more than 100 to 150 µg/ml. The use of 20 g/day of ampicillin in combination with gentamicin was effective in one patient with vancomycin-resistant *E. faecium* IE (with susceptibility to ampicillin ≤ 64 µg/ml) (43). To date, there has been little toxicity when employing high doses of ampicillin but more experience is still needed with such doses (42). In 1999, the FDA approved QD as the first antibacterial drug to treat infections associated with vancomycin-resistant *Enterococcus faecium* (VREF) bacteremia when no alternative treatment is available. However, QD alone is unlikely to be curative in VREF IE because the antibacterial is not usually bactericidal against *E. faecium*. Endocarditis models do suggest that the association of QD with ampicillin may be beneficial (42). It is important to note that *E. faecalis* is not susceptible to QD.

Linezolid has a bacteriostatic effect against VRE (*E. faecalis* or *E. faecium*) and it is not recommended as a first line therapy to treat VRE IE. However, cases with successful outcomes (one patient with prolonged bacteremia, and another with IE caused by VRE *faecium*) have been reported (44,45). In 2000, the FDA approved linezolid to treat infections associated with vancomycin-resistant *E. faecium* (VREF), including cases with bloodstream infection.

Daptomycin, a novel lipopeptide antibiotic, displays *in vitro* efficacy against *E. faecium* in pharmacodynamic models with simulated endocardial vegetations, however, clinical experience is not yet available (46,47).

The experience with combinations such as chloramphenicol plus minocycline is anecdotal (48).

23.2.3.2 Prosthetic Valves

Patients with enterococcal PVE can be treated with antibiotics alone if no other indications for surgery are present (e.g., unstable prosthesis, heart failure, new or progressive paravalvular prosthetic leak, perivalvular extension or persistent infection after 7 to 10 days of appropriate antibiotic therapy) (16).

23.2.4 HACEK Microorganisms

HACEK organisms, including *Haemophilus* spp. (*Haemophilus parainfluenzae, H. aphrophilus,* and *H. paraphrophilus*), *Actinobacillus actinomycetemcomitans, Cardiobacterium hominis, Eikenella corrodens,* and *Kingella kingae* account for 5 to 10% of native valve endocarditis in non-IVDA (3). A characteristic of the group is their fastidious growth characteristics when cultured without modern microbiologic techniques. Thus, when standard microbiological techniques are used, incubation for 2 to 3 weeks is recommended for those cases in which endocarditis is suspected and the initial blood cultures are negative. Third generation cephalosporins such as ceftriaxone and cefotaxime are the drugs of choice for the treatment of HACEK endocarditis. The recommended duration is 3 to 4 weeks for native valves and 6 weeks for PVE (3). Ampicillin monotherapy is no longer recommended because many strains produce β-lactamase. HACEK microorganisms are susceptible *in vitro* to fluoroquinolones, however, since clinical data is still lacking they should be used as an alternative therapy only in patients who can not tolerate β-lactams. The same principles apply for the use of aztreonam and thrimethoprim-sulfamethoxazole (3).

23.2.5   *Pseudomonas aeruginosa*

Pseudomonas aeruginosa causes endocarditis primarily in IVDA. Most of our information about this disease comes from the experience at one medical center in Detroit, in the 1970s and 1980s. When the disease affects the right side of the heart antibiotics alone can be curative in 50 to 75% of the cases. Medical therapy is occasionally curative for left-sided endocarditis (49) though in most cases surgical treatment along with antimicrobial therapy is mandatory (50,51). The usual antibiotic treatment regimen consists of two antimicrobials such as antipseudomonal penicillin in large dose (e.g., piperacillin 18 g/day) along with a high dose aminoglycoside (e.g., tobramicin 5 to 8 mg/kg/day to achieve levels of 8 to 10 µg/ml) for 6 weeks (49,50). Alternative regimens can be used as long as they are supported by *in vitro* susceptibility results. Examples of such alternative regimens are the combination of imipenem plus an aminoglicoside or a quinolone based-regimen with the addition of a second drug (52,53). However the clinical experience is limited with both and they are not recommended as first line therapy.

23.2.6   *Coxiella burnetti*

Q fever is caused by *Coxiella burnetti*, a strict intracellular pathogen. Though rare in most parts of the world, in selected locations (e.g., southern France) infection with this organism is a relatively common cause of IE on native or prosthetic valves. The intracellular location of the microorganism is associated with frequent relapses and makes eradication extremely difficult. To achieve cure, valve replacement is commonly required along with an extended course of antibiotics.

Some experts favor antibiotic treatment for a minimum of 3 years once IgG antibody titers drop below 1:400 and IgA phase I antibodies are undetectable (50). Several regimens are recommended including doxycycline with trimethoprim/sulfamethoxazol, rifampin or fluoroquinolones. However, recent evidence suggests that combination of doxycycline and hydroxychloroquine allows therapy to be shortened and decreases relapses (54).

Surgical valve replacement is indicated for prosthetic-valve infection, heart failure or uncontrolled infection (50).

23.3   SURGICAL TREATMENT

The reduction of mortality in patients with IE during the last three decades may be explained by the introduction of echocardiographic techniques, allowing earlier diagnosis as well as more aggressive surgical management in those patients with congestive heart failure (CHF), perivalvular abscesses or prosthetic valves. The later approach has been supported by the finding that the shortening of antibiotic therapy preoperatively does not increase operative mortality (55). In addition, though some studies have found a higher risk of persistence or relapse associated with valve replacement surgery during the acute phase of endocarditis, others did not confirm this association, particularly after surgery for mitral valve endocarditis (56–58). Recently a retrospective study utilizing propensity score methodology to adjust for confounders, found that patients undergoing surgery for left-sided IE had a lower mortality at 6 months when compared with those matched for clinical characteristics who received medical therapy alone (59). Interestingly the study did not find significant differences in mortality for many of the recognized clinical indications of surgery beyond the presence of moderate to severe congestive heart failure. While the findings of this study are debatable, they underline the fact that indications for surgery most probably will evolve depending on future research (60).

23.4   SURGICAL INDICATIONS

Accepted indications for surgery in infective endocarditis can be seen in Table 23.2. Here we will discuss selected indications. For further information a comprehensive review is recommended (55).

TABLE 23.2
Accepted Indications for Surgery in Patients with IE[a]

Emergency Indication

Acute AR with early closure of mitral valve
Rupture of a sinus of Valsalva aneurysm into a right heart chamber
Rupture into the pericardium

Urgent Indication (within 1 to 2 days)

Valvular obstruction
Unstable prosthesis
Acute or worsening AR or MR with heart failure, NYHA III–IV
Septal perforation
Presence of annular or aortic abscess, sinus, fistula

Elective Indication (earlier is better)

Staphylococcal PVE[b]
New or progressive paravalvular prosthetic leak
Persistent infection after 7 to 10 days of appropriate antibiotic therapy[b]
Fungal endocarditis[c]
Pseudomonas aeruginosa (left sided-disease)

[a] Modified from Olaison et al. (55)
[b] Not absolute.
[c] Not absolute. Based on the fungal pathogen and patient response to the medical treatment.
  New anti-fungal agents may change prognosis.

23.4.1   Congestive Heart Failure

Among patients with IE, congestive heart failure (CHF) is the complication with biggest impact on prognosis (50). In patients with native valve endocarditis, CHF occurs more often in aortic valve infections (29%) than in mitral (20%) or tricuspid valve disease (8%) (61). Prognosis without surgical therapy is poor, and surgical delay increases operative mortality by up to 25% (50). Since the incidence of reinfection of newly implanted valves is estimated to be only 2 to 3% extending preoperative antibiotic treatment is not recommended. Therefore surgery should be performed before frank ventricular decompensation occurs. Echocardiographic techniques are invaluable for such assessment. Valve position also plays an important role in prognosis. Acute mitral regurgitation is usually better tolerated and carries a better prognosis than acute aortic regurgitation (55) while tricuspid or pulmonic IE rarely requires emergency surgery. Preexisting valvular disease is an important consideration when contemplating surgery. Patients with IE and preexisting moderate or severe aortic valve disease usually require surgery, whereas other patients who remain compensated may be treated with medical therapy alone (55). However, high one year mortality rates in this population suggest the need for earlier intervention in both groups. Severe valvular dysfunction usually requires valve replacement. However, selected patients with ruptured mitral chordae or perforated leaflets can be treated effectively with valvular repair alone (50).

23.4.2   Periannular Extension of Infection

Periannular extension is a common complication of IE, occurring in 10 to 40% of patients with native valve infections, and in 56 to 100% of patients with prosthetic valves (50). Extension beyond the valve annulus increases the risk of CHF secondary to paravalvular regurgitation, heart block and death (50). TEE is particularly useful in detecting such extension of the infection.

Medical therapy alone is inadequate and virtually all patients with macroscopic periannular extension should undergo cardiac surgery. A small number of patients with significant comorbidities can be treated without surgical intervention, especially those without heart block, echocardiographic evidence of progression, or valvular dehiscence or insufficiency (55).

23.4.3   Risk of Embolization

Clinical evidence of systemic embolization occurs in one quarter to one half of patients with IE (50). Two thirds of these events affect the central nervous system, most commonly in the distribution of the middle cerebral artery. Most emboli occur during the first two weeks of effective antimicrobial therapy (62). The incidence of systemic emboli appears to be higher with *S. aureus*, *Candida* species, HACEK and *Abiotrophia* organisms than with *S. viridans*. Importantly, mitral vegetations, regardless of size, are associated with higher rates of embolic events (25%) than aortic valves (10%) (50). Interestingly, the incidence seems to be higher when the anterior mitral leaflet is affected (37%) (50). Studies of vegetation mobility and size have shown conflicting results (63). We feel that vegetation size is an important predictor of thromboembolic events but does not precisely identify a high-risk cohort of patients (64).

Some experts recommend surgical therapy for those patients with ≥2 embolic events, but this approach does not consider many of the risk factors for recurrent embolization or time of the event within the antibiotic treatment regimen. A more rational approach may be to recommend surgery for those patients with a major embolic event within two weeks of initiation of appropriate antimicrobial therapy, with other predictors of a complicated course such as a large vegetation (>1 cm) remaining on the mitral or aortic valve, CHF, PVE, or infection of the mitral valve caused by aggressive or resistant organisms (50).

In the absence of complications *S. aureus* IE on a native valve does not have a clear indication for surgery. One retrospective study showed that early surgery (within 14 days) had higher survival rates when compared with delayed intervention; the findings were mostly due to the outcomes in the *S. aureus* group (65). However, given the retrospective design of the study and the fact that numbers of intracardiac devices were significantly different between groups, the results should be considered with caution.

23.4.4   PROSTHETIC VALVE ENDOCARDITIS

As discussed in previous sections of this chapter PVE commonly requires surgical intervention along with combined antibimicrobial therapy. However, there is a subset of patients in whom medical therapy alone may be effective to cure PVE. Such patients usually have late-onset infections (≥12 months after valve placement), infection produced by HACEK microorganisms (*Haemophilus* spp., *Actinobacillus acinemycetomcomitans*, *Cardiobacterium hominis*, *Eikinella corrodens* and *Kingella kingae*), *viridans* streptococci, or enterococci in the absence of invasive infection (23,55). Accepted indications for surgery in PVE can be seen in Table 23.2.

23.5   TREATMENT OF INFECTED ELECTROPHYSIOLOGIC CARDIAC DEVICES

Cardiac prosthetic devices such as permanent implantable pacemakers or implantable cardioverter-defibrillators have increasingly become part of modern cardiovascular medicine. Infection of cardiac prosthetic devices is a devastating complication whose incidence seems to be rising. Among Medicare beneficiaries the implantation rate of intracardiac devices increased by 42% from 3.26/1,000 in 1990 to 4.64/1,000 in 1999. Remarkably, infection rates among those patients increased by 124% from 0.94/1,000 to 2.11/1,000 patients over the same period of time (2). Among patients with *S. aureus* bacteremia and intracardiac devices, 75% of them had evidence of cardiac device infection (CDI) when the bloodstream infection occurred within one year of placement or surgical modification (66).

Unfortunately, conservative therapy has a limited role in the management of CDI and complete explantation of the device along with appropriate antimicrobial therapy is recommended (23,67,68). Higher mortality was reported in patients with pacemaker infection treated with antibiotics alone compared to those treated with antibiotics plus removal of the complete device system (41% vs. 19% respectively) (68). Intracardiac removal of transvenous devices was more efficacious and associated with fewer complications utilizing laser techniques (68).

Successful reimplantation of a new device is usually accomplished when the patient is afebrile and blood cultures are negative after the initial removal. Reimplantation is recommended utilizing a new site. In a retrospective study the only relapse among 117 patients in whom complete explantation was performed occurred in a patient with reimplantation in the old pocket (67). In this study the mean time from explantation to reimplantation was 7 days.

Antimicrobial treatment for CDI is usually given for 4 to 6 weeks based on the microorganism, susceptibility tests as well as the echocardiographic findings. Antibiotic therapy without full explantation of the device is rarely successful (67).

REFERENCES

1. Cabell, C.H., Jollis, J.G., and Peterson, G.E., et al., Changing patient characteristics and the effect on mortality in endocarditis. *Arch. Intern. Med.* 162, 90–94. 2002.
2. Cabell, C.H., Heidenreich, P.A., Chu, V.H., et al., Increasing rates of cardiac device infections among Medicare beneficiaries, 1990–1999. *Am. Heart. J.* 147:582–586, 2004.
3. Wilson, W., Karchmer, A.W., Dajani, A.S., et al., Antibiotic treatment of adults with infective endocarditis due to Streptococci, Enterococci, Staphylococci, and HACEK microorganisms. *JAMA* 274, 1706–1713, 1995.
4. Korzeniowski, O., and Sande, M.A., The National Collaborative Endocarditis Study Group. Combination antimicrobial therapy for *Staphylococcus aureus* endocarditis in patients addicted to parenteral drugs and in nonaddicts: a prospective study. *Ann. Intern. Med.* 97, 496–503. 1982.
5. Nannini, E.C., Singh, K.V., and Murray, B.E., Relapse of type A β-lactamase-producing *Staphylococcus aureus* native valve endocarditis during cefazolin therapy: revisiting the issue. *Clin. Infect. Dis.* 37, 1194–1198, 2003.
6. Steckelberg, J.M., Rouse, M.S., Tallan, B.M., Henry, N.K., and Wilson, W.R., Relative efficacies of broad-spectrum cephalosporins for treatment of methicillin-susceptible *Staphylococcus aureus* experimental infective endocarditis. *Antimicrob. Agents Chemother.* 37, 554–558, 1993.
7. Levine, D.P., Fromm, B., and Reddy, R., Slow response to vancomycin plus rifampin in methicillin-resistant *Staphylococcus aureus* endocarditis. *Ann. Intern. Med.* 115, 674–680, 1991.
8. Small, P., and Chambers, H., Vancomycin for *Staphylococcus aureus* endocarditis in intravenous drug users. *Antimicrob. Agents Chemother.* 34, 1227–1231, 1990.
9. Fortun, J., Navas, E., Martinez-Beltran, J., et al., Short-course therapy for right-side endocarditis due to *Staphylococcus aureus* in drug abusers: cloxacillin versus glycopeptides in combination with gentamycin. *Clin. Infect. Dis.* 33, 120–125, 2001.
10. Ruiz, M.E., Guerrero, I.C., and Tuazon, C., Endocarditis caused by methicillin-resistant *Staphylococcus aureus* endocarditis: treatment failure with linezolid. *Clin. Infect. Dis.* 35, 1018–1020, 2002.
11. Eliopulos, G.M., Quinupristin/Dalfopristin and Linezolid: evidence and opinion. *Clin. Infect. Dis.* 36, 473–481, 2003.

12. Drew, R.H., Perfect, J.R., and Srinath, L., et al., Treatment of methicillin-resistant *Staphylococcus aureus* infections with quinupristin-dalfopristin in patient intolerant or failing prior therapy. *J. Antimicrob. Chemother.* 46, 775–784, 2000.

13. Olsen, K.M., Rebuck, J.A., and Rupp, M.E. Arthralgias and myalgias related to Quinupristin/Dalfopristin administration. *Clin. Infect. Dis.* 32, e83–86, 2001.

14. Gilbert, D.N., Wood, C.A., and Kimbrough, R.C., The Infectious Diseases Consortium of Oregon. Failure of treatment with teicoplanin at 6 mg/kg/day in patients with *Staphylococcus aureus* intravascular infections. *Antimicrob. Agents Chemother.* 35, 79–87, 1991.

15. Sakoulas, G., Eliopoulos, G.M., Alder, J., and Thauvin-Eliopoulos, C., Efficacy of Daptomycin in experimental endocarditis due to methicillin-resistant *Staphylococcus aureus*. *Antimicrob. Agents Chemother.* 47, 1714–1718, 2003.

16. Mylonakis, E., and Calderwood, S.B., Infective endocarditis in adults. *N. Eng. J. Med.* 345, 1318–1330, 2001.

17. Chambers, H.F., Miller, T., and Newman, M.D., Right-sided *Staphylococcus aureus* endocarditis in intravenous drug abusers: two week combination therapy. *Ann. Intern. Med.* 109, 619–624, 1988.

18. DiNubile, M.J., Short-course antibiotic therapy for right-sided endocarditis caused by *Staphylococcus aureus* in injection users. *Ann. Intern. Med.* 121, 873–876, 1994.

19. Ribera, E., Gomez-Jimenez, J., Cortes, E., et al., Treatment of right-sided *Staphylococcus aureus* endocarditis. *Ann. Intern. Med.* 125, 969–974, 1996.

20. Dworkin, R.J., Lee, B.L., Sande, M.A., et al., Treatment of right-sided *Staphylococcus aureus* endocarditis in intravenous drug users with ciprofloxacin and rifampin. *Lancet* 2, 1071–1073, 1989.

21. Petti, C.A., and Fowler, V.G., *Staphylococcus aureus* bacteremia and endocarditis. *Infect. Dis. Clin. N. Am.* 16, 413–435, 2002.

22. Williams, I., Venables, W.A., Lloyd D., et al., The effects of adherence to silicone surfaces on antibiotic susceptibility in *Staphylococcus aureus*. *Microbiology* 143, 2407–2413, 1997.

23. Karchmer, A.W., and Longworth, D.L., Infections of intrcardiac devices. *Infect. Dis. Clin. North Am.* 16, 477–505, 2002.

24. Rouse, M.S., Wilcox, R.M., Henry, N.K., Steckelberg, J.M., and Wilson, W.R., Ciprofloxacin therapy of experimental endocarditis caused by methicillin-resistant. *Staphylococcus epidermidis*. *Antimicrob. Agents Chemother.* 34, 273–276, 1990.

25. Stanbridge, T.N., and Isalska, B.J., Aspects of prosthetic valve endocarditis. *J. Infect.* 35, 1–6. 1997.

26. John, M.D.V., Hibberd, P.L., and Karchmer, A.W., *Staphylococcus aureus* prosthetic valve endocarditis: optimal management and risk factors for death. *Clin. Infect. Dis.* 26, 1302–1309, 1998.

27. Calderwood, S.B., Swinski, L.A., Karchmer, A.W., Waternaux, C.M., and Buckley, M.J., Prosthetic valve endocarditis: analysis of factors affecting outcome therapy. *J. Cardiovasc. Surg.* 92, 776–783, 1986.

28. Caputo, G.M., Archer, G.L., Calderwood, S.B., DiNubile, M.J., and Karschmer, A.W., Native valve endocarditis due to coagulase-negative staphylococci. Clinical and microbiologic features. *Am. J. Med.* 83, 619–625, 1987.

29. Rocha, J.L., Janoff, E.N., Ellingson, L.A., and Crossley, K.B., Consequences of coagulase-negative staphylococcal endocarditis of native valves. Abstract 276. 41st Annual Meeting of the Infectious Disease Society of America. San Diego, CA, 2003.

30. Drinkovi, D., Morris, A.J., Pottumarthy, S., MacCulloch, D., and West, T., Bacteriological outcome of combination versus single-agent treatment for staphylococcal endocarditis. *J. Antimicrob. Chemother.* 52, 820–825, 2003.

31. Arber, N., Militianu, A., Ben-Yehuda A., et al., Native valve *Staphylococcus epidermidis* endocarditis: report of seven cases and review of the literature. *Am. J. Med.* 90, 758–762, 1991.

32. Wilson, W.R., Thompson, R.L., Wilkowske, C.J., et al., Short-term therapy for streptococcal infective endocarditis: combined intramuscular administration of penicillin and netilmicin. *JAMA* 245, 360–363, 1981.

33. Siegel, M., and Timpone, J., Penicillin-resistant *Streptococcus pneumoniae* endocarditis: a case report and review. *Clin. Infect. Dis.* 32, 972–974, 2001.

34. Martinez, E., Miro, J.M., Almirante, B., et al., Effect of penicillin-resitance of *Streptococcus pneumoniae* on the presentation, prognosis, and treatment of pneumococcal endocarditis in adults. *Clin. Infect. Dis.* 35, 130–139, 2002.

35. Levy, C.S., Kogulan, P., Gill, V.J., et al., Endocarditis caused by penicillin-resistant *viridans streptococci*: 2 cases and controversies in therapy. *Clin. Infect. Dis.* 33, 577–579, 2001.

36. Krogstad, D.J., and Parquette, A.R., Defective killing of enterococci: a common property of antimicrobial agents acting on the cell wall. *Antimicrob. Agents Chemother.* 17, 965, 1980.

37. Le, T., and Bayer, A., Combination antibiotic therapy for infective endocarditis. *Clin. Infect. Dis.* 36, 615–621, 2003.

38. Megran, D.W., Enterococcal endocarditis. *Clin. Infect. Dis.* 15, 63–71. 1992.

39. Watanakunakorn, C., and Bakie, C., Synergism of vancomycin-gentamycin and vancomycin–streptomycin against enterococci. *Antimicrob. Agents Chemother.* 4, 120–124, 1973.

40. Brandt, C.M., Rouse, M.S., Laue, N.W., et al., Effective treatment of multidrug-resistant enterococcal experimental endocarditis with combination of cell wall-active agents *J. Infect. Dis.* 173, 909–913, 1996.

41. Galvada, J., Miro, J., Torres C., et al., Efficacy of ampicillin plus ceftriaxone or cefotaxime in treatment of endocarditis due to *Enterococcus faecalis* [abstract L1342]. In: *Programs and abstracts of the 41st Interscience Conference on Antimicrobial Agents and Chemotherapy* (Chicago). Washington, D.C.: American Society of Microbiology, 2001.

42. Murray, B.E., Vancomycin-resistant enterococcal infections. *N. Engl. J. Med.* 342, 710–721, 2000.

43. Mekonen, E.T., Noskin, G.A., Hacek, D.M., and Peterson, L.R., Successful treatment of persistent bacteremia due to vancomycin-resistant, ampicillin-resistant *Enterococcus faecium. Microb. Drug Resist.* 1, 249–253, 1995.

44. McNeil, S.A., Clark, N.M., Chandrasekar, P.H., and Kauffman, C.A., Successful treatment of vancomycin-resistant *Enterococcus faecium* bacteremia with linezolid after failure of treatment with Synercid (Quinupristin-Dalfopristin). *Clin. Infect. Dis.* 30, 403–404, 2000.

45. Babcock, H.M., Ritchie, D.J., Christiansen E., et al., Successful treatment of vancomycin-resistant *Enterococcus* endocarditis with oral linezolid. *Clin. Infect. Dis.* 32, 1373–1375, 2001.

46. Akins, R.L. and Rybak, M.J., Bactericidal activities of two daptomycin regimens against clinical strains of glycopeptide intermediate-resistant *Staphylococcus aureus*, vancomycin-resistant *Enterococcus faecium* and methicillin-resistant

Staphylococcus aureus isolates in an *in vitro* pharmacodynamic model with simulated endocardial vegetations. *Antimicrob. Agents Chemother.* 45, 454–459, 2001.

47. Cha, R., and Rybak, M.J., Daptomycin against multiple drug-resistant *Staphylococcus* and *Enterococcus* isolates in an *in vitro* pharmacodynamic model with simulated endocardial vegetations. *Diag. Microbiol. and Infect. Dis.* 47, 539–546, 2003.

48. Safdar, A., Bryan, C.S., Stinson, S., and Saunders, D.E., Prosthetic valve endocarditis due to vancomycin-resitant Enterococcus faecium: Treatment with chloramphenicol plus minocycline. *Clin. Infect. Dis.* 34, e61–63, 2002.

49. Levine, D.P., Crane, L.R., and Zervos, Z.M., Bacteremia in narcotic addicts at the Detroit Medical Center II. Infectious endocarditis: a prospective comparative study. *Rev. Infect. Dis.* 8, 374–396, 1986.

50. Bayer, A.S., Bolger, A.F., Taubert, K.A., et al., Diagnosis and management of infective endocarditis and its complications. *Circulation* 98, 2936–2948, 1998.

51. Mammana, R.B., Levitsky, S., Sernaque, D., et al., Valve replacement for left-sided endocarditis in drug addicts. *Ann. Thorac. Surg.* 35, 436, 1983.

52. Fitchtenbaum, C.H., and Smith, M.J., Treatment of endocarditis due to *Pseudomonas aeruginosa* with imipenem. *Clin. Infect. Dis.* 14, 353, 1992.

53. Daikops, G.L., Kathopalia, S.B., Lolans, V.T., et al., Long-term oral ciprofloxacin in the treatment of incurable infective endocarditis. *Am. J. Med.* 84, 786–790, 1988.

54. Raoult, D., Houpikian, P., Tissot Dupont, H., et al., Treatment of Q fever endocarditis. *Arch. Intern. Med.* 159, 167–173, 1999.

55. Olaison, L., and Pettersson, G., Current best practices and guidelines. Indications for surgical intervention in infective endocarditis. *Infect. Dis. Clin. N. Am.* 16, 453–475, 2002.

56. Chastre, J., and Trouillet, J.L., Early infective endocarditis on prosthetic valves. *Eur. Heart J.* 16(Suppl. B), 32–38, 1995.

57. Varheul, H., Renee, B., Vanden Brink, A., et al., Effect of changes in management of active infective endocarditis on outcome in a 25-years period. *Am. J. Cardiol.* 72, 682–687, 1993.

58. Aranki, S.F., Adams, D.H., Rizzo, R.J., et al., Determinants of early mortality and late survival in mitral valve endocarditis. *Circulation* 92(SII), 143–149, 1995.

59. Vikram, H.R., Buenconsejo, J., Hasbun, R., and Quagliarello, V.J., Impact of valve surgery on 6-month mortality in adults with complicated left-sided endocarditis. A propensity analysis. *JAMA* 290, 3207–3214, 2003.

60. Durak, D.T., Evaluating and optimizing outcomes of surgery for endocarditis. *JAMA* 290, 3250–3251.

61. Mills, J., Utley, J., and Abbott, J., Heart failure in infective endocarditis: predisposing factors, course and treatment. *Chest* 66, 151–157, 1974.

62. Steckelberg, J.M., Murphy, J.G., Ballard, D., et al., Emboli in infective endocarditis: the prognostic value of echocardiography. *Ann. Intern. Med.* 114, 635–640, 1991.

63. Mugge, A., Daniel, W.G., Frank, G., and Litchtlen, P.R., Echocardiography in infective endocarditis: reassessment of prognostic implications of vegetation size determined by transthoracic or transesophageal approach. *J. Am. Coll. Cardiol.* 14, 631–638, 1989.

64. Cabell, C.H., and Fowler, V.G. Jr., Vegetations in endocarditis: big is bad, but is there more to it? *Am. Heart J.* 146, 189–190, 2003.

65. Jihad, B., Lebovici, L., Gartman-Israel D., et al., Long-term outcome of infective endocarditis: the impact of the surgical intervention. *Clin. Infect. Dis.* 33, 1636–1643, 2001.

66. Chamis, A.L., Peterson, G.E., Cabell, C.H., et al., *Staphylococcus aureus* bacteremia in patients with permanent pacemakers or implantable cardioverter-defibrillators. *Circulation* 104, 1029–1033, 2001.

67. Chua, J.D., Wilkoff, B.L., Lee, I., et al., Diagnosis and management of infections involving implantable electrophysiologic devices. *Ann. Intern. Med.* 133, 604–608, 2000.

68. Wilkoff, B.L., Byrd, C.L., Love, C.J., et al., Pacemaker lead extraction with laser sheat: results of the pacing lead extraction with the excimer sheath (PLEXES) trial. *J. Am. Coll. Cardiol.* 33, 1671–1676, 1999.

24 Treatment Protocol of Infections of Orthopedic Devices

Vera Antonios, Elie Berbari, and Douglas Osmon

CONTENTS

24.1 INTRODUCTION

Prosthetic devices have become a cornerstone in many orthopedic surgeries. Their use has dramatically impacted patients' quality of life, providing symptom relief, restoration of a limb or a joint function, improved mobility and independence. Orthopedic device infections, although uncommon, remains one of the most devastating complications, and may lead to significant morbidity. This event often implies the need for subsequent surgeries, a prolonged course of antimicrobial therapy, functional limitation, amputation in some instances, and occasionally death. It constitutes a heavy burden not only to the patient, but also to the health care system in the United States. Incidence rates range from 0.5% to more than 10%, and vary with the type of procedure and the availability of dedicated orthopedic services (1). Management of these infections increase costs of surgery several fold compared to the uncomplicated procedure (2).

This chapter will address the management of infections involving prosthetic joints, fracture fixation devices and vertebral devices. Despite significant progress in this field, and multiplicity of published papers and reports, many questions pertaining to the diagnosis and management of these infections remain unanswered, and the physician is often confronted with challenging decisions: first deciding if surgery is required, and selecting the most appropriate surgical option; then choosing the most effective antibiotic, and finally the correct duration of treatment and follow-up. An essential component of this therapeutic approach is the strong collaboration between surgical and medical caregivers, which should guide all decisions in achieving optimal results.

24.2 PROSTHETIC JOINT INFECTIONS

Over the last 50 years, the use of prophylactic antimicrobial therapy, along with the improvement in aseptic measures, surgical techniques and operating rooms have significantly reduced the risk of prosthetic joint infections (PJI). The incidence rate is currently reported to be less than 2.5% after primary total hip arthroplasty (THA) or total knee arthroplasty (TKA), but increases to 5.6% after revision surgeries. The rate is highest in the first two postoperative years (5.9/1,000 joint-years) and continues to decline afterwards (2.3/1,000 joint-years) (3). Mortality rates range between 1% and 18%. The estimated cost of treating an episode of PJI is >$50,000 (3).

Identifying patients at high risk for developing PJI enabled clinicians to provide an intensified and focused care for susceptible patients. Berbari et al. (4) conducted a case-control study to define risk factors for PJI. Development of a postoperative surgical site infection, National Nosocomial Infection Surveillance (NNIS) score of >2, history of systemic malignancy, and history of prior total joint arthroplasty were significant predictors of infection. NNIS score, as defined by the Centers for Disease Control and Prevention, takes into account the duration of the procedure, the American Society of Anesthesiologists' preoperative assessment score, and the surgical wound classification. No increased risk was associated with either rheumatoid arthritis or diabetes mellitus in the multivariate analysis (4). These results allow clinicians to optimize prevention efforts in high-risk patients, increase their index of suspicion for infection, and provide adequate and prolonged follow-up.

Understanding the pathogenesis of PJI helps to clarify some aspects of the therapeutic approach. Colonization of the prosthesis by microorganisms at the time of implantation is believed to be the most common mechanism of infection.

Strong supportive evidence to this hypothesis is the significant decline in PJI rates that accompanied antimicrobial prophylaxis (5). Infection can also occur via hematogenous seeding from a distant site following a bacteremia, or through direct spreading from a contiguous source. However, these mechanisms are thought to be rare and account for a small number of PJI cases.

Biofilm formation is the hallmark characteristic of PJI. Bacterial adherence to implanted biomaterial results from the production of extracapsular glycocalyx (slime) (6–8). This provides a mechanical barrier that allows bacteria to withstand host defense mechanisms and to evade antimicrobial therapy. Gristina et al. (9) showed bacterial growth in glycocalyx-enclosed biofilms in 59% of orthopedic biomaterial-related infections. Using electron microscopy, they were able to identify organisms not previously recovered by routine culture methods (9). Using routine culturing techniques may not detect slime-producing microorganisms. Biofilm formation accounts for the difficulty in eradicating PJI or other device-related infection with antimicrobial therapy alone (10). There is an obvious need for improved diagnostic methods that could provide the answer to recurrent and culture-negative PJI. Soaking the biomaterial without any prior treatment appears to have the lowest yield in the detection of microorganisms (14). Vortexing prior to sonication appears to be superior to simple sonication and scraping for bacterial removal. Cultures following biomaterial sonication and various isolation techniques are currently being evaluated (11–13).

Gram-positive cocci account for >60% of PJI cases. *Staphylococcus aureus* and coagulase-negative staphylococci are the most commonly reported organisms in both early (within 2 years following surgery) and late (beyond 2 years) infections, in both THA and TKA. Two or more species of microorganisms are reported in 14 to 19% of PJI cases. Aerobic Gram-negative bacilli and anaerobic organisms are less frequently encountered. The organism remains unknown in up to 10% of cases. Unusual microorganisms such as *Candida* species, *Brucella*, and various mycobacteria have been reported. The type of infecting organism does affect the therapeutic approach (3).

In order to define prognostic parameters and to establish treatment guidelines, different staging systems for PJI have been proposed (15,16). The classification by McPherson et al. (16) is based on three major variables: type of infection, patient's immune and medical status, and local tissue factors. This staging system was evaluated in infected TKA and THA cases. Results were promising in some reports and inconclusive in others, reflecting the small number of cases (15–17). Other classifications are being used in different institutions, and take into account clinical presentation as proposed by Tsukayama et al. (18,19), or local and systemic host factors as proposed by Cierny et al. (20). A multicenter collaborative study is needed to make the necessary modifications to establish an optimized staging system.

24.2.1 MANAGEMENT STRATEGIES

Therapeutic approach to PJI should take into account several factors, including symptoms duration, joint age, infecting pathogen and its susceptibility pattern, prosthesis stability, patient's immune and medical conditions, and soft tissue status. Early and acute infections are less likely to be associated with a biofilm, which usually is a slow and long process (21). Therefore, the chance of cure without prosthesis removal is higher in these cases. The type of microorganism has affected the outcome of each surgical modality differently. Choosing an aggressive versus a conservative approach is also guided by the patient's ability to tolerate surgical procedures. The ultimate goal of therapy is to have adequate mobility through a functional and pain-free joint. Eradicating the infection may not be achievable in all cases, and therefore suppressive antimicrobial therapy may become an alternative that provides the least morbidity in some patients (22).

24.2.1.1   Suppressive Antimicrobial Therapy

Antimicrobial therapy without concomitant surgical intervention may be an acceptable alternative in frail patients with a severe medical condition or terminal illness (22). The goal of such approach is to provide symptomatic relief and to maintain a functioning joint. When this strategy is used, the success rate is reported as 10 to 25% (23–25). In selected cases and when initial surgical debridement with prosthesis retention was possible, results were more encouraging (26). This modality should not be considered the standard of care, and should only be used when the following criteria are met: (a) curative surgery is not feasible, (b) the prosthesis is well fixed and stable, (c) the pathogen is identified, of low virulence and amenable to suppression with oral antimicrobials, (d) no evidence of systemic infection, and (e) the patient is compliant and able to tolerate oral antimicrobial regimen (27). Clinicians should be aware of the possible long-term toxicity of antimicrobials (see Section 24.2.2), and the risk of developing resistant microorganisms in future relapses.

24.2.1.2   Debridement with Retention

This option offers the advantage of a single surgery that preserves both prosthesis and bone stock (22). On the other hand, it carries the risk of leaving an infected foreign body in place. It has been suggested that this therapeutic modality could be applied in patients with early postoperative and hematogenous infections, short duration of symptoms (28,29), stable prosthesis and in the absence of sinus tracts. Zimmerli et al. (30) reported favorable results using this method for patients with staphylococcal orthopedic implant-related infections (TKA, THA and fracture fixation devices) who received rifampin-based antimicrobial regimen. Of importance, these patients had a short duration of symptoms and stable implants (30). Other studies reported a success rate of 50% and 70% for TKA and THA respectively. These were early postoperative (within 1 month after implantation) or acute hematogenous infection. In addition to surgical debridement, antibiotic-impregnated cement spacer and beads in TKA infections, and removal of the polyethylene insert in both THA and TKA cases were done in most reported cases (18,19,31). *S. aureus* PJI treated with debridement and retention at our institution resulted in treatment failure in more than 60% of cases. A higher probability of failure was reported when debridement was performed after 2 days from onset of symptoms (32). Similar findings were noted when considering all cases of infected THA (33). Results were, however, beneficial in patients with penicillin-susceptible streptococcal PJI, presenting within 1 month after implantation, with a well-fixed prosthesis and a short duration of symptoms (34). In one report by Marculescu et al. on 99 PJI episodes treated with this modality, factors associated with treatment failure included *S. aureus* infections, presence of a sinus tract, and duration of symptoms of >7 days (35). Loosening and age of the prosthesis and rheumatoid arthritis did not seem to affect the outcome. In summary, there seems to be a convergence toward the necessity of careful patient selection for this surgical modality. Suggested criteria include: (a) short duration of symptoms prior to debridement, (b) absence of a sinus tract, (c) well fixed prosthesis, and (d) early infection. Some authors would even suggest that this modality could be used for treatment of *S. aureus* PJI (36). In early postoperative or acute hematogenous PJI cases, defining the onset of symptoms is possible, and therefore fulfilling the first criterion is more likely than in chronic infections. Although implant loosening was not a risk factor for infection recurrence, it may affect the functional outcome of a prosthesis. Therefore, stable prosthesis would still be preferable for this type of treatment. Failure of this method in late chronic infections is probably expected from their biofilm-based pathogenesis.

24.2.1.3 Resection Arthroplasty

Once considered the standard therapeutic modality, this procedure currently has limited indications. It involves removal of all infected components including prosthesis, cement, bone and soft tissues, with no subsequent implantation. It is usually followed by intravenous administration of antimicrobials for 4 to 6 weeks.

Using this modality, eradication of infection was achievable in 60 to 100% of THA cases in multiple reports. However, discrepant results were noted regarding patients' satisfaction and symptoms relief (25,37–40). In one study of TKA infections treated with resection arthroplasty, success rate of 89% was reported. Subsequent arthrodesis was performed in 21% of patients who were initially unsatisfied with the results (41).

Obviously, a major limitation of this procedure is the loss of joint function and adequate mobility. It may be an acceptable alternative in nonambulatory patients. Otherwise, indications should be limited to situations where reimplantation is not feasible (e.g., patients with major bone loss, recurrent infections, highly resistant organisms or medical condition precluding major surgery).

24.2.1.4 One-Stage Replacement

This approach involves excision of all prosthetic components, meticulous debridement of devitalized bone and soft tissues, and immediate implantation of a new prosthesis during the same surgery. Intravenous antimicrobial therapy is usually administered for a variable period of time. Potential advantages of this single exchange procedure result from saving patient and healthcare system an additional surgery, and include lower morbidity rate and lower cost.

Success rate for THA infections treated with one-stage replacement ranges from 80 to 90% in most reports (42–45). The benefit of using antibiotic-impregnated cement remains controversial. Suggested selection criteria for this procedure include: (a) a relatively healthy patient with adequate bone stock and soft tissues, (b) a low virulence, antibiotic-sensitive organism identified preoperatively, and (c) inability of patient to tolerate a second major procedure. Fulfilling these criteria is difficult in most situations, and most authorities favor two-stage exchange that allows adequate identification of the infecting microorganism (46). In a few studies, there was reported success in treating PJI associated with Gram-negative bacilli or with draining sinuses by this method (47,48). However, this practice is discouraged by most authorities. Appropriate antimicrobial therapy is essential in this situation where a new foreign material is placed before achieving "prolonged sterilization" of the joint. Intact soft tissues are considered a prerequisite for such procedure by most investigators. Draining sinuses may become the portal of entry that maintains infection and may increase the risk of failure.

Treating TKA infections with one stage placement yielded more controversial result. Reported success rates range between 25 and 100% with an average of 78% (49). Factors that seem to contribute to a successful outcome include intact soft tissues, use of antibiotic-impregnated cement, Gram-positive organisms, and prolonged antimicrobial therapy (12 weeks) (50). Comparative studies of single vs. two-stage exchange are inconclusive, because of the small number of cases in these reports. Most authorities advocate two-stage exchange for treatment of TKA infections. This approach allows multiple debridements and confirmation of sterilization prior to prosthesis reimplantation.

24.2.1.5 Two-Stage Replacement

This is the procedure most often used in the United States for PJI treatment (46). It involves an initial removal of prosthetic components, debridement of infected tissue, followed by a delayed stage where a second prosthesis is reimplanted. The time interval between the two surgical procedures is variable. Antimicrobials are administered following resection for a 4 to 6 weeks period (Table 24.1). Confirmation of a successful joint sterilization is usually required prior to reimplantation. Therefore, multiple debridements may be needed. Patients should have adequate bone stock, be medically fit, and willing to undergo at least two surgeries. Patients with sinus tracts or with virulent organisms qualify for this procedure as well. Brandt et al. (51) reported success rate of over 97% when this modality was used for treatment of *S. aureus* PJI. None of the patients had evidence of infection at time of reimplantation, based on histopathologic and microbiologic criteria.

In chronic THA infections, two-stage exchange is considered standard treatment by most authorities. Success rate has been reported to range from 68 to 100% (51–53). Antimicrobials are usually given for 4 to 6 weeks. The benefit of antibiotic-impregnated cement fixation remains controversial. Local antibiotics have been traditionally considered part of the treatment, but recent studies on cementless reconstruction report a success rate of 92% (52,53). Cementless technique offers the advantages of preserving bone stock and avoiding the use of foreign material that may have a deleterious effect on the immune system. Long-term effects of both techniques are to be determined.

Most authorities consider two-stage exchange the treatment of choice for TKA infections. Early implantation (within 3 weeks after excision) resulted in unsatisfactory results (54), whereas delayed implantation (>4 weeks after excision) was associated with a success rate averaging 90% (31,55–57). Antimicrobial therapy is usually

TABLE 24.1
Antimicrobial Treatment of Common Microorganisms Causing PJI

Microorganism	First choice[a]	Alternative[a]	Comments
Staphylococcus spp., coagulase-negative, oxacillin-susceptible	Nafcillin sodium 1.5–2 g IV q4 hours or Cefazolin 1 to 2 g IV q8 hours	Vancomycin IV 15 mg/kg q12 hours or Levofloxacin 500 to 750 mg PO or IV q24 hours + Rifampin 300–450 mg PO q12 hours	4 to 6 weeks. Vancomycin only in case of allergy. Levofloxacin/Rifampin combination for patients treated with debridement and retention; duration could be extended up to 6 months.[b]
Staphylococcus spp., coagulase-negative, oxacillin-resistant	Vancomycin 15 mg/kg IV q12 hours	Linezolid 600 mg PO or IV q12 hours or Levofloxacin 500 to 750 mg PO or IV q24 hours + Rifampin 300–450 mg PO q12 hours	4 to 6 weeks. Levofloxacin/Rifampin combination for patients treated with debridement and retention; duration could be extended up to 6 months.[b]
Staphylococcus aureus, Oxacillin-susceptible	Nafcillin sodium 1.5-2 g IV q4 hours or Cefazolin 1 g IV q8 hours	Vancomycin 15 mg/kg IV q24 hours or Levofloxacin 500 to 750 mg PO or IV q24 hours + Rifampin 300–450 mg PO q12 hours	4-6 weeks. Vancomycin only in case of allergy. Levofloxacin/Rifampin combination for patients treated with debridement and retention; duration could be extended up to 6 months.[b]
Staphylococcus aureus, Oxacillin-resistant	Vancomycin 15 mg/kg IV q12 hours	Linezolid 600 mg PO or IV q12 hours or Levofloxacin 500-750 mg PO or IV q24 hours + Rifampin 300–450 mg PO q12 hours	4-6 weeks. Levofloxacin/Rifampin combination for patients treated with debridement and retention; duration could be extended up to 6 months.[b]
Enterococcus spp., penicillin-susceptible	Penicillin G 20 million units IV q24 hours continuously or in six divided doses or Ampicillin sodium 12 g IV q24 hours continuously or in six divided doses	Vancomycin 15 mg/kg IV q12 hours	4 to 6 weeks. Aminoglycoside optional. Vancomycin only in case of allergy.

TABLE 24.1—Cont'd
Antimicrobial Treatment of Common Microorganisms Causing PJI

Microorganism	First choice[a]	Alternative[a]	Comments
Enterococcus spp., penicillin-resistant	Vancomycin 15 mg/kg IV q12 hours	Linezolid 600 mg PO or IV q12 hours	4 to 6 weeks. Aminoglycoside optional
Pseudomonas aeruginosa	Cefepime 1–2 g IV q12 hours or Meropenem 1 g IV q8 hours	Ciprofloxacin 750 mg PO or 400 mg IV q12 hours or Ceftazidime 1–2g IV q8 hours	4 to 6 weeks. Aminoglycoside optional. Double coverage optional.
Enterobacter spp.	Cefepime 1 g IV q12 hours or Meropenem 1 g IV q8 hours	Ciprofloxacin 750 mg PO or 400 mg IV q12 hours	4 to 6 weeks.
β-Hemolytic streptococci	Penicillin G 20 million units IV q24 hours continuously or in 6 divided doses or Ceftriaxone 1–2 g IV q24 hours	Vancomycin 15 mg/kg IV q12 hours	4 to 6 weeks. Vancomycin only in case of allergy.
Propionibacterium acnes	Penicillin G 20 million units IV q24 hours continuously or in six divided doses or Ceftriaxone 1–2 g IV q24 hours	Clindamycin 600 to 900 mg IV q8 hours or Vancomycin 15 mg/kg IV q12 hours	4 to 6 weeks. Vancomycin only in case of allergy.

[a] Listed antimicrobial dosage are for patients with normal renal and hepatic function and need to be adjusted based on patients' creatinine clearance and hepatic function.
[b] In patients undergoing debridement and retention, Levofloxacin/Rifampin duration is 3 months for THA infection and fracture fixation device infection or 6 months for TKA infection.

given for 4 to 6 weeks. McPherson et al. (55) recommend a period of observation off therapy, during which patient is evaluated for residual infection by clinical and laboratory parameters (sedimentation rate, C reactive protein, and knee aspirate). In case of suspected infection, another debridement is performed (55). In one cohort study, prerevision aspirate cultures have been shown to improve clinical outcome (58). The use of antibiotic-impregnated cement and spacers has not been evaluated in randomized controlled trials. Data from retrospective reports showed benefit in some reports (59) and was inconclusive in others (60–62). Currently, no report has shown any harm related to the use of antibiotic-impregnated cement or spacers. Moreover, the FDA has recently approved two premixed antibiotic-impregnated bone cements for use in the reimplantation stage of a two-stage revision (see Section 24.2.3). It is recommended to choose antimicrobials that are active against the infecting pathogen and have a low risk of hypersensitivity.

24.2.1.6 Arthrodesis

Arthrodesis is the treatment of choice in TKA infections, when subsequent joint reimplantation is not feasible because of poor bone stock (e.g., recurrent PJI) (63–65). Bone loss plays a critical role in predicting outcome of this procedure (66). Adequate bone apposition and rigid fixation are essential to achieve successful bony fusion. External fixation has been used in active TKA infection, with variable success rate. Fusion was reported in 93 to 100% of cases when the Ilizarov technique was used. Unfortunately, external fixators are usually cumbersome and may result in a number of complications (pin tract infection, bone fracture at a pin site, etc.). Intramedullary nails result in fusion rate of 80 to 100% (67). Better outcome has been reported when arthrodesis was performed as part of a two-stage procedure (65,68). Lower fusion rates were associated with Gram-negative or mixed infections. Complications from internal fixation include nail migration or breakage, distal tibial fractures, and related events.

24.2.1.7 Amputation

Amputation may be required in selected cases such as the presence of a life-threatening infection, recurrent and uncontrollable PJI where other therapeutic alternatives have failed, intractable pain, and severe bone loss precluding other surgical procedures. In most cases, amputation is used after multiple revision arthroplasties have failed and arthrodesis is not technically feasible. Therefore, it may be reasonable, in selected cases, to consider arthrodesis early on, rather than multiple revision attempts that could exhaust the bone stock (69,70).

24.2.2 ANTIMICROBIALS

Appropriate antimicrobial therapy is an essential part of PJI treatment. The role of the infectious diseases specialist is to provide optimal choice of antimicrobials, adequate treatment duration and focused follow-up. Clinicians should be aware of the importance of establishing a microbiological diagnosis prior to administering empiric antimicrobial therapy. Unless PJI is presenting as overwhelming sepsis, antimicrobials should be held until intraoperative or aspiration cultures are obtained.

Initial therapy should cover the most common pathogens, i.e., staphylococcal species. A first-generation cephalosporin (Cefazolin) is an adequate first-line empirical treatment in the majority of patients. In patients with penicillin allergy, this can be substituted with vancomycin. Results of intraoperative cultures and *in vitro* antimicrobial susceptibility testing should guide further adjustment in the antibiotic regimen. Table 24.1 summarizes suggested therapeutic options for each pathogen.

Vancomycin use should be restricted to methicillin-resistant staphylococcal infections, and to penicillin allergic patients. Quinolones provide an oral alternative because of the excellent bioavailability of these drugs. Rifampin-based regimens have been widely used in Europe. In vitro studies and animal models of device-related infections showed efficacy of rifampin on adherent and stationary-phase microorganisms (71,72). Rifampin monotherapy often leads to the emergence of resistance. The ability of both rifampin and quinolones to achieve excellent concentrations in bone and soft tissues makes this combination an attractive regimen for PJI treatment (73). Clinical studies showed a success rate of 82% to 100% when rifampin-quinolone combination was used in selected cases of staphylococcal orthopedic implant-related infections treated with retention and debridement (30,74). Because of the limitations of these studies and the emergence of quinolone-resistant staphylococcal species, this practice has not been widely accepted yet in the United States. A French study showed promising results with the use of rifampin–fusidic acid combination (75).

Linezolid has an excellent bioavailability. Razonable et al. (76) reported 20 patients with orthopedic infections treated with linezolid. Success rate was 90%, and side effects were reversible myelosuppression in 40% and irreversible peripheral neuropathy in 5% (76). Daptomycin is a new antimicrobial, the first in the class of lipopeptides, that was recently approved for the treatment of complicated skin and soft tissue infections. Daptomycin role in the treatment of orthopedic infections has been limited to animal studies and a few case reports (77,78). Its good activity against resistant Gram-positive cocci makes it a promising antimicrobial in the setting of MRSA or VRE orthopedic infections, or when allergy or side effects precludes the use of vancomycin and linezolid. There is a clear need for large randomized controlled multicenter trials of different antimicrobial regimens, including the newer quinolones. The optimal duration of treatment varies with the surgical procedure and the pathogen. Most authorities favor a minimum of 4 weeks.

Treating clinicians should be familiar with the Infectious Diseases Society of America guidelines pertaining to the use of community-based parenteral antiinfective therapy (CoPAT) (79). Table 24.2 summarizes the recommendations of the laboratory parameters that should be monitored when using CoPAT. Side effects of selected antimicrobials commonly used for chronic suppression are listed in Table 24.3.

24.2.3 ANTIBIOTIC-IMPREGNATED DEVICES

Since the 1970s, local antibiotic-impregnated devices have been commonly used in the treatment of PJI and other orthopedic devices infection. In a survey by Heck et al. (80), over 80% of orthopedic surgeons in the US reported use of antibiotic-impregnated cement more than two-thirds of the time in septic hip and knee revision arthroplasty. There was significant variability in the type of cement and antimicrobial used, reflecting the lack of standardization in this practice (80).

TABLE 24.2

Laboratory Parameters to be Monitored on a Weekly Basis During Community-Based Parenteral Anti-Infective Therapy. (Adapted from *Clin. Infect. Dis.* with permission)

Antimicrobial	Complete blood count (CBC)	Creatinine level	Potassium level	Magnesium level	Other
Penicillins	1	1			LFTs[a] with nafcillin and oxacillin. Potassium level with Ticarcillin.
Cephalosporins	1	1			LFTs with ceftriaxone.
Carbapenems	1	1			
Aminoglycosides	1	2			Serum drug levels. Consider audiogram.
Vancomycin	1	2			Serum drug levels.
Linezolid	1	1			CBC twice weekly for patients at risk[b].
Daptomycin	1	1			CPK[c].
Clindamycin	1	1			
Trimethoprim-sulfamethoxazole	1	1	1		
Fluconazole	1	1			LFTs.
Itraconazole	1	1			LFTs. Serum drug levels when capsule form is used.
Voriconazole	1	1			LFTs.
Amphotericin B	1	2	2	2	

[a] LFTs: Liver function tests.

[b] Patients at risk: preexisting myelosuppression, concomitant use of other myelosuppressive drugs.

[c] CPK: Creatine phosphokinase.

Antibiotic-impregnated cement is either premixed or mixed by the surgeon in the operating room (81). Two premixed antibiotic-impregnated bone cements have been recently approved by the FDA for use in the second stage of a two-stage revision in total joint arthroplasty, one using tobramycin and the other gentamicin (82,83). This will help in standardizing the dose of antimicrobial used for prosthesis implantation.

Aminoglycosides are the most commonly used antimicrobials in antibiotic-impregnated devices (mainly tobramycin), followed by vancomycin. Penicillin and cephalosporins are generally avoided because of their potential allergenicity and problems with stability. The use of quinolones is currently experimental. A limiting factor may be the potential interference with bone and soft tissue healing that was

TABLE 24.3
Long-Term Side Effects of Common Antimicrobials Used for Chronic Suppression

Antimicrobial	Long-term side effects/Cautions
Minocycline	Phototoxicity. Discoloration of skin, sclerae, teeth, and bone. Drug-induced lupus. Dizziness. Pseudotumor cerebri (rarely). Should not be taken with antacids. May decrease contraceptive efficacy.
Trimethoprim-sulfamethoxazole	Myelosuppression. Elevated creatinine without nephrotoxicity. Nephrotoxicity/crystalluria. Possible disulfiram-like reactions. May enhance hypoglycemic effects of sulfonylureas.
Rifampin	Orange discoloration of urine, tears, sweat, saliva. Elevated LFTs. Should not be used as monotherapy. Significant drug–drug interactions.
Linezolid	Myelosuppression. Peripheral/optic neuropathy. Avoid tyramine-containing food.
Levofloxacin	Arrhythmias (very rarely). Tendon rupture. Alteration in blood glucose. Should not be taken with antacids.
Metronidazole	Peripheral neuropathy. Ataxia. Disulfiram-like reaction.
Fluconazole	Hepatotoxicity. Drug–drug interactions.

observed in recent animal studies (84,85). In vitro studies showed that fluconazole and amphotericin B retain their efficacy when mixed with cement (86).

Antibiotic-impregnated devices offer the advantage of achieving high concentrations of antibiotic locally (87), and therefore potentially reducing the risk of recurrence. In addition, spacers reduce dead space, providing joint stability while awaiting reimplantation in a two-stage procedure (88,89). Emerson et al. reported a better range of motion with articulating spacers but found no difference in reinfection rate when compared to static spacers (90). The PROSTALAC (prosthesis of antibiotic-loaded acrylic cement) is a temporary hip prosthesis used in two-stage exchange, in which both acetabulum and femoral head articulate through antibiotic-impregnated cement. It can provide early mobilization and shorter hospitalization. In one report, 94% of patients were infection-free 2 years after reconstruction (91). Clinical trials failed to show conclusive results on the benefit of antibiotic-impregnated devices in one-stage (as cement) and two-stage (as spacer or beads) exchange. High doses of antibiotic may affect the mechanical properties of bone cement when used for implant fixation. However, it may still be used in spacers or beads (81). Systemic toxicity was addressed in a recent report by Springer et al. (92). There was no evidence of renal insufficiency in any of the 34 TKA infections treated with high dose vancomycin and gentamycin antibiotic spacers (92). Antibiotics used should be chosen to cover the infecting organism.

24.2.4   SPECIAL SITUATIONS

24.2.4.1   Culture-Negative PJI

Prior antimicrobial exposure probably plays a major role in most culture-negative cases (Table 24.4). As mentioned earlier, antibiotics should be held until appropriate

---

TABLE 24.4
Common Causes for Culture-Negative PJI

Culture-negative PJI

1. Prior use of antimicrobials
2. Not optimal culture specimens
3. Local antibiotic release
4. Slime-producing microorganisms
5. Fastidious microorganisms (anaerobes, small-colony mutants of *S. aureus*, etc.)
6. Non-infectious etiologies (RA, SLE, etc.)

---

cultures are taken. The minimum duration for which a patient should be off antimicrobials prior to surgery is unknown. A recent report suggested discontinuation of antimicrobial therapy >2 weeks prior to surgery in order to improve culture sensitivity (93). Therapy should be directed against pathogens isolated from previous intraoperative or joint aspirate cultures if available. Otherwise, it should cover the most common culprits, i.e., Gram-positive cocci. First-generation cephalosporin is an adequate empirical treatment in the majority of these cases.

Local antibiotic release from fractured cement during surgery has been shown to inhibit growth of microorganisms in culture specimens (94). Therefore, it is recommended to obtain samples early in the procedure, before the cement is disturbed if possible. In general, multiple culture specimens are recommended for better yield and accuracy.

Slime-producing microorganisms may not grow in routine cultures. Ultrasonication, followed by immunofluorescence, PCR amplification of bacterial 16S rRNA or electron microscopy, may be a better diagnostic tool in identifying these bacteria (11–13). In the right context, uncommon pathogens that need specific growth media should be suspected, e.g., anaerobes, fungi, mycobacteria, and fastidious bacteria. Clinical history, intraoperative findings and histopathologic features may provide suggestive diagnostic clues. These patients often present with recurrent culture-negative PJI, unresponsive to commonly used antimicrobials. When pathology reveals granulomatous inflammation, fungal and mycobacterial cultures should be done. All available histopathology specimens should have an auramine-rhodamine and Giemsa stains.

Small-colony mutants of *S. aureus* are slow-growing variants that could easily be missed on routine solid media cultures (95). They have atypical characteristics probably likely related to defective electron transport, but they remain as infective as their parent strain. Although their role in PJI is not well described, they have been associated with persistent and relapsing infections in patients with chronic osteomyelitis (96,97). They may be induced by aminoglycosides and quinolones. Resistance pattern may be unusual and susceptibility data may be misleading if these mutants were missed in cultures. They are resistant to aminoglycosides and may be resistant to trimethoprim-sulfamethoxazole (98). Stepwise resistance to quinolones has been shown in-vitro (99). The optimal diagnostic methods and treatment of this infection are not yet defined. Routine cultures from PJI specimens are usually followed for 5 to 7 days. Longer incubation time may be needed to identify these organisms.

Finally, non-infectious etiologies could mimic culture-negative PJI, such as the inflammatory arthritidis secondary to rheumatoid arthritis (RA) and systemic lupus erythematosus (SLE).

24.2.4.2 Positive Cultures and/or Pathology at Reimplantation

Two-stage exchange procedures imply placing a new prosthesis in a sterile milieu. This is also the case in revision arthroplasty for presumed aseptic loosening. However, declaring the joint space infection-free is not always straightforward. In the operating room, two parameters guide the surgeon's decision in whether to proceed with reimplantation or not: intraoperative findings and frozen section analysis. Gram stain is not a sensitive diagnostic tool (sensitivity of 14.7% in one study) (100). Cultures should be awaited for further management decisions. When microbiology and pathology provide contradictory results, the situation becomes even more complicated.

Several studies compared these two diagnostic methods. Variability in defining the gold standard diagnostic tool makes it difficult to interpret these results. Criteria for infection in pathology specimens also varied between reports (101,102). Lonner et al. (103) showed that using an index of 10, rather than 5 polymorphonuclear leukocytes per high-power field to define infection, increased positive predictive value of frozen section (gold standard being culture results). It did not however affect sensitivity or specificity (103). Most authors agree that frozen section analysis has an excellent negative predictive value (97 to 100%) and a good specificity (89 to 96%) (103–106). Data on sensitivity is more controversial. Banit et al. (107) showed an acceptable positive predictive value and sensitivity in TKA but not in THA revision. It can be concluded that, in the absence of infection on frozen section, reimplantation is considered safe if surgical findings are otherwise normal. When cultures come back unexpectedly positive, the main question is whether this represents contamination or infection. Atkins et al. (108) found that three or more positive cultures yielding the same organism is a definite diagnosis of infection. Most authors use a minimum of two positive cultures as diagnostic, and recommend multiple specimens to be sent. Management of pathology-negative, culture-positive cases is not well defined. According to Tsukayama et al. (18,19) these cases belong to a separate category (class I infection) and may be treated without surgical intervention. Six weeks of antibiotics have been recommended. Success rate of 90% was reported in THA (31 patients) (19), and 100% in TKA cases (5 patients) (31). Marculescu et al. (109) reported 16 cases of unsuspected PJI diagnosed by positive cultures at revision arthroplasty. Pathogens were of low virulence. Outcome was excellent regardless of the treatment modality used (chronic suppression versus short course of antimicrobials). The five-year cumulative probability of success was 89% (109).

When frozen section analysis shows evidence of inflammation in the presence of normal intraoperative findings, the surgeon may choose to delay reimplantation or proceed with caution. In case of delayed replacement, further antimicrobial course may be warranted. The surgeon's clinical judgment may become his only guide in these difficult situations. Inflammatory reaction was thought to be related to the presence of foreign material (beads, cement, etc.). However, a recent report in our institution addressed the significance of acute inflammation in joint tissue at reimplantation arthroplasty. In the 23 cases that were analyzed, the presence of acute inflammation appeared to decrease the probability of success. Furthermore, chronic suppression seemed to carry a better outcome. The two-year cumulative probability of success with and without chronic suppression was 100% and 60% respectively (110). Although this was a small study, the results suggest that a positive pathology carries a realistic risk of failure.

24.2.4.3   Recurrent PJI

Therapeutic approach to reinfection differs in some aspects from primary PJI treatment. Choosing the best salvage option can be difficult. Risks of losing bone stock and soft tissues may outweigh the benefits of another revision arthroplasty. Few reports in the literature discuss management and outcome of recurrent PJI. Bengston et al. (111) reported nine recurrent TKA infections, eight of which were treated with placement of a hinged knee design resulting in a poor outcome. The remaining one was successfully treated with another reimplantation (111). Hanssen et al. (112) described course and outcome of 24 TKA reinfection cases. An average of 3.7 procedures was performed on the affected knee. There were 10 patients with successful arthrodesis, five with suppressive antibiotics, four amputations, three arthrodesis nonunions, one resection arthroplasty, and one uninfected total knee prosthesis. Aspiration followed by antibiotic suppression failed in all four cases initially treated with this method (112). The authors concluded that this option should be used only when surgery is not feasible. They recommended arthrodesis with either external fixation device (for patients with more preserved bone stock) or long intramedullary nail (for patients with bone loss compromising >50% of the tibial and femoral cancellous bone surfaces). Repeated attempts at reimplantation should be avoided in order to prevent dismal outcome, e.g., amputation. Revision arthroplasty by itself is a definite risk factor for PJI. Success rate of two-stage reimplantation has been reported to be as low as 41% in patients with multiple previous knee operations (56,113).

24.2.4.4   Unusual Microorganisms

A wide variety of unusual microorganisms has been reported in the literature as a cause of PJI. It is not possible to outline treatment protocol for each pathogen, but we will summarize what is known about some of them.

In a recent report on pneumococcal PJI, 8/13 patients were cured with long-term antibiotics and drainage; four of them remained on indefinite suppression. Of the remaining patients, three died of pneumococcal sepsis, and two required two-stage exchange (114). Candidal PJI was successfully treated in 10 cases with delayed reimplantation arthroplasty (8.6 months in average for THA and 2.3 months for TKA) after appropriate antifungal therapy (115). Among seven cases of *Brucella* PJI reported, three underwent two-stage replacement, four received medical treatment alone, one of which had to undergo hip replacement. All patients had a favorable outcome. Doxycycline–rifampin combination for a minimum of 6 weeks has been recommended (116). Tuberculous PJI requires both medical and surgical treatment, although the best surgical option remains unknown (117). Prosthesis removal seems necessary in treatment of *Mycobacterium fortuitum* cases (118). Other examples of unusual PJI include *Listeria monocytogenes* (119), *Haemophilus parainfluenzae* (120), *Yersinia enterocolitica* (121), *Campylobacter fetus* (122), *Tropheryma whippeli* (123), *Pasteurella multocida* (124), and *Clostridium difficile* (125). This diversity of possible pathogens undermines the importance of an accurate microbiological diagnosis in PJI.

24.3   FRACTURE FIXATION DEVICES INFECTION

Recent technological advances have helped in creating and improving a wide variety of devices used in internal and external fixation of orthopedic fractures. Incidence rate of infection in fracture devices ranges from 0% up to 30% (126–130). This variability is attributed to several factors, including type of fracture, device, procedure, vascular supply, and soft tissue integrity. Open fractures carry a significant risk of infection that correlates with the severity of skin, soft tissue and vascular injury. There is a higher infection rate in open compared to closed fractures. Comminuted fractures have an infection rate of 10.3% compared to 2.1% in torsional tibial fractures treated with internal fixation (131).

Fixation with plates and screws implies soft tissue stripping and therefore may increase the risk of infection (127). Closed and open femoral shaft fractures treated with this method had an infection rate of 3% and 7% respectively (126,132). In one report on open tibial shaft fractures, open reduction internal fixation performed on the day of injury resulted in an infection rate of 20%. When surgery was delayed six days, the rate decreased to 6.6% (133). Another contradictory report showed increased infection rate with delayed intramedullary nailing with reaming for treatment of open tibial fractures (134). For closed tibial fractures, infection was reported in 0% of cases treated with percutaneous plating, i.e., using small incisions, and plating associated fibular fractures (135).

Intramedullary fixation seems to have a low infection risk. Locked intramedullary nailing is considered the treatment of choice for most femoral and tibial shaft fractures. Infection occurred in 0.9% of closed femoral fractures treated with closed nailing (136), compared to 13% with open reduction using simple nails and cerclage (137).

External fixators may result in pin site infection. Ring fixators have lower infection rate compared to hybrid and unilateral fixators (138). Infection has also been reported in femoral and tibial fractures treated with external fixation followed by nailing (139,140). Reaming is thought to disturb cortical blood flow and potentially increase the risk of infection. High infection rates have been reported in open tibial fractures treated with reamed nailing (141). Although unreamed nailing has received increased attention in both tibial and femoral fractures, comparative studies yielded controversial results in term of infection and union rates (142–146).

Most authorities believe that bone alignment, integrity of blood supply, adequacy of debridement and wound coverage are more significant predictors of infection than the device or procedure itself.

Pathogenesis involves biofilm formation (see pathogenesis of PJI). Bone healing plays an important role in this context. It is thought that achieving union helps controlling infection. Hence, maintaining bone stability and soft tissue viability is essential. Gram-positive cocci have been reported as the most common pathogens, namely *S. aureus* (147–151). Gram-negative bacilli are not infrequent. In one report, *Pseudomonas* was more common than *S. aureus* (152).

24.3.1 MANAGEMENT STRATEGIES

Infection of a fixation device is often associated with nonunion. Although foreign body removal is usually necessary to eradicate infection, device retention may be required for fracture stability and bone union. Furthermore, some authors believe that reamed nailing may reestablish medullary vascular continuity and promote healing. Two different approaches have been suggested: conventional and active. Conventional treatment includes an initial stage of device removal, tissue debridement, and fixation using external devices or intramedullary nailing. Soft tissue coverage is performed at a later stage, as well as bone grafting if needed. Active treatment implies debridement and retention of the device if required for fracture stability, as long as sepsis does not develop and drainage is controlled. Systemic antibiotics with or without antibiotic-impregnated devices are usually used for various periods of time.

Seligson and Klemm (153–155) suggested a therapeutic approach for infections complicating intramedullary nailing. In their protocol, stability at the fracture site is considered more important than the implant itself. With interlocking nails and solid fixation, device retention is advocated, along with drainage control and suppression with antibiotics (156). In the presence of sepsis, persistent drainage or loose simple nails, hardware removal is required, usually followed by installation of antibiotic-impregnated implant. External fixation may be used at a later stage, versus "secondary nailing" if infection is suppressed (153,154). Court-Brown et al. (149) proposed a similar protocol and used exchange nailing when persistent drainage was present. Patzakis et al. (157) recommended nail retention for femoral and early tibial (within 6 weeks of surgery) intramedullary nail infection. Late tibial infections should be treated by nail removal and fracture stabilization. This can be achieved by external fixation (for unstable fractures and poor soft tissues) or bone grafting (for stable fractures and adequate soft tissues) (148).

Klemm et al. (155) treated 64 cases of infected tibial and femoral pseudarthrosis with interlocking nailing, after removal of the primary device. Success rate was 89.5% in femoral and 62.5% in tibial cases (155). The authors recommended treating pseudarthrosis with external fixation and local implantation of antibiotic chains, and using intramedullary nailing when union is not achieved after these measures. Other investigators reported higher success rate when interlocking nailing was combined with open wound management. Healing rates reached 100% when using this method for infected nonunion of the tibia (156). It was emphasized however that this approach should be reserved for cases where other methods cannot be used, and when healthy granulation tissue can be stimulated. Shahcheraghi et al. (151,152) reported a success rate of 100% with bone grafting and intramedullary nailing of infected tibial nonunion, compared to 84% with bone grafting and compression plates. Patzakis et al. (157) used external fixation in 32 patients with infected tibial nonunion without substantial bone loss. There was no infection in any of these cases (157). Device removal and debridement followed by open bone grafting was successful in 100% of cases of infected tibial nonunion in one report (150).

Ilizarov technique uses coricotomy and the application of a circular external fixator in order to improve vascularity. It is based on the bone ability to generate itself when exposed to tensile stress. The device is designed in a manner to preserve the blood supply and stabilize the limb, in order to achieve distraction osteogenesis. It is beneficial when substantial bone loss or limb length discrepancy is present. It has been used in infected pseudarthrosis with 100% success rate in some reports (151). Ring et al. (158) compared external fixation using Ilizarov technique with bone grafting under adequate soft tissue coverage, for treatment of infected tibial nonunion. Failure was reported in 0/17 patients in the bone graft group, and 4/10 in the Ilizarov group (150).

Antibiotic-impregnated beads have been used after device removal. Technically, they preclude external fixator use, provide no mechanical support and are difficult to be removed after two weeks. In one study on intramedullary nailing infection, intraoperative custom-made antibiotic cement rod was used after nail removal (147). Advantages include dead space management, mechanical support, easy removal, and adequate local antibiotic concentration for long period of time. There was no recurrent infection in any of the nine cases treated with this method. Results are promising and more data will be needed for further recommendation.

Pin tract infection may complicate external fixation (159). There are no reports addressing management of this entity in particular. Pin removal and antibiotic administration are usually recommended (140). It is important to keep in mind that infection may involve the medullary canal at the pin's inner end. Therefore, subsequent nailing for fracture stabilization is discouraged. Intramedullary infections have been reported in this context.

Antimicrobial therapy of fracture fixation device infection follows the same principles cited in PJI treatment. However, an important note should be mentioned regarding quinolones use in this setting. Animal studies showed delayed fracture healing in presence of therapeutic serum concentrations of ciprofloxacin, during early stages of fracture repair (85). Levofloxacin and Trovafloxacin produced the same effect (84). There are currently no data on quinolones effect on bone healing in humans. Awaiting further studies, it may be reasonable to avoid quinolones use if fracture healing has not been completely achieved. In one animal study, systemic use of gentamicin and vancomycin did not seem to affect fracture healing (160).

24.4   SPINAL DEVICES INFECTION

Introduction of instrumentation in the operative management of various spinal pathologies (e.g., injury, deformity, and instability) has increased infection rates several fold. In the 1980s, initiation of preoperative antimicrobial prophylaxis has decreased the infection rate significantly (161). Nonetheless, spinal infection remains a dreaded complication where optimal management is still controversial.

The average cost of treatment of postoperative spine infection is estimated to be >$100,000 (162).

Incidence rate varies from <1% in adolescent idiopathic scoliosis to as high as 9.4% in spinal injury (163) and 9.7% in cerebral palsy and myelodysplasia (84,164). Posterior instrumentation has been associated with higher risk of infection in multiple reports (165). Various risk factors have been suggested based on observational studies, and include smoking, malnutrition, age, cognitive impairment, allograft use, bulky devices, and miscellanous others (166).

Pathogenesis is thought to be similar to other orthopedic device-related infections. Intraoperative contamination is the most common route, responsible for the majority of both early and late infections. Hematogenous seeding can also occur (167). Richards et al. (168) advocate extending incubation time of cultures to 7 to 14 days to improve cultures yield. When using this method, Clark et al. (169) had in fact shown significant improvement. Dubousset et al. (170) described 18 patients with Cotrel–Dubousset (CD) instrumentation who presented with symptoms suggestive of delayed infection. All patients had pathologic evidence of inflammation with granuloma formation, but cultures were positive in only two patients. The authors attributed the inflammatory response to micromotion between components and metal fretting corrosion (170). Whereas some investigators support this theory (171), others believe that delayed drainage is always the result of a bacterial infection. Granuloma formation is not typical of bacteria causing orthopedic hardware infections, and may be attributable to metal corrosion. However, eliminating an infectious process is essential, especially when indolent bacteria may use the irritated soft tissue as a growth medium. As stated before, routine cultures may not be sensitive in biofilm-based infections. On the other hand, extending incubation time may result in higher contamination rates. This is further complicated by the fact that true infection and contamination are caused by the same type of organisms. As with other orthopedic device related infections, we recommend sending multiple specimens, and following cultures for at least 5 days.

Staphylococcus spp. are the most common isolates (172,173). *Staphylococcus epidermidis* and *Propionibacterium acnes* are significant pathogens in late infections. Higher incidence of Gram-negative bacilli has been associated with bowel and bladder incontinence (174,175). In one case-controlled study, risk factors for methicillin-resistant staphylococcal species were lymphopenia, history of chronic infections, alcohol abuse, recent hospitalization, and prolonged postoperative wound drainage (172,176,177).

Most authors classify postoperative spinal infections as early or late (178). Thalgott et al. (177) proposed a clinical staging system based on Cierny's osteomyelitis classification. A treatment protocol was proposed for each group of patients. Further evaluation of this classification is needed to define risk categories and appropriate treatment guidelines.

24.4.1 MANAGEMENT STRATEGIES

Management of postoperative spinal infections is one of the most controversial topics in orthopedic infections. Prompt exploration of the wound is an essential initial step. It is difficult to define the depth of the infectious process on clinical grounds only. Once infection is suspected, the wound should be opened in the operating room. Aggressive debridement of all necrotic tissue and irrigation should be performed. Multiple cultures should be taken intraoperatively before the initiation of antimicrobials. Hardware removal may compromise the mechanical integrity of the spine and often is not done.

Most authorities differentiate between early, i.e., occurring within 3 to 12 months after surgery, and late infections (178). In the early postoperative period, instrumentation is necessary to maintain spine stability. Successful outcome has been reported when aggressive, repeated debridements were performed, while retaining hardware in place (164,172,173,175). A course of antibiotics that varies between 4 weeks and 6 months is usually given. Primary (179) vs. secondary (164,180) closure is dictated by the amount of necrosis and the surgeon's preference. Simple drains and inflow-outflow suction irrigation systems (165,177) have been used, as well as antibiotic-impregnated beads (181) and chains (182). There are no clinical trials comparing these different techniques.

In general, outcome is encouraging when using this approach in early-occurring infections. Bulky implants such as CD devices were associated with a higher failure rate (183). Hardware removal remains necessary in patients with persistent drainage despite multiple debridements or with loose instrumentation. Advocates of this approach emphasize that soft tissue appear affected during revision, but rarely the underlying bone. For this reason, a defined course of antibiotics is thought to be curative. However, Abbey et al. (184) recommend using suppressive antibiotics until bone grafts are solid. Two different options are then considered: either implant removal followed by a final course of antibiotics, or observation and follow-up for recurrence, after discontinuing antibiotics (184). Data is limited to support this conservative versus the "curative" approach. It would be difficult however to argue with the success reported in multiple studies with adequate follow-up, when the latter approach was used. Treatment should be at least continued until resolution of fever, and normalization of inflammatory parameters. At this point, we do not recommend prolonged antimicrobial therapy.

In delayed infections, hardware removal is usually necessary. At the time of diagnosis, bony fusion is already achieved. Even when pseudarthrosis is present, implant removal is considered part of the treatment. Debridement and primary closure of the wound are performed, followed by a course of culture-directed antibiotics. Pseudarthrosis is treated with immediate or delayed instrumentation. Most authors report excellent results when using this method for treatment of delayed infections (168,169,185,186).

REFERENCES

1. Allen, D.M., Orthopaedic implant infections: current management strategies. *Ann. Acad. Med.* Singapore 26, 687–690, 1997.
2. Jansen, B., and Peters, G., Foreign body associated infection. *J. Antimicrob. Chemother.* 32 Suppl A, 69–75, 1993.
3. Steckelberg, J., and Osmon, D.R., Prosthetic joint infections. In: Bisno, A.L., and Waldvogel, F.A., eds., *Infections Associated with Indwelling Medical Devices.* 3rd edition. Washington: American Society for Microbiology, 2000:173–209.

4. Berbari, E.F., Hanssen, A.D., Duffy, M.C., et al., Risk factors for prosthetic joint infection: case-control study. *Clin. Infect. Dis.* 27, 1247–1254, 1998.
5. Carlsson, A.K., Lidgren, L., and Lindberg, L., Prophylactic antibiotics against early and late deep infections after total hip replacements. *Acta. Orthop. Scand.* 48, 405–410, 1977.
6. Dougherty, S.H., Pathobiology of infection in prosthetic devices. [comment]. *Rev. Infect. Dis.* 10, 1102–1117, 1988.
7. Gristina, A.G., and Costerton, J.W., Bacterial adherence to biomaterials and tissue. The significance of its role in clinical sepsis. *J. Bone Joint Surg. Am.* 67, 264–273, 1985.
8. Costerton, J.W., Stewart, P.S., and Greenberg, E.P., Bacterial biofilms: a common cause of persistent infections. *Science* 284, 1318–1322, 1999.
9. Gristina, A.G., and Kolkin, J., Current concepts review. Total joint replacement and sepsis. *J. Bone Joint Surg. Am.* 65, 128–134, 1983.
10. Khardori, N., and Yassien, M., Biofilms in device-related infections. *J. Ind. Microbiol.* 15, 141–147, 1995.
11. Trampuz, A.P.K., Mandrekar, J., Greenleaf, J.F., Steckelberg, J.M., and Patel, R., Comparison of scraping, sonication and vortexing for removal of *Staphylococcus epidermidis* biofilms from polycarbonate coupons. American Society for Microbiology Conference on Biofilms, Nov. 1–6, 2003, Victoria, Canada,
12. Nguyen, L.L., Nelson, C.L., Saccente, M., Smeltzer, M.S., Wassell, D.L., and McLaren, S.G., Detecting bacterial colonization of implanted orthopaedic devices by ultrasonication. *Clin. Orthop.* 403, 29–37, 2002.
13. Tunney, M.M., Patrick, S., Curran, M.D., et al., Detection of prosthetic hip infection at revision arthroplasty by immunofluorescence microscopy and PCR amplification of the bacterial 16S rRNA gene. *J. Clin. Microbiol.* 37, 3281–3290, 1999.
14. Tunney, M.M., Patrick, S., Gorman, S.P., et al., Improved detection of infection in hip replacements. A currently underestimated problem. *J. Bone Joint Surg. Br.* 80, 568–572, 1998.
15. Hanssen, A.D., and Osmon, D.R., Evaluation of a staging system for infected hip arthroplasty. *Clin. Orthop.* 403, 16–22, 2002.
16. McPherson, E.J., Tontz, W., Jr., Patzakis, M., et al., Outcome of infected total knee utilizing a staging system for prosthetic joint infection. *Am. J. Orthop.* 28, 161–165, 1999.
17. McPherson, E.J., Woodson, C., Holtom, P., Roidis, N., Shufelt, C., and Patzakis, M., Periprosthetic total hip infection: outcomes using a staging system. *Clin. Orthop.* 403, 8–15, 2002.
18. Tsukayama, D.T., Goldberg, V.M., and Kyle, R., Diagnosis and management of infection after total knee arthroplasty. *J. Bone Joint Surg. Am.* 85-A Suppl 1, S75–S80, 2003.
19. Tsukayama, D.T., Estrada, R., and Gustilo, R.B., Infection after total hip arthroplasty. A study of the treatment of one hundred and six infections. *J. Bone Joint Surg. Am.* 78, 512–523, 1996.
20. Cierny, G., 3rd, and DiPasquale, D., Periprosthetic total joint infections: staging, treatment, and outcomes. *Clin. Orthop.* 23–28, 2002.
21. Trampuz, A., Osmon, D.R., Hanssen, A.D., Steckelberg, J.M., and Patel, R., Molecular and antibiofilm approaches to prosthetic joint infection. *Clin. Orthop.* 69–88, 2003.
22. Fisman, D.N., Reilly, D.T., Karchmer, A.W., and Goldie, S.J., Clinical effectiveness and cost-effectiveness of 2 management strategies for infected total hip arthroplasty in the elderly. [comment]. *Clin. Infect. Dis.* 32, 419–430, 2001.

23. Johnson, D.P., and Bannister, G.C., The outcome of infected arthroplasty of the knee. *J. Bone Joint Surg. Br.* 68, 289–291, 1986.

24. Tsukayama, D.T., Wicklund, B., and Gustilo, R.B., Suppressive antibiotic therapy in chronic prosthetic joint infections. *Orthopedics* 14, 841–844, 1991.

25. Canner, G.C., Steinberg, M.E., Heppenstall, R.B., and Balderston, R., The infected hip after total hip arthroplasty. *J. Bone Joint Surg. Am.* 66, 1393–1399, 1984.

26. Segreti, J., Nelson, J.A., and Trenholme, G.M., Prolonged suppressive antibiotic therapy for infected orthopedic prostheses [comment]. *Clin. Infect. Dis.* 27, 711–713, 1998.

27. Goulet, J.A., Pellicci, P.M., Brause, B.D., and Salvati, E.M., Prolonged suppression of infection in total hip arthroplasty. *J. Arthroplasty* 3, 109–116, 1988.

28. Tattevin, P., Cremieux, A.C., Pottier, P., Huten, D., and Carbon, C., Prosthetic joint infection: when can prosthesis salvage be considered? [comment]. *Clin. Infect. Dis.* 29, 292–295, 1999.

29. Mont, M.A., Waldman, B., Banerjee, C., Pacheco, I.H., and Hungerford, D.S., Multiple irrigation, debridement, and retention of components in infected total knee arthroplasty. *J. Arthroplasty* 12, 426–433, 1997.

30. Zimmerli, W., Widmer, A.F., Blatter, M., Frei, R., and Ochsner, P.E., Role of rifampin for treatment of orthopedic implant-related staphylococcal infections: a randomized controlled trial. Foreign-Body Infection (FBI) Study Group [comment]. *JAMA* 279, 1537–1541, 1998.

31. Segawa, H., Tsukayama, D.T., Kyle, R.F., Becker, D.A., and Gustilo, R.B., Infection after total knee arthroplasty. A retrospective study of the treatment of eighty-one infections. *J. Bone Joint Surg. Am.* 81, 1434–1445, 1999.

32. Brandt, C.M., Sistrunk, W.W., Duffy, M.C., et al., Staphylococcus aureus prosthetic joint infection treated with debridement and prosthesis retention. *Clin. Infect. Dis.* 24, 914–919, 1997.

33. Crockarell, J.R., Hanssen, A.D., Osmon, D.R., and Morrey, B.F., Treatment of infection with debridement and retention of the components following hip arthroplasty. *J. Bone Joint Surg. Am.* 80, 1306–1313, 1998.

34. Meehan, A.M., Osmon, D.R., Duffy, M.C., Hanssen, A.D., and Keating, M.R., Outcome of penicillin-susceptible streptococcal prosthetic joint infection treated with debridement and retention of the prosthesis. *Clin. Infect. Dis.* 36, 845–849, 2003.

35. Marculescu, C., Berbari, E., Hanssen, A., Harmsen, S., Mandrekar, J., Steckelberg, J., and Osmon, D., Outcome of PJI treated with debridement and retention of components. Infectious Diseases Society of America, Oct. 9–12, 2003, San Diego, California.

36. Zimmerli, W., and Ochsner, P.E., Management of infection associated with prosthetic joints. *Infection* 31, 99–108, 2003.

37. Kantor, G.S., Osterkamp, J.A., Dorr, L.D., Fischer, D., Perry, J., and Conaty, J.P., Resection arthroplasty following infected total hip replacement arthroplasty. *J. Arthroplasty* 1, 83–89, 1986.

38. Grauer, J.D., Amstutz, H.C., O'Carroll, P.F., and Dorey, F.J., Resection arthroplasty of the hip. *J. Bone Joint Surg. Am.* 71, 669–678, 1989.

39. McElwaine, J.P., and Colville, J., Excision arthroplasty for infected total hip replacements. *J. Bone Joint Surg. Br.* 66, 168–171, 1984.

40. Bittar, E.S., and Petty, W., Girdlestone arthroplasty for infected total hip arthroplasty. *Clin. Orthop.* 83–87, 1982.

41. Falahee, M.H., Matthews, L.S., and Kaufer, H., Resection arthroplasty as a salvage procedure for a knee with infection after a total arthroplasty. *J. Bone Joint Surg. Am.* 69, 1013–1021, 1987.

42. Elson, R., One-stage exchange in the treatment of the infected total hip arthroplasty. *Semin. Arthroplasty* 5, 137–141, 1994.

43. Callaghan, J.J., Katz, R.P., and Johnston, R.C., One-stage revision surgery of the infected hip. A minimum 10-year followup study. *Clin. Orthop.* 139–143, 1999.
44. Ure, K.J., Amstutz, H.C., Nasser, S., and Schmalzried, T.P., Direct-exchange arthroplasty for the treatment of infection after total hip replacement. An average ten-year follow-up. *J. Bone Joint Surg. Am.* 80, 961–968, 1998.
45. Raut, V.V., Siney, P.D., and Wroblewski, B.M., One-stage revision of total hip arthroplasty for deep infection. Long-term followup. *Clin. Orthop.* 202–207, 1995.
46. Hanssen, A.D., and Osmon, D.R., Assessment of patient selection criteria for treatment of the infected hip arthroplasty. *Clin. Orthop.* 91–100, 2000.
47. Raut, V.V., Orth, M.S., Orth, M.C., Siney, P.D., and Wroblewski, B.M., One stage revision arthroplasty of the hip for deep gram negative infection. *Int. Orthop.* 20, 12–14, 1996.
48. Raut, V.V., Siney, P.D., and Wroblewski, B.M., One-stage revision of infected total hip replacements with discharging sinuses. *J. Bone Joint Surg. Br.* 76, 721–724, 1994.
49. Goksan, S.B., and Freeman, M.A., One-stage reimplantation for infected total knee arthroplasty. *J. Bone Joint Surg. Br.* 74, 78–82, 1992.
50. Silva, M., Tharani, R., and Schmalzried, T.P., Results of direct exchange or debridement of the infected total knee arthroplasty. *Clin. Orthop.* 125–131, 2002.
51. Brandt, C.M., Duffy, M.C., Berbari, E.F., Hanssen, A.D., Steckelberg, J.M., and Osmon, D.R., Staphylococcus aureus prosthetic joint infection treated with prosthesis removal and delayed reimplantation arthroplasty. *Mayo. Clin. Proc.* 74, 553–558, 1999.
52. Haddad, F.S., Muirhead-Allwood, S.K., Manktelow, A.R., and Bacarese-Hamilton, I., Two-stage uncemented revision hip arthroplasty for infection. *J. Bone Joint Surg. Br.* 82, 689–694, 2000.
53. Fehring, T.K., Calton, T.F., and Griffin, W.L., Cementless fixation in 2-stage reimplantation for periprosthetic sepsis. *J. Arthroplasty* 14, 175–181, 1999.
54. Rand, J.A., and Bryan, R.S., Reimplantation for the salvage of an infected total knee arthroplasty. *J. Bone Joint Surg. Am.* 65, 1081–1086, 1983.
55. McPherson, E., Periprosthetic total knee infection. In: Calhoun, J.H., and Mader, J.T., eds., *Musculoskeletal Infections.* New York: Marcel Dekker, 2003:293–324.
56. Hirakawa, K., Stulberg, B.N., Wilde, A.H., Bauer, T.W., and Secic, M., Results of 2-stage reimplantation for infected total knee arthroplasty. *J. Arthroplasty,* 13, 22–28, 1998.
57. McPherson, E.J., Patzakis, M.J., Gross, J.E., Holtom, P.D., Song, M., and Dorr, L.D., Infected total knee arthroplasty. Two-stage reimplantation with a gastrocnemius rotational flap. *Clin. Orthop.* 73–81, 1997.
58. Mont, M.A., Waldman, B.J., and Hungerford, D.S., Evaluation of preoperative cultures before second-stage reimplantation of a total knee prosthesis complicated by infection. A comparison-group study [comment]. *J. Bone Joint Surg. Am.* 82-A, 1552–1557, 2000.
59. Hanssen, A.D., Rand, J.A., and Osmon, D.R., Treatment of the infected total knee arthroplasty with insertion of another prosthesis. The effect of antibiotic-impregnated bone cement. *Clin. Orthop.* 44–55, 1994.
60. Teeny, S.M., Dorr, L., Murata, G., and Conaty, P., Treatment of infected total knee arthroplasty. Irrigation and debridement versus two-stage reimplantation. *J. Arthroplasty* 5, 35–39, 1990.
61. Rosenberg, A.G., Haas, B., Barden, R., Marquez, D., Landon, G.C., and Galante, J.O., Salvage of infected total knee arthroplasty. *Clin. Orthop.* 29–33, 1988.
62. Jhao, C., and Jiang, C.C., Two-stage reimplantation without cement spacer for septic total knee replacement. *J. Formos. Med. Assoc.* 102, 37–41, 2003.
63. Bengston S.K.K., The infected knee arthroplasty. A 6-year follow-up of 357 cases. *Acta. Orthop. Scand.* 62, 301–311, 1991.

64. Rand, J.A., Bryan, R.S., Morrey, B.F., and Westholm, F., Management of infected total knee arthroplasty. *Clin. Orthop.* 75–85, 1986.

65. Bengston, S., Knutson, K., and Lidgren, L., Treatment of infected knee arthroplasty. *Clin. Orthop.* 173–178, 1989.

66. Rand, J.A., Bryan, R.S., and Chao, E.Y., Failed total knee arthroplasty treated by arthrodesis of the knee using the Ace–Fischer apparatus. *J. Bone Joint Surg. Am.* 69, 39–45, 1987.

67. Donley, B.G., Matthews, L.S., and Kaufer, H., Arthrodesis of the knee with an intramedullary nail [comment]. *J. Bone Joint Surg. Am.* 73, 907–913, 1991.

68. Kaufer, H., and Matthews, L.S., Resection arthroplasty: an alternative to arthrodesis for salvage of the infected total knee arthroplasty. *Instr. Course Lect.* 35, 283–289, 1986.

69. Sierra, R.J., Trousdale, R.T., and Pagnano, M.W., Above-the-knee amputation after a total knee replacement: prevalence, etiology, and functional outcome. *J. Bone Joint Surg. Am.* 85-A, 1000–1004, 2003.

70. Isiklar, Z.U., Landon, G.C., and Tullos, H.S., Amputation after failed total knee arthroplasty. *Clin. Orthop.* 173–178, 1994.

71. Widmer, A.F., Frei, R., Rajacic, Z., and Zimmerli, W., Correlation between in vivo and in vitro efficacy of antimicrobial agents against foreign body infections. *J. Infect. Dis.* 162, 96–102, 1990.

72. Blaser, J., Vergeres, P., Widmer, A.F., and Zimmerli, W., In vivo verification of in vitro model of antibiotic treatment of device-related infection. *Antimicrob. Agents Chemother.* 39, 1134–1139, 1995.

73. Drancourt, M., Stein, A., Argenson, J.N., Zannier, A., Curvale, G., and Raoult, D., Oral rifampin plus ofloxacin for treatment of *Staphylococcus*-infected orthopedic implants. *Antimicrob. Agents Chemother.* 37, 1214–1218, 1993.

74. Widmer, A.F., Gaechter, A., Ochsner, P.E., and Zimmerli, W., Antimicrobial treatment of orthopedic implant-related infections with rifampin combinations. *Clin. Infect. Dis.* 14, 1251–1253, 1992.

75. Drancourt, M., Stein, A., Argenson, J.N., Roiron, R., Groulier, P., and Raoult, D., Oral treatment of *Staphylococcus* spp. infected orthopaedic implants with fusidic acid or ofloxacin in combination with rifampicin. *J. Antimicrob. Chemother.* 39, 235–240, 1997.

76. Razonable, R.R., Osmon, D.R., and Steckelberg, J.M., Linezolid therapy for orthopedic infections. *Mayo Clin. Proc.* 70, 1137–1144, 2004.

77. Mader, J.T., and Adams, K., Comparative evaluation of daptomycin (LY146032) and vancomycin in the treatment of experimental methicillin-resistant *Staphylococcus aureus* osteomyelitis in rabbits. *Antimicrob. Agents Chemother.* 33, 689–692, 1989.

78. Juthani-Mehta M.V.H., Gorelick, J., and Rappeport, J.M., Successful therapy of Methicillin-resistant *Staphylococcus aureus* (MRSA) vertebral osteomyelitis (VOM) with Daptomycin. Infectious Diseases Society of America, Oct. 9–12, 2003, San Diego, California,

79. Williams, D.N., Rehm, S.J., Tice, A.D., Bradley, J.S., Kind, A.C., and Craig, W.A., Practice guidelines for community-based parenteral anti-infective therapy. ISDA Practice Guidelines Committee. *Clin. Infect. Dis.* 25, 787–801, 1997.

80. Heck, D., Rosenberg, A., Schink-Ascani, M., Garbus, S., and Kiewitt, T., Use of antibiotic-impregnated cement during hip and knee arthroplasty in the United States. *J. Arthroplasty* 10, 470–475, 1995.

81. Joseph, T.N., Chen, A.L., and Di Cesare, P.E., Use of antibiotic-impregnated cement in total joint arthroplasty. *J. Am. Acad. Orthop. Surg.* 11, 38–47, 2003.

82. www.stryker.com/orthopaedics/sites/bonecement/

83. www.biomet.com/products/index.cfm?p=090F06
84. Perry, A.C., Prpa, B., Rouse, M.S., et al., Levofloxacin and trovafloxacin inhibition of experimental fracture-healing. *Clin. Orthop.* 95–100, 2003.
85. Huddleston, P.M., Steckelberg, J.M., Hanssen, A.D., Rouse, M.S., Bolander, M.E., and Patel, R., Ciprofloxacin inhibition of experimental fracture healing. *J. Bone Joint Surg. Am.* 82, 161–173, 2000.
86. Silverberg, D., Kodali, P., Dipersio, J., Acus, R., and Askew, M., In vitro analysis of antifungal impregnated polymethylmethacrylate bone cement. *Clin. Orthop. Relat. Res.* 403, 228–231, 2002.
87. Hendriks, J.G., Neut, D., van Horn, J.R., van der Mei, H.C., and Busscher, H.J., The release of gentamicin from acrylic bone cements in a simulated prosthesis-related interfacial gap. *J. Biomed. Mater. Res.* 64B, 1–5, 2003.
88. Hofmann, A.A., Kane, K.R., Tkach, T.K., Plaster, R.L., and Camargo, M.P., Treatment of infected total knee arthroplasty using an articulating spacer. *Clin. Orthop.* 45–54, 1995.
89. Magnan, B., Regis, D., Biscaglia, R., and Bartolozzi, P., Preformed acrylic bone cement spacer loaded with antibiotics: use of two-stage procedure in 10 patients because of infected hips after total replacement. *Acta. Orthop. Scand.* 72, 591–594, 2001.
90. Emerson, R.H., Jr., Muncie, M., Tarbox, T.R., and Higgins, L.L., Comparison of a static with a mobile spacer in total knee infection. *Clin. Orthop.* 132–138, 2002.
91. Haddad, F.S., Masri, B.A., Garbuz, D.S., and Duncan, C.P., The treatment of the infected hip replacement. The complex case. *Clin. Orthop.* 144–156, 1999.
92. Springer B.J.D., Lee, G.C., Osmon, D., Haiukewych, G., Hanssen, A., and Jacofsky, D.J., Systemic safety of high dose antibiotic loaded cement spacers after resection of an infected total knee arthroplasty. *Clin. Orthop. Relat. Res.* 427, 47–51, 2004.
93. Trampuz A.H.A., Osmon, D.R., Piper, K.E., Rouse, M.S., Steckelberg, J.M., and Patel, R., The role of preoperative antimicrobial therapy on culture sensitivity of joint fluid, periprosthetic tissue, and explant sonicate in the diagnosis of prosthetic joint infection. Musculoskeletal Infection Society, 2003, Snowmass, Colorado.
94. Powles, J.W., Spencer, R.F., and Lovering, A.M., Gentamicin release from old cement during revision hip arthroplasty. *J. Bone Joint Surg. Br.* 80, 607–610, 1998.
95. von Eiff, C., Proctor, R.A., and Peters, G., *Staphylococcus aureus* small colony variants: formation and clinical impact. *Int. J. Clin. Pract. Suppl.* 44–49, 2000.
96. von Eiff, C., Proctor, R.A., and Peters, G., Small colony variants of *Staphylococci:* a link to persistent infections. *Berl. Munch. Tierarztl.* Wochenschr. 113, 321–325, 2000.
97. Proctor, R.A., van Langevelde, P., Kristjansson, M., Maslow, J.N., and Arbeit, R.D., Persistent and relapsing infections associated with small-colony variants of *Staphylococcus aureus*. *Clin. Infect. Dis.* 20, 95–102, 1995.
98. Looney, W.J., Small-colony variants of *Staphylococcus aureus*. *Br. J. Biomed. Sci.* 57, 317–322, 2000.
99. Pan, X.S., Hamlyn, P.J., Talens-Visconti, R., Alovero, F.L., Manzo, R.H., and Fisher, L.M., Small-colony mutants of *Staphylococcus aureus* allow selection of gyrase-mediated resistance to dual-target fluoroquinolones. *Antimicrob. Agents Chemother.* 46, 2498–2506, 2002.
100. Della Valle, C.J., Scher, D.M., Kim, Y.H., et al., The role of intraoperative Gram stain in revision total joint arthroplasty. *J. Arthroplasty* 14, 500–504, 1999.
101. Pandey, R., Drakoulakis, E., and Athanasou, N.A., An assessment of the histological criteria used to diagnose infection in hip revision arthroplasty tissues. *J. Clin. Pathol.* 52, 118–123, 1999.

102. Feldman, D.S., Lonner, J.H., Desai, P., and Zuckerman, J.D., The role of intraoperative frozen sections in revision total joint arthroplasty [comment]. *J. Bone Joint Surg. Am.* 77, 1807–1813, 1995.

103. Lonner, J.H., Desai, P., Dicesare, P.E., Steiner, G., and Zuckerman, J.D., The reliability of analysis of intraoperative frozen sections for identifying active infection during revision hip or knee arthroplasty. *J. Bone Joint Surg. Am.* 78, 1553–1558, 1996.

104. Della Valle, C.J., Bogner, E., Desai, P., et al., Analysis of frozen sections of intra-operative specimens obtained at the time of reoperation after hip or knee resection arthroplasty for the treatment of infection. *J. Bone Joint Surg. Am.* 81, 684–689, 1999.

105. Abdul-Karim, F.W., McGinnis, M.G., Kraay, M., Emancipator, S.N., and Goldberg, V., Frozen section biopsy assessment for the presence of polymorphonuclear leukocytes in patients undergoing revision of arthroplasties. *Modern Pathology* 11, 427–431, 1998.

106. Fehring, T.K., and McAlister, J.A. Jr., Frozen histologic section as a guide to sepsis in revision joint arthroplasty. *Clin. Orthop.* 229–237, 1994.

107. Banit, D.M., Kaufer, H., and Hartford, J.M., Intraoperative frozen section analysis in revision total joint arthroplasty. *Clin. Orthop.* 230–238, 2002.

108. Atkins, B.L., Athanasou, N., Deeks, J.J., et al., Prospective evaluation of criteria for microbiological diagnosis of prosthetic-joint infection at revision arthroplasty. The OSIRIS Collaborative Study Group. *J. Clin. Microbiol.* 36, 2932–2939, 1998.

109. Marculescu, C., Berbari, E., Steckelberg, J., Hanssen, A., and Osmon, D., Medical treatment and outcome of clinically unsuspected prosthetic joint infection diagnosed by multiple positive cultures at the time of revision arthroplasty. Infectious Diseases Society of America, Oct. 9–12, 2003, San Diego, California.

110. Marculescu, C., Berbari, E., Hanssen, A., Steckelberg, J., and Osmon, D., Significance of acute inflammation in joint tissue at reimplantation arthroplasty in patients with prosthetic joint infectio treated with two-stage exchange. Infectious Diseases Society of America, Oct. 9–12, 2003, San Diego, California.

111. Bengston, S.K.K., and Lidgren, L., Revision of infected knee arthroplasty. *Acta. Orthop. Scand.* 57, 484–494, 1986.

112. Hanssen, A.D., Trousdale, R.T., and Osmon, D.R., Patient outcome with reinfection following reimplantation for the infected total knee arthroplasty. *Clin. Orthop.* 55–67, 1995.

113. Pagnano, M.W., Trousdale, R.T., and Hanssen, A.D., Outcome after reinfection following reimplantation hip arthroplasty. *Clin. Orthop.* 192–204, 1997.

114. Ross, J.J., Saltzman, C.L., Carling, P., and Shapiro, D.S., Pneumococcal septic arthritis: review of 190 cases. *Clin. Infect. Dis.* 36, 319–327, 2003.

115. Phelan, D.M., Osmon, D.R., Keating, M.R., and Hanssen, A.D., Delayed reimplantation arthroplasty for candidal prosthetic joint infection: a report of 4 cases and review of the literature. *Clin. Infect. Dis.* 34, 930–938, 2002.

116. Weil, Y., Mattan, Y., Liebergall, M., and Rahav, G., Brucella prosthetic joint infection: a report of 3 cases and a review of the literature. *Clin. Infect. Dis.* 36, e81–86, 2003.

117. Berbari, E.F., Hanssen, A.D., Duffy, M.C., Steckelberg, J.M., and Osmon, D.R., Prosthetic joint infection due to *Mycobacterium tuberculosis*: a case series and review of the literature. *Am. J. Orthop.* 27, 219–227, 1998.

118. Herold, R.C., Lotke, P.A., and MacGregor, R.R., Prosthetic joint infections secondary to rapidly growing *Mycobacterium fortuitum*. *Clin. Orthop.* 183–186, 1987.

119. Weiler, P.J., and Hastings, D.E., *Listeria monocytogenes*—an unusual cause of late infection in a prosthetic hip joint. *J. Rheumatol.* 17, 705–707, 1990.

120. Jellicoe, P.A., Cohen, A., and Campbell, P., *Haemophilus parainfluenzae* complicating total hip arthroplasty: a rapid failure. *J. Arthroplasty* 17, 114–116, 2002.

121. Iglesias, L., Garcia-Arenzana, J.M., Valiente, A., Gomariz, M., and Perez-Trallero, E., *Yersinia enterocolitica* O:3 infection of a prosthetic knee joint related to recurrent hemarthrosis. *Scand. J. Infect. Dis.* 34, 132–133, 2002.

122. Bates, C.J., Clarke, T.C., and Spencer, R.C., Prosthetic hip joint infection due to *Campylobacter fetus* [comment]. *J. Clin. Microbiol.* 32, 2037, 1994.

123. Fresard, A., Guglielminotti, C., Berthelot, P., et al., Prosthetic joint infection caused by *Tropheryma whippelii* (Whipple's bacillus). *Clin. Infect. Dis.* 22, 575–576, 1996.

124. Maradona, J.A., Asensi, V., Carton, J.A., Rodriguez Guardado, A., and Lizon Castellano, J., Prosthetic joint infection by *Pasteurella multocida*. *Eur. J. Clin. Microbiol. Infect. Dis.* 16, 623–625, 1997.

125. McCarthy, J., and Stingemore, N., *Clostridium difficile* infection of a prosthetic joint presenting 12 months after antibiotic-associated diarrhoea. *J. Infect.* 39, 94–96, 1999.

126. Magerl, F., Wyss, A., Brunner, C., and Binder, W., Plate osteosynthesis of femoral shaft fractures in adults. A follow-up study. *Clin. Orthop.* 62–73, 1979.

127. Bach, A.W., and Hansen, S.T. Jr., Plates versus external fixation in severe open tibial shaft fractures. A randomized trial. *Clin. Orthop.* 89–94, 1989.

128. Lambiris, E., Tyllianakis, M., Megas, P., and Panagiotopoulos, E., Intramedullary nailing: experience in 427 patients. *Bull. Hosp. Joint Dis.* 55, 25–27, 1996.

129. Whittle, A.P., Russell, T.A., Taylor, J.C., and Lavelle, D.G., Treatment of open fractures of the tibial shaft with the use of interlocking nailing without reaming. *J. Bone Joint Surg. Am.* 74, 1162–1171, 1992.

130. Brumback, R.J., Ellison, P.S., Jr., Poka, A., Lakatos, R., Bathon, G.H., and Burgess, A.R., Intramedullary nailing of open fractures of the femoral shaft. *J. Bone Joint Surg. Am.* 71, 1324–1331, 1989.

131. Johner, R., and Wruhs, O., Classification of tibial shaft fractures and correlation with results after rigid internal fixation. *Clin. Orthop.* 7–25, 1983.

132. Ruedi, T.P., and Luscher, J.N., Results after internal fixation of comminuted fractures of the femoral shaft with DC plates. *Clin. Orthop.* 74–76, 1979.

133. Smith, J.E., Results of early and delayed internal fixation for tibial shaft fractures. A review of 470 fractures. *J. Bone Joint Surg. Br.* 56B:469–477, 1974.

134. Fischer, M.D., Gustilo, R.B., and Varecka, T.F., The timing of flap coverage, bone-grafting, and intramedullary nailing in patients who have a fracture of the tibial shaft with extensive soft-tissue injury. *J. Bone Joint Surg. Am.* 73, 1316–1322, 1991.

135. Collinge, C.A., and Sanders, R.W., Percutaneous plating in the lower extremity. *J. Am. Acad. Orthop. Surg.* 8, 211–216, 2000.

136. Winquist, R.A., Hansen, S.T., Jr., and Clawson, D.K., Closed intramedullary nailing of femoral fractures. A report of five hundred and twenty cases. *J. Bone Joint Surg. Am.* 66, 529–539, 1984.

137. Johnson, K.D., Johnston, D.W., and Parker, B., Comminuted femoral-shaft fractures: treatment by roller traction, cerclage wires and an intramedullary nail, or an interlocking intramedullary nail. *J. Bone Joint Surg. Am.* 66, 1222–1235, 1984.

138. Parameswaran, A.D., Roberts, C.S., Seligon, D., and Voor, M., Pin tract infection with contemporary external fixation: how much of a problem? *J. Orthop. Trauma* 17, 503–507, 2003.

139. Marshall, P.D., Saleh, M., and Douglas, D.L., Risk of deep infection with intramedullary nailing following the use of external fixators [comment]. *J. R. Coll. Surg. Edinb.* 36, 268–271, 1991.

140. Tornqvist, H., Tibia nonunions treated by interlocked nailing: increased risk of infection after previous external fixation. *J. Orthop. Trauma* 4, 109–114, 1990.

141. Bone, L.B., and Johnson, K.D., Treatment of tibial fractures by reaming and intramedullary nailing. *J. Bone Joint Surg. Am.* 68, 877–887, 1986.

142. Tornetta, P., 3rd, and Tiburzi, D., Reamed versus nonreamed anterograde femoral nailing. *J. Orthop. Trauma.* 14, 15–19, 2000.

143. Tornetta, P., 3rd, and Tiburzi, D., The treatment of femoral shaft fractures using intramedullary interlocked nails with and without intramedullary reaming: a preliminary report. *J. Orthop. Trauma* 11, 89–92, 1997.

144. Moed, B.R., and Watson, J.T., Retrograde nailing of the femoral shaft. *J. Am. Acad. Orthop. Surg.* 7, 209–216, 1999.

145. Moed, B.R., Watson, J.T., Cramer, K.E., Karges, D.E., and Teefey, J.S., Unreamed retrograde intramedullary nailing of fractures of the femoral shaft. *J. Orthop. Trauma* 12, 334–342, 1998.

146. Court-Brown, C.M., Will, E., Christie, J., and McQueen, M.M., Reamed or unreamed nailing for closed tibial fractures. A prospective study in Tscherne C1 fractures [comment]. *J. Bone Joint Surg. Br.* 78, 580–583, 1996.

147. Paley, D., and Herzenberg, J.E., Intramedullary infections treated with antibiotic cement rods: preliminary results in nine cases. *J. Orthop. Trauma* 16, 723–729, 2002.

148. Patzakis, M.J., Wilkins, J., and Wiss, D.A., Infection following intramedullary nailing of long bones. Diagnosis and management. *Clin. Orthop.* 182–191, 1986.

149. Court-Brown, C.M., Keating, J.F., and McQueen, M.M., Infection after intramedullary nailing of the tibia. Incidence and protocol for management. *J. Bone Joint Surg. Br.* 74, 770–774, 1992.

150. Emami, A., Mjoberg, B., and Larsson, S., Infected tibial nonunion. Good results after open cancellous bone grafting in 37 cases. *Acta. Orthop. Scand.* 66, 447–451, 1995.

151. Morandi, M., Zembo, M.M., and Ciotti, M., Infected tibial pseudarthrosis. A 2-year follow up on patients treated by the Ilizarov technique. *Orthopedics* 12, 497–508, 1989.

152. Shahcheraghi, G.H., and Bayatpoor, A., Infected tibial nonunion. *Can. J. Surg.* 37, 209–213, 1994.

153. Seligson, D., and Klemm, K., Treatment of infection following intramedullary nailing. In: Browner, B.D., ed., The science and practice of intramedullary nailing. Media, PA: Williams and Wilkins, 1996:317–333.

154. Klemm, K.H.S, and Seligson, D., The treatment of infection after interlocking nailing. *Techniques Orthop.* 3, 54–61, 1988.

155. Klemm, K.W., Treatment of infected pseudarthrosis of the femur and tibia with an interlocking nail. *Clin. Orthop.* 174–181, 1986.

156. Miller, M.E., Ada, J.R., and Webb, L.X., Treatment of infected nonunion and delayed union of tibia fractures with locking intramedullary nails. *Clin. Orthop.* 233–238, 1989.

157. Patzakis, M.J., Scilaris, T.A., Chon, J., Holtom, P., and Sherman, R., Results of bone grafting for infected tibial nonunion. *Clin. Orthop.* 192–198, 1995.

158. Ring, D., Jupiter, J.B., Gan, B.S., Israeli, R., Yaremchuk, M.J., Infected nonunion of the tibia. *Clin. Orthop. Rel. Res.* 369, 302–311, 1999.

159. Mahan, J., Seligson, D., Henry, S.L., Hynes, P., and Dobbins, J., Factors in pin tract infections. *Orthopedics* 14, 305–308, 1991.

160. Haleem, A.A., Rouse, M.S., Lewallen, D.G., Hanssen, A.,D., Steckelberg, J.M., and Patel, R. Gentamicin and vancomycin do not impair experimental fracture healing. *Clin. Orthop. Relat. Res.* 427, 22–24, 2004.

161. Lonstein, J., Winter, R., Moe, J., and Gaines, D., Wound infection with Harrington instrumentation and spine fusion for scoliosis. *Clin. Orthop.* 96, 222–233, 1973.

162. Calderone, R.R., Garland, D.E., Capen, D.A., and Oster, H., Cost of medical care for postoperative spinal infections. *Orthop. Clin. North Am.* 27, 171–182, 1996.

163. Blam O.G., Vaccaro, A.R., Vanichkachorn, J.S., Albert, T.J., Hilibrand, A.S., Minnich, J.M., and Murphey, S.A., Risk factors for surgical site infection in the patient with spinal injury. *Spine* 28, 1475–1480, 2003.

164. Theiss, S.M., Lonstein, J.E., and Winter, R.B., Wound infections in reconstructive spine surgery. *Orthop. Clin. North Am.* 27, 105–110, 1996.

165. Levi, A.D., Dickman, C.A., and Sonntag, V.K., Management of postoperative infections after spinal instrumentation. *J. Neurosurg.* 86, 975–980, 1997.

166. Stambough, J.L., and Beringer, D., Postoperative wound infections complicating adult spine surgery. *J. Spinal Disord.* 5, 277–285, 1992.

167. Heggeness, M.H., Esses, S.I., Errico, T., and Yuan, H.A., Late infection of spinal instrumentation by hematogenous seeding. *Spine* 18, 492–496, 1993.

168. Richards, B.R., and Emara, K.M., Delayed infections after posterior TSRH spinal instrumentation for idiopathic scoliosis: revisited. *Spine* 26, 1990–1996, 2001.

169. Clark, C.E., and Shufflebarger, H.L., Late-developing infection in instrumented idiopathic scoliosis. *Spine* 24, 1909–1912, 1999.

170. Dubousset. J.S.H., and Wenger, D., Late "infection" with Cotrel-Dubousset instrumentation. *Orthop. Trans.* 18, 121, 1994.

171. Aydinli, U., Karaeminogullari, O., and Tiskaya, K., Postoperative deep wound infection in instrumented spinal surgery. *Acta. Orthop. Belg.* 65, 182–187, 1999.

172. Massie, J.B., Heller, J.G., Abitbol, J.J., McPherson, D., and Garfin, S.R., Postoperative posterior spinal wound infections. *Clin. Orthop.* 99–108, 1992.

173. Weinstein, M.A., McCabe, J.P., and Cammisa, F.P. Jr., Postoperative spinal wound infection: a review of 2, 391 consecutive index procedures. *J. Spinal Disord.* 13, 422–426, 2000.

174. Sponseller, P.D., LaPorte, D.M., Hungerford, M.W., Eck, K., Bridwell, K.H., and Lenke, L.G., Deep wound infections after neuromuscular scoliosis surgery: a multicenter study of risk factors and treatment outcomes. *Spine* 25, 2461–2466, 2000.

175. Perry, J.W., Montgomerie, J.Z., Swank, S., Gilmore, D.S., and Maeder, K., Wound infections following spinal fusion with posterior segmental spinal instrumentation. *Clin. Infect. Dis.* 24, 558–561, 1997.

176. Klekamp, J., Spengler, D.M., McNamara, M.J., and Haas, D.W., Risk factors associated with methicillin-resistant staphylococcal wound infection after spinal surgery. *J. Spinal Disord.* 12, 187–191, 1999.

177. Thalgott, J.S., Cotler, H.B., Sasso, R.C., LaRocca, H., and Gardner, V., Postoperative infections in spinal implants. Classification and analysis—a multicenter study. *Spine* 16, 981–984, 1991.

178. Wimmer, C., and Gluch, H., Management of postoperative wound infection in posterior spinal fusion with instrumentation. *J. Spinal Disord.* 9, 505–508, 1996.

179. Dernbach, P.D., Gomez, H., and Hahn, J., Primary closure of infected spinal wounds. *Neurosurgery* 26, 707–709, 1990.

180. Picada, R., Winter, R.B., Lonstein, J.E., et al., Postoperative deep wound infection in adults after posterior lumbosacral spine fusion with instrumentation: incidence and management. *J. Spinal Disord.* 13, 42–45, 2000.

181. Glassman, S.D., Dimar, J.R., Puno, R.M., and Johnson, J.R., Salvage of instrumental lumbar fusions complicated by surgical wound infection. *Spine* 21, 2163–2169, 1996.

182. Dietze, D.D., Jr., and Haid, R.W., Jr., Antibiotic-impregnated methylmethacrylate in treatment of infections with spinal instrumentation. Case report and technical note. *Spine* 17, 981–987, 1992.

183. Harle, A., and van Ende, R., Management of wound sepsis after spinal fusion surgery. *Acta. Orthop. Belg.* 57 Suppl 1, 242–246, 1991.
184. Abbey, D.M., Turner, D.M., Warson, J.S., Wirt, T.C., and Scalley, R.D., Treatment of postoperative wound infections following spinal fusion with instrumentation. *J. Spinal Disord.* 8, 278–283, 1995.
185. Soultanis, K., Mantelos, G., Pagiatakis, A., and Soucacos, P.N., Late infection in patients with scoliosis treated with spinal instrumentation. *Clin. Orthop.* 116–123, 2003.
186. Viola, R.W., King, H.A., Adler, S.M., and Wilson, C.B., Delayed infection after elective spinal instrumentation and fusion. A retrospective analysis of eight cases. *Spine* 22, 2444–2450. Discussion, 2450–2441, 1997.

Index

Page numbers in italics indicate figures and tables.